Scientific Basis for Nuclear Waste Management

Volume 3

Scientific Basis for Nuclear Waste Management

Volume 3

Edited by

John G. Moore
Oak Ridge National Laboratory
Oak Ridge, Tennessee

Associate Editors

Ernest A. Bryant
Los Alamos, Scientific Laboratory
Los Alamos, New Mexico

Lawrence D. Ramspott
Lawrence Livermore National Laboratory
Livermore, California

James O. Duguid
Office of Nuclear Waste Isolation
Columbus, Ohio

Wayne A. Ross
Battelle-Pacific Northwest Laboratories
Richland, Washington

Clyde J. M. Northrup, Jr.
Sandia National Laboratories
Albuquerque, New Mexico

James G. Steger
Los Alamos Scientific Laboratory
Los Alamos, New Mexico

Stephen V. Topp
Savannah River Laboratory
Aiken, South Carolina

Editorial Assistants

Amy L. Harkey, Imogene G. Loope, and Sue R. Damewood
Oak Ridge National Laboratory
Oak Ridge, Tennessee

SPRINGER SCIENCE+BUSINESS MEDIA, LLC

Library of Congress Cataloging in Publication Data

International symposium on the Scientific Basis for Nuclear Waste Management (3rd: 1980: Boston, Mass.)
Scientific basis for nuclear waste management.

Proceedings of the symposium held as part of the annual meeting of the Materials Research Society.
Bibliography: p.
Includes index.
1. Radioactive waste disposal — Congresses. I. Moore, John G. II. Bryant, Ernest A., 1931- . III. Harkey, Amy L. IV. Materials Research Society. V. Title.
TD898.I57 1980 621.48'38 81-10663
AACR2

ISBN 978-1-4684-4042-3 ISBN 978-1-4684-4040-9 (eBook)
DOI 10.1007/978-1-4684-4040-9

Proceedings of the Third International Symposium on
the Scientific Basis for Nuclear Waste Management
held in Boston, Massachusetts, on November 17-20, 1980,
as part of the Annual Meeting of the Materials Research Society.

Originally published by Plenum Press New York in 1981
Softcover reprint of the hardcover 1st edition 1981

PREFACE

The third International Symposium on the Scientific Basis for Nuclear Waste Management was held in Boston, Massachusetts, on November 17-20, 1980, as part of the Annual Meeting of the Materials Research Society. The purpose of this Symposium was to provide an interdisciplinary forum for the discussion of scientific research dealing with all levels and types of radioactive wastes and their management. Since its inception in 1978, this annual Symposium has provided a unique opportunity for scientists of widely differing backgrounds to share in such discussions. The proceedings of the first two meetings were published as Volumes 1 and 2 in this series. The fourth Symposium is scheduled to be held in the autumn of 1981.

The efforts of many people went into making this meeting a success. The scope of the 1980 Symposium was guided by the following Steering Committee:

K. J. Notz (Chairman), Oak Ridge National Laboratory, USA
G. H. Daly, Department of Energy, USA
D. E. Ferguson, Oak Ridge National Laboratory, USA
R. H. Flowers, Atomic Energy Research Establishment, UK
F. Girardi, Ispra Establishment, Italy
T. Ishihara, Radioactive Waste Management Center, Japan
R. W. Lynch, Sandia Laboratories, USA
S. A. Mayman, Atomic Energy of Canada Ltd., Canada
G. J. McCarthy, North Dakota State University, USA
E. Merz, Kernforschunganlage Jülich, FRG
L. Nilsson, KBS Project, Sweden
D. M. Rohrer, Nuclear Regulatory Commission, USA
R. Roy, Pennsylvania State University, USA
T. E. Scott, Ames Laboratory, USA
C. Sombret, Centre d'Etudes Nucleaires, Marcoule, France
W. S. Twenhoeffel, U.S. Geological Survey, USA
V. I. Spitsyn, Academy of Sciences, USSR

The Program Committee, responsible for program organization, included:

J. G. Moore (Chairman), Oak Ridge National Laboratory, USA
E. A. Bryant, Los Alamos Scientific Laboratory, USA
J. O. Duguid, Office of Nuclear Waste Isolation, USA
C. J. Northrup, Jr., Sandia Laboratories, USA
L. D. Ramspott, Lawrence Livermore Laboratory, USA
W. A. Ross, Battelle Pacific Northwest Laboratory, USA
J. G. Steger, Los Alamos Scientific Laboratory, USA
S. V. Topp, Savannah River Laboratory, USA

The Session Cochairmen helped to maintain the Symposium at a high level of performance throughout the meeting; in addition to the members of the Program and Steering committees, the following served in this capacity:

R. C. Ewing, University of New Mexico, USA
N. J. Hubbard, Office of Nuclear Waste Isolation, USA
J. F. Kircher, Office of Nuclear Waste Isolation, USA
P. W. Levy, Brookhaven National Laboratory, USA
J. D. Tewhey, Lawrence Livermore Laboratory, USA

The reports presented this year represent research programs from a large number of universities and government institutions in eight countries. The 77 papers published in these proceedings have been divided into 11 chapters. These encompass various aspects of high- and non-high-level radioactive waste management ranging from repository characterization and waste form production to product and performance assessment. All of the contributions have been refereed. The authors and the editorial staff have attempted to make the documents free from errors; thus the publication should be an excellent source of current information in the field.

The preparation of these proceedings was made possible by the support of the Oak Ridge National Laboratory, operated by Union Carbide Corporation, the Department of Energy, and the Nuclear Regulatory Commission.

John G. Moore
Chemical Technology Division
Oak Ridge National Laboratory
Oak Ridge, Tennessee 37830
USA
March 1981

UNPUBLISHED PAPERS

For a variety of reasons, the following papers presented at the Symposium are not published in this volume:

"Immobilization of Savannah River Plant Sludge Waste by Consolidation with Calcium Titanate," A. W. Lynch, Sandia National Laboratories, Albuquerque, New Mexico.

"Synthesis, Characterization and Soil Interactions of Ethylenediaminetetraacetic Acid (EDTA) and Diethylenetriaminepentaacetic Acid (DTPA) Complexes of Technetium-99," L. Y. Martin, D. Rai, and J. A. Franz, Pacific Northwest Laboratory, Richland, Washington.

"Selecting Zeolites for Adsorption and/or Fixation of Cesium and Strontium," S. Komarneni and R. Roy, Materials Research Laboratory, Pennsylvania State University, University Park, Pennsylvania.

"Cleanup Problems at Three Mile Island," R. E. Brooksbank, Oak Ridge National Laboratory, Oak Ridge, Tennessee.

"Activities of the State Planning Council on Low-Level Radioactive Waste Management," J. J. Stucker, Executive Director of the State Planning Council, Washington, D.C.

"EPA's Environmental Standards for Management and Disposal of High-Level Radioactive Waste," D. J. Egan, Jr., Environmental Protection Agency, Washington, D.C.

CONTENTS

REPOSITORY CHARACTERIZATION

HIGH-LEVEL WASTE FORMS

Assessment and Head End Processing

Vitreous Materials

Crystalline Materials

NON-HIGH-LEVEL WASTE

NATURAL ANALOGUES

LEACH STUDIES

RADIATION EFFECTS

RADIONUCLIDE MIGRATION

ENGINEERED BARRIERS

PERFORMANCE ASSESSMENT

DEVELOPMENT OF REFERENCE CONDITIONS FOR GEOLOGIC REPOSITORIES FOR NUCLEAR WASTE IN THE USA†

G. E. Raines*, L. D. Rickertsen**, H. C. Claiborne***, J. L. McElroy****, and R. W. Lynch*****

INTRODUCTION

This paper summarizes activities to determine interim reference conditions for temperatures, pressure, fluid, chemical, and radiation environments that are expected to exist in commercial and defense high-level nuclear waste and spent fuel repositories in salt, basalt, tuff, granite, and shale. These interim conditions are being generated by the Reference Repository Conditions Interface Working Group (RRC-IWG), an ad hoc IWG established by the National Waste Terminal Storage Program's (NWTS) Isolation Interface Control Board (I-ICB). Members of the RRC-IWG are:

G. E. Raines, ONWI, Chairman
N. E. Bibler, SRL, Defense High-Level Waste
H. C. Claiborne, ORNL, Salt Repository Environment
K. R. Hoopingarner, Rockwell (BWIP), Basalt Repository Environment
N. Hubbard, ONWI, Geochemical Environment
T. O. Hunter, SNL (WIPP), Salt Repository Environment
R. W. Lynch, SNL (NNWSI), Tuff Repository Environment
J. L. McElroy, PNL, Commercial High-Level Waste
J. D. Osnes, RE/SPEC, Granite Repository Environment
L. D. Rickertsen, SAI, Shale Repository Environment

---

†This work was funded by the U.S. Department of Energy.
*Office of Nuclear Waste Isolation
**Science Applications, Inc.
***Oak Ridge National Laboratory
****Battelle Pacific Northwest Laboratories
*****Sandia National Laboratories

The reference repository conditions being developed are intended to serve as a guide for: a) scientists conducting material performance tests; b) engineers preparing the design of repositories; c) the technically conservative conditions to be used as a basis for DOE license applications; and d) scientists and engineers developing waste forms. Present plans call for the completion of generic reference repository conditions for salt, basalt, tuff, and granite, by December, 1981. Shale has been assigned a lower priority and RRC work on that rock type has been discontinued. However, interim conditions for all five rock types will be published as ONWI reports in the near future.

REFERENCE CANISTERED HIGH-LEVEL WASTES

Three types of waste are being considered in this effort: spent fuel (SF) from light water reactors, commercial high-level waste (CHLW) that would result from reprocessing of light water reactor fuel, and defense high-level waste (DHLW). Pressurized Water Reactor (PWR) spent fuel was chosen over Boiling Water Reactor (BWR) spent fuel because of its greater thermal power. CHLW resulting from a 3:1 mix of wastes from fresh UO_2 and MOX fuel was chosen as a very probable candidate. The maximum thermal output DHLW described by Savannah River Laboratory[1] was selected over Hanford or other defense wastes. Characteristics of the canistered reference wastes are given in Tables 1 and 2.

GEOLOGIC MEDIA INVESTIGATED

The geologic media for which reference conditions are being established include salt, basalt, tuff, granite, and shale. These media were selected because of present NWTS emphasis and ongoing repository projects.

REFERENCE REPOSITORY DESCRIPTION

The bases selected for calculation of reference repository conditions in salt, basalt, tuff, granite, and shale are given in Table 3. These descriptions are based on the standard room and pillar mined repository concept. In this concept, storage rooms are excavated deep in the rock and vertical emplacement holes in rows are drilled in the floor. Waste packages are emplaced in the holes and the holes are then backfilled and plugged with concrete or other shielding plugs. The variation in design parameters selected for the various repositories in the five rock types reflects the variation in heat dissipation and rock strength properties for the different media as well as the different heat generation rates for the different waste types. There will undoubtedly be some evolution in these bases before the final reference conditions are selected. The effects of changes in these

TABLE 1. RELATIVE HEAT-GENERATION RATES OF REFERENCE SF[a], CHLW[b], and DHLW[c]

Years After Emplacement[d]	SF	CHLW	DHLW
0	1.000	1.000	1.000
5	.838	.810	.886
10	.750	.692	.789
15	.681	.600	.705
20	.622	.529	.630
30	.525	.402	.505
40	.449	.313	.407
50	.387	.246	.330
70	.301	.157	.191
100	.238	.0864	.128
190	.137	.0296	.032
290	.108	.0215	.013
390	.0919	.0163	.0072
490	.0806	.0145	.0047
990	.0466	.00810	.0021
1990	.0247	.00404	.0013
5990	.0148	.00230	.0009
9990	.0114	.00175	.0008

a. Pressurized Water Reactor fuel; 33,000 MWd/MTU burnup.
b. See Y/OWI/TM-34, "Nuclear Waste Projections and Source Term Data for FY 1977." The CHLW decay rates correspond to waste arising from fuel which is a 3:1 mix of fresh UO_2 and MOX fuels.
c. E. I. DuPont de Nemours and Co., "Preliminary Technical Data Summary No. 3," DPSTD-77-13-3 (1980).
d. Assumes commercial waste (SF or CHLW) is 10 years out of reactor at emplacement in repository; DHLW is 15 years out of reactor at emplacement.

design parameters or of other repository design concepts will be reported in future communications of the working group.

It should also be noted that since final designs do not exist for engineered barriers for the repositories in the several rock types, relatively simple geometries were assumed for the waste package configuration in the repository environments calculations. These geometries were not the same for all rock types and one of the major modifications planned in the final reference repository conditions reports will be the incorporation of more realistic waste packages in the calculations.

RESULTS

The full complement of interim reference conditions for all rock types being investigated cannot be presented in this brief paper. Interim reference conditions for a CHLW and an SF repository in salt are discussed in References 2 and 3. Interim reference conditions for repositories in the four other rock types will be published as ONWI reports in the near future. Nevertheless, some illustrative examples are warranted and we have drawn largely on Reference 3 for these.

a. Thermal Environments. The maximum emplacement hole wall, canister surface, and waste centerline temperature histories for the specified SF and CHLW repositories in salt are shown in Figures 1 and 2.[3] These temperature histories were calculated based on a temperature of 34°C for the undisturbed formation at repository depth. It can be seen that for either repository, the maximum salt temperature is less than 160°C or about 90°C below the lowest reported[4] decrepitation temperature for salt samples. In the CHLW repository the temperature of the canister peaks at 260°C a few years after emplacement and decreases to less than 120°C at 100 years. The centerline temperature of the glass peaks at about 320°C at 1-2 years after emplacement and decreases to about 120°C at 100 years. In the SF repository the canister

TABLE 2. DESCRIPTION OF REFERENCE CANISTERED WASTE

Characteristics	SF	CHLW	DHLW
Waste Description			
Active Length (m)	3.7	2.4	2.3
Active Volume (m^3)	NA	0.18	0.63
Age of Waste (yr)	10[a]	10[a]	15[a]
Thermal Loading (kW/canister)	0.55[b]	2.16[c] or 1.0[d]	0.31[e]
Canister Dimensions			
Outer Diameter (cm)	35.6[f]	32.4[g]	61.0[h]
Inner Diameter (cm)	33.7	30.5	59.1
Length (m)	4.7	3.0	3.0
Materials			
Waste	UO_2[j]	Glass[j]	Glass[j]
Filler in Canister	Helium	Air	Air
Canister	Carbon Steel	304L Stainless Steel	304L Stainless Steel

a. At emplacement (years after discharge from reactor).
b. Heat generation rate for a single PWR assembly 10 years out of the reactor. BWR assemblies would have a lower heat generation rate but this value has been chosen as its use results in predicting maximum temperatures in repository.
c. Heat generation rates for CHLW used in salt and tuff studies to date.
d. Heat generation rate for CHLW used in granite and shales studies to date.
e. Maximum expected thermal loading for Savannah River Plant wastes; many canisters will have a lower loading.
f. Nominal 14-inch schedule 30 carbon steel pipe.
g. Nominal 12-inch schedule 40s 304L stainless steel pipe.
h. Nominal 24-inch schedule 20 304L stainless steel pipe.
j. The choice of waste form for these calculations was based on their advanced state of engineering development. The calculated environments outside the waste forms are insensitive to the details of the waste forms themselves other than heat output and physical dimensions.

TABLE 3. REFERENCE REPOSITORY DESCRIPTION

	Host Rock--Salt			Host Rock--Basalt			Host Rock--Tuff			Host Rock--Granite			Host Rock--Shale		
Repository Characteristics	SF	CHLW	DHLW	SF	CHLW	DHLW	SF	CHLW	DHLW	SF	CHLW	DHLW	SF	CHLW	DHLW
Repository Depth Below Surface (m)	600	600	600	1000	(c)	(c)	800	800	(c)	1000	1000	1000	600	600	600
Storage Room Width (m)	5.5	5.5	5.5	4.3	(c)	(c)	7.5	5.0	(c)	7.5	7.5	7.5	5.5	5.5	5.5
Storage Room Height (m)	6.4	5.5	5.5	6.1	(c)	(c)	7.0	5.0	(c)	7.0	7.0	7.0	6.4	5.5	5.5
Adjacent Pillar Width (m)	21.3	18.3	18	32.3	(c)	(c)	30	20	(c)	22.5	22.5	22.5	18	18	18
Canister Rows per Room	2	1	2	1	(c)	(c)	2	1	(c)	2	2	2	2	1	2
Row Separation (m)	1.67	--	2.29	--	(c)	(c)	2.5	--	(c)	2.5	2.5	2.5	1.67	--	2.29
Hole Pitch (along row) (m)	1.67	3.66	2.29	3.66[d] or 1.22	(c)	(c)	1.19	3.50	(c)	1.83	2.67	1.53	2.34	2.85	2.70
Canisters per Hole	1	1	1	1	(c)	(c)	1	1	(c)	1	1	1	1	1	1
Canister Thermal Loading (kW)[a]	0.55	2.16	0.31	1.65[d] or 0.55	(c)	(c)	0.55	2.16	(c)	0.55	1.0	0.31	0.55	1.0	0.31
Local Areal Thermal Loading (W/m^2)[a]	25	25	11.6	12.3	(c)	(c)	25	25	(c)	20	25	13.5	10	10	10
Average Areal Thermal Loading (W/m^2)[a]	15	<25	<11.6	8.2	(c)	(c)	<25	<25	(c)	<20	<25	<13.5	8	8	8
Emplacement Hole Depth (m)	7.0	5.5	5.5	6.4	(c)	(c)	8.0	6.0	(c)	6.7	5.0	5.0	7.0	5.5	5.5
Emplacement Hole Diameter (m)	0.54	0.54	0.76	1.15	(c)	(c)	0.41	0.37	(c)	0.56	0.52	0.81	0.51	0.51	0.76
Hole Liner Dimensions															
Outer Diameter (cm)	53.3	53.3	--	114.3[e]	(c)	(c)	(f)	(f)	(c)	(g)	(g)	(g)	(h)	(h)	(h)
Inner Diameter (cm)	50.8	50.8	--	94.0	(c)	(c)	(f)	(f)	(c)	(g)	(g)	(g)	(h)	(h)	(h)
Length (m)	6.25	5.5	--	6.4	(c)	(c)	(f)	(f)	(c)	(g)	(g)	(g)	(h)	(h)	(h)
Backfill Dimensions															
Thickness (cm)	5.1	5.1	7.6	15.2[e]	(c)	(c)	2.5	2.5	(c)	10	10	10	5.1	5.1	7.6
Length (m)	5.5	4.0	4.0	5.4	(c)	(c)	4.7	3.0	(c)	6.7	5.0	5.0	5.5	4.0	4.0
Overpack Dimensions															
Outer Diameter (cm)	40.6[b]	40.6[b]	--	53.3[e]	(c)	(c)	(f)	(f)	(c)	(g)	(g)	(g)	40.6[b]	40.6[b]	(h)
Inner Diameter (cm)	38.1	38.1	--	52.8	(c)	(c)	(f)	(f)	(c)	(g)	(g)	(g)	38.1	38.1	(h)
Length (m)	5.1	3.4	--	4.9	(c)	(c)	(f)	(f)	(c)	(g)	(g)	(g)	5.1	3.4	(h)
Materials															
Overpack	CSt	CSt	--	Ti	(c)	(c)	(f)	(f)	(c)	(g)	(g)	(g)	CSt	CSt	(h)
Backfill	CSa	CSa	CSa	TBf	(c)	(c)	air	air	(c)	CGr	Bt	CGr	CSh	CSh	CSh
Hole Liner	CSt	CSt	CSt	Grt	(c)	(c)	(f)	(f)	(c)	(g)	(g)	(g)	(h)	(h)	(h)
Emplacement Hole Plug	Ct	Ct	Ct	Ct	(c)	(c)	Ct	Ct	(c)	CGr	Bt	CGr	Ct	Ct	Ct

a. At emplacement of wastes.
b. Nominal 16-inch Schedule 40 carbon steel pipe.
c. Not yet available.
d. For a repository in basalt, two configurations have been calculated; three PWR or seven BWR elements per canister, and one PWR or three BWR elements per canister.
e. The BWIP project has developed advanced concepts for waste packages. Their thinking is not adequately treated in this table. There is an additional 5-cm air gap between the titanium overpack and the tailored backfill in this design.
f. Hole liner, backfill, and overpack were not used in the calculations to determine interim reference conditions for the tuff repositories. A more realistic representation of the waste package is planned for the final reference conditions report.
g. Hole liner and overpack were not included in the calculations to determine interim reference conditions for the granite repositories.
h. A hole liner was not included in the calculations to determine interim reference conditions for the shale repositories. An overpack was not included for the DHLW calculations.

CST = Carbon Steel; CSa = Crushed Salt; Ct = Concrete; Ti = Titanium; TBf = Tailored Backfill; Grt = Grout; Bt = Bentonite; CGr = Crushed Granite; CSh = Crushed Shale.

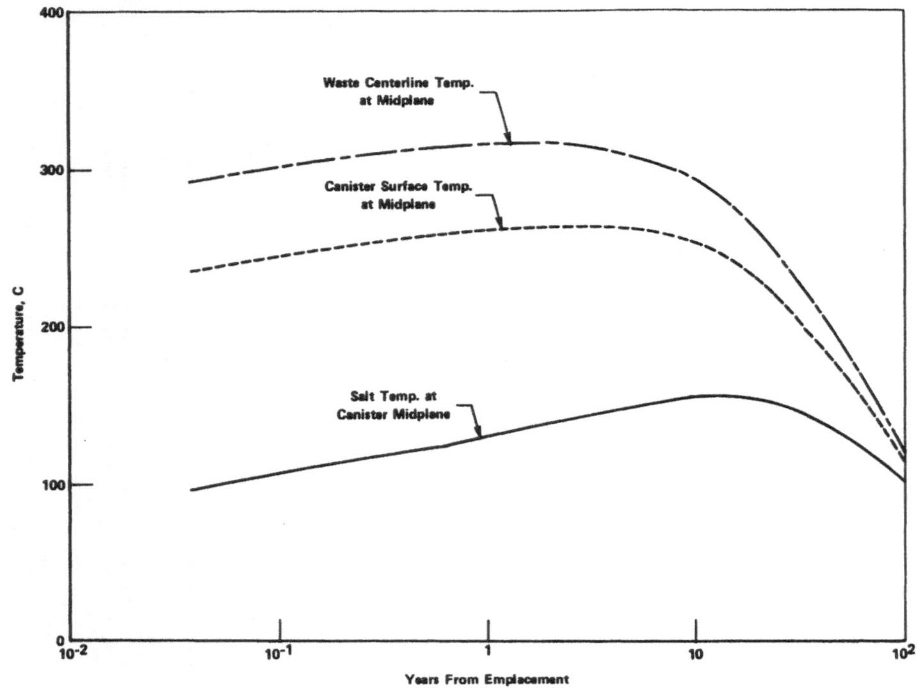

FIGURE 1. SALT TEMPERATURE HISTORIES FOR CHLW 25 W/m^2 (100 KW/ACRE)

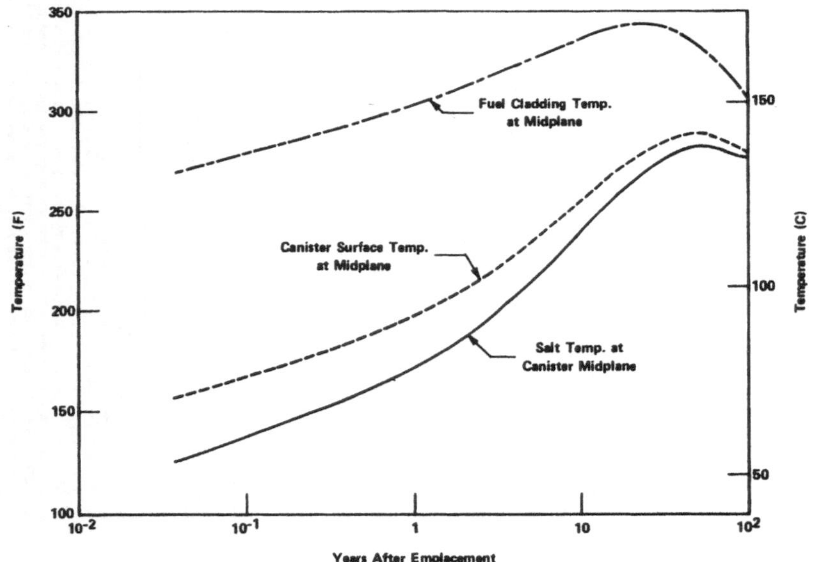

FIGURE 2. SALT TEMPERATURE HISTORIES FOR SF 25 W/m^2 (100 KW/ACRE)

reaches a maximum temperature of about 145°C after 40-50 years. The maximum fuel cladding temperature reaches about 175°C after approximately 25 years and decreases to less than 155°C by 100 years. Peak temperatures for the salt repositories as well as for repositories in the four other rock types are summarized in Table 4

Temperatures resulting in the rock above the various reference repositories will also be available in the Reference Repository Conditions IWG reports. The expected temperatures rise at the earth's surface was calculated as well and was always less than 0.1°C.

b. Fluid Environment. For the salt repositories, the fluid environment was calculated for normal operating conditions only, i.e., the only fluid assumed to contact the waste was that resulting from migration of brine inclusions in the host rock.* The volume of water migrating to each emplacement hole was calculated using the predicted thermal gradients, an equation developed by Jenks[5] which relates the velocity of the brine inclusion to the temperature and temperature gradients, and the MIGRAIN code[2] which solves the equation for mass continuity numerically. The

TABLE 4. REFERENCE PEAK NEAR-FIELD TEMPERATURES (°C)[a]

Host Rock	Location	SF[b]	CHLW	DHLW
Salt	Host Rock	140	160	80
	Canister Wall	145	260	90
	Waste[c]	175	320	100
Basalt	Host Rock	165	---	---
	Canister Wall	170	---	---
	Waste[c]	185	---	---
Tuff	Host Rock	185	225	---
	Canister Wall	190	260	---
	Waste[c]	230	295	---
Granite	Host Rock	150	165	105
	Canister Wall	170	205	115
	Waste[c]	190	225	120
Shale	Host Rock	125	140	125
	Canister Wall	140	210	135
	Waste[c]	165	235	140

a. Assumes initial formation temperature of 34°C for salt, 57°C for basalt, 35°C for tuff, 20°C for granite, and 38°C for shale.
b. Results for 1 PWR element per canister waste package configuration. The BWIP project is also calculating results for a 3 PWR elements per canister waste package configuration.
c. Maximum centerline temperature for CHLW and DHLW; maximum cladding temperature for SF.

*The average initial concentration of brine inclusions was taken to be 0.5% by volume. This figure is typical of bedded salt, but it is a factor of 2-10 too high for dome salt and is, therefore, conservative.

TABLE 5. REFERENCE COMPOSITIONS FOR INTRUDING WATERS

Constituent	Salt Brine A[a]	Salt Brine B[b] (mg/l)	Basalt[c]	Tuff[d]	Granite[e]	Shale[f]
Lithium	20	---	---	0.05	---	
Sodium	42,000	115,000	250	51	125	
Potassium	30,000	15	1.9	4.9	0.4	
Rubidium	20	1	---	---	---	
Cesium	1	1	---	---	---	
Magnesium	35,000	10	0.04	2.1	0.5	
Calcium	600	900	1.3	14	59	
Strontium	5	15	---	0.05	---	
Barium	---	---	---	0.003	---	
Iron	2[g]	2[g]	---	0.04	0.02	
Aluminum	----	---	---	0.03	---	
Silica	---	---	92	61	11	
Fluoride	---	---	37	2.2	3.7	
Chloride	190,000	175,000	148	7.5	283	
Bromide	400	400	---	---	---	
Iodide	10	10	---	---	---	
Carbonate	---	---	25	0.0	3	
Bicarbonate	700	10	21	120	13	
Sulfate	3,500	3,500	108	22	19	
Nitrate	---	---	---	5.6	---	
Borate	1,200	10	---	---	---	
Phosphate	---	---	---	0.12	---	
pH	6.5	6.5	9.7	7.1	9.	
Eh (volts)	mildly oxidizing	mildly oxidizing	-0.50	mildly reducing	+0.17	

a. Brine A is based on analyses of brine inclusions in the McNutt potash zone in the Salado formation near Los Medanos, New Mexico.[6]
b. Brine B results from dissolving a core sample from the Salado formation at the WIPP site in local ground water.[6] This latter brine represents that formed by a hypothetical mine flooding accident and would be the case if water flow were sufficiently slow to achieve saturation.
c. Based on analyses of water from holes drilled in the basalts of the Hanford Reservation by the BWIP project.[7]
d. Based on reported analyses of water from well J-13 in the southwest corner of the Nevada Test Site.[8,9,10]
e. The RRC-IWG experienced considerable difficulty in selecting a reference groundwater composition for granite. That reported for the Stripa Granite[11] was selected.
f. The RRC-IWG has not yet selected a reference groundwater composition for shale.
g. Iron as Fe^{+3}.

expected in-flow of brine per hole is about 8-9 liters for CHLW and 3-4 liters for SF in 1000 years. The reference composition of the accumulated brine is given in Table 5. It is worth noting that the total in-flow would only fill a small fraction of the emplacement hole if it were all to accumulate.

The situation for the repositories in hard rock (basalt, tuff, granite, and shale) is quite different. After closure, one would expect any such repository located below the water table to slowly fill with water. The rate of filling will depend on hydrologic conditions and heat output from the wastes. The ultimate composition of the water interacting with the waste packages depends on a variety of factors including rock and engineering materials compositions, radiation field, initial ground-water composition, temperature, pressure, and water replenishment rate. For the interim, the RRC-IWG has used available literature to identify representative compositions for the intruding ground waters and established a set of reference compositions (Table 5). Insufficient data exist to specify reference compositions for ground waters equilibrated with host rocks and engineering materials at high temperatures and pressures in a radiation environment.

c. Pressure Environment. For salt a model was developed to predict pressure in the emplacement hole as a function of time based on the volume of the hole, its temperature, the volume of air in the hole, and the brine inflow rate through brine migration. Estimates of the pressure in the hole were made for two limiting cases.[2,3] In the first case, it was assumed that the hole was perfectly sealed. In the second, it was assumed that the hole was poorly sealed so that the pressure in the hole was essentially equal to that in the room.

In the first case, pressure peaks at about 3.2 MPa at 10 years for CHLW and 0.45 MPa at 50 years for SF. For CHLW pressure quickly subsides, decreasing to about 0.3 MPa at 100 years. For SF pressure decreases at a slower rate, reaching 0.4 MPa at 100 years. In the second, more probable, case, pressures remain near atmospheric (0.1 MPa) for the entire 100-year period.

For hard rock, one may obtain the maximum possible vapor pressure in the emplacement hole during the operation phase of the repository by assuming the hole is perfectly sealed and calculating the saturation pressure corresponding to the maximum temperature of any surface exposed to the vapor. The maxima are of the order of 2-3 MPa depending on the type of waste. Clearly this is an overprediction in that the pressure in the hole is more likely determined by the minimum temperature of any surface exposed to the vapor. Further, it is not likely that any hole would be perfectly sealed, and all of them certainly would not be. If a hole is not well sealed, pressure will not rise significantly in it until after backfilling and sealing of the particular room in which it is located. In this case, pressure would remain essentially atmospheric until ground water ingress occurred in the underground excavations. Pressure would then rise at some, as yet, unknown rate until the local hydrostatic pressure was reached. At that point, pressure would remain constant, totally dependent on depth below the water table, and usually less than 7 MPa.

d. Nuclear Radiation Environment. Radiation dose rates and maximum integrated dose have been calculated for the several rock types for the reference SF and CHLW waste packages. For the case of salt, the maximum dose, which occurs at the inner edge of the crushed salt backfill, is roughly a factor of 15 greater for CHLW than for SF. At 10^4 years, these maximum doses are $\sim 1.5 \times 10^{10}$ rads for CHLW and $\sim 9.5 \times 10^8$ rads for SF. Results are not greatly dependent on rock type; thus, these results may be considered within an order of magnitude of results for other rocks.

REFERENCES

1. E. I. duPont de Nemours and Co., "Preliminary Technical Data Summary No. 3.," DPSTD-77-13-3 (1980).
2. H. C. Claiborne, L. D. Rickertson, and R. F. Graham, "Expected Environments in High-Level Nuclear Waste and Spent Fuel Repositories in Salt," Oak Ridge National Laboratory Report ORNL/TM-7201, August 1980.
3. Office of Nuclear Waste Isolation, "Interim Reference Repository Conditions for Spent Fuel and Commercial and Defense High-Level Nuclear Waste Repositories in Salt, Draft," NWTS-3, to be published.
4. R. L. Bradshaw et al., "Properties of Salt in Radioactive Waste Disposal," The Geological Society of America, Inc., Special Paper 88, 1968.
5. G. H. Jenks, "Effects of Temperature, Temperature Gradients, Stress, and Irradiation of Brine Inclusions in a Salt Repository," Oak Ridge National Laboratory Report ORNL-5526, July 1979.
6. R. G. Dosch and A. W. Lynch, "Interaction of Radionuclides with Geomedia Associated with the Waste Isolation Pilot Plant (WIPP) Site in New Mexico," Sandia National Laboratories Report SAND 78-0297, June 1978; J. W. Braitwaite and M. A. Molecke, "Nuclear Waste Canister Corrosion Studies Pertinent to Geologic Waste Isolation," Nucl. and Chem. Waste Management 1, No. 1, p. 39-50 (1980).
7. R. E. Gephart et al., "Hydrologic Studies within the Columbia Plateau, Washington: An Integration of Current Knowledge," Rockwell Hanford Operations Report RHO-BWI-ST-5, October 1979.
8. I. J. Winograd, "Hydrogeology of Ash-Flow Tuff: A Preliminary Statement," Water Resources Research 7, No. 4, p. 994-1006 (1971).
9. K. Wolfsberg et al., "Sorption-Desorption Studies on Tuff, I. Initial Studies with Samples from the J-13 Drill Site, Jackass Flats, Nevada," Los Alamos Scientific Laboratory Information Report LA-7480-MS, April 1979.
10. E. N. Vine, Los Alamos Scientific Laboratories, Private Communications, October 9, 1980.
11. P. Frits et al., "Geochemistry and Isotope Hydrology of Groundwater in the Stripa Granite--Results and Preliminary Interpretation," Lawrence Berkeley Laboratory Report LBL-8285, 1979.

BASALT WASTE ISOLATION PROJECT DATA PACKAGE FOR REFERENCE PHYSICOCHEMICAL CONDITIONS

William E. Coons

Rockwell Hanford Operations, P.O. Box 800
Richland, Washington 99352

ABSTRACT

This package defines the physicochemical conditions expected to prevail in a closed nuclear waste repository mined from Columbia River Basalt. Estimates have been derived from experimental data, in-situ measurements, or thermodynamic calculations. Equations are provided from estimation of fO_2, Eh and pH as a function of temperature. Expected conditions are summarized below.

	Temperature (°C)	Equilibrium pH	log Oxygen Fugacity (atm)	Eh(V)
Period of Geologic Controls	65	9.4	-64.5 to -67.15	-0.50 to -0.54
Thermal Period	100	8.7	-57.8 to -60.0	-0.49 to -0.53
	150	7.9	-49.6 to -51.8	-0.48 to -0.53
	200	7.2	-43.3 to -45.4	-0.47 to -0.52
	250	6.7	-38.3 to -40.2	-0.48 to -0.52
	300	6.2	-34.1 to -35.9	-0.45 to -0.51

INTRODUCTION

The reference conditions listed in this data package are based upon field measurements, experimental testing, and thermodynamic calculations. A more complete discussion is contained within the draft Rockwell Hanford Operations document, RHO-BWI-ST-7, "Engineered Barriers Development for a Nuclear Waste Repository Located in Basalt: An Integration of Current Knowledge".

Groundwater compositions and pH were determined by in-house Rockwell staff and are in good agreement with a previous determination made by the United States Geological Survey. Hydrothermal data were obtained through experiment by personnel at the Pennsylvania State University and have been presented at a poster session of the 1979 Annual Meeting of the Geological Society of America (M. W. Barnes and B. E. Scheetz, 1979). Calculations were made using the compilation of thermodynamic equations contained in Research Techniques for High Pressure and High Temperature (G. C. Ulmer ed., Springer-Verlag, 1971) and fundamental physics. The resulting values from all of the above result in the current best estimate of physicochemical conditions likely to obtain in a nuclear waste repository in basalt.

The design and evaluation of engineered barrier systems for nuclear waste containment are directly linked to the range of physicochemical conditions expected to occur in a geologic repository. The parameters which are expected to affect barrier stability included temperature, pressure, oxygen fugacity (or Eh), pH, and dominant groundwater species. Over most of repository life, these conditions will be controlled by repository depth, ambient geothermal gradient, and reaction between groundwater and basalt. Thus, the current ambient conditions are directly relevant to long-term stability of the waste package. In the short term, however, repository excavation and emplacement of heat producing waste are certain to cause deviation from the natural norm.

PERIOD OF GEOLOGIC CONTROL

Presently, the most likely horizon for repository construction at Hanford is located at a depth near one kilometer. As indicated above, current ambient conditions in this region will be similar to the long-term repository environment. At one kilometer depth, within a likely repository horizon, the maximum ambient temperature has been determined as 65°C by in-situ well measurements. This corresponds to a geothermal gradient near 45°C/km. The groundwater pH has also been determined by in-situ measurement and is in the range 9.4 to 9.9; groundwater compositon is shown in Table 1.

TABLE 1. Hanford Groundwater Compositions (Grande Ronde)

Temperature $\approx$ 45°C

	Na^+	K^+	Ca^{2+}	Mg^{2+}	CO_3^{2-}	HCO_3^-	OH^-	$H_3SiO_4^-$	Cl^-	F^-	SO_4^-
(mg/l)	250	1.9	1.3	0.04	21	25	10	145	148	37	108

At a depth of one kilometer, the weight of overlying rock exerts a pressure near 300 bars (4,410 PSI). Similarly, the weight of a one kilometer column of water results in a hydrostatic pressure near 100 bars (1470 PSI).

If the water in the repository is directly connected to the surface, the water pressure available to hydrothermal reaction is equal to the hydrostatic pressure (100 bars). If the water in the repository is isolated from the surface, then the water pressure can be no greater than the load pressure of the confining rock (300 bars). The water pressure in a closed repository must therefore range between 100 and 300 bars. To be conservative, BWIP assumes 300 bars.

The active oxygen content in the geologic environment directly affects barrier corrosion, waste solubility and radionuclide migration. In geologic systems, available oxygen is controlled by oxygen exchange among mineral phases comprising the host geology. For basalt, oxygen fugacity (active oxygen concentration) is customarily taken to be in the range defined by the quartz - fayalite - magnetic buffer (QFM) and the nickel - nickel oxide (NNO) buffer. The equilibrium oxygen fugacities for these reactions are temperature and pressure dependent and may be calculated from the equations below:

Eq (1) $\log FO_2 = 9.36 - \frac{24930}{T^\circ} + 0.046 \left[\frac{p-1}{T^\circ} \right]$ NNO Where T° = temperature in degrees Kelvin and P = pressure in atmospheres

Eq (2) $\log FO_2 = 9.00 - \frac{25738}{T^\circ} + 0.092 \left[\frac{p-1}{T^\circ} \right]$ QFM (see Ulmer, 1971)

Under ambient repository conditions, these reactions constrain oxygen fugacity near 10^{-67} atmospheres (QFM).

Oxygen - reduction potential (Eh) is another parameter used to describe the corrosive properties of an aqueous solution. Eh is related to oxygen fugacity by the dissociation of water. For instance:

Eq (3) $$2\,H_2O \rightarrow 4\,H^+ + O_2 + 4e$$

Where e refers to an electron in aqueous solution.

Water dissociation is dependent on hydrogen ion activity (or pH); thus oxygen fugacity and temperature do not define Eh uniquely. However, Eh may be calculated at any temperature provided oxygen fugacity and pH are known, from the following equation:

Eq (4) $Eh = 1.23 + 4.96 \times 10^{-5}(T^\circ K)(\log fO_2) - 1.984 \times 10^{-4}$

Using equations (1), (2) and (4), ambient geologic Eh is calculated to be in the range -0.50 to -0.54 volts.

THERMAL PERIOD

Emplacement of waste will increase temperature in the repository environment during the early (or thermal period) of repository history. For a dry environment, using the pre-conceptual design for burial of spent fuel at Hanford, the maximum near-canister temperature is calculated as 275°C (Hardy and Hocking, 1978). To be conservative, the Basalt Waste Isolation Project takes 300°C as the thermal maximum. The energy related to this elevated temperature will drive chemical reactions that affect oxygen fugacity (and Eh), pH and concentrations of major groundwater species.

Groundwater pH displays the most complex variation of the parameters under discussion (see Figure 1). This behavior results from two processes which drive pH in opposite directions while proceeding at different rates. The two processes are precipitation and solid-liquid reaction, and in either case, clay is the product.

Precipitation of clay from dissolved salts in groundwater occurs as temperature is increased. This reaction results in addition of hydrogen ion to the residual solution, causing pH to decrease rapidly. Precipitation will be the controlling reaction when the system is dominated by water rather than rock. For instance:

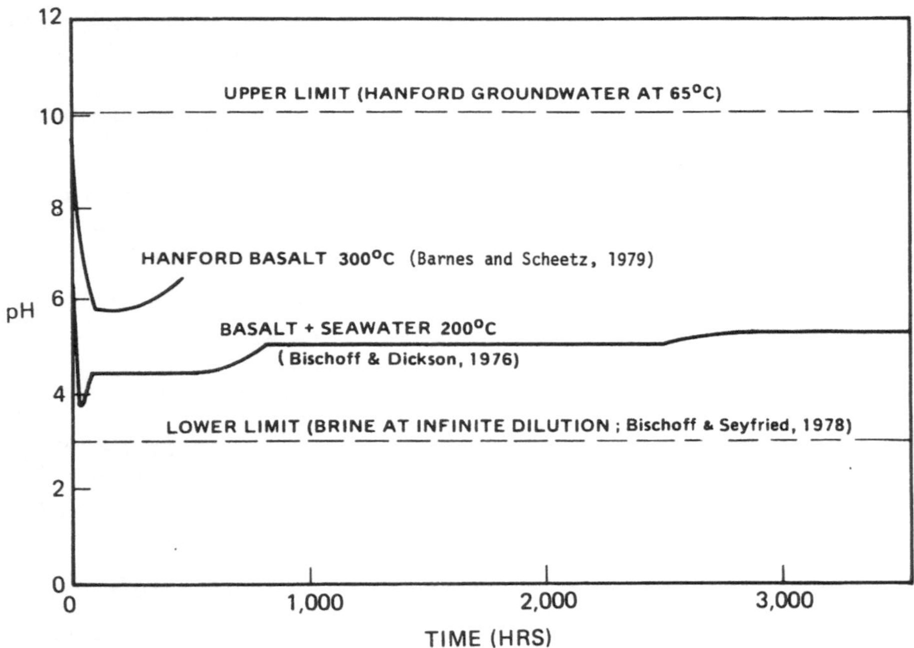

Fig. 1. Variation trend of solution pH as a function of time.

Eq (5) $$6\ Mg^{2+} + 8H_4SiO_4 = Mg_6Si_8O_{20}(OH)_4 + 8H_2O + 12H^+$$

dissolved material = smectite + water + hydrogen ion

Hydrothermal reaction of water and basalt also results in clay. In contrast to precipitation, the hydrolysis of silicates consumes hydrogen ion causing the residual solution to be enriched in base. This reaction causes a slow but steady increase in groundwater pH, and it will control the solution composition when the system is dominated by rock rather than water. For instance:

Eq (6) $$4NaAlSi_3O_8 + 6H_2O = Al_4Si_4O_{10}(OH)_8 + 8SiO_2 + 4Na^+ + 4OH^-$$

Albite + water = kaolinite + silica + base (sodium hydroxide)

The combined effect of precipitation and silicate hydrolysis is depicted in Figure 1. The characteristic trend in geologic systems is an early depression of pH followed by recovery to a steady state determined by the water - rock equilibrium. The extent of pH depression is determined by the water:rock ratio. In a freely flowing repository the absolute value for hydrothermal pH is difficult to predict; it might, however, be as low as pH = 4. Under more normal near-stagnant conditions pH is controlled by the steady-state portion of Figure 1 and can be predicted from:

Eq (7) $$pH = 1.21 + 2.81\ (1000/T°K).$$

This result is illustrated in Figure 2 which also shows some field measurements made on the Hanford Site. Extrapolation of the experimental data is in good agreement with the field data. The coincidence validates the experimental result.

In Figure 3, the experimental trend is compared with the dissociation of the silicic acid. The near perfect agreement indicates that steady-state groundwater pH is controlled by silica speciation. While unusual, silicic acid control results from the composition of Hanford groundwater which is of low ionic strength and saturated in silica.

In summary, field measurements and thermodynamic calculations have been used to define conditions existing in the present geologic environment. These conditions are appropriate to late repository history when perturbations caused by waste burial have subsided. Experimental measurements and additional calculations have been used to predict the repository environment as it will be during the thermal period. A wider range in physicochemical conditions will occur in the thermal period than will occur during geologic control because of thermal and kinetic effects.

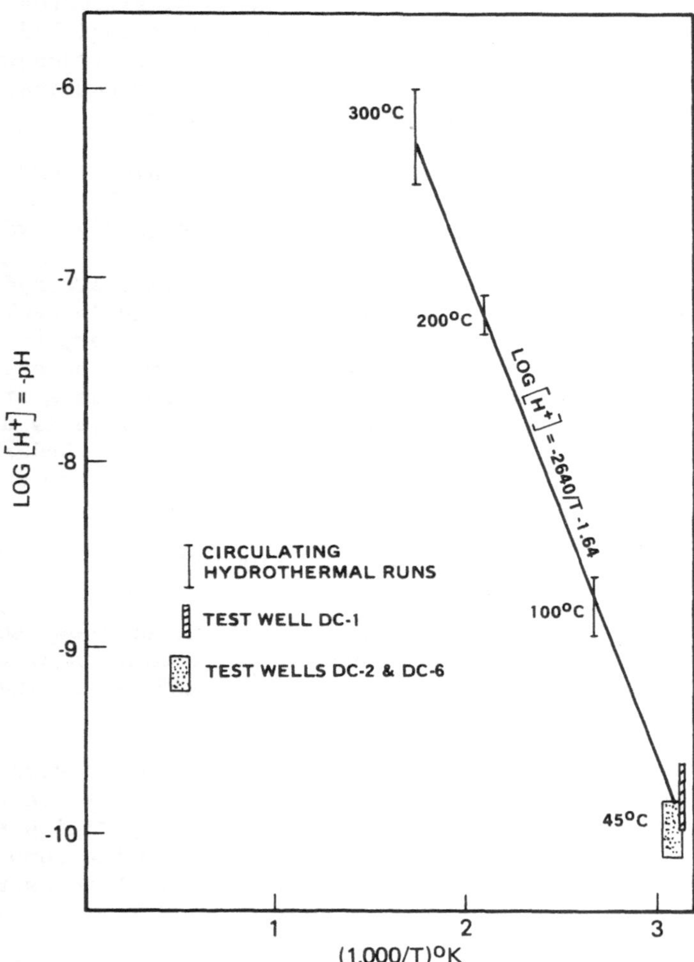

Fig. 2. pH versus 1/T for groundwater in equilibrium with Umtanum basalt, from Barnes and Scheetz (1979).

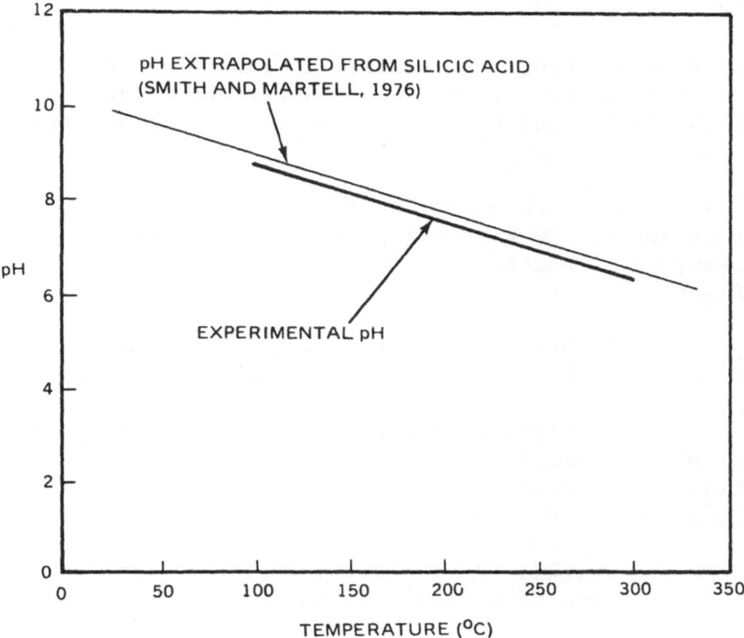

Fig. 3. Control of hydrothermal groundwater pH by silicic acid.

For easy reference, equilibrium conditions for a closed repository are summarized in Table 2. These values have been obtained at a series of temperatures using equations (1), (2), (4) and (7).

TABLE 2. Summary of Repository Equilibrium Conditions

	Temperature (°C)	Equilibrium pH	log Oxygen Fugacity (atm)	Eh(V)
Period of Geologic Controls	65	9.4	-64.5 to -67.15	-0.50 to -0.54
Thermal Period	100	8.7	-57.8 to -60.0	-0.49 to -0.53
	150	7.9	-49.6 to -51.8	-0.48 to -0.53
	200	7.2	-43.3 to -45.4	-0.47 to -0.52
	250	6.7	-38.3 to -40.2	-0.48 to -0.52
	300	6.2	-34.1 to -35.9	-0.45 to -0.51

REFERENCES

Barnes, M. W. and Scheetz, B. E. (1979) "Laboratory Alteration of a Columbia River Basalt by Hot Groundwater: An Application to Deep Geological Disposal of Nuclear Waste, Geological Society of America Bulletin, No. 11, p. 384.

Bischoff, J. L. and Dickson, F. W. (1976) "Seawater Basalt Interactions at 200°C, 500 Bars - Implications for the Origin of Sea-Floor Heavy Metal Deposits and Regulation of Seawater Chemistry", Earth Planet Scie. Lett. 25, pp. 385-97.

Bischoff, J. L. and Seyfried, W. E. (1978) "Hydrothermal Chemistry of Seawater from 25° to 350°C. Amer. J. Sci. 278, pp. 838-60.

Hardy, M. P. and Hocking, G. (1978) "Numerical Modeling of Rock Stresses with a Basalt Nuclear Waste Repository: Phase II - Parametric Design Studies: RHO-BWI-C-23, Rockwell Hanford Operations, Richland, Washington, 303 p.

Smith, R. M. and Martell, A. E. (1976) "Critical Stability Constants, Volume 4: Inorganic Complexes", Plenum Press, N.Y., 255 p.

Ulmer, G. C. (1971) "Research Techniques for High Pressure and High Temperature", Springer-Verlag, N.Y., 367 p.

EVALUATION OF PRODUCT SPECIFICATIONS WITH A SAFETY ANALYSIS FOR A DISPOSAL MINE

Ernst Warnecke and Heinrich Illi

Physikalisch-Technische Bundesanstalt (PTB)

Bundesallee 100, D-3300 Braunschweig

INTRODUCTION

The Physikalisch-Technische Bundesanstalt (PTB) became responsible for the long-term storage[1] and final disposal[2] of radioactive wastes in the Federal Republic of Germany with the 4th amendment of the Atomic Energy Act (Atomgesetz) in 1976. Thus the PTB is the authorized applicant for the licensing procedure for repositories and has to prove to the licensing authority that a repository can be operated safely.

This can be done with a safety analysis analogously to the procedure in nuclear power plant licensing. The normal operation and accident conditions have to be taken into account in the operational phase of the repository. Additionally the stability of the geologic formation has to be investigated for the post operational phase of the mine after backfilling and sealing. The radioactive waste form is the main variable parameter after the selection of a geologic formation and a site for a repository. Thus a safety analysis directly results in requirements for the radioactive waste products to guarantee the overall safety of the total system.

It is intended to come to quantitative product specifications on a scientific basis in an iterative process. This means that the waste forms have to be adapted step by step in such a way that the safety requirements finally can be met. These quantitative requirements result from the limits for radionuclide releases and radiation exposure of the Radiation Protection Order (Strahlenschutzverordnung).

In mid 1978 the PTB came to an agreement on this procedure[3] with the waste producers and started a safety analysis for the planned disposal mine in the Gorleben salt dome, to demonstrate the applicability of this method as a first step towards this goal. Conclusions on the actual radiation exposure cannot be drawn yet.

BASIC MODEL DATA FOR THE SAFETY ANALYSIS

Radioactive Wastes

The different types of radioactive wastes arising from a 1400-t/a reprocessing plant and Pu-fuel element fabrication have been considered and were condensed into seven categories of model wastes (Table 1).

Table 1. Radioactive Wastes

Category	Waste type	Number/a	Transport	Storage
A-1	High active waste (HAW), fission product concentrate, fixed in glass, 70-L glass in a steel canister, 900 W/canister	1760	Cask	Borehole
A-2	Medium active waste (MAW), cladding hulls, dissolution residues, etc., fixed in concrete; 400-L drums; average heat production 158 W/drum	4331	Cask	Borehole
A-3	Medium active liquid aqueous wastes, mainly $NaNO_3$ fixed in cement, 400-L drums	8849	Cask or concrete shielding (A-3 to A-5)	Top loading chamber or staple technique (A-3 to A-5)
A-4	Medium active solid wastes, filters, ion exchange resins, etc., fixed in concrete, 400-L drums	3116		
A-5	Medium active liquid organic waste, TBP fixed in PVC, 400-L drums	499		
A-6	Low active wastes (LAW), all kinds of solid and liquid wastes, fixed in cement; 400-L drums	5385		Tumble down technique
A-7	Low active α-bearing solid wastes, burnable α-wastes from glove boxes, 10-L cans placed in a 400-L drum, space filled with concrete	5175		Tumble down technique

Geology and Hydrology at the Gorleben Site

In 1978 knowledge of the Gorleben salt dome resulted mainly from seismic exploration of that region. A model for the salt dome and the surrounding rocks has been constructed for the safety analysis (Fig. 1). Additionally, a simple hydrologic model has been assumed.

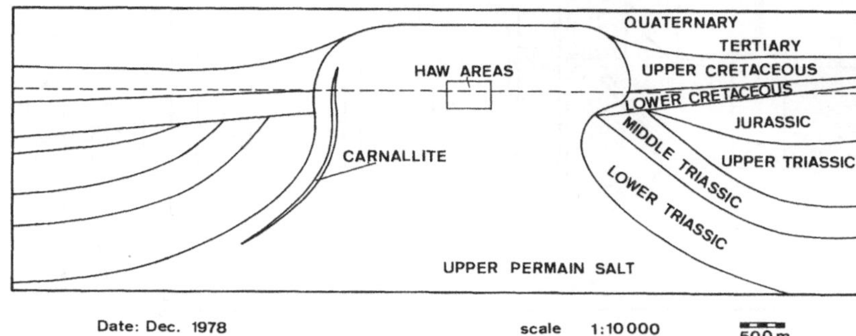

Fig. 1. Model of the Gorleben salt dome.

Layout of the Mine and Transports

A one-floor disposal model mine with two shafts has been planned at 830 m below the surface on the assumption that large amounts of pure halite are available in the salt dome (Fig. 2). The heat producing wastes A-1 and A-2 are stored in boreholes separately from the other wastes which are planned for disposal using tumble down techniques, stacking techniques, and top loading chamber techniques.

Salt mining and waste emplacement are separated as well as the transports of radioactive wastes and mined salt. A one-way transport system is used to avoid accidents.

The thermal layout in the borehole fields is for ≤200°C surface temperature for HAW and ≤100°C central temperature for the 400-L MAW drums.[4]

Excavated salt is used as a backfill material for the disposal chambers, boreholes, and drifts. A filled disposal field is sealed by dams.

SAFETY ANALYSIS

A preliminary detailed safety analysis of the repository under normal and accident conditions was carried out to find out if the

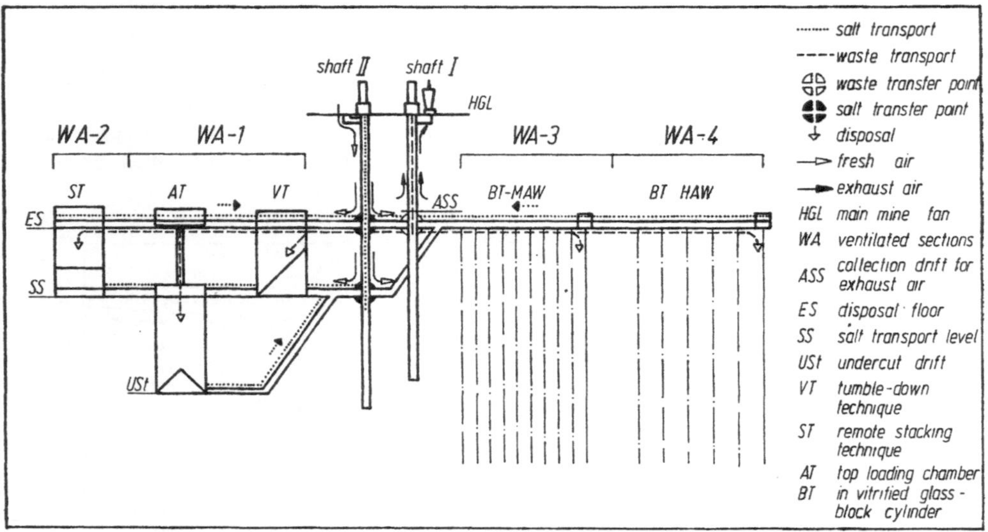

Fig. 2. Flow diagram of waste disposal and salt transport.

mine can be operated safely. The possibilities and consequences of radionuclide releases have been investigated quantitatively. The releases of airborne radionuclides into the environment have been judged with a simplified model by introducing a release criterion. A site specific diffusion calculation must finally prove that exposures are within the limits of the Radiation Protection Order.

Gaseous Radionuclides and Aerosols

The relevant releases from radioactive wastes in the normal operation of the mine could result from gaseous radionuclides and aerosols. Possibly volatile radioactive elements such as radon, iodine, or ruthenium were found to be fixed well and releases could only occur if the waste form is damaged, for example, by accidents.

However, it was found that the release rates for H-3 from Zircaloy fuel element hulls (A-2) fixed in concrete possibly could exceed the legal limits for the H-3 concentration within the air. As this result is based on only a few experimental data, the highest and the lowest estimate differ by several orders of magnitude. A value for the H-3 releases can be taken from Münzel[5] for Zircaloy hulls to be 100 Ci/a after 50 years of mine operation.

A decision whether this waste form is suitable for disposal or needs another fixation can only be made when more experimental results are available. Nevertheless, a maximum permissible average release rate for H-3 from waste category A-2 can be calculated to be under 80 mCi/a per drum for the actual planning data of the repository.

The exposure rates from γ-radiation of the disposed wastes and from radionuclide releases (except possible H-3 releases) were found to be below the legal limits in the mine and in the environment.

In addition to radionuclide releases, the generation of hydrogen due to corrosive or radiolytic effects cannot be excluded. Its impact on limiting requirements for the waste forms has to be studied in detail.

Accidents

Airborne releases. The total disposal system has been analyzed by a deterministic analysis ("release tree analysis"). Sixty-one events have been identified for the operation of the mine and the 35 most severe ones have been investigated.

Some examples of possible accidents in the various parts of the disposal system are as follows:

- Shafthouse
  fire, airplane crash, fall of a waste package from a truck, explosion of a pressure cylinder
- Shaft
  fall of a waste package down the shaft
- Pit bottom and drifts
  fire, salt fall, explosion of a truck with blasting materials
- disposal areas
  (a) borehole technique: fall of a HAW/canister or a drum into a borehole

  (b) top loading chamber: fall of a drum into the chamber

  (c) stacking technique: fall of a waste package from the crane or from top of the staple into the chamber, fall down of the transport vehicle, fall of the crane upon the waste

  (d) tumble down technique: salt fall, fall of a transport vehicle into a chamber, fire in the loading area, fall of a crane upon a transport vehicle.

These events represent a mechanical or a thermal impact or a combination of both. It was difficult to quantify the radioactive releases under these conditions due to a lack of experimental data on some of the disposal-specific parameters. For example, only a few data are available for the generation of airborne particles by mechanical impacts.[6]

Several accidents were found to exceed the release criterion. Practically all accidents within the mine were found not to have exceeded the environmental exposures, due to the mine's good barrier function.

In most events the transgression of the release limit results from the burnable Pu-waste (A-7) if a mechanical impact is followed by a fire because Pu on organic matrices has a high resuspension factor in the case of fire.

The radiological consequences of these accidents, however, can be eliminated by using a suitable shipping cask, a concrete container or a better treatment and fixation of the waste, for example by the fixation of its ashes in concrete.

Within the iterative process an advanced waste form will be used for a second run of the safety analysis to test its suitability for disposal.

Critical releases from other waste forms were found to be possible, especially for HAW, in the case of the accidents (a) fall into the shaft, (b) fall into a borehole, or (c) a collision within a drift. Provisions have been suggested to avoid accidents, to reduce the exposure rates, and to consider or introduce additional barriers.

These measures could, for example, be (a) to consider an appropriate sump, (b) to give up the ball headed gudgeon for the various containers because of its penetration potential in collisions, and (c) shock absorbers for drums or canisters used in the borehole technique.

The other waste forms lead to lower release rates under accident conditions. Nevertheless the organic waste form A-5 (TBP in PVC) seems to be less suitable for disposal because of possible segregation of the constituents.

A comparison of the fall of a container with HAW into the shaft with that of unreprocessed fuel results in an anticipated higher consequence for the fuel element.[7] A rough estimate is about one order of magnitude.

Releases on the water path. Water-borne releases of radionuclides into the environment could occur if the mine were actually flooded by an inrush of water into the shaft. This scenario is a licensing requirement in the Federal Republic of Germany. A first test of the methodology to deal with such an event showed the need for detailed data especially for the kinetics of the reaction between water and salt, the transport mechanisms within the mine, and the geosphere.

Thus, it is necessary to wait for the results of the site investigations before product specifications can be obtained; e.g., the requirements for corrosion and leach rates depend on the actual transport of radionuclides into the biosphere.

Postoperational Phase

The long-term integrity of the salt dome has to be studied mainly because of the thermal load due to the HAW. The temperature distribution and the effects from temperature rise in the model dome have been simulated with a numerical computer program. It could be shown that the thermal expansion of the salt leads to an uplift of 1.2 m in 450 years. Induced diapirism of the salt, failures of the geologic formation, and the surrounding strata could not be identified. From this, no arguments could be found against the thermal output of the HAW canisters and the total heat load of the salt dome.

CONCLUSION

This work was done on the basis of preliminary data for radioactive wastes generated in a future reprocessing plant and assumptions of the geology and hydrology of the salt dome at Gorleben. It showed that it is possible to come to quantitative requirements for type specific radioactive waste forms with an iterative safety analysis for a repository if a suitable data base is available.

ORIENTATION OF FUTURE INVESTIGATIONS

For improved basic data it is necessary

- to incorporate the results of the site investigation program[8] into the safety analysis
- to proceed in the research and development on the characterization of the waste forms due to the specific waste disposal parameters
- to develop a complete quality assurance program for radioactive waste that is able to guarantee the basic data for the safety analysis in practice.

A second run of a safety analysis has to be conducted with the real data from the site and the waste forms. The second run must incorporate the experience from the first run and include the waste from nuclear power plants, research centers, etc., that are currently being generated.

In this way final conclusions on product specifications for radioactive waste can be evaluated that meet the safety requirements and that are flexible enough to be used in practice.

ACKNOWLEDGMENTS

This paper is based on the results of a preliminary safety analysis prepared for PTB by Elektrowatt Ingenieurunternehmung A.G., Zurich, Switzerland (operational phase) and by Battelle-Institut e.V., Frankfurt, F.R.G. (postoperational phase).

REFERENCES

1. E. Warnecke, S. Ahner
Air-Cooled Krypton-85 Storage Facility with Natural Convection International Symposium on Management of Gaseous Wastes from Nuclear Facilities, Vienna, 18-22 February, 1980, IAEA-SM-245/11

2. H. Röthemeyer
Site Investigations and Conceptual Design for the Repository in the Nuclear "Entsorgungszentrum" of the Federal Republic of Germany - International Symposium on Underground Disposal of Radioactive Wastes, Helsinki 2-5 July, 1979, Vol. I, p. 297-310.

3. Seminar über Forschungs- und Entwicklungsarbeiten und Untersuchungen zur Sicherstellung und Endlagerung radioaktiver Abfälle, PTB-Bericht-SE 1, p. 32-39 (Oct. 1979).

4. G. Delisle
Berechnung zur raumzeitlichen Entwicklung des Temperaturfeldes um ein Endlager für mittel- und hochaktive Abfälle in einer Salzformation
Z. dt. geol. Ges. 131: 461 (1980).

5. H. Münzel
Abgabe von Tritium aus Zircaloy - Private communication (April 1980).

6. T. H. Smith, W. A. Ross
Impact Testing of Vitreous Simulated High-Level Waste in Canisters, BNWL - 1903 (May 1975).

7. Vergleich der verschiedenen Entsorgungsalternativen und Beurteilung ihrer Realisierbarkeit. Studie "Entsorgungsalternativen". KfK 3000 (Sept. 1980).

8. H. Röthemeyer, K. D. Closs
High-Level Waste Disposal
ANS/ENS International Conference on World Nuclear Energy - Accomplishments and Perspectives, Washington, D.C., 17-21 November, 1980.

PREDICTING THE REACTION STATE OF BRINES IN PROPOSED REGIONS OF NUCLEAR WASTE ISOLATION*

R. L. Bassett, M. E. Bentley, E. A. Duncan,
and J. A. Griffin

Bureau of Economic Geology
The University of Texas at Austin
Austin, Texas 78712

INTRODUCTION

A number of independent investigations are being conducted nationally to evaluate the feasibility of storing radioactive waste in salt formations. Thorough evaluation of any potential host geologic system must include site-specific assessment of the volume and type of fluids present, the direction and magnitude of flow, and the reaction potential. This paper describes the preliminary investigation of the Bureau of Economic Geology (BEG) into the reaction state of subsurface fluids in the Texas Panhandle.

Permian salt sequences in the Palo Duro and Dalhart Basins have been selected by the Department of Energy for further study as potential host media for high-level nuclear waste (fig. 1). A multidisciplinary study has been initiated by BEG, employing simultaneous investigation of surface processes, structure, resource analysis, stratigraphy, hydrology, and geochemistry.[1] The Palo Duro and Dalhart Basins are Late Paleozoic intracratonic subbasins of the Permian Basin. Pre-Permian strata are dominated by shelf carbonates and peripheral nearshore sandstones. In Permian time the environment was dominated by coastal sabkha systems with well-defined cyclicity. Sabkha lithofacies range (landward) from shelf carbonates and supratidal dolomite and anhydrite, halite salt pan, upper sabkha salt/mudstone, to wadi plain mudstone.[1]

The basins are now stable, and gravity-driven fluids are moving through them on a regional scale. Except for the shallow

*Publication authorized by the Director, Bureau of Economic Geology, The University of Texas at Austin.

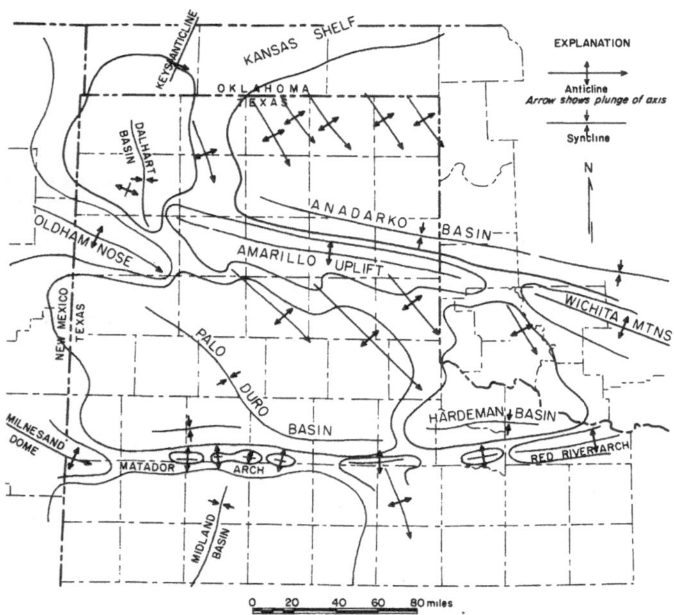

Fig. 1. Structural elements and general index map of the Texas Panhandle (from Nicholson, 1960, The University of Texas at Austin, Bureau of Economic Geology Publication No. 6017, p. 51-64.)

potable unconfined aquifers, the fluids have brine compositions (>35,000 ppm TDS). Three areas of uncertainty affect efforts to quantify the behavior of waste in this hydrologic system: (1) location, quantity, and mobility of brine, (2) chemical composition of the brines, and (3) theoretical limitations in describing complex and concentrated electrolyte solutions.

SPATIAL DISTRIBUTION OF BRINES

The Palo Duro and Dalhart Basins, like most basins containing evaporites, have aqueous fluid distributed in at least five locations: (1) shallow fresh ground water, (2) brine from dissolution of halite, (3) regional deep flow systems, (4) brine as fluid inclusions in evaporite minerals, and (5) water of hydration on clay minerals, oxyhydroxides, and hydrous phases. Shallow potable aquifers in this region have rapid through-put and average Darcian velocities of up to 10 meters per year. There is a hydraulic potential for downward flow of this water through the evaporite section; however, the permeability is low, and the flux is probably very small under current conditions.

Shallow ground-water flow above the evaporites is essentially horizontal. Fluid moving along the base of aquifers in the eastern part of the area encounters halite, which is continually dissolved and transported out of the region. Stratigraphic and geochemical evidence indicates that dissolution is an active process, and substantial collapse and subsidence have resulted from removal of salt.[1] Dense sodium chloride brines emerge as springs and seeps along eastward-draining tributaries to several major rivers.

The regional deep flow system has been described in preliminary fashion.[2] Hydraulic properties are being derived from analysis of petroleum drill-stem tests and core tests. There appears to be an eastward movement of deep basin fluids with interstitial velocities ranging from 1.3 to 13 cm per year. The porous shelf carbonates underlying the evaporite sequences provide a permeable hydraulic conduit through the basin. Composition of the brines is regionally variable, but sodium and chloride are dominant.

Brine present as fluid inclusions occurs almost exclusively in the halite crystals. This brine is saturated with sodium chloride but also must contain significant quantities of calcium and probably magnesium, as indicated by freezing point determinations (E. Roedder, pers. comm., 1980). Water present as fluid inclusions is isolated, is relatively immobile in the absence of a strong thermal gradient, and is not connected hydraulically to the regional flow field.

Hydrous phases do not provide mobile water unless perturbed by heat or pressure. Water is normally released when clays transform during diagenesis; however, this source will not be discussed here.

AVAILABILITY OF DATA

Shallow ground-water systems and brine-emission areas are easily sampled and the fluid compositions are well known. Data are available from historical monitoring of water supplies as well as from samples recently collected by the authors.

Chemical analyses of deep basin fluids are available almost exclusively from the oil industry as a by-product of their exploration program. Samples are frequently contaminated with drilling fluid or hydrocarbons produced during drill-stem testing.

Composition of fluid inclusions in relatively insoluble minerals like quartz is obtained by crushing the mineral, leaching the inclusions, then diluting and analyzing the fluid. In soluble minerals like halite, we are currently limited to heating and freezing measurements until microanalytical techniques are perfected. No chemical analyses of fluid inclusions are currently available for these salt formations. All data will be derived from experimental work currently ongoing at the Bureau of Economic Geology.

THEORETICAL CONSTRAINTS ON ELECTROLYTE CHEMISTRY

In recent years there have been attempts to develop equations that describe ionic behavior in concentrated electrolyte solutions. The fundamental expression by Debye and Hückel[3] and subsequent extensions presented by others generally have not performed adequately in the evaluation of concentrated brines, or in the investigation of solubilities of highly soluble minerals. The exclusion of an accurate description of short range interionic forces in the computation of potentials apparently has been a major source of error in fitting the experimental data. Pitzer,[4] Scatchard,[5] and others have developed a rigorous and comprehensive theory for electrolyte solutions which can be approximated by a series of semiempirical equations. The equations though somewhat cumbersome, especially for mixed electrolyte systems, are rather easily implemented with the aid of a digital computer.

A rigorous test of Pitzer's equations to model mineral solubilities has been conducted by Harvey and Weare.[6,7] These workers found excellent agreement with published solubility data and have used the approach of free energy minimization to compute mineral sequences observed in the Permian Zechstein evaporites.[7]

A computer model, AQ/SALT, has been developed by the authors for applications in radioactive waste disposal and the geochemistry of deep formation brines. We have done little to improve the equations of Pitzer, used in the model, or the computationally more appropriate expressions given by Harvey and Weare.[7] The model is discussed in more detail elsewhere and will be only briefly mentioned here.[8] AQ/SALT is table driven, will accommodate any number of species, and is readily expanded as new data and equation parameters become available. The model computes activity coefficients, activities of ions, and the activity of water. Temperature dependence is available for only a few species; however, additional parameterization is currently underway.

By comparing calculated activity products with thermodynamic equilibrium constants, the reaction state of a brine with respect to the minerals of interest can be determined. The model lists the possible phase assemblages and determines whether they are stable or have the potential to dissolve in the brine or if new phases will be precipitated as the brine moves through the system. Activity products computed from solubility data for halite and gypsum (fig. 2) are compared with the equilibrium constants to illustrate the accuracy of the model for this assemblage.

REACTION POTENTIAL OF FLUIDS

Deep Basin Formations

Long-term risk-assessment studies of regions like the Palo Duro Basin will eventually evaluate the most likely path for nuclide movement away from a repository in the event that a breach in the storage facility should occur. Most hydraulic connections in the deep basin will contain brine, and assessments of the potential for transport or retention must consider the effects of the brine and host-rock interactions.

The origin of deep formation brine compositions has always been an enigma for geochemists due in part to inaccurate chemical data and the lack of a model for electrolyte chemistry. The approach generally taken is simply to compare ionic ratios or discuss chemical processes that modify seawater during burial of the

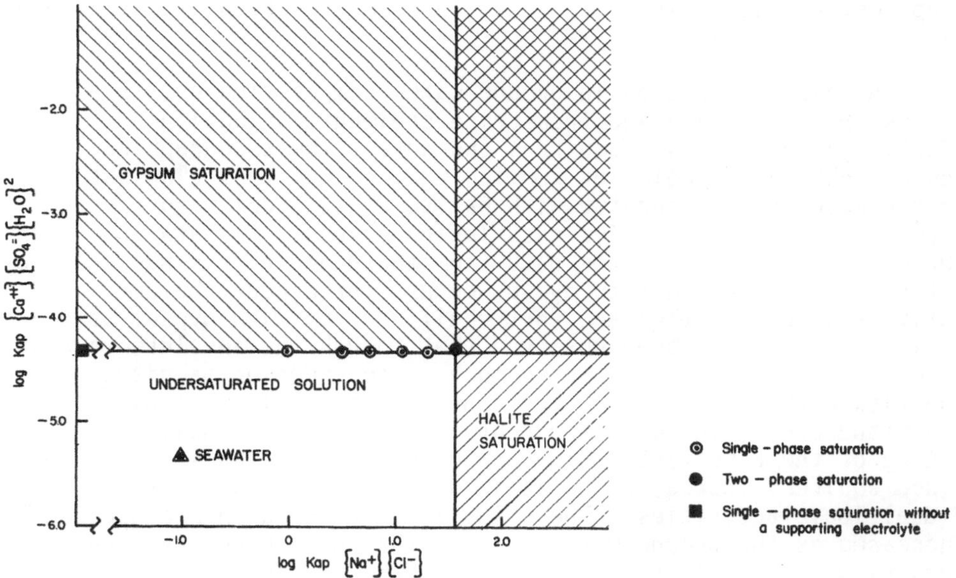

Fig. 2. Stability field diagram based on computed log activity products (log Kap) for gypsum and halite solubility data (from Linke, 1965, Am. Chem. Soc., p. 681).

sediments. Evaluation of the brine reaction states would be particularly instructive when unraveling diagenetic changes, mixing of fluids, or predicting mass transfer along a flow path.

In the deep Palo Duro Basin, the absolute mass of calcium and sulfate ion increases with increasing ionic strength, and the concentration ratio $Ca^{++}/SO_4^{=}$ ranges from 0.1 to 11.8. Preliminary computation by the AQ/SALT model indicates that 85 analyzed deep-basin brines in a seven-county region are all very near equilibrium with anhydrite but are undersaturated with respect to halite (fig. 3). Halite has not been reported in these deep carbonate aquifers, but saline waters from overlying evaporites may mix with the deeper fluids and affect the brine composition. Secondary anhydrite has been reported in carbonate rocks in the northern Palo Duro Basin and ongoing research will investigate the relationship between brine reaction states and the occurrence of anhydrite cements. A discussion of deep-basin geochemistry and hydrology is given in detail elsewhere.[3]

Shallow Zones of Salt Dissolution

Substantial dissolution of evaporite lithologies has occurred around the margins of the Palo Duro Basin. Brines near saturation with sodium chloride are known to exist below potable ground-water supplies, particularly in the eastern counties of the Texas Panhandle.

A major alteration in ground-water composition occurs as it leaves the fluviatile and lacustrine sediments of Ogallala (Tertiary) and Triassic aquifers and enters the topographically lower sabkha lithologies of Permian age. Along this flow path ground water undersaturated with halite and anhydrite enters the sabkha sediments and active dissolution ensues. Computations with AQ/SALT, using ground-water data from these aquifers, indicate that saturation with respect to gypsum occurs throughout the area of active salt dissolution and that equilibration with gypsum is rapid. It has also been observed that though the absolute mass of calcium sulfate may increase as much as an order of magnitude along the flow path, the ground water continues to follow the gypsum phase boundary (fig. 4). Apparently dissolution of halite and mixing of the resultant saline ground water with shallower unconfined aquifers increased the ionic strength and lowered the individual ionic activities. The solubility of gypsum has consequently increased as the ground water follows the phase boundary. For comparison, a sample from Bristol Dry Lake, a continental sabkha in California, is actually at the point of two-phase equilibrium with halite and gypsum (fig. 4). The reason for apparent equilibrium with gypsum rather than anhydrite at those ionic strengths is unclear at this time and further work is in progress.

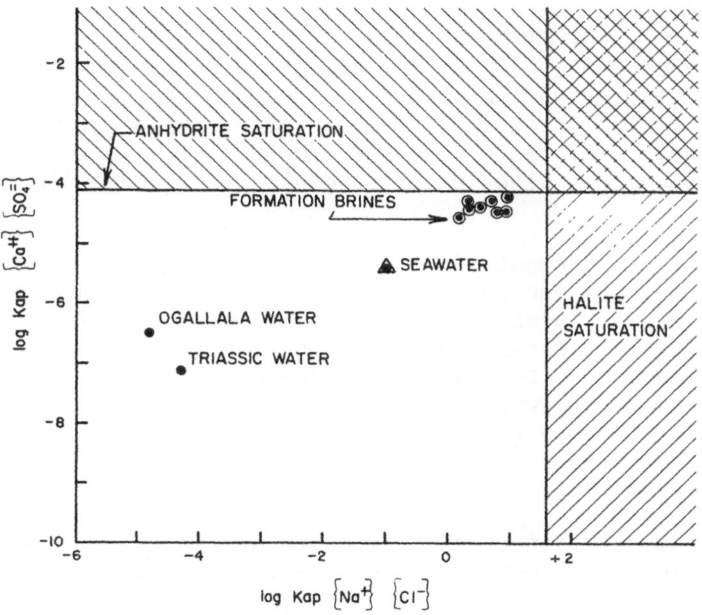

Fig. 3. Stability field diagram based on log activity products (log Kap) for anhydrite and halite (county-wide averages).

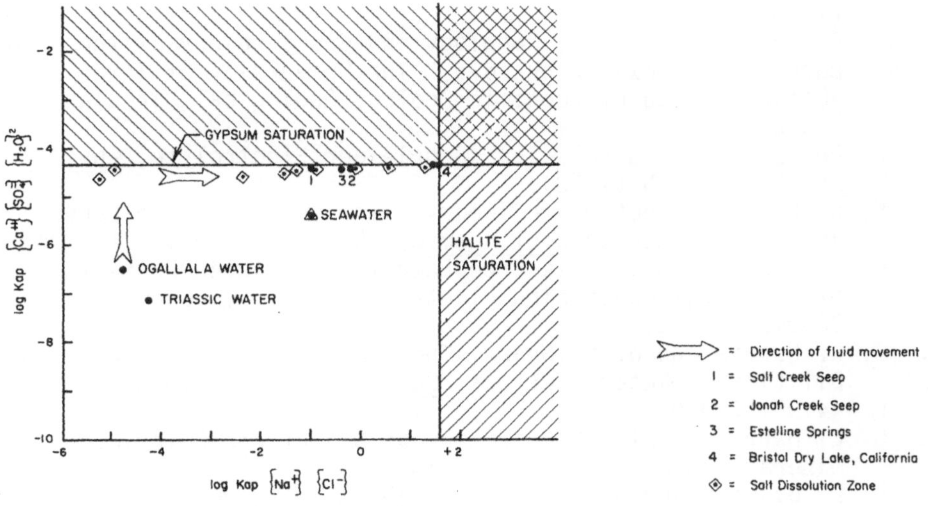

Fig. 4. Stability field diagram based on log activity products (log Kap) for gypsum and halite.

CONCLUSIONS

The fluids most likely to interact with nuclear waste materials buried in evaporite strata are brines. An accurate characterization of the potential for these fluids to react with the host media or the buried waste requires a mathematical model that describes the ionic behavior in concentrated aqueous electrolyte solutions.

Preliminary computations with the new computer model AQ/SALT indicate that equilibrium between anhydrite and the deep formation brines may be widespread. The shallow brines resulting from active salt dissolution are derived from undersaturated ground water, which dissolves both halite and anhydrite, then quickly comes to equilibrium with gypsum. Calcium sulfate is being removed from the basin largely because of increasing ionic strength from halite dissolution, even though the fluids are following the anhydrite or gypsum phase boundary.

Characterization of the carbonate system will require additional chemical and thermodynamic data, which should be available soon.

REFERENCES

1. S. P. Dutton et al., "Geology and Geohydrology of the Palo Duro Basin, Texas Panhandle," The University of Texas at Austin, Bureau of Economic Geology Geological Circular 79-1, 99 p. (1979).
2. T. C. Gustavson et al., "Geology and Geohydrology of the Palo Duro Basin, Texas Panhandle," The University of Texas at Austin, Bureau of Economic Geology Geological Circular (in press).
3. P. Debye and E. Hückel, Zur Theorie der Electrolyte, Phys. Z. 24:185; 24:305 (1923).
4. K. S. Pitzer, Electrolyte Theory--Improvements Since Debye and Hückel, Accounts of Chemical Research, 19:371 (1979).
5. G. Scatchard, The Excess Free Energy and Related Properties of Solutions Containing Electrolytes, J. Am. Chem. Soc., 90:3124 (1968).
6. C. E. Harvie and J. H. Weare, The Prediction of Mineral Solubilities in Natural Waters, Geochimica et Cosmochimica Acta, 44:981 (1980).
7. C. E. Harvie et al., Evaporation of Seawater: Calculated Mineral Sequences, Science, 208:498 (1980).
8. R. L. Bassett and J. A. Griffin, AQ/SALT (in review).

LABORATORY INVESTIGATION OF WATER CONTENT WITHIN ROCK SALT AND ITS BEHAVIOR IN A TEMPERATURE FIELD OF DISPOSED HIGH-LEVEL WASTE

Norbert Jockwer

Gesellschaft für Strahlen- und Umweltforschung mbH
Institut für Tieflagerung, Wissenschaftliche Abteilung
Theodor-Heuss-Str. 4
D-3300 Braunschweig, Federal Republic of Germany

In the FRG it is planned to dispose high-level radioactive waste from reprocessing fuel elements in rock salt formations. The rock salt of these formations contains small amounts of water. In order to develop the disposal concept and to optimize the long-term safety, it is important to know the following details concerning the water content:

1. How much water is in the rock salt?
2. In what physical and chemical states does the water exist?
3. What happens to the water in the disposed area of the predicted temperatures and temperature gradients?
4. What interaction between the water and the waste containers, the solidification products, and the radioactive waste can occur?
5. Can gas be produced by radiolysis?
6. Can the water be a transport medium for released radionuclides?

The investigation of rock salt from different North German salt mines showed that at least three components of water exist within the rock salt. The majority of the water is from water of hydration of the minor minerals in the salt and from intergranular water adsorbed to the crystal boundaries. The amount of water from fluid inclusions in the salt of North German mines is comparatively low.

Disposal of high-level radioactive waste is planned in the center of a salt anticline at a depth greater than 800 m. At this depth the minor minerals that are more than 0.1 wt % are anhydrite ($CaSO_4$), sylvite (KCl), polyhalite [$Ca_2K_2Mg(SO_4)_4 \cdot 2H_2O$], and kieserite ($MgSO_4 \cdot H_2O$). Gypsum does not exist at a depth greater than 600 m. Therefore the only minerals with water of hydration in a disposal area are polyhalite and kieserite. Rock salt formations which contain carnallite ($KMgCl_3 \cdot 6H_2O$) are excluded from consideration for the disposal of radioactive waste. But it might be possible that formations of carnallite contact the potential disposal areas and that this carnallite would be heated indirectly by the rock salt formation of the disposal area. Therefore it must be guaranteed by the repository layout that carnallite will not be heated to the temperature where water is released. Kainite ($KMgClSO_4 \cdot 11/4H_2O$) is a mineral that is found only at the top of a salt anticline where radioactive waste disposal would not take place; hence this mineral is not important relative to this paper. Other minerals do not exist in salt formations where radionuclides are to be disposed or they are only trace minerals with an amount much less than 0.1 wt %, so that it is not necessary to take them into consideration in evaluating the water content of the rock salt.

In order to investigate the thermal liberation of the different water components within the rock salt, the loss of weight as a function of time at different constant temperatures between 90 and 630°C has been measured. Some typical curves are shown in Fig. 1. This figure indicates that over nearly the same temperature intervals the distance between the curves are not constant. When the loss of weight after 50 hours is plotted as a function of the temperature, it shows that increasing temperature frees the water within the rock salt in steps. A relationship between these steps and different water components does exist. Figure 2 shows the liberation of these components from three salt samples of different mineralogical compositions.

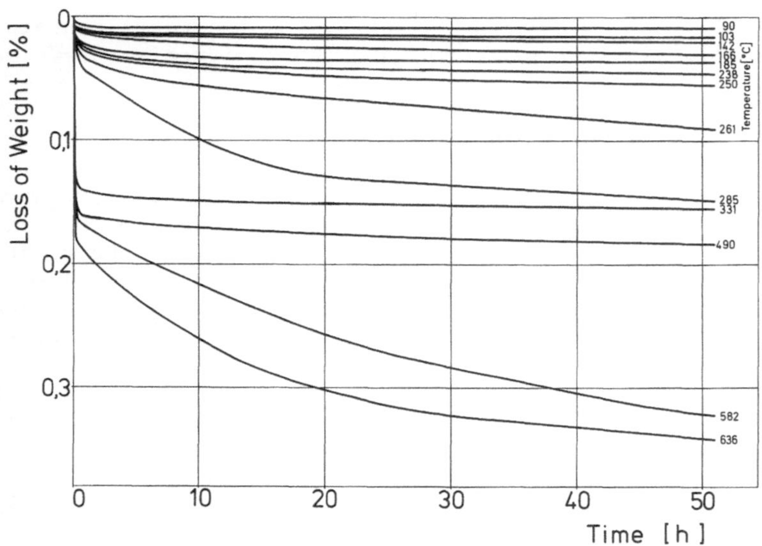

Fig. 1. Loss of weight as a function of time for various temperatures.

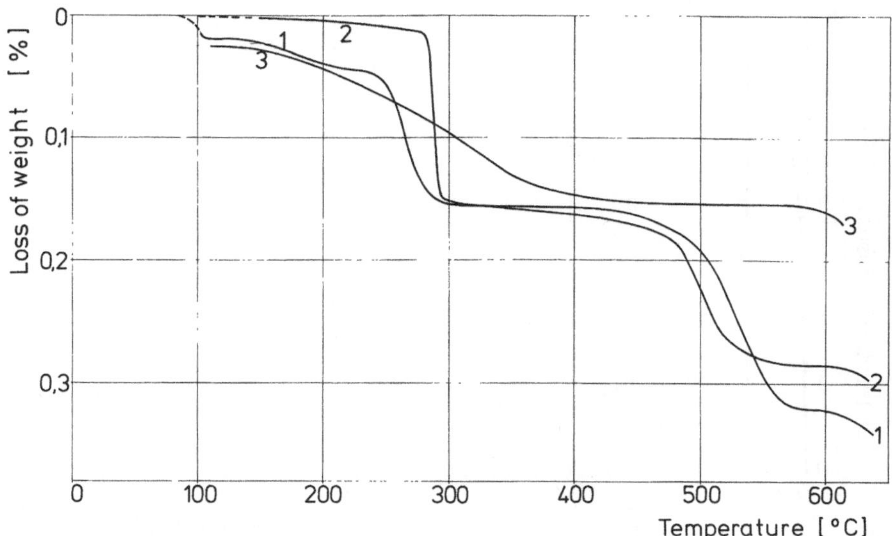

Fig. 2. Loss of weight from three salt samples as a function of temperature. Curve 1: Polyhalitic salt with 0.12% hydration water from polyhalite and 0.035% adsorbed water. Curve 2: Kieseritic salt with 0.147% hydration water from kieserite and no adsorbed water. Curve 3: Synthetic salt with all water adsorbed at the crystal boundaries.

To measure the loss of weight, the oven of the thermoanalyser was rinsed with dried nitrogen. Figure 2 shows that in an absolute dry gas stream no minimum temperature for the liberation of the water exists. The different steps are caused by different release velocities. The loss of weight above 300°C is caused by thermal cracking of some of the minor and trace minerals, the liberation of gas components, and submicroscopic fluid inclusions. Above 600°C rock salt began to sublime.

One of the most important questions concerning the disposal of radioactive waste is the amount of water within the rock salt. Hence 202 salt samples from the Asse salt mine from different depths and different stratigraphic layers have been analysed. The water content has been measured by at least two different methods. Figure 3 shows the distribution frequency of the water content of these 202 samples. The distribution maximum is at about 0.04 wt %, showing that the water content is not constant but depends on the amount of minor minerals present containing water of hydration. Fifty-five percent of these salt samples had a water content less than 0.1%, and 75% were less than 0.2% water.

The investigation of salt samples from core drillings in different stratigraphic layers indicated that the water content varies by a factor of 10 within a distance of less than 1 m.

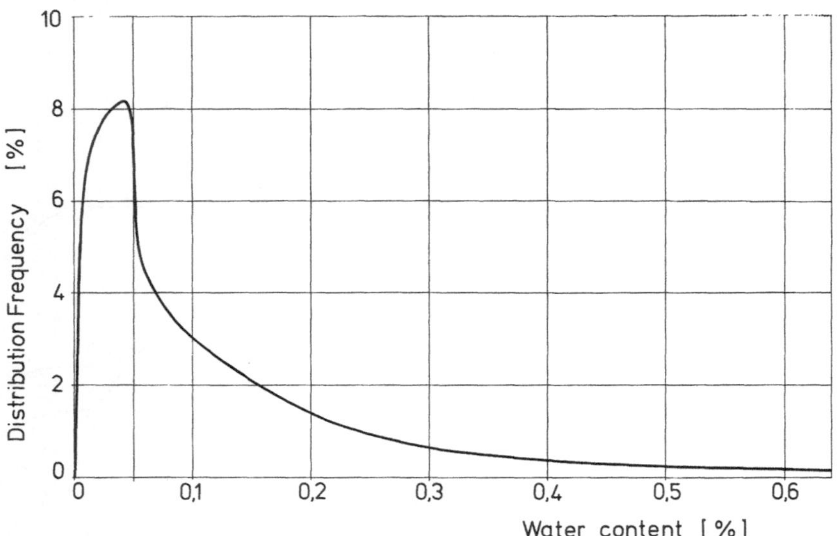

Fig. 3. Distribution frequency of the water content of 202 rock salt samples.

As the majority of the water within the rock salt is associated with the minerals polyhalite and kieserite, the thermal behavior and the liberation of the hydration water from these minerals were analysed. In this analysis the carnallite was included. Though the carnallite is not a minor mineral within the rock salt of a disposal area, such formations may border the disposal area and be heated indirectly.

The thermal investigation of the hydrated minerals indicated that the liberation temperature of water depends on the humidity of the air above the sample. Therefore in succeeding examinations the oven of the thermoanalyser was rinsed with air of a specific humidity. Figure 4 shows the temperature at which the liberation of the hydration water begins for carnallite versus the absolute humidity. With absolute dry air above the sample, no minimum liberation temperature exists.

Figure 5 shows the release velocity on a logarithmic scale as a function of the reciprocal absolute temperature for different humidities of the air above the sample.

Here the release velocity is

$$g = \frac{G}{t \cdot E},$$

where G is weight loss by liberation of hydration water, t is time, and E is total sample weight.

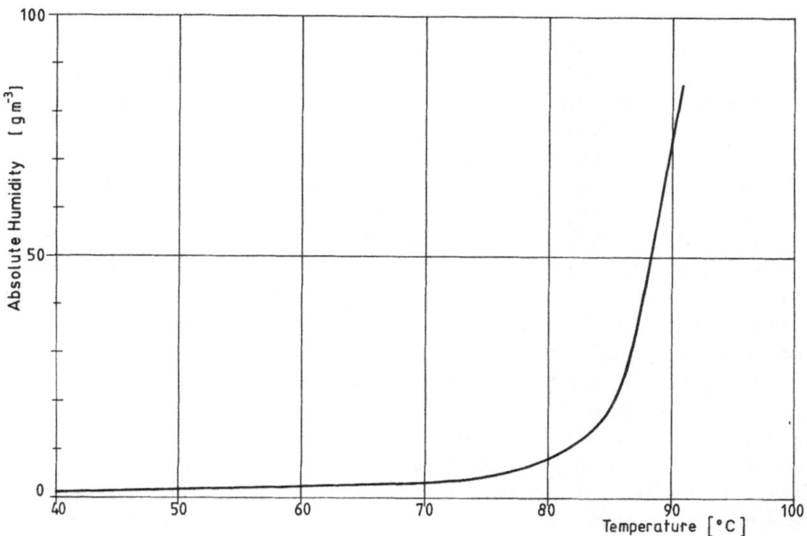

Fig. 4. Temperature at which the liberation of the hydration water of carnallite begins versus the absolute humidity.

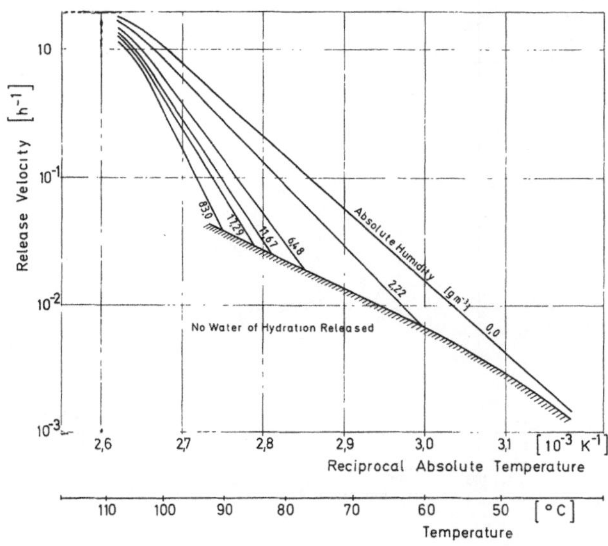

Fig. 5. Release velocity of the hydration water from carnallite at absolute humidities between 0 and 83.0 g/m^3 versus the reciprocal absolute temperature.

This figure indicates that at a given temperature, equilibrium between liberation of hydration water and resorption occurs if the air above the sample has a specific humidity. With absolutely dry air above the sample, this equilibrium occurs only after essentially all of the water of hydration has been released.

Figures 6 and 7 show the release velocities of the hydration water of the minerals polyhalite and kieserite. They show the same phenomenon as the carnallite; the only difference is that the release range for the carnallite in this diagram is between 50 and 100°C whereas the range for the polyhalite is between 200 and 250°C and for the kieserite between 250 and 325°C.

As the minerals with hydration water are found in the rock salt of a disposal area in amounts up to several percent, this liberation will also take place in the temperature field of high-level radioactive waste. The liberated water may migrate along grain boundary surfaces and through microfissures and intergranular spaces as a result of a vapor pressure gradient, temperature gradient, or concentration gradient. In order to develop a mathematical model of the mass transport and to calculate how much water will be released into a borehole, it is important to know what happens to the water

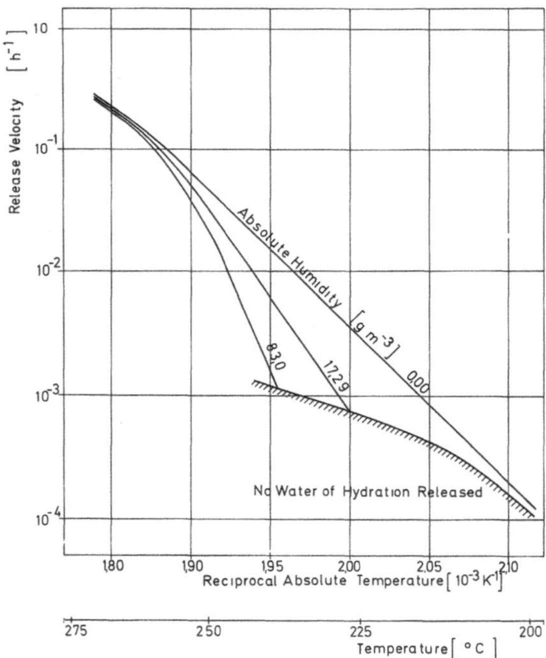

Fig. 6. Release velocity of the hydration water from polyhalite at absolute humidities between 0 and 83.0 g/m^3 versus the reciprocal absolute temperature.

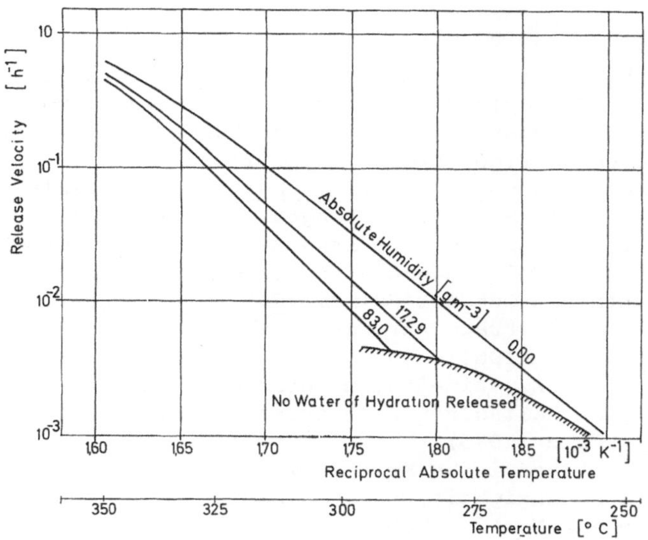

Fig. 7. Release velocity of the hydration water from kieserite at absolute humidities between 0 and 83.0 g/m^3 versus the reciprocal absolute temperature.

that migrates to the surface of emplaced high-level radioactive waste and to know the diffusivity and permeability of rock salt in the area surrounding the emplacement borehole.

The diffusivity range for rock salt samples from the Asse salt mine have been measured at different temperatures and pressure gradients. Values varied between 0.2 and 2.0 x 10^{-4} cm^2/sec. The diffusivity decreases with increasing temperature, which means that it is inversely proportional to the viscosity of the water vapor within the intergranular spaces.

The permeability and porosity of the rock salt are being measured at the present time. Values and bandwidth do not exist yet.

ACKNOWLEDGEMENT

This investigation is being made within the research contracts 058-78-1 WASD of the European Atomic Energy Community "Management and Storage of Radioactive Waste," Action No. 7: "Disposal of Radioactive Waste in Geological Formations."

REFERENCES

1. T. R. Anthony and H. E. Cline, "The Thermomigration of Biphase Vapour-Liquid Droplets in Solids," Acta Metall. Vol. 20 (1972).
2. E. Hofrichter, "Zur Frage der Porosität und Permeabilität von Salzgesteinen," Erdöl Erdgas 92(3) (1976).
3. O. Braitsch, Entstehung und Stoffbestand der Salzlagerstätten, Springer-Verlag, Berlin, Gottingen, Heidelberg (1962).
4. J. Leonhardt, "Der Kleinste Laugentropfen als Grenzgröße Zwischen van't Hoffschen Gleichgewichtssystemen und Laugen-freien Synthesen Ozeaner Salze," Naturwissenschaft 39:20 (1951).
5. H. Trollert, "Beiträge zur Porosität von Salzgesteinen, Kali Steinsalz No. 2 (1964).
6. S. Vielhauer, "Experimentelle Untersuchungen über die Wasser-bestimmung in Kalisalzen nach er Karl-Fischer-Methode," Kali Steinsalz No. 8 (1970).
7. N. Jockwer, "Thermische Kristallwasserfreisetzung des Car-nallits in Abhängigkeit von der Absoluten Luftfeuchtigkeit," Kali Steinsalz No. 2 (1980).

THERMAL CONDUCTIVITY OF SELECTED REPOSITORY MINERALS*

M. J. Skvarla, J. W. Vandersande,** M. L. Linvill
and R. O. Pohl

Laboratory of Atomic and Solid State Physics
Cornell University, Ithaca, New York 14853

For the assessment of the safety of nuclear waste disposal in a geologic repository, knowledge of the near- and far-field temperature profile is of utmost importance. For a given thermal loading, this profile is determined largely by the thermal conductivity of the rock in which the repository is constructed. It is the purpose of the present investigation to explore the range of conductivity that may be expected in granite, basalt and rock salt, and to search in particular for their lower limits. Knowledge of such a limit would be useful in determining the maximum temperature rise that may occur for a given repository loading.

The thermal conductivity Λ was determined below 150 K with the steady-state heat flow technique,[1] and above 100 K with the Angstrom method.[2] In the latter, Λ was calculated from the measured thermal diffusivity, $D = \Lambda/C_v$, using the known specific heat, C_v. The sample sizes were typically 6 cm x 5 mm x 5 mm. All measurements were performed in a vacuum.

In electrical insulators, heat is carried by elastic waves or, in the quantum picture, by phonons. Lattice defects can act as scattering centers for these waves and thus lower the thermal conductivity. In the highly disordered amorphous solids or glasses, for example, the conductivity is much smaller than in the ordered,

*Work supported by the U.S. Nuclear Regulatory Commission, Contract No. NRC-0-78-261.

**Present address: Physics Department, University of the Witwatersrand, Johannesburg, South Africa.

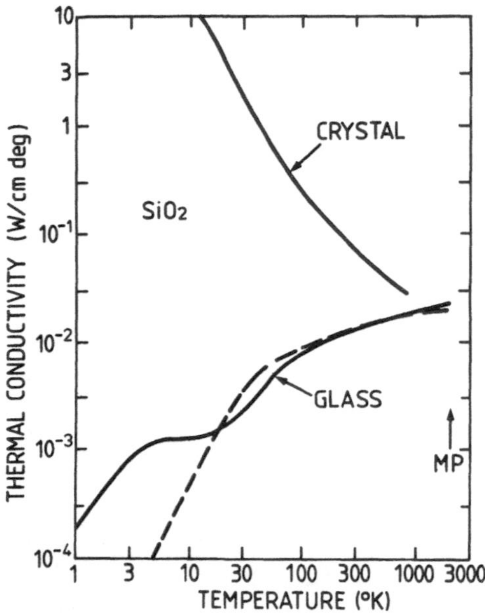

Fig. 1. Thermal conductivity of crystalline SiO_2 (quartz) and of amorphous SiO_2 (silica). The dashed line is the theoretical minimum thermal conductivity of SiO_2, after Slack (ref. 3). MP: melting point of quartz.

crystalline phase, as is shown in Fig. 1 for SiO_2. It has been shown that in the amorphous phase the mean free path traveled by the waves between collisions is only one wavelength. An example is shown in Fig. 1, in which the dashed curve, calculated by Slack,[3] assuming a mean free path of one wavelength agrees well with the measured conductivity above 100 K. It has been pointed out by Kittel[4] that this mean free path near room temperature is of the order of a few Angstrom units (Å = 10^{-8} cm), i.e. the interatomic spacing. In this case, the heat transport is better described as a random walk of the thermal, vibrational energy, from one atom to a neighboring one, with the residence time of the energy at each atom being of the order of one oscillatory period. Such a process leads to the lowest heat transport, i.e., the lowest conceivable thermal conductivity of the solid. This conductivity is determined by the period of oscillation, which is related to the Debye frequency, and the average atomic or ionic radius.

Figure 2 shows a comparison of the conductivities of samples of granite and basalt with those of SiO_2 in both the crystalline and the amorphous phases. At low temperatures, the conductivities of the two rocks differ markedly from those of either quartz or silica, as would be expected in these highly impure crystalline solids.

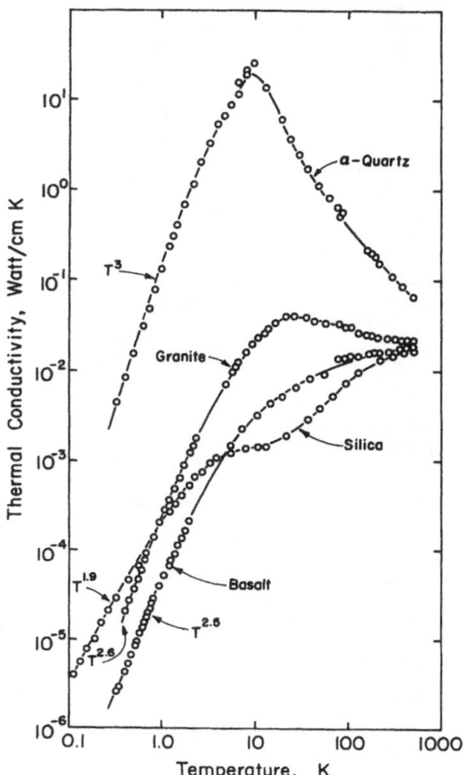

Fig. 2. Thermal conductivity of a sample of granite (quartz monzonite, Climax Stock) and of basalt (Hanford), compared with that of quartz and silica.

Near room temperature, however, the conductivities of granite and basalt approach that of the amorphous silica. The phonon mean free path determined from these data has also been shown to become equal to that of silica (i.e., a few Å).[5] This observation leads to the important conclusion that in these two rock samples at least, the thermal conductivity is almost as low already as it could possibly be. The cause for such a low conductivity in these largely crystalline, though highly disordered solids, is presently unknown.

In an attempt to test our conclusion that the thermal conductivity of silica represents a lower limit of the conductivity of all basaltic and granitic rocks, we measured samples from different sources, and also compiled data from different studies.[6-12] The data shown in Fig. 2 are those shown with open circles in Figs. 3 and 4, and do indeed represent essentially lower limits for the conductivities measured on all other samples. The measured range

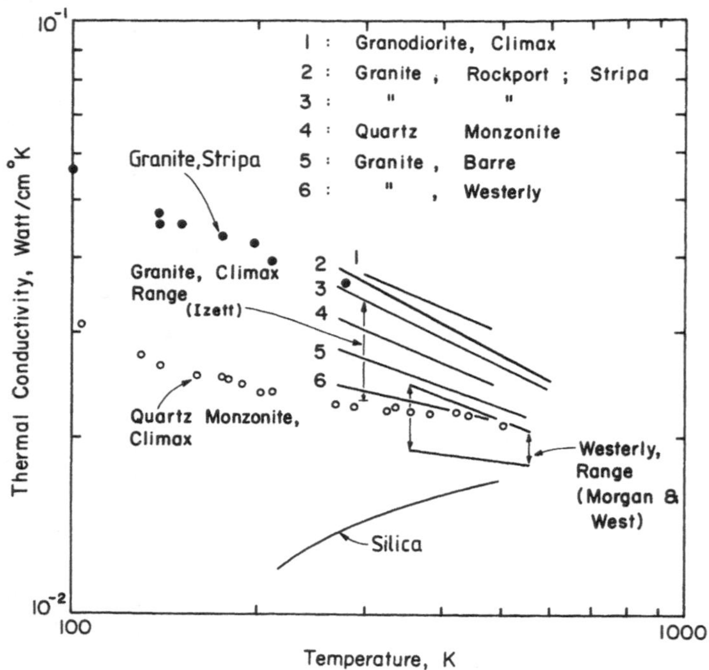

Fig. 3. Summary of thermal conductivity of granite from a variety of sources. Curve 1: Climax granodiorite, ref. 7; curves 2 through 6: ref. 6; curve 2 also: Stripa granite, ref. 7; double arrow, labeled Izett: range of room temperature data from one drill hole into the Climax stock, ref. 8; Westerly range: ref. 9; data points, present investigation: closed circles, Stripa granite; open circles, quartz monzonite, Climax.

of conductivity for all granitic or all basaltic rocks is quite large. Note that the granitic samples measured by Izett[8] had been obtained from the same drill hole, and yet show a wide range in conductivity. In inspecting the Figs. 3 and 4, there appears to be no reason why conductivities much lower than those measured should not also exist in these rock types. Yet, based on the arguments presented here, it follows that a lower limit of the conductivity does exist, which is less than a factor of two smaller than the thermal conductivities known to exist in these materials, on the basis of the existing measurements.

Our attempt to extend the concept of the minimum thermal conductivity to rock salt turned out to be less fruitful. Its minimum conductivity has been estimated by Slack[3] to lie between 3 and 5×10^{-3} watt cm^{-1} K^{-1} near and above 300 K, much lower than that of silicon dioxide. By contrast, the lower limit of the range of conductivity previously measured on different samples of rock salt

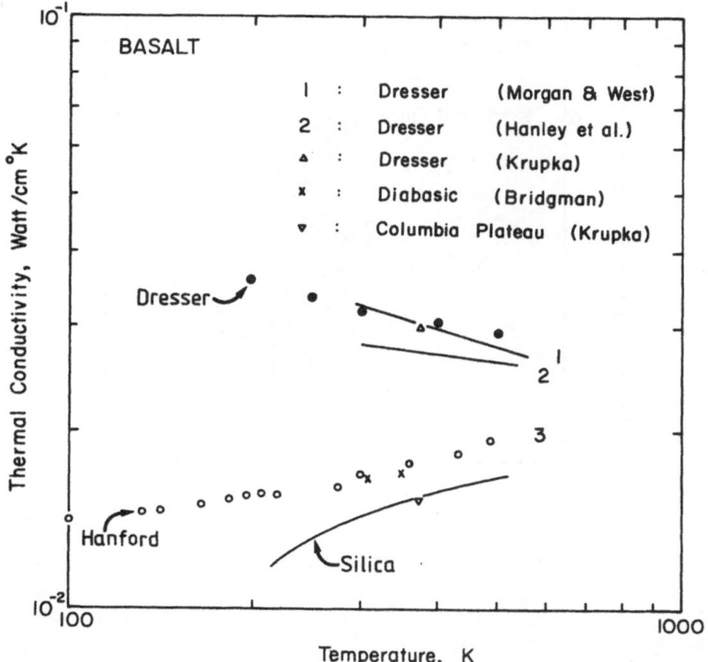

Fig. 4. Summary of thermal conductivity of basalt from different sources. Dresser basalt: Morgan and West, ref. 9; Hanley et al., ref. 10; Krupka, ref. 11; closed circles: present investigation. Two data points for diabasic rock after Bridgman, ref. 12; one point labeled "Columbia Plateau" after ref. 11; open circles: present investigation, Hanford basalt.

lies near 3×10^{-2} watt cm^{-1} K^{-1} at room temperature (see Fig. 5), almost ten times higher than this (theoretical) lower limit. Amorphous sodium chloride, which we would expect to have such a low conductivity, does not exist. It seems hardly justifiable to assume, for reasons of safety, such a low conductivity, particularly since there is no indication as of yet that even in the most highly disordered or contaminated rock salt such a low conductivity will ever be found.

In exploring the range of conductivity of rock salt samples taken at the WIPP site, however, we made the remarkable observation that the conductivity of certain samples was irreversibly decreased as a result of thermal cycling. Figure 5 shows an example of such behavior. The rock salt sample was taken from a core labeled WIPP # 10 by Sweet and McCreight.[14] It contained between approximately 10 and 20 wt % polyhalite ($K_2SO_4 \cdot MgSO_4 \cdot 2CaSO_4 \cdot 2H_2O$).

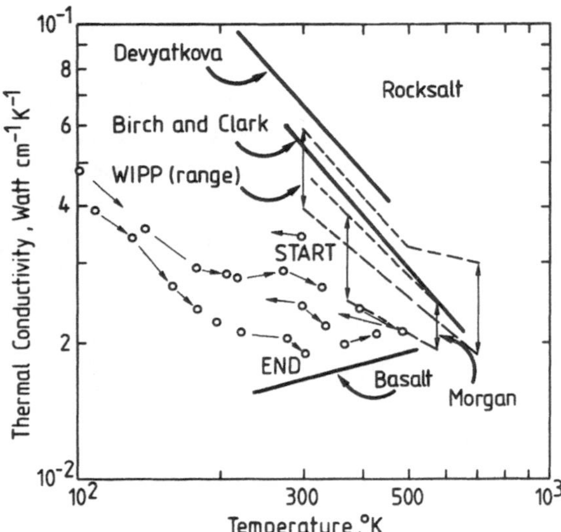

Fig. 5. Thermal conductivity of rock salt. Devyatkova: pure synthetic NaCl, ref. 13. Birch and Clark: Natural rock salt, ref. 6. WIPP (range): Range of rock salt from WIPP site, ref. 14. Morgan: Range of salt from salt domes in Louisiana, ref. 15. Open circles: Irreversible conductivity changes on a sample of rock salt from WIPP # 10, containing polyhalite. Arrows mark the sequence in which data were taken.

At the beginning of the thermal cycling ("Start") the conductivity fell in the range of previously reported conductivities; after repeated cooling and heating, however, the conductivity at 300 K had decreased by almost a factor of two ("End"), with no indication that a saturation had been reached. Visual inspection of the sample showed numerous small cracks. These measurements have not yet been repeated on a sample which was only heated, and not also cooled, although similar cracks have been observed after heat treatment alone. Cracking due to expansion of water inclusions could be an explanation, although in other samples in which water inclusions had been observed, but which contained no polyhalite, no irreversible changes of the conductivity have been found. It might perhaps be speculated that the polyhalite inclusions weaken the rock salt to the point that expanding water (and ice) can cause the cracking. Clearly, more work is required to determine the cause of the low thermal conductivity, and of the irreversibility, so that such critical areas can be avoided in a repository.

If, as we have speculated, cracks in the salt are important, we should explore their role also in the silicate based rocks. Small ($\sim$10%) irreversible decreases of the conductivity had indeed been observed by Morgan and West following heat treatment,[9] and had also been explained by cracks. We have confirmed these observations. However, we would expect that any cracks in granite and basalt in a nuclear waste repository would soon be filled with water. Water has a relatively low conductivity ($\sim 5 \times 10^{-3}$ watt cm^{-1} K^{-1}), but in a thin layer would not significantly affect the overall thermal conductivity. This point must be further explored through field tests.

In conclusion, the concept of the minimum thermal conductivity has been suggested as a means of estimating an upper limit of the temperature rise around a waste disposal site located in granitic or basaltic rock. For rock salt, however, this concept is less useful, and more systematic studies will be needed to determine the range of conductivity to be expected, and to identify the regions which have either unacceptably low thermal conductivities, or those in which irreversible change may occur as a result of the emplacement of the nuclear waste.

REFERENCES

1. See, e.g., G. K. White in "Thermal Conductivity," Vol. 1, R. P. Tye, ed., Academic Press, London and New York, 1969.
2. J. W. Vandersande and R. O. Pohl, A simple apparatus for the measurement of thermal diffusivity between 80 and 500 K using the modified Angstrom method, Rev. Sci. Instr., to be published.
3. G. A. Slack, The thermal conductivity of nonmetallic crystals, in "Solid State Physics," Vol. 34, Academic Press, New York, 1979, p. 1-71. H. Ehrenreich et al., eds.
4. C. Kittel, Interpretation of the thermal conductivity of glasses, Phys. Rev. 75: 972 (1949).
5. R. O. Pohl and J. W. Vandersande, A review of heat dissipation in geologic media, to be published in "Proceedings of the workshop on alternate nuclear waste forms and interactions in geologic media," Gatlinburg, Tenn., May 13-15, 1980, Lynn A. Boatner, ORNL, Chairman.

6. F. Birch and H. Clark, The thermal conductivity of rocks and its dependence upon temperature and composition, Am. J. of Science, 238:529 (1940).
7. H. R. Pratt et al., Terratek, Inc., Salt Lake City, Utah, Rept. TR-77-92 (Oct. 1977) and TR-78-47 (Aug. 1978).
8. G. A. Izett, U.S. Geological Survey Rept. TEM-836-C (1960), Interim Report, Part C, Physical Properties.
9. M. T. Morgan and G. A. West, Oak Ridge National Laboratory, Oak Ridge, Tenn., Rept. ORNL/TM-7052 (Jan. 1980).
10. E. J. Hanley, D. P. De Witt, and R. E. Taylor, Proc. Sympos. on Thermophysical Properties, Vol. 7, p. 386 (1977), and E. J. Hanley, D. P. De Witt, and R. F. Roy, Eng. Geology (Amsterdam) 12:31 (1978).
11. M. C. Krupka, Los Alamos Laboratories, Rept. No. LA-5540-MC (1974).
12. P. W. Bridgman, Am. J. of Science, 7:81 (1924).
13. E. D. Devyatkova and I. A. Smirnov, Soviet Phys. - Solid State 4 (7):1445 (1963).
14. J. M. Sweet and J. E. McCreight, SAND 79-1665, March 19, 1980.
15. M. T. Morgan, Oak Ridge National Laboratory, Oak Ridge, Tenn., Rept. ORNL/TM-6809 (June 1979).

PHASE CHEMISTRY OF THE UMTANUM BASALT, A REFERENCE REPOSITORY HOST IN THE COLUMBIA PLATEAU

Albert F. Noonan, Kurt Fredriksson* and Joseph Nelen*

Rockwell Hanford Operations, P.O. Box 800, Richland, Washington 99352 and *Smithsonian Institution, Washington, D.C. 20560

ABSTRACT

Kilogram quantities of reference Umtanum basalt collected from a surface exposure are compared with Umtanum basalt DC-2 core samples extracted from the central Pasco Basin. The two materials are similar in texture and phase composition, suggesting that sorption, waste/rock/water interaction experiments, and radiation effects studies using reference basalt will successfully simulate the proposed Umtanum repository environment. Textural studies of DC-2 core samples also show that dissolution of mesostasis is restricted to flow top breccia with precipitation of secondary minerals in the flow top, entablature, and colonnade sealing fissures and vesicles.

INTRODUCTION

The Pasco Basin is a structural and topographic depression in the Columbia Plateau of southeastern Washington which contains an accumulation of tholeiitic basalt in excess of 1,500m. The basalts were deposited during an extended period of lava extrusion (∿6.0-16.5 million years before the present) and were subsequently overlain by up to 200m of Quanternary fluvial, glaciofluvial, and lacustrine sediments.

Primarily on the basis of chemistry and paleomagnetic properties, the basalt units have been divided into three major formations; the Grande Ronde, the Wanapum, and the Saddle Mountains Formations. The Grande Ronde Basalt occupies the basin below about 650m and accounts for nearly 70% of the total basalt accumulation in the Pasco Basin.

The Umtanum flow of the Grande Ronde Formation is one of several deep basalt units being considered to host a nuclear waste repository. It is a dense, continuous flow of nearly uniform thickness (∿70m) occurring at a depth of 900-1,000m in the central Pasco Basin. The basalt is characterized by relatively high glass content that is susceptible to alteration to smecitite clay and clinoptilolite, particularly so at elevated temperatures in the presence of groundwater. It is essential to chemically and mineralogically characterize reference Umtanum basalt before and after planned waste/rock/water interaction studies in order to successfully predict changes imposed on a basalt host by the natural decay of radioactive waste. It is also important to ensure that reference materials used in natural barrier studies are similar to the basalt equivalents found at depth in the central Pasco Basin. In this report the phase chemistry of reference Umtanum basalt is compared with that of the Umtanum basalt samples from core hole DC-2 at a depth in excess of 900m.

SAMPLING AND ANALYTICAL PROCEDURES

Because of the need for kilogram quantities of basalt for experimental studies, reference Umtanum colonnade (RUC,1), entablature (RUC,1), and flow top breccia (RUFT,1) were collected from an outcrop on the north slope of Umtanum Ridge, an east-west trending anticline on the northwest flank of the Pasco Basin. These are the principal basalt intraflow structural types recognized in the Grande Ronde (Long, 1978). Small representative samples of Umtanum basalt were also taken from core hole DC-2 in the central Pasco Basin and designated DC-2-UC, DC-2-UE, and DC-2-UFT for colonnade, entablature, and flow top material, respectively. The reference flow top material is highly weathered, cannot be compared with DC-2 equivalent basalt, and may be used only to determine its sorption capacity for certain key radionuclides. The reference colonnade and entablature will be used in sorption studies and hydrothermal experiments to test the effects imposed on the host rock in a repository environment.

The major mineral phases and glass (here referred to as mesostasis) in the reference and DC-2 equivalent Umtanum colonnade and entablature samples were analyzed with an ARL-SEMQ electron microprobe using a standard accelerating potential of 15kV and a specimen current of 0.015 MA. The analyses were corrected on line against mineral and glass standards using the procedures of Albee and Ray, 1970. Standards analyzed as samples were used to provide the basis for internal drift corrections. A broad beam (30-50 microns) was used to analyze "glass" to minimize volatile loss and achieve an average composition of material that actually is a mixture of microcrystalline oxides, silicates, and true glass. This material will be referred to as mesostasis in the text. Compositions reported are means based on 4 to 5 ten-second analyses per grain or mesostasis area

PHASE CHEMISTRY

The compositions of major silicate phases in reference Umtanum colonnade and entablature are very similar to DC-2 equivalents (Tables 1 and 2). The clinopyroxenes pigeonite and augite show the chemical variations normally found in basaltic rocks. Most pyroxene is in the augite compositional range with the exception of minor pigeonite, subcalcic augite, and ferroaugite. Pigeonite is less abundant than augite. Plagioclase is the most abundant silicate mineral in all reference and DC-2 equivalent samples and is very constant in composition (AN_{49-51}). The major opaque phase in all Umtanum samples is titaniferous magnetite which is nearly constant in composition for all samples with 28-32% TiO_2.

Broad beam analyses of mesostasis in the reference and core basalt samples are presented in Table 3. Colonnade mesostasis appears very similar in composition in both materials for all major oxides with the exception of the alkalies. Potassium is higher in the reference colonnade and lower in Na_2O relative to the DC-2 equivalent. In general, mesostasis in the colonnade is higher in K_2O and lower in Na_2O and FeO, similar in composition to fractionated rhyolitic glass. The entablature mesostasis is more variable in composition; the reference material apparently containing more microcrystalline magnetite (high-FeO) relative to the DC-2 equivalent.

PETROFABRIC ANALYSIS

Textural differences noted between reference Umtanum colonnade and entablature are consistent with DC-2 equivalent samples. Entablature in both materials is characterized by abundant mesostasis and titaniferous magnetite dendrites whereas colonnade typically appears more coarsely crystalline and contains larger equant opaque grains and less mesostasis (Figure 1). These differences are usually reflected in the composition of the mesostasis (Table 3) with entablature mesostasis containing higher iron (i.e., microcrystalline magnetite) relative to colonnade whereas the high-K colonnade "glass" appears more highly fractionated.

Preliminary studies of DC-2 Umtanum flow top breccia samples suggest that preferential dissolution of mesostasis by groundwater may be restricted to this relatively porous zone. Reflected light photomicrographs of DC-2 and reference entablature and colonnade samples (Figure 1) exhibit very little dissolution pitting whereas a DC-2 flow top sample shows dissolution and vesicle filling with silica, euhedral clinoptilolite, and clay (Figure 2a). Precipitation of secondary minerals takes place below this zone where entablature and colonnade fissures and vesicles appear sealed (Figure 2b).

Table 1. Composition of major silicate minerals in reference (RUC,1) and DC-2 (DC-2-UC, 963m) Umtanum colonnade

	Pigeonite		Augite				Plagioclase			
	* RUC,1	DC-2-UC	RUC,1		DC-2-UC		RUC,1		DC-2-UC	
	(1)	(1)	(5)	σ	(3)	σ	(9)	σ	(7)	σ
SiO_2	50.78	51.39	50.11	0.35	49.82	0.34	56.33	1.85	57.34	2.44
TiO_2	0.57	0.52	1.05	0.19	0.94	0.18	0.11	0.02	0.09	0.01
Al_2O_3	0.90	0.80	2.53	0.60	1.71	0.61	26.97	1.20	26.62	1.89
Cr_2O_3	<0.05	<0.05	<0.05	----	<0.05	----	N.D.	----	N.D.	----
FeO	24.52	21.66	15.56	1.86	16.11	0.52	0.79	0.09	0.84	0.07
MnO	0.67	0.50	0.36	0.01	0.39	----	N.D.	----	N.D.	----
MgO	17.76	20.42	15.45	0.74	14.83	1.24	0.09	0.04	0.06	0.02
CaO	5.91	4.81	15.31	1.68	15.03	1.10	9.43	1.38	9.38	2.15
Na_2O	0.07	0.09	0.27	0.04	0.23	0.02	5.28	0.59	5.05	0.55
K_2O	N.D.	N.D.	N.D.	----	N.D.	----	0.69	0.32	0.78	0.37
P_2O_5	N.D.	N.D.	N.D.	----	N.D.	----	N.D.	----	N.D.	----
TOTAL	101.25	100.24	100.69		99.11		99.69		100.16	
Mol.%	En 50 Fs 38 Wo 12	En 57 Fs 34 Wo 9	En 44 Fs 25 Wo 31		En 43 Fs 26 Wo 31		An 49 Ab 47 Or 4		An 50 Ab 45 Or 5	

N.D. = Not Detected
*( ) = Number of Analyses

Table 2. Composition of major silicate minerals in reference (RUE,1) and DC-2 (DC-2-UE, 954.4m) Umtanum entablature

	Pigeonite			Augite				Plagioclase			
	* RUE,1		DC-2-UE	RUE,1		DC-2-UE		RUE,1		DC-2-UE	
	(3)	σ	(1)	(5)	σ	(4)	σ	(9)	σ	(5)	σ
SiO_2	50.90	0.78	51.48	49.42	1.82	50.09	0.43	55.51	1.22	56.11	0.67
TiO_2	0.58	0.10	0.61	1.15	0.57	0.94	0.16	0.11	0.03	0.13	0.02
Al_2O_3	0.82	0.15	0.72	1.89	0.55	2.03	0.46	27.15	0.78	26.64	0.59
Cr_2O_3	<0.05	----	<0.05	<0.05	----	<0.05	----	N.D.	----	N.D.	----
FeO	23.47	1.09	22.51	17.94	2.68	15.25	0.99	0.70	0.14	0.90	0.06
MnO	0.53	----	0.48	0.46	0.06	0.48	----	N.D.	----	N.D.	----
MgO	19.09	1.48	18.10	13.89	1.77	14.78	1.24	0.11	0.02	0.08	0.02
CaO	5.16	0.94	6.27	15.18	1.04	15.05	1.54	9.67	0.19	9.76	0.64
Na_2O	0.07	0.02	0.09	0.25	0.02	0.28	0.04	4.97	0.47	5.09	0.27
K_2O	N.D.	----	N.D.	N.D.	----	N.D.	----	0.68	0.13	0.71	0.14
P_2O_5	N.D.	----	N.D.	N.D.	----	N.D.	----	N.D.	----	N.D.	----
TOTAL	100.67		100.31	100.23		98.95		98.90		99.42	
Mol.%	En 53 Fs 37 Wo 10		En 51 Fs 36 Wo 13	En 40 Fs 29 Wo 31		En 43 Fs 25 Wo 32		An 51 Ab 44 Or 5		An 51 Ab 45 Or 4	

N.D. = Not Detected
*( ) = Number of Analyses

Table 3. Composition of mesostasis in reference and Umtanum basalt

	(RUC,1)		(DC-2-UC)		(RUE,1)		(DC-2-UE)	
	*(7)	σ	(7)	σ	(5)	σ	(7)	σ
SiO_2	72.20	1.49	72.88	1.20	61.22	1.78	69.37	0.87
TiO_2	0.69	0.14	0.82	0.13	1.53	0.60	0.86	0.14
Al_2O_3	12.47	1.44	12.71	0.60	12.49	0.63	13.74	0.43
Cr_2O_3	N.D.	----	N.D.	----	N.D.	----	N.D.	----
FeO	1.59	0.16	2.16	0.34	10.49	1.08	4.49	0.88
MnO	<0.05	----	<0.05	----	0.16	0.02	<0.05	----
MgO	<0.05	----	N.D.	----	0.59	0.13	<0.05	----
CaO	0.51	0.19	0.62	0.20	3.87	0.47	1.90	0.24
Na_2O	4.01	0.45	6.62	1.10	7.14	0.37	7.45	0.10
K_2O	6.47	0.27	4.07	1.04	2.68	0.30	2.74	0.24
P_2O_5	0.12	0.09	0.10	0.43	0.62	0.02	0.36	0.11
TOTAL	98.16		100.03		100.79		101.01	

N.D. = Not Detected

* = Number of Analyses

(a)

(b)

Fig. 1 Reflected light photomicrographs of (a) reference Umtanum colonnade (RUE,1) and (b) reference Umtanum entablature (RUE,1). Note the greater volume percent mesostasis (dark-gray) in entablature relative to colonnade and the lack of dissolution pitting in both samples. White phase=titaniferous magnetite; light-gray=clinopyroxene; medium-gray=plagioclase; and dark-gray=mesostasis.

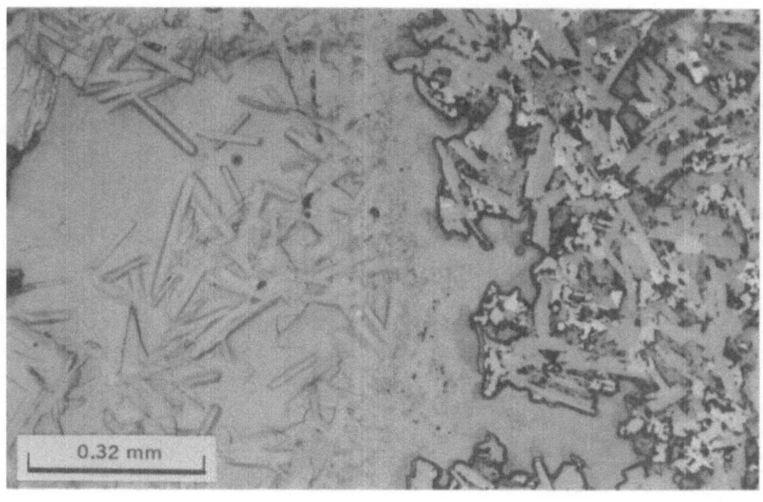

(a)

(b)

Fig. 2 (a) Reflected light photomicrograph of Umtanum flow top sample (DC-2-UFT). Note the dissolution of mesostasis between crystals in basalt clasts (right side) and secondary mineral vesicle filling (left). (b) Oil immersion reflected light photomicrograph of DC-2 Umtanum entablature (DC-2-UE) sample which shows clinoptilolite fracture filling.

CONCLUSION

Umtanum basalt selected as reference material to be used in sorption and hydrothermal tests required to qualify a nuclear waste repository in basalt are similar in texture and phase composition to equivalent material taken from DC-2 core samples. Dissolution of glass in flow top material with subsequent precipitation of secondary minerals (e.g., silica, zeolite, smectite clay) in fissures and vesicles in the underlying entablature and colonnade is an important process. The swelling clays tend to seal channels and fissures inhibiting groundwater movement and together with the zeolites, which exhibit high cation exchange properties, provide a natural barrier to radionuclide migration. The chemical composition of the mesostasis thus becomes important for successfully modeling reactions in the repository environment, including radionuclide precipitation and/or transport away from the proposed repository.

REFERENCES

Albee, A. L. and Ray, L. 1970, Correction factors for electron probe microanalysis of silicates, oxides, carbonates, phosphates and sulfates, Analytical Chem., 42:1408.

Long, P.E., 1978, Characterization and recognition of intraflow structures, Grande Ronde basalt: RHO-BWI-LD-10, Rockwell Hanford Operations, Richland, Washington.

GEOSCIENTIFIC EVALUATION OF RADIOACTIVE WASTE ISOLATION IN JAPAN

Kazumi Doi

Radioactive Waste Management Center

2-8-10, Toranomon, Minatoku, Tokyo, Japan

INTRODUCTION

Because Japan is located in the circum-Pacific structural zone, geological conditions in that country are not always favorable for radioactive waste isolation. The author has investigated the feasibility of isolating radioactive waste in geologic formations in Japan; this investigation is based on criteria in ref. 1.

GEOSCIENTIFIC SITUATION OF JAPAN

General Geology

A geological outline of Japan is shown in Table 1; Cenozoic sedimentary and volcanic rocks cover over half of the territory. Japan is divided into three blocks by two large faults — namely, the Fossa Magna and the median dislocation line (Fig. 1). These and numerous other faults and fractures have cut formations in most areas of the country.

Lithology

Hard rocks distributed in the region include Palaeozoic and Mesozoic sediments, plutonic rock, metamorphic rock, and some volcanic rock. Although these rocks normally have low porosity and permeability, structural fractures have caused these formations to become quite porous and permeable.

Table 1. Geological Outline of Japan — The Lithology and Distribution (Compiled after Ref. 2)

	Distribution		Lithology*
	km^2	%	
Quaternary sediments	76,500	20.7	ss, m, c, s, ga
Tertiary sediments	69,900	18.9	ss, m, c, sh
Mesozoic sediments	34,400	9.3	ss, sh, sl, c
Palaeozoic sediments	45,200	12.2	ss, sl, t, l, c
Cenozoic volcanic rocks	75,400	20.4	a, b, r, t
Acidic plutonic rocks	49,300	13.3	gr, q
Basic plutonic rocks	5,800	1.6	g, d
Metamorphic rocks	13,300	3.6	gn, sc
Total	369,800	100.0	

*a, andesite; b, basalt; c, conglomerate; d, diabase; g, gabbro; ga, gravel; gn, gneiss; gr, granite; l, limestone; m, mudstone; q, quartz porphyry; r, rhyolite; s, sand; sc, crystalline schist; sh, shale; sl, slate; ss, sandstone; t, tuff and tuff breccia

Cenozoic sedimentary and pyroclastic rocks are generally not well consolidated and consequently have high porosity and permeability. Argillaceous and pyroclastic rocks of that period, however, have low porosity and permeability. It is thought that the plasticity of these formations has minimized the influence of structural movements by absorbing the geological stress. Generally, quaternary sediments do not have enough mechanical strength to be used for radioactive waste isolation.

Volcanism

Active volcanism has occurred repeatedly in Japan, especially since the Neogene era. Over 70 volcanoes are reported to still be active (Fig. 2).

Precipitation and Groundwater

Precipitation in Japan is high as compared with other parts of the world (Table 2). The average value is 1777 mm/year.

Fig. 1. Main techtonic lines in Japan. (1) Fossa Magna; (2) median dislocation line.

Groundwater is abundant, and flowing groundwater is common in most parts of Japan. This situation is a result of high precipitation and the areal distribution of permeable rocks, such as Cenozoic sediments and pyroclastics.

Earthquakes and Tidal Waves

In the past 100 years, Japan has withstood more than 50 earthquakes measuring over 7 on the Richter scale, and this trend is assumed to continue (Figs. 3 and 4). A listing of the big (>10 m high) tidal waves that have hit Japan is given in Table 3. Most of these tidal waves were caused by earthquakes in or near Japan.

EVALUATION OF THE FEASIBILITY OF RADIOACTIVE WASTE ISOLATION

Japan has been hit by several kinds of calamities related to geologic events, and the recurrence of such events is still a possibility in most parts of the country. Accordingly, it would be difficult to find a favorable site for the permanent storage of radioactive waste, but the author believes that a site suitable for retrievable storage could possibly be found.

Fig. 2. Main active volcanoes in Japan (after Ref. 2).

Table 2. Annual Precipitation in Japan (1941-1970) and in Other Countries (1931-1960) (After Ref. 2)

		Annual Precipitation (mm)	Locality
Japan (80 samples)	Maximum	4,158	Owase, Mie Pref.
	Median	1,689	Ooita, Ooita Pref.
	Minimum	848	Abashiri, Hokkaido
	Average	1,777	
Other countries (440 samples)	Maximum	7,410	Quibdo, Columbia
	Median	760	Toronto, Canada
	Minimum	2	Aswan, Egypt
	Average	993	

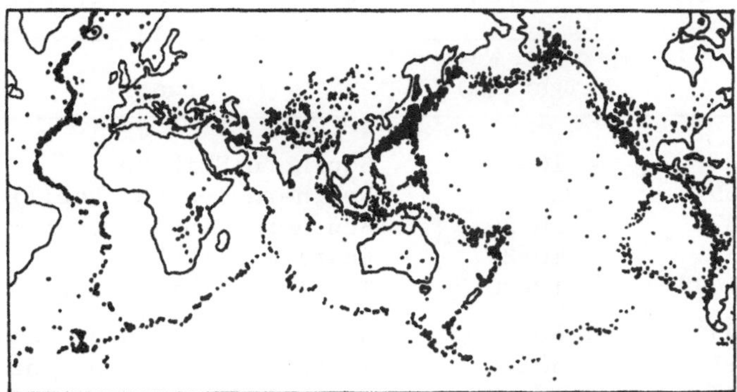

Fig. 3. Distribution of earthquakes in the world. Plotted earthquakes are of magnitude over 4 on the Richter scale and at a depth less than 100 km (after Ref. 2).

Fig. 4. Distribution of earthquakes in the vicinity of Japan. Plotted earthquakes are of magnitude over 7 on the Richter scale and include only those that occurred from 1874 to 1974.

Table 3. Historical Tidal Waves of Over 10 Meters that Occurred in Japan from 684 to 1946 (After Ref. 2)

Date	Maximum Wave Height (m)	Area
Nov. 29, 684	10-20	Southern Shikoku
July 13, 869	30	Northern Tohoku, Pacific coast
Aug. 26, 887	10-20	Southern Kinki, Pacific coast
Aug. 3, 1361	10-20	Eastern Shikoku
Sept. 20, 1498	10-20	Tokai, Pacific coast
Feb. 3, 1605	30	Western half Japan, Pacific coast
Dec. 2, 1611	30	Northern Tohoku and Hokkaido
Nov. 4, 1677	10-20	Central Japan, Pacific coast
Dec. 31, 1703	10-20	Kanto, Pacific coast
Oct. 28, 1707	26	Western Japan, Pacific coast
Aug. 28, 1741	10-20	Northern Japan, Japan sea coast
Apr. 24, 1771	40	Southern Kyushyu
May 21, 1792	10-20	Western Kyushyu
Dec. 23, 1854	10-20	Tokai, Pacific coast
Dec. 24, 1854	30	Western Japan, Pacific coast
June 15, 1896	30	Northern Tohoku, Pacific coast
March 3, 1933	25	Northern Tohoku, Pacific coast
Dec. 7, 1944	10-20	Tokai, Pacific coast
July 21, 1946	10-20	Western Japan, Pacific coast

In this region of a high degree of faulting and relatively large groundwater flows, it is necessary to satisfy the following factors to derive a favorable area for radioactive waste isolation.

(a) areas with little or no volcanic or recent tectonic activity, such as active faults or active volcanoes;

(b-1) geologic formations with little or no flowing groundwater;

(b-2) formations where all or most of the groundwater is easily controlled by artificial drainage.

Potential Formation for the Repository

Some of the Neogene argillaceous sediments may be favorable because they satisfy the factor (b-1). Some Neogene argillaceous sediments have fewer faults and fractures; this is because tectonically induced stress has been absorbed by the plasticity of the rocks. Generally, the argillaceous sediments in the region contain water, but most of the water is bound by clay minerals contained in the rocks and is not free to migrate.

Palaeozoic calcareous formations potentially satisfy the factor (b-2). In several areas of the region, dry limestone mines exist in these formations. The groundwater in limestones would have to be gathered into a drainage system below the repository by draining either natural fractures or cave systems.

REFERENCES

1. _Site Selection Factors for Repositories of Solid High-Level and Alpha-Bearing Wastes in Geological Formations_, IAEA, Technical Report Series No. 177 (1977).

2. K. Doi, "Geo-Scientific Considerations on Evaluation of Possible Sites for Radioactive Waste Isolation in Japan," _Atomic Energy Soc. Japan_ (in Japanese), 22:543 (1980).

PRELIMINARY EVALUATION OF ALTERNATIVE FORMS FOR IMMOBILIZATION OF HANFORD HIGH-LEVEL WASTES

W. W. Schulz, M. M. Beary, S. A. Gallagher, B. A. Higley, R. G. Johnston, M. J. Kupfer, and R. A. Palmer

Rockwell Hanford Operations
Richland, Washington, USA 99352

INTRODUCTION

A preliminary assessment of the applicability of 19 different waste forms and 20 different processes to immobilization of Hanford high-level defense waste was performed.[1] The high-level waste addressed in this study included salt cake (95,000 m^3) and sludge (49,000 m^3) in single-shell tanks and residual liquid (45,000 m^3) in double-shell tanks. Hanford salt cake is an admixture of large amounts of $NaNO_3$, minor amounts of other sodium salts, and small amounts of ^{90}Sr, ^{137}Cs, ^{99}Tc and other radionuclides. Highly alkaline (4 to 6<u>M</u> NaOH) residual liquid resulting from production of salt cake contains large amounts of soluble sodium salts, ^{90}Sr, and ^{137}Cs. Hanford sludge is all the water-insoluble solids stored in tanks irrespective of their origin or composition and contains most of the actinides and ^{90}Sr in the high-level wastes.[2]

Because key decisions affecting long-term management of the Hanford wastes will not be made until the mid-1980s, the waste forms and processes were evaluated and rated with respect to five alternative management schemes:

Alternative 1	Near-term retrieval of all in-tank wastes (Radionuclide Removal option).
Alternative 2	Same as Alternative 1 but without Radionuclide Removal.
Alternative 3	In-situ disposal of in-tank wastes.

Alternative 4 Near-term retrieval of residual liquid only (Radionuclide Removal option).

Alternative 5 Same as Alternative 4 but without Radionuclide Removal.

Early and separate immobilization of residual liquid would require that salt cake and sludge be treated by Alternative 1, 2, or 3. The Radionuclide Removal option involves removal of all long-lived radionuclides ($t_{1/2} \geq 10y$) from Hanford salt cake and residual liquid to partition these high-level wastes into a small volume of highly radioactive waste requiring immobilization for long-term storage or disposal and a large volume of low-level ($\sim$10 nCi/g) wastes that can be safely stored by relatively inexpensive means.

CANDIDATE WASTE FORMS

These five management alternatives give rise to four different waste fractions: Sludge plus Radionuclide Concentrate (Alternative 1), Blended Wastes (Alternatives 2 and 3), Residual Liquid (Alternatives 3 and 5), and Radionuclide Concentrate from Residual Liquid (Alternative 4). A list of the candidate waste forms and their compatibilities with these waste fractions is given in Table 1. The Sol Gel process currently being developed at Oak Ridge National Laboratory was also evaluated as a method of preparing ceramic and other waste forms from Hanford waste.

TABLE 1. Compatibility of Candidate Waste Forms with Hanford High-Level Wastes.

Candidate Waste Forms	Compatible With			
	Sludge and Radionuclide Concentrate	Blended Waste	Residual Liquid	Radionuclide Concentrate From Residual Liquid
Borosilicate Glass	Yes	Yes	Yes	Yes
Porous Glass Matrix	Yes	No	No	No
Glass-Ceramics	Yes	No	No	Yes
Supergrout Concrete	No	Yes	Yes	No
Sludge In Concrete	Yes	No	No	Yes
FUETAP Concrete	Yes	Yes	Yes	Yes
Hot-Pressed Concrete	Yes	No	No	Yes
Aqueous Silicate	No	Yes	Yes	No
Supercalcine Ceramic	Yes	No	No	No
SYNROC Ceramic	Yes	No	No	No
Tailored Ceramics	Yes	No	No	No
Clay Ceramics	Yes	Yes	Yes	Yes
Cermet (Urea Process)	Yes	No	No	No
Glass in Metal Matrix	Yes	Yes	Yes	Yes
Ceramic Pellets in Metal Matrix	Yes	No	No	No
Ceramic in Concrete	Yes	No	No	No
Coated Ceramic	Yes	No	No	No
Bitumen	Yes	Yes	Yes	Yes
Calcine	Yes	No	No	No

BASES FOR RATING

Evaluation of the alternative waste forms was based on the following items:

- When specific data were not available, the assessment was based upon published statements and judgements of the proponents of that particular waste form

- Waste form and process characteristics are of equal importance in the overall rating of the waste form

- Disposal of Hanford high-level wastes on site either in caverns mined in Columbia Plateau Basalt or in suitable near-surface facilities is assumed

- The waste form will come into contact with water or other aqueous solutions present in the disposal site

- Protection of the waste form by canisters, overpacks, backfills, and other engineered barriers is not taken into account

- Research and development to determine the feasibility of providing engineered barriers for safe, direct geologic disposal of $^{137}CsCl$ and $^{90}SrF_2$ capsules has a higher current priority than developing alternative forms for these Hanford high-level wastes

- Hanford residual liquid and any Purex and B plant sludges containing organic complexants will be processed for complexant removal before immobilization.

RATING PROCEDURE

Each waste form and/or process was rated by five different raters for the characteristic listed on the standard rating form (Table 2) on a scale of 1 to 5 (5 = highest ranking). The weighting factors were assigned by a procedure described by Gelpi.[3] In this procedure, each characteristic is compared to every other characteristic. For each comparison, a value of "1" is assigned to that characteristic which is judged to be more important, and a value of "0" is assigned to that characteristic judged to be less important. The positive decisions ("1's" are summed for each characteristic and divided by the total number of possible decisions or $n(n-1)/2$ where n = number of items being compared. The weighting factors were then arbitrarily multiplied by a factor of 10 to put the ratings on a scale of 1 to 100. The determination of the weighting factors for the waste form characteristics is illustrated in Table 3.

TABLE 2. Standard Rating Form.

WASTE FORM: NAME COATED CERAMIC APPENDIX A-17
RATER: NAME U. B. Rater* DATE July 1, 1980

WASTE FORM EVALUATION	Numerical Rating	x	Factor Weight	=	Product	PROCESS EVALUATION	Numerical Rating	x	Factor Weight	=	Product
Waste Loading	4	x	2.382	=	9.528	Status of Development	2	x	0.476	=	0.952
Leachability	5	x	2.857	=	14.285	Simplicity	1	x	2.857	=	2.857
Thermal Stability	5	x	0.952	=	4.76	Scale-Up Potential	4	x	1.905	=	7.62
Radiation Stability	5	x	0.952	=	4.76	Remotability	2	x	1.905	=	3.81
Repository Environment/ Waste Form Interaction	5	x	1.905	=	9.525	Quality Assurance	4	x	0.476	=	1.904
Mechanical Stability	4	x	0.476	=	1.904	Rework Capacity	2	x	0.476	=	0.952
Fire Resistance	5	x	0.476	=	2.38	Process Safety	3	x	1.905	=	5.715
			Sum	=	47.142				Sum	=	23.81

OVERALL RATING:

WASTE FORM EVALUATION	=	47.142
PROCESS EVALUATION	=	23.81
SUM	=	71

* Un Biased Rater

TABLE 3. Weighting of Waste Form Rating Factors.

Factors	1	2	3	4	5	6	7	8	9	10	11	12	13	14	15	16	17	18	19	20	21	Positive Decisions (N)	Weighting Factor (x10) (N/21) (x10)
Waste Loading	0	1	1	1	1	1																5	2.382
Leachability	1						1	1	1	1	1											6	2.857
Thermal Stability		0					0					1	0	0	1							2	0.952
Radiation Stability			0					0				0				0	1	1				2	0.952
Repository Environment/Waste Form Interaction				0					0				1			1			1	1		4	1.905
Mechanical Stability					0					0				1			0		0		0	1	.476
Fire Resistance						0					0				0			0		0	1	1	.476
																						21	10.000

The relative ranking of each waste form/process compatible with a particular waste fraction is shown in Figures 1 through 4. Confidence levels (75%) around the mean overall rating of each waste form were calculated according to standard statistical procedure. All those waste forms whose confidence levels were not overlapped by the confidence levels around the mean for the second highest ranked waste form were rejected from further consideration. Additional screening factors (e.g., minimum acceptable leachability, maximum number of canisters which could reasonably be produced per day, etc.) were also imposed to shorten the list of candidate waste forms/processes deemed most suitable for immobilization of Hanford high-level wastes. The screened list of waste forms to be considered for immobilization of each waste fraction is given in Table 4. Additional factors (e.g., costs, geologic repository availability and location, engineered barriers, etc.) which could impact the final choice of waste forms/processes were not considered.

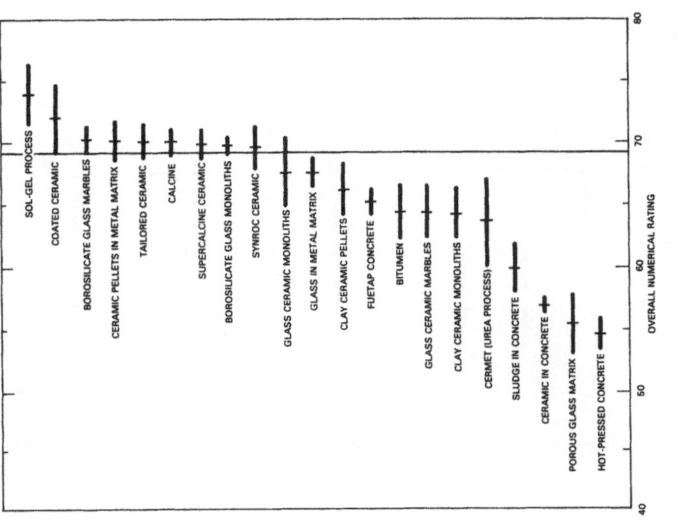

FIGURE 1. Relative Ranking of Waste Forms for Immobilization of Hanford Sludge Plus Radionuclide Concentrate.

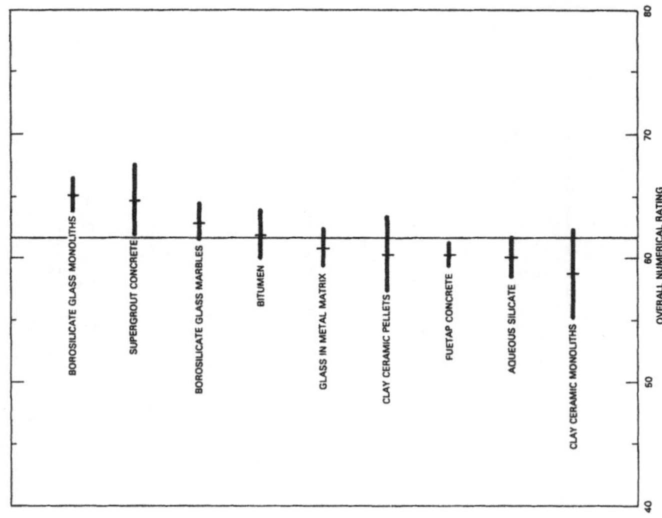

FIGURE 2. Relative Ranking of Waste Forms for Immobilization of Hanford Blended Wastes.

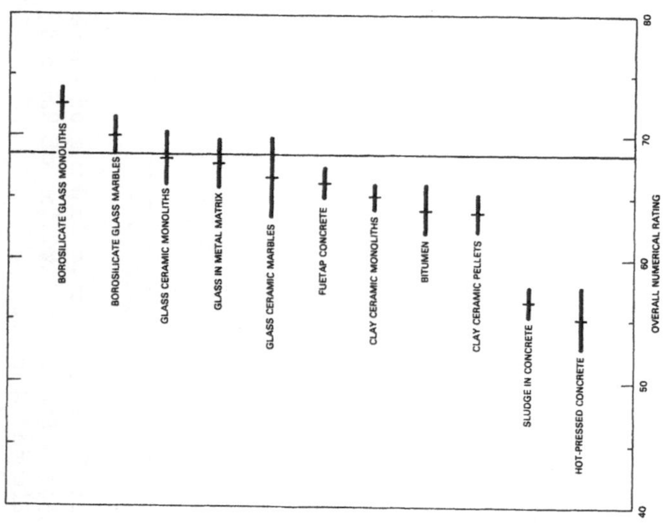

FIGURE 4. Relative Ranking of Waste Forms for Immobilization of Hanford Radionuclide Concentrate from Residual Liquid.

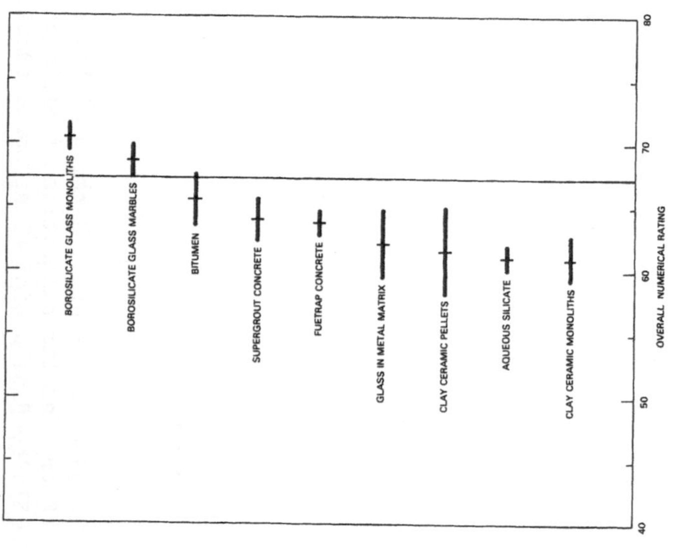

FIGURE 3. Relative Ranking of Waste Forms for Immobilization of Hanford Residual Liquid.

TABLE 4. Screened List of Waste Forms to be Considered Further for Immobilization of Hanford High-Level Waste.

Sludge and Radionuclides Concentrate	Blended Wastes(b)	Residual Liquid	Radionuclide Concentrate From Residual Liquid
Coated Ceramic	Supergrout Concrete	Borosilicate Glass-Monoliths	Borosilicate Glass-Monoliths
Borosilicate Glass-Marbles	Borosilicate Glass-Marbles	Borosilicate Glass-Marbles	Borosilicate Glass-Marbles
Ceramic Pellets in Metal Matrix	Bitumen	Bitumen	Glass in Metal Matrix
Tailored Ceramics	Clay Ceramic-Pellets		
Supercalcine Ceramic			
Borosilicate Glass-Monoliths			
SYNROC Ceramic			

(a) For each Hanford waste fraction, waste forms are listed in descending order of numerical ranking.
(b) For bulk disposal only; no waste form qualifies for containerized disposal under rating criteria used in this assessment.

CONCLUSIONS AND RECOMMENDATIONS

Some important conclusions and recommendations for research and development were based on the selection of the candidate waste forms listed in Table 4. They are as follows:

- Borosilicate Glass Marbles and/or monoliths were rated among the top three waste forms for immobilization of all types of Hanford high-level waste. Borosilicate Glass research and development at Rockwell Hanford including bench- and pilot plant-scale melter tests with actual waste should be continued on a high priority basis.

- Supergrout Concrete and Bitumen, low temperature processes, are judged to be particularly suitable for immobilization and bulk disposal of high sodium blended wastes and/or residual liquid. Research and engineering studies should be initiated at Rockwell Hanford to develop the technology needed for possible large-scale application of these waste forms for immobilization of Hanford wastes.

- This preliminary assessment indicates that certain ceramic waste forms (e.g., Tailored Ceramics, Supercalcine Ceramic, and SYNROC Ceramic) are equal to or superior to Borosilicate Glass waste forms for immobilization of Hanford sludges and radionuclides removed from salt cake and residual liquid. These ceramic waste forms can be made by the Sol Gel process. It is recommended that research and development programs currently underway at Oak Ridge National Laboratory and other DOE laboratories be broadened to include scouting studies to develop a Sol Gel process for preparation of a generic ceramic waste form from Hanford sludges and associated radionuclide concentrate.

• Some multibarrier waste forms (e.g., Coated Ceramics, Ceramic Pellets in Metal Matrix, and Glass in Metal Matrix) are judged to be superior waste forms for immobilization of Hanford sludges and/or radionuclide concentrate. If funding is available, it is recommended that developers and proponents of these multibarrier waste forms perform appropriate tests to determine the feasibility of preparing such waste forms from simulated Hanford wastes. Extensive development of multibarrier waste forms for Hanford high-level wastes should be deferred, however, pending detailed evaluation of the single barrier glass, ceramic, and concrete waste forms referred to previously.

LITERATURE CITED

1. W. W. Schulz, M. M. Beary, S. A. Gallagher, B. A. Higley, R. G. Johnston, F. M. Jungfleisch, M. J. Kupfer, R. A. Palmer, R. A. Watrous, G. A. Wolf, "Preliminary Evaluation of Alternative Forms for Immobilization of Hanford High-Level Defense Wastes," RHO-ST-32, Rockwell Hanford Operations, Richland, Washington, (September 1980).
2. "Alternatives for Long-Term Management of Defense High-Level Radioactive Waste, Hanford Reservation," ERDA 77-44, U.S. Energy Research and Development Administration, Washington, D. C. (1977).
3. M. J. Gelpi, "Forcing a Good Decision," Westinghouse Engineer, 25, 25 (1967).

DEVELOPMENT OF AN IMPROVED ION-EXCHANGE PROCESS FOR REMOVING CESIUM AND STRONTIUM FROM HIGH-LEVEL RADIOACTIVE LIQUID WASTES

R. M. Wallace and R. B. Ferguson

E. I. du Pont de Nemours & Co.
Savannah River Laboratory
Aiken, South Carolina 29808

Processes are being developed to solidify and isolate the biologically hazardous radionuclides from approximately 23 million gallons of alkaline high-level waste accumulated at the Savannah River Plant. The waste consists mainly of a liquid supernate, a damp salt cake, and a gelatinous, insoluble sludge. The reference solidification process involves separation of the water soluble fraction (supernate) from the insoluble fraction, removal of cesium and traces of strontium from the supernate, incorporation of the sludge and the radionuclides from the supernate in glass, and incorporation of the residual salt in concrete.

A previously reported process for treating the supernate consisted of sorbing the cesium and strontium separately on different ion exchange resins.[1] Cesium was sorbed on Duolite* ARC-359, a phenol methylene sulfonate resin. Strontium was sorbed on Amberlite** IRC-718, an iminodiacetate resin. The cesium was eluted with a mixture of ammonium carbonate and ammonium hydroxide, and the strontium was eluted with an EDTA solution. The ammonium carbonate in the eluate was thermally decomposed, leaving a solution that was principally sodium carbonate containing the radionuclides. These were further concentrated by sorption in a zeolite column. The zeolite and the EDTA-strontium eluate were combined with the sludge prior to calcination and incorporation into glass.

* Trademark of Diamond Shamrock.
** Trademark of Rohm and Haas.

A new process, now being developed, involves sorbing cesium on phenolic resins that contain no strongly acidic sulfonate groups. These resins can then be eluted with formic acid, which is not possible with Duolite ARC-359. Duolite CS-100, a phenol-carboxylate resin, and Duolite S-761, a phenolic resin containing methylol groups, were studied initially. CS-100 was chosen for further development because of its greater breakthrough capacity and because it also sorbs strontium to some extent. Strontium sorption by CS-100 was not sufficient to eliminate the need for Amberlite IRC-718. However, the latter resin can also be eluted with formic acid because its functional groups are weakly acidic.

Formic acid elution permits several options to be considered. The preferred option consists simply of mixing the eluate with sludge prior to calcination. Sodium formate, which is formed when sodium is eluted from the resin, decomposes rapidly between 450°C and 500°C and will be destroyed in either the calciner or the melter. The resulting sodium oxide would be incorporated into glass. About 60% of the required sodium for glass production would come from this source. Another attractive option is to elute most of the sodium prior to elution of cesium and strontium. This is possible because only small amounts of the radionuclides are eluted until most of the sodium has been removed. They are then eluted rapidly (~99% within 1 column volume of eluate). The sodium formate would be recycled to feed. Both of the above options would eliminate the ammonium carbonate recovery step, would eliminate the zeolite sorption step, and simplify the process by using the same elutriant for both resins. Another possibility is the destruction of formic acid and sodium formate with ozone or hydrogen peroxide prior to vitrification. The separation of sodium formate by sorption of cesium and strontium on zeolite prior to vitrification is also possible. One of these steps might be necessary if the presence of too much formate should cause difficulties with operation of the calciner or melter.

Advantages of the New Process

The principal advantage of the new process is the elimination of a number of process steps. This reduces the amount of canyon space required for ion-exchange to one-third to one-half of its original value and thereby may significantly reduce the cost. A further advantage is that it is more compatible with the production of alternative waste forms than is the present process. Several of these waste forms require incorporation of the radionuclides into a mineral matrix; large amounts of sodium and silica interfere with the production of the most desirable minerals. The product of the new process contains no silica (from zeolite) and potentially much less sodium than the older one.

Basis of the Process

The sorption of cesium on Duolite CS-100 and Duolite S-761 was studied in connection with a continuing investigation of the causes of the high specificity of phenolic resins for cesium. Roberts and Holcomb[2,3] have studied the sorption of cesium and strontium by CS-100 from low-level, alkaline (pH 12) waste waters and their elution with dilute nitric or hydrochloric acid. The sorption of these fission products from highly alkaline high-level waste by weak-acid phenolic resins and their elution with weak acids, such as formic acid, have not previously been studied. The following sections describe the sorption and elution characteristics of Duolite CS-100 and Duolite S-761 for cesium, the sorption and elution characteristics of CS-100 for strontium, and the elution characteristics of Amberlite IRC-718 for strontium.

Absorption Studies

Equilibrium distribution coefficients (K_d) for cesium between the resins and solutions containing 4.75M $NaNO_3$, 1M NaOH are shown in Table 1. The dry H^+ form was used as the mass basis for K_d. The sorption characteristics of Duolite CS-100 and S-761 are very similar to the sorption characteristics of ARC-359; the distribution coefficients were a very sensitive function of cesium concentration. Both weak acid resins were as good for cesium sorption as the best samples of Duolite ARC-359 previously tested when K_d's were compared at a cesium concentration of 2×10^{-4}M. At cesium concentrations lower than 2×10^{-4}M, the sorption characteristics of CS-100 and S-761 were better than ARC-359.

Table 1

DISTRIBUTION COEFFICIENTS (K_d) OF Cs^+ BETWEEN 4.75M $NaNO_3$ - 1M NaOH AND VARIOUS RESINS AS A FUNCTION OF Cs^+ CONCENTRATION

Duolite CS-100		Duolite S-761	
Equilibrium Cs^+ Conc., M	K_d	Equilibrium Cs^+ Conc., M	K_d
8.67×10^{-2}	9.41	8.65×10^{-2}	10.4
7.31×10^{-3}	15.9	7.18×10^{-3}	25.2
3.74×10^{-4}	54.0	5.90×10^{-4}	57.4
2.85×10^{-5}	159	3.22×10^{-5}	150
1.84×10^{-6}	265	2.25×10^{-6}	212
0	402	0	317

Breakthrough curves were run using columns containing 25 mL of resin when in equilibrium with 1M NaOH. The feed was designed to simulate the supernate expected in actual plant operation. The feed contained: 5.65M Na^+, 2.36 x 10^{-4}M Cs^+ (with ^{137}Cs tracer), 1.67M NO_3^-, 0.94M NO_2^-, 1.67M OH^-, 0.52M AlO_2^-, 0.21M $CO_3^{2-}$and 0.22M SO_4^{2-}. Columns were fed at a rate of 1.25 column volumes per hour. The results of these tests are shown in Figure 1. The ratio of the ^{137}Cs activity in an effluent sample to that in the feed (C/C_o) is plotted as a function of throughput in column volumes (CV). Results of the tests show that CS-100 is better than S-761 as far as breakthrough capacity is concerned. The principal reason for the superiority of CS-100 over S-761 is its greater bulk density in equilibrium with feed solutions.

Figure 2 shows the results of column tests in which $^{85}Sr^{2+}$ was sorbed on CS-100 from simulated supernate run at 1.25 CV/hour. Although strontium sorbs well as indicated by the midpoint of the breakthrough curve at greater than 30 CV, the rate of sorption appears to be too slow for this resin to be effective in removing strontium in the plant process; the initial breakthrough occurred after only 5 CV.

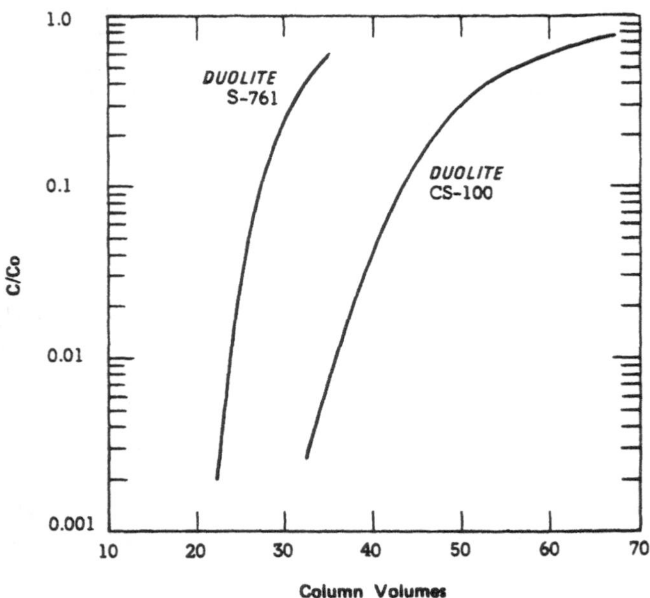

Fig. 1. Absorption of Cs^+ on Duolite CS-100 and S-761 from simulated supernate.

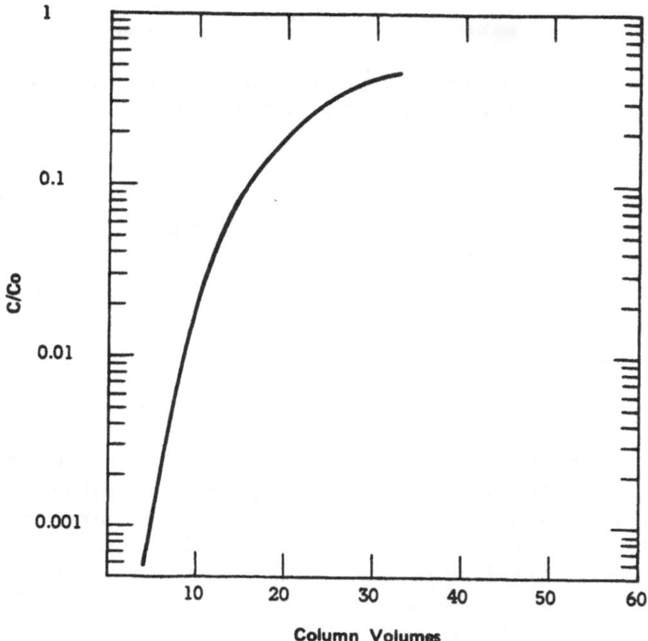

Fig. 2. Absorption of Sr^{2+} on Duolite CS-100 from simulated supernate.

Elution Studies

Although the sorption characteristics of Duolite CS-100 and S-761 are not greatly different from ARC-359, they offer a distinct advantage in elution. Because these resins are weak acids, elution can be effected by a stronger acid. However, ARC-359, which contains a strongly acidic sulfonate group, can be eluted only with a high concentration of an ion, such as ammonium ion, that tends to replace cesium. Formic acid was chosen for study as an elutriant because: (1) it is a weak acid that is relatively strong and therefore should convert the resin to the hydrogen form; (2) it should not damage the resin as would nitric acid if it were used; (3) it would have a smaller impact on subsequent operations in the vitrification plant than any other organic acid; (4) it should not interfere with the ion exchange process if some were recycled; and (5) it would be fairly easy to destroy in aqueous solution if that were necessary.

Table 2 gives the results of measurements of distribution coefficients of $^{137}Cs^+$ tracer between several resins and a solution containing 2M $(NH_4)_2CO_3$ - 1M NH_4OH and one containing 2M formic acid. These studies were made to indicate conditions that

Table 2

DISTRIBUTION COEFFICIENTS (K_d) OF Cs^+ AT TRACER LEVEL BETWEEN VARIOUS RESINS AND ELUTRIANTS

Resin	Cs^+ K_d	
	2M $(NH_4)_2CO_3$ - 1M NH_4OH	2M Formic Acid
Duolite ARC-359	6.95	535
Duolite CS-100	2.13	1.45
Duolite S-761	1.24	0.78

will be favorable for elution; low K_d's indicate favorable conditions. The results indicated that: cesium should elute from CS-100 and S-761 more efficiently with ammonium carbonate than from ARC-359. Cesium should elute from CS-100 and S-761 with 2M formic acid more efficiently than with ammonium carbonate, but it should not elute well from ARC-359 with formic acid. The elution performance of S-761 should be slightly better than CS-100 with both elutriants.

Tests of the elution performance of the resins were run in 1 cm ID columns with the feed introduced down-flow. The columns were eluted down-flow at a rate of 1.25 CV/hour. Down-flow elution was used because it was easier experimentally and because it is a more severe test of elution performance than the up-flow elution used in the reference process.[1] The eluates were collected in 0.25 CV fractions. Figure 3 shows the results of a test of cesium elution from CS-100 with 2M formic acid. One curve gives the % of the total cesium activity originally on the column eluted with each of the 0.25 CV fractions, the other is the cumulative % eluted. Virtually all of the activity was removed within 4 CV, but very little was removed in the first few fractions (1.75 CV); most of the activity is then removed in the next 1 CV. It appears that very little cesium is removed until after the bulk of the sodium has been eluted and the resin is converted to the hydrogen form. It should therefore be possible to separate much of the sodium from the cesium should it prove desirable. The elution curve for cesium from Duolite S-761 with 2M formic acid was nearly the same as that from the CS-100. Elution of Cs from Duolite CS-100 with 2M $(NH_4)_2CO_3$ - 1M NH_4OH resulted in a broader curve than that shown in Figure 3.

Elution curves were also run for $^{85}Sr^{2+}$ from CS-100 and Amberlite IRC-718 with 2M formic acid as an elutriant. Strontium elutes well from both resins with formic acid and the shapes of

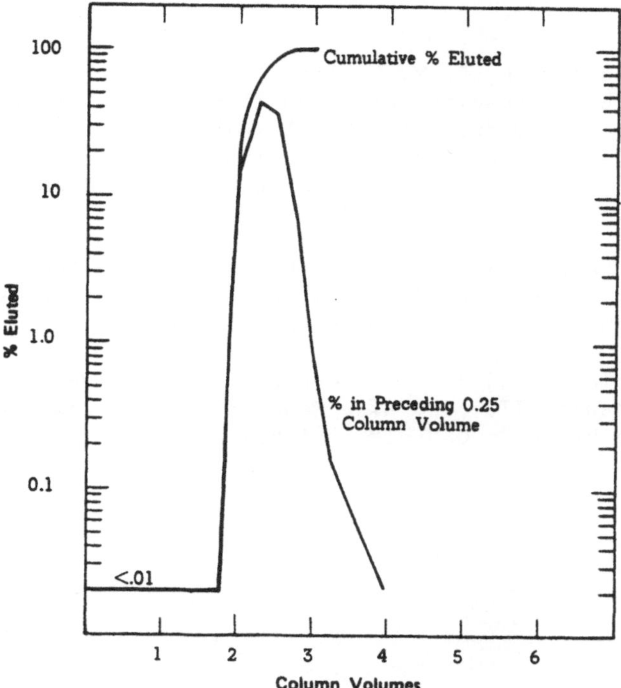

Fig. 3. Elution of Cs^+ from Duolite CS-100 with 2 M formic acid.

the curves are similar to those obtained with cesium. This suggests that a sodium/strontium separation is also possible. Because formic acid can be used to elute both cesium and strontium from Duolite CS-100, and to elute strontium from Amberlite IRC-718, there does not appear to be any reason why these resins could not be mixed together in the same column, should that prove desirable.

Future Program

The gross features of the new process have been demonstrated but a few questions remain. These include: (1) Will plutonium be sorbed by CS-100? (2) How well can sodium and cesium be separated during formic acid elution? (3) How will formic acid or sodium formate react in the vitrification process? (4) What will be the effect of sodium formate on the ion exchange sorption of strontium and cesium if some formate must be recycled? (5) How can formate ion best be destroyed in aqueous solution if the need should arise? These problems are currently under investigation.

REFERENCES

1. P. K. Baumgarten, R. M. Wallace, D. A. Whitehurst and J. M. Steed, "Development of an Ion-Exchange Process for Removing Cesium from High-Level Radioactive Liquid Waste." Paper presented at "International Symposium on the Scientific Basis of Nuclear Waste Management," Boston, Massachusets, November 1979.

2. J. T. Roberts and R. R. Holcomb, "A Phenolic Resin Ion Exchange Process for Decontaminating Low-Radioactivity-Level Process Water Wastes." ORNL-3036 (1961).

3. R. R. Holcomb and J. T. Roberts, "Low-Level Waste Treatment by Ion Exchange, II. Use of a Weak Acid, Carboxylic-Phenolic Ion Exchange Resin." ORNL-TM-5 (1961).

A SMALL-SCALE INTEGRATED DEMONSTRATION OF HIGH-LEVEL RADIOACTIVE WASTE PROCESSING AND VITRIFICATION USING ACTUAL SRP WASTE

R. B. Ferguson, G. B. Woolsey, R. M. Galloway, P. K. Baumgarten, and R. E. Eibling

E. I. du Pont de Nemours & Co.
Savannah River Laboratory
Aiken, SC 29808

Introduction

The Savannah River Plant (SRP) produces nuclear materials for national defense programs and civilian use. These materials are produced by irradiation in nuclear reactors and are recovered by chemical separations processes. The plant is operated by Du Pont for the U.S. Department of Energy. In 26 years of SRP operations, some 68 million gallons of high-level radioactive wastes have been generated as a byproduct of the nuclear materials production processes, and new waste is continuing to be generated at a rate of about 1.5 million gallons per year. This waste has been evaporated to 23 million gallons and is now stored in 33 underground carbon-steel tanks.

A research and development program has been underway at Savannah River since 1973 to develop and demonstrate appropriate technology for immobilizing this high-level waste in forms suitable for permanent disposal. This paper reports results from one of these programs - a small-scale demonstration of the reference SRP solidification process using actual high-level radioactive waste.

Description of SRP Waste and the Reference Waste Immobilization Process

Most wastes from the SRP radiochemical separations processes are initially acidic solutions of heavy metal nitrates. However, before transfer to the carbon-steel storage tanks, these solutions are neutralized with sodium hydroxide. This causes precipitation of insoluble oxides and hydroxides of elements present in the fuel or added in reprocessing (primarily Fe, Al, Mn, and Hg) along with most of the fission products and small amounts of actinides and uranium not recovered in reprocessing. After these insoluble materials, referred to as sludge, have settled to the bottom of the waste tank, the supernatant liquid is removed and concentrated by evaporation. The concentrate from the evaporator is transferred to tanks where, on cooling, salts crystallize out. Residual supernate is recycled to the evaporators until these tanks contain only a wet salt cake. The salt consists primarily of sodium nitrate and nitrite. A large quantity of supernate also remains in the gelatinous sludge layer as interstitial liquid. The principal radionuclides of biological concern are ^{90}Sr and ^{239}Pu, which are found almost exclusively in the sludge, and ^{137}Cs present in the salt cake and supernate solution.

Figure 1 shows a simplified flowsheet for the reference waste immobilization process. In this process, the salt cake in waste tanks will be redissolved in fresh water, while the washed sludge will be removed from the tanks by hydraulic slurrying techniques. These two feed streams will then be fed separately into a large shielded building, referred to as the Defense Waste Processing Facility (DWPF), for remote processing to incorporate the wastes into borosilicate glass.

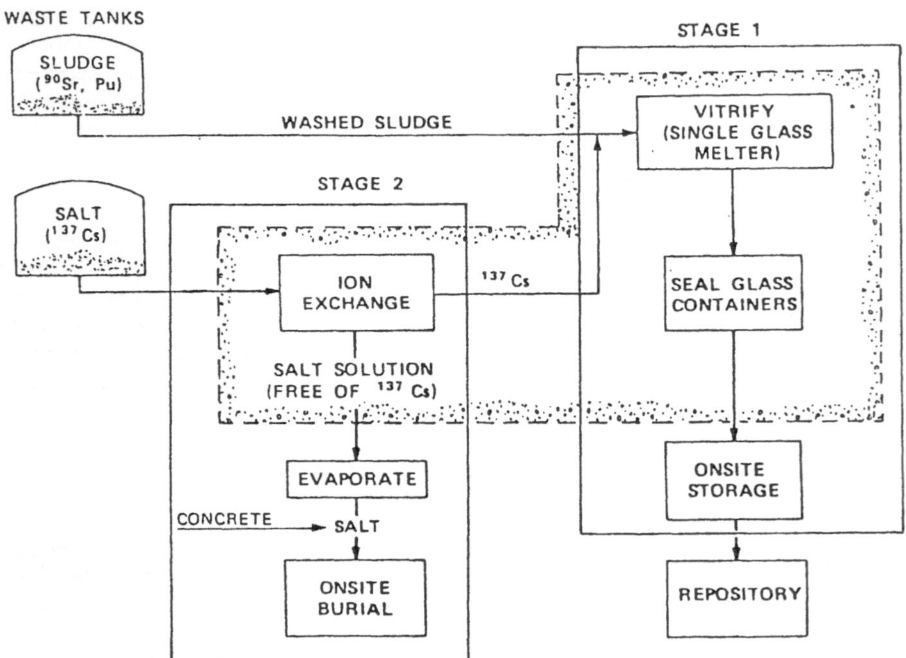

Figure 1. Reference Process for SRP Waste

The staged approach for design and construction of the DWPF was selected for the main effort on the basis of lowest initial investment and competitive overall cost after evaluating several alternatives. The staged approach consists of a first stage which provides a vitrification facility to incorporate the sludge portion of SRP waste containing about 90% of the total long-lived (>100 yrs) curie inventory of fission products into borosilicate glass. The second stage will provide for decontamination of the supernate, or salt solution, with transfer of the removed radioactive isotopes to the vitrification facility and incorporation of the decontaminated salt solution into concrete (saltcrete) for onsite burial as low-level waste.

The ion exchange process removes the cesium as well as trace amounts of soluble strontium and actinides from the dissolved salt solution. The decontaminated salt solution is then evaporated and stored as low-level waste. The cesium and strontium eluted from the ion exchange columns are mixed with the washed sludge and glass frit, forming a slurry for feed to the vitrification step. The slurry is sprayed onto the molten glass surface in a joule-heated ceramic glass melter. Borosilicate waste glass is poured from the melter into stainless-steel canisters which are cleaned and welded shut in preparation for interim storage on-site and eventual storage in a suitable geologic repository. By decontaminating the salt cake and incorporating only the sludge and the cesium and strontium from the salt cake into the borosilicate glass, the total volume of waste glass which must be melted and stored is reduced by a factor of about 20.

Much of the work to develop and demonstrate the SRP reference waste soldification process has been conducted with synthetic non-radioactive waste rather than actual waste. This substitution is essential to reduce the cost and difficulty of carrying out the development work. However, a small-scale demonstration of the process is being performed with actual waste retrieved from SRP waste tanks for the purposes of:

- Providing a comparison of synthetic and actual waste processing behavior and product quality.
- Confirming the detailed chemistry of the flowsheet.
- Obtaining basic data on behavior of minor constituents in the ion exchange and vitrification off-gas treatment steps.

The following sections of this paper discuss the results obtained in processing actual waste with small-scale equipment in shielded hot cells.

Sludge Washing

The soluble salts entrained within the settled sludge at the bottom of the waste tanks must be removed before a high durability borosilicate waste glass can be produced. The sodium sulfate present in the salts is nearly insoluble in borosilicate glass and would form a separate cesium-rich sulfate phase susceptible to rapid leaching if not washed from the sludge before vitrification.

The washing operation consists of contacting the sludge with fresh water in the waste tank and then allowing the resulting slurry to separate into two phases by gravity settling.

Supernate Decontamination

The salts in the spent wash water from sludge washing are concentrated by evaporation and combined with dissolved salt cake to be decontaminated. Cesium will be removed from the clarified solution by ion exchange on "Duolite" ARC-359 resin or CS-100 resin (trademark of Diamond Shamrock Corporation). Strontium will then be removed from the solution by ion exchange on "Amberlite" IRC-718 resin (trademark of Rohm & Haas) or on sodium titanate. For both cesium and strontium ion exchange, the solution will flow downward through fixed resin beds. When very small amounts of cesium and strontium begin to appear or "break through" in the effluent solution, column operation will be halted. The cesium and strontium sorbed on the beds will be eluted and the beds will be regenerated, thus permitting the cycle to begin again.

Ion exchange processes are conventionally characterized by breakthrough curves in which the concentration of the ion being removed from the solution is plotted as a function of time or column throughput. Figure 2 shows typical cesium breakthrough curves experimentally obtained for both synthetic and actual dissolved salt solutions. Because of the difference in scale of the two tests, the data are plotted in dimensionless form. The ordinate shows the ratio of the outlet cesium concentration to the inlet concentration (also referred to as the decontamination factor), while the abcissa is the throughput in column volumes. The inlet concentration of cesium in both cases is about 10^{-4}M, while the residence time of the solution in the bed is held constant at 2 column volumes/hour. Figure 3 shows typical strontium breakthrough column throughput.

The test with synthetic supernate was conducted in a plant-height, 8-foot-tall column, while the actual dissolved salt was decontaminated in 8-cm-long laboratory glassware. Because the column residence time was held constant for the two tests, the ratio of fluid linear velocity to column length (V/L) was also constant. Thus, the linear velocity of synthetic supernate through its column was about 30 times faster than that of the actual supernate. Nevertheless, the breakthrough curves for the two tests are very similar.

The radioisotope content of decontaminated salt cake is being thoroughly characterized to ensure the acceptability of storing it as low-level radioactive waste. Table 1 indicates the specific radioactivity present in the salt (about 250 nCi/g). Figure 4 places this level of radioactivity in perspective with that present in other natural and commercial sources.

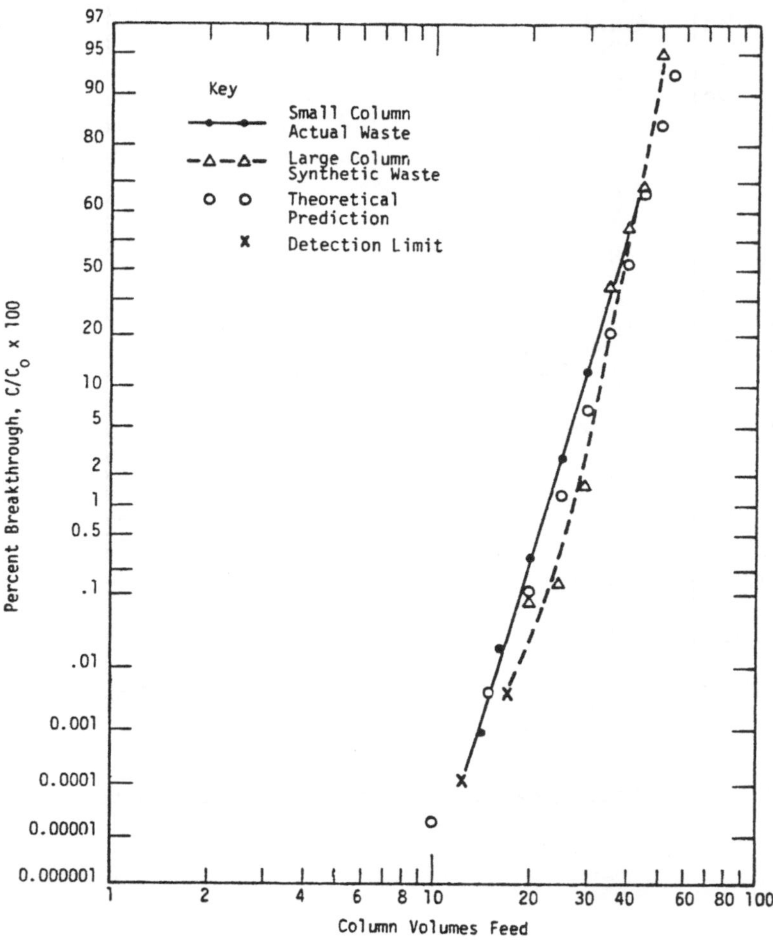

Figure 2. Performance of Cesium Ion Exchange Columns

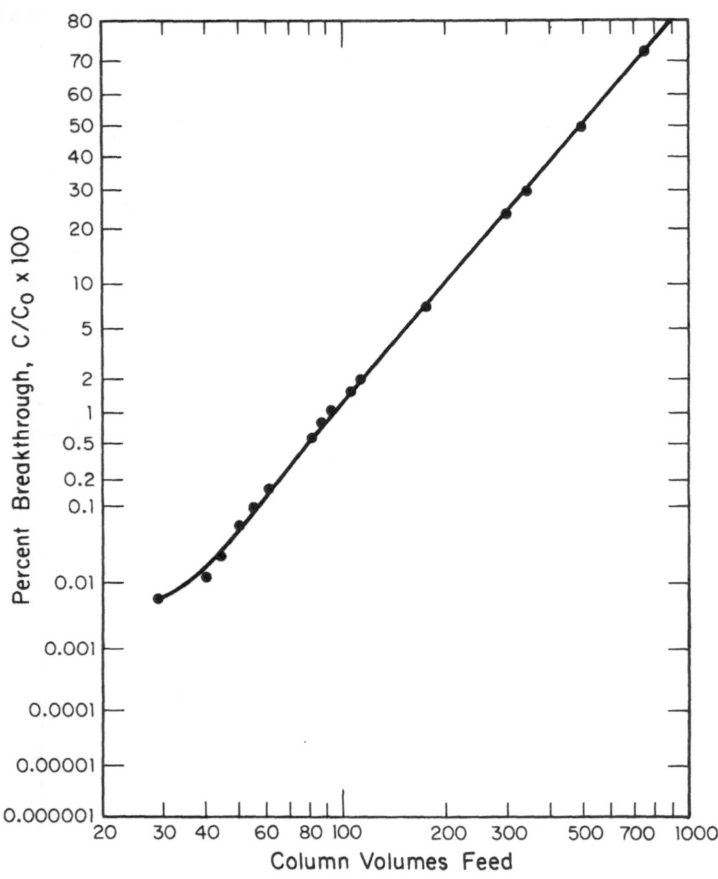

Figure 3. Performance of Strontium Ion Exchange Columns

TABLE 1. Specific Radioactivity of Decontaminated Saltcake

Radionuclide	Conc., nCi/g	Radionuclide	Conc., nCi/g
Cs^{137}	31	Ag^{110m}	<0.0085
Ba^{137m}	31	Ce^{144}-Pr^{144}	<1.10
Cs^{134}	0.28	Eu^{152}	<0.021
Sr^{90}	0.4	Eu^{154}	<0.0075
Y^{90}	0.4	Zr^{95}-Nb^{95}	<0.013
Ru^{106}	28	H^{3}	
Rh^{106}	28	Tc^{99}	
Sb^{125}	3.5	I^{129}	100 total
Sb^{126m}	0.24	Pm^{147}	
Sn^{126}	0.24	Sm^{151}	
Pu^{238}	1.9	Sn^{121n}	
Co^{60}	<0.0015		

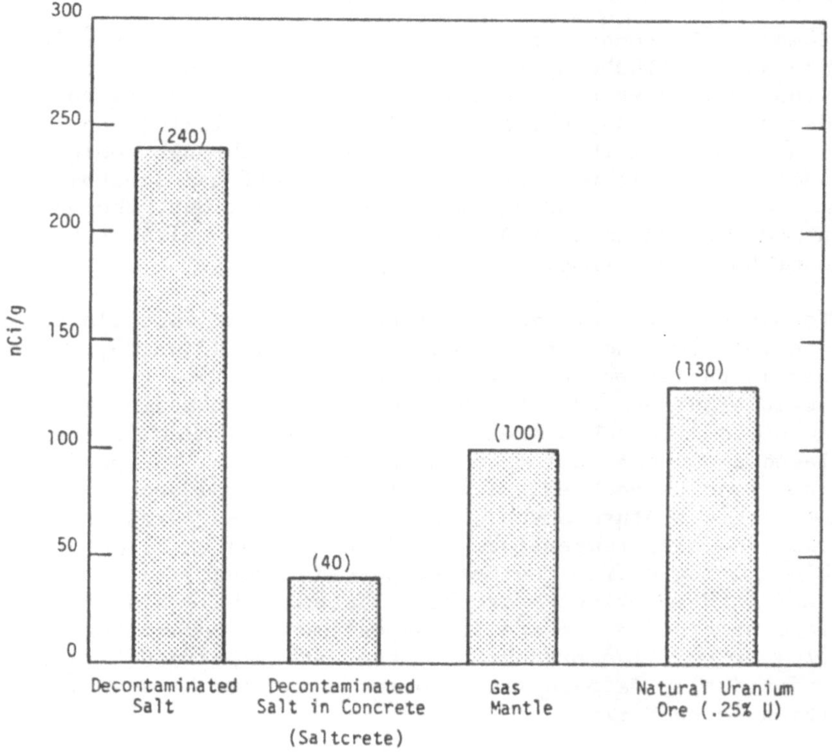

Figure 4. Residual Radioactivity of Decontaminated Salt

Glass Melting

In the reference small-scale waste vitrification flowsheet, washed sludge plus cesium and strontium eluted from the ion exchange columns are slurry-fed to a joule-heated ceramic melter. The main melt chamber is rectangular in shape and about 23-cm-long with a pair of "Inconel" 690 (trademark of International Nickel Corp.) electrodes at the ends of the chamber. The resistance (joule-effect) heating generated by passing an electric current through the molten glass provides the necessary energy to maintain the glass at an operating temperature of 1150°C as well as to melt the frit-waste mixture fed to the melter. Normal power requirements of the small-scale melter are about 2-1/2 kW. The molten glass exits the main melt chamber through a small channel or throat at the bottom of the chamber and flows into an overflow chamber. Glass is poured from the overflow chamber into 500 mL stainless steel beakers by tilting the melter at a 6-degree angle.

Off-Gas Characteristics

Some of the components of the waste are at least partially volatile at the 1150°C operating temperature of the melter. Elements that will partially volatilize and are of special concern include Ru, I, Cs, Hg, Cl, and F. After volatilizing from the melt, the vapors of these elements will pass through cooler sections of the melter before reaching the off-gas scrubbing system. As they pass through these cooler sections, the volatile elements may condense or react and plate out on melter surfaces and contribute to corrosion.

To begin characterizing the behavior of the volatile components generated during vitrification of actual SRP waste, a vapor trap was installed on the small-scale melter. This trap consists of a stainless-steel tube (9/16-inch diameter) containing a stack of concentric 1/2-inch-diameter rings. The entrance of the trap was placed about one inch from the melt surface. Thermocouples were installed at each end of the ring stack, and each ring was assigned a temperature based on an assumed linear gradient between the measured inlet temperature of 800°C and the exit temperature of 130°C. Off-gas from the melter passed through the stacked rings and into an off-gas scrubbing system. The inside walls of the rings provided a surface for deposition of condensing volatiles or reaction products. After each experiment, the rings were individually leached with aqueous solutions to quantify deposition on each ring.

Typical deposition patterns of semivolatiles from dry sludge tests are shown in Figs. 5, 6, and 7.

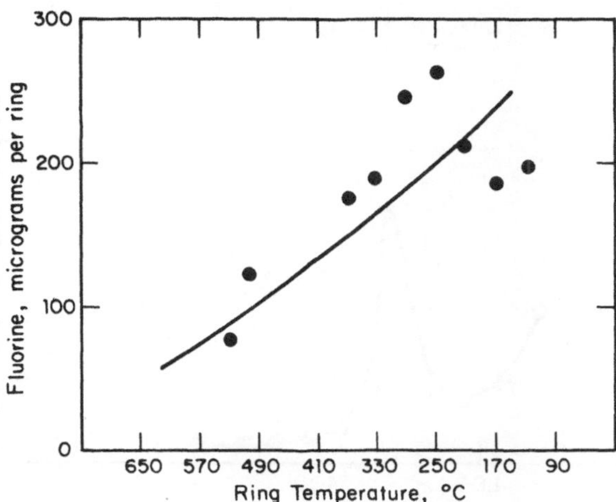

Figure 5. Deposition of Fluorine

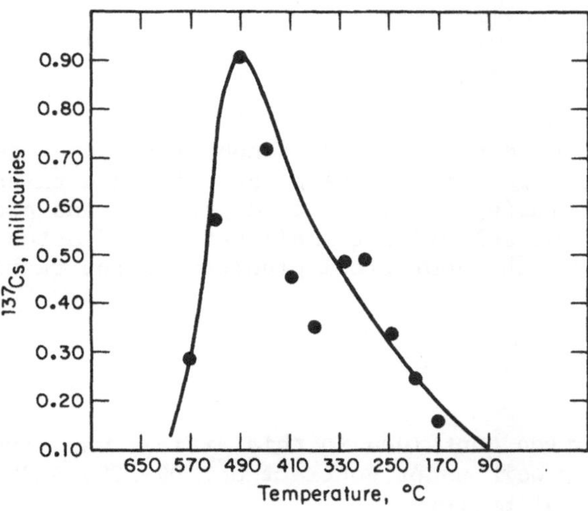

Figure 6. Deposition of Cesium

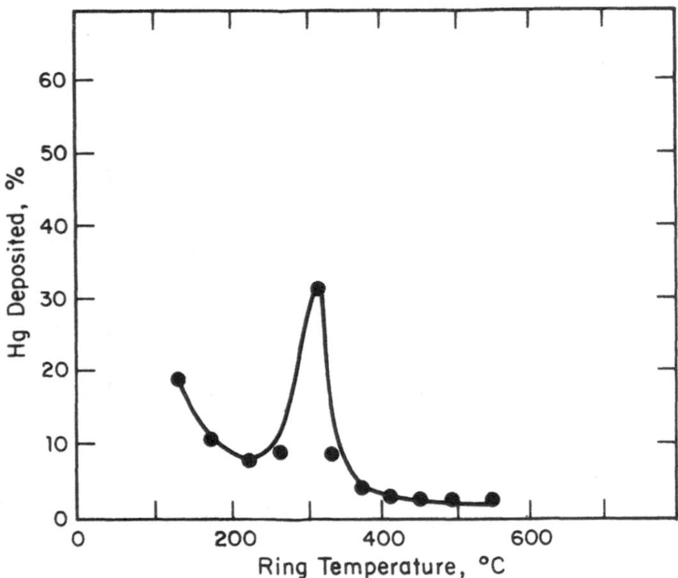

Figure 7. Deposition of Mercury

Conclusions

Experiments have been made to demonstrate the feasibility of immobilizing SRP high-level waste in borosilicate glass. Results to date are encouraging. Equipment performance and processing characteristics for solidifying small batches of actual SRP waste have agreed well with theoretical predictions and tests using synthetic waste.

Acknowledgement

The information contained in this article was developed during the course of work under Contract No. DE-AC09-SR00001 with the U. S. Department of Energy.

ADVANCED METHOD FOR MAKING VITREOUS WASTE FORMS*

J. M. Pope and D. E. Harrison

Westinghouse Research and Development Center

Pittsburgh, Pennsylvania 15235

ABSTRACT

A process is described for making waste glass that circumvents the problems of dissolving nuclear waste in molten glass at high temperatures. By this process, nuclear waste and glass formers are chemically combined at low temperatures prior to melting. Separation of the mixing and melting operations permits any glass composition to be prepared and it enables novel glass fabrication methods to be employed.

INTRODUCTION

Glass is the most extensively studied solid waste form for the immobilization of high-level nuclear waste. Usual practice is to dissolve waste oxide in molten glass at temperatures in excess of 1000°C. In order to achieve homogenization of the melt in reasonable times, glass forming systems are selected that have low viscosities at these temperatures. This constraint is best met by borosilicate glass compositions. Conversely, highly durable glass compositions, such as natural obsidian and nepheline syenite, require such high forming temperatures (∿1350°C) when made by conventional waste glass processing methods that they have been largely eliminated from consideration as a solid waste form.

Westinghouse has developed a process for making nuclear waste glass that largely eliminates the problems of volatilization, foaming, slagging and dusting by chemically combining the waste sludge and glass additives at low temperatures prior to melting.[1] Because the reactive mixing process is independent of the inherent viscosity of the melt, waste glass compositions that are high in alumina can

*Prepared for the U.S. Department of Energy, Contract DE-AC-08-76NV00597.

be prepared with ease. Separation of the mixing and melting operations permits novel glass fabrication techniques to be employed.

PROCEDURE

Reactive Mixing Process

Overall the reactive mixing process consists of four steps: (1) hydrolysis of the individual metal-alkoxide glass additives; (2) blending of the alkoxide glass additives; (3) mixing of the glass additives and waste sludge to develop the required glass composition; and (4) heating of this blend to promote chemical reaction by removal of alcohol and water. In order to avoid gross precipitation during blending of the alkoxides, the individual alkoxides must be hydrolyzed to the correct degree. Once this is accomplished the alkoxides and waste sludge will readily blend into a homogeneous mixture that is insensitive to variations in composition, temperature and time.

Two variations of the reactive mixing process have been developed. One is named the reference alkoxide process which employs all metal alkoxide glass additives. The other, called the simplified process, uses only silicon and boron alkoxide, the other additives being in the form of hydroxides.

Glass Fabrication

After the water and alcohol are removed, the reactively mixed material containing 30 wt % waste may be transformed into a glass by melting at temperatures between 1000 and 1200°C. Typically, glasses were formed by melting 50 cc batches of either wet or dry reactively mixed material in Al_2O_3-998 crucibles. A few 1-kg glass ingots were prepared by intermediate feeding of a larger alumina crucible. The nominal firing schedule involved heating at 5°C/min, holding at temperature for 1 hour (for convenience) and furnace cooling at about 20°C/min. Furnace cooling could be programmed to be as low as 2°C/min.

RESULTS

Reactive Mixing Evaluation

X-ray diffraction and IR absorption spectra of reactively mixed glass additives alone and with waste additions were used to assess the degree of homogeneity achieved by means of the alkoxide process. The diffraction patterns of the glass additives after drying and after melting are amorphous. Likewise, the diffraction patterns of the dried waste glass mixtures are generally amorphous except for small peaks which are due to $Al(OH)_3$ and U_3O_8. After melting the waste glass can be completely amorphous or, depending on composition, it can contain crystals of an iron-manganese-nickel spinel phase.

Surprisingly, calcined (500°C) glass additives alone and those containing 30% waste have nearly identical IR spectra between 6 and 40 μm. An apparent difference between 2 and 6 μm is undoubtedly largely due to the difference in refractive index between the transparent glass additives and the highly colored waste glass.

Waste Glass Compositions

Table 1 lists the calculated low-, medium- and high-alumina glass additive compositions normalized at the 30 wt % waste level. Table 2 gives selected compositions of the "average," "high-ferric oxide" and "high-alumina" simulated waste sludge (supplied by Southwestern Analytical Chemicals, Inc.) normalized to the 70% glass additive level. Thus, the complete waste glass composition is determined by the sum. For example, the total alumina content of the high-alumina glass added to a high-alumina waste is 27.7 wt %.

Confirmation of the calculated waste glass compositions by chemical analysis gives general agreement except for consistently higher values for the Al_2O_3 and lower values for the Fe_2O_3. Melting in alumina crucibles may account for some of this discrepancy.

Compositions containing less than 15 wt % Al_2O_3 melt satisfactorily at 1100°C. Compositions containing 28 wt % Al_2O_3 and 11 wt % B_2O_3 melt nicely at 1200°C, but if they contain 11% CaO rather than B_2O_3 the temperature required to form a good glass increases to 1300°C. The high-alumina glasses were much too viscous to be poured from a crucible, yet they formed a smooth meniscus. Even the low-

Table 1. Calculated Melted Compositions (wt %) of Alkoxide Derived Glasses Compared to that of a Typical SRL Glass

Glass Designation	SiO_2	Al_2O_3	B_2O_3	Na_2O	K_2O	Total Waste, Oxides
Low-alumina	45.5	3.5	10.5	10.5		30.0
Medium-alumina	38.5	10.5	10.5	10.5		30.0
High-alumina[a]	35.5	14.0	10.9[b]	9.0		30.6
Reduced Viscosity #1	40.8	3.5	10.5	10.5	4.7	30.0
Reduced Viscosity #2	40.8	3.5	10.5	15.2		30.0
Reduced Viscosity #3	40.8	3.5	7.8	14.4[c]		30.0
Ref. SRL 211	40.8	d	7.8	14.4		30.0

[a]Gives "nepheline syenite" when combined with "high-alumina" waste.
[b]CaO instead of B_2O_3 also studied.
[c]In addition, 3.9% CaO.
[d]3.1% Li_2O + 3.9% CaO.

Table 2. Waste Compositions (wt %)

Waste Designation	Fe_2O_3	Al_2O_3	MnO_2	U_3O_8	CaO	NiO	SiO_2	Na_2O	Na_2SO_4	Ion-Siv
Avg + Na_2CO_3	13.4	2.7	3.8	1.2	1.0	1.7	1.2	1.8	0.4	2.8
Avg	14.4	2.9	4.1	1.3	1.1	1.8	1.3	0.1	-	3.0
Hi Fe + Na_2CO_3	16.6	0.3	1.2	3.9	1.0	2.8	-	1.4	0.1	2.7
Hi Fe	17.5	0.3	1.2	4.1	1.0	3.0	-	-	-	2.9
Hi Al + Na_2CO_3	5.5	12.9	3.2	2.4	0.4	0.6	0.4	1.6	0.1	2.9
Hi Al	5.8	13.7	3.4	2.5	0.4	0.7	0.4	-	-	3.1

alumina waste glasses had higher viscosities than Savannah River Reference Glass 211. Consequently, the reduced viscosity glasses listed in Table 1 were prepared by increasing the alkali content.

Observation of one kilogram ingots being made by successive feeding of the melt through a tube clearly shows that the feed is readily assimilated by the melt within a few minutes without foaming or any other difficulty. Air quenching or furnace cooling always gives a bubble-free glass monolith of either reference or simplified process material.

Volatilization

TGA and chromatographic determination of the volatile species at temperatures up to 600°C show them to be the alcohols and water expected from the condensation reactions of the various alkoxides.

Thermobalance results indicate the weight loss between 600 and 1300°C for the reference alkoxide process waste glasses is small for the low-alumina composition and negligible for higher-alumina glasses:

Aluminosilicate glass	% wt loss from 600 to 1300°C
Low-alumina	0.68
High-alumina	0
Nepheline syenite	0

These results are essentially the same for all of the waste types given in Table 2, including those with Na_2CO_3. Although the weight loss for the waste glasses produced by the simplified alkoxide route is slightly higher, it is still less than 1 wt % absolute.

Leaching Measurements

Leach tests were performed in a Pyrex soxhlet apparatus operating at 100°C with "circulating" water for 24 days. DC plasma and atomic absorption methods were used to determine the amount of the species leached. Leach rates were calculated using the method of Kelley and Wallace[2]

$$L = f\left(\frac{M}{A}\right)\frac{1}{t}$$

where f is the fraction of ions leached during the leaching period, t, and M/A is the ratio of mass to surface area of the sample. The ratio M/A was estimated to be 0.00583 g/cm^2 by assuming all particles to be spheres with a diameter of 250 μm (the smallest particle that will not pass through a 60-mesh sieve), a density of 2.8 g/cm^3, and a roughness factor of 2. The upper sieve was 40-mesh through which all of the sample passed.

The leach rates were corrected for impurities in the deionized water and for species leached from the Pyrex apparatus during "blank" tests of the same duration. These results indicate that the nepheline syenite glass has a hundred-fold lower leach rate than the low-alumina glass composition.

Waste glass microstructure. At the 30 wt % loading level, average waste and especially high-iron waste compositions yield glasses that exhibit cubic or occasionally tetrahedral-shaped crystals of an iron-nickel-manganese spinel as determined by EDAX and x-ray diffraction. Increasing the proportion of either soda or alumina in the glass composition decreases the number of spinel crystallites. High-alumina sludges, on the other hand, result in glasses that are free of crystalline phases that can be detected by either petrographic, electron optic, or x-ray techniques.

DISCUSSION AND CONCLUSIONS

Reactive Mixing

The melting behavior of material made by the reactive mixing process indicates that the glass additives and nuclear waste sludge are combined on essentially a molecular level by the time the alcohol and water are evaporated. Whether this is the result of a chemical polymerization reaction between the alkoxides and waste sludge or a high degree of reactivity that promotes homogenization at modest temperatures is not clear. The best indicators of a chemical reaction are gelation and the unique role of boron alkoxide. If the alkoxide-waste mixture is allowed to remain quiescent, a weak gel will form. Since stable gel formation signifies chemical polymerization in alkoxide processing, even a weak gel which is destroyed on mixing is suggestive of localized polymerization. This supposition is supported by the markedly different behavior of boric acid and boron alkoxide. Glass additives plus waste sludge made using boric acid exhibit upon heating the bubbling and copious vapors that are typical of compositions containing boric acid. However, when trimethylborate is used instead, these characteristics disappear, again signifying that the boron is chemically combined. Other evidence supporting a chemical polymerization reaction is the effect of a slight change in pH. A few drops of nitric acid will cause the calcined mixture to form a cake rather than a fine powder, which suggests an electrolyte effect.

That chemical polymerization may only partially explain the results is indicated by the tolerance of the process to variations in temperature, time and composition. The glass additives alumina and soda may be added as either alkoxides or hydroxides, and a wide range of waste compositions can be treated with equal facility. This insensitivity to process and compositional variables contrasts with the close control required when stable gels are produced using metal alkoxides exclusively in the conventional manner. A high reactivity resulting from the use of alkoxides may also contribute to the homogenization of the mixture at the modest temperatures needed to drive off the alcohol and water.

Diagnostic examination of the reactive mixing process has thus far revealed little. X-ray analysis shows the dried waste-glass precursor mixture to be generally amorphous with minor amounts of a few discrete phases such as $Al(OH)_3$, U_3O_8 and occasionally $CaCO_3$. More surprising is the absence of any major shifts in the IR spectra of the glass additives when 30 wt % waste is incorporated.

Melting Characteristics

Furnace melting. The most notable characteristic of the reactively mixed material is the ease with which it melts. Melting progresses rapidly without foaming or slagging and little bubbling. Withdrawing a melt as soon as it reaches furnace temperature always gives a uniform, bubble-free glass.

Since no thermal driving force is required for homogenization, the apparent melting temperature of the waste glass is decreased. For example, a nepheline syenite waste glass composition containing 28% Al_2O_3 and 11% B_2O_3 will melt into bubble-free glass at 1200°C, which is about 150°C lower than the temperature required by conventional waste glass processes. This melting behavior suggests that the waste glass composition is mixed on a "molecular scale." Waste glass compositions prepared by the reactive mixing process have the characteristics of high melting rates, short residence times, and low volatility. They exhibit neither foaming, reboil nor slagging, and they may be tailored to the desired waste glass composition. In contrast, the conventional process of using glass as a high temperature solvent gives low melting rates, long residence times for melt homogenization, high volatility, persistent foaming, reboil, slagging, and permits minimal compositional control of the melt.

Compositions high in Fe_2O_3 and low in either Al_2O_3 or alkali exhibit cubic or tetrahedral crystals of an iron-nickel-manganese spinel. These crystals will disappear as either the alumina or alkali content is increased. Because these two components strongly affect the viscosity in opposite directions, the indication is that the spinel crystals developed under equilibrium conditions rather than as a result of devitrification during cooling.

Remote heat source melting. The feasibility of glass making by localized melting was explored using a continuous 18 kW/cm^2 CO_2 laser, a pulsed 150 watt/cm^2 YAG-Nd laser, and a 200 watt/cm^2 xenon lamp source. The localized melting produced by either the CO_2, YAG or xenon lamp sources appears to depend mainly on the magnitude of the lattice absorption and the low thermal diffusivity of the oxide material. As would be expected from the absorption spectra of the waste glass material, a 10.2 μm CO_2 laser coupled extremely well and gave rise to high surface temperatures and copious volatilization (smoke). In contrast, a 1.0 μm YAG-Nd laser gave excellent melting characteristics with negligible smoke formation, and it required less than one percent of the energy density. This behavior indicates that the radiant energy is being absorbed in a larger volume of material while generating lower surface temperatures. The melting characteristics of the xenon lamp are surprising in that it behaved more like a CO_2 laser in terms of the amount of smoke produced while most of its emission is at wavelengths below one micron. The reasons for this phenomenon and the significantly lower melting rates at comparable power densities may be revealed in the details of the emission spectra.

Waste Glass Composition

A wide variety of waste glass compositions can be prepared by the reactive mixing process because it is independent of the inherent viscosity of the melt. Compositions ranging from the Savannah River 211 Glass, which contains less than 3% alumina, to nepheline syenite, which contains over 25% alumina, have been produced. Each of these waste glass compositions contains 30 wt % simulated waste, and they span the range from high Fe_2O_3 to high Al_2O_3 waste sludges (see Table 2). All of these compositions form bubble-free glasses, over the range of compositions from the viscous, high-alumina to the fluid "reduced viscosity" glasses. Of the four glass additives (silica, alumina, boron oxide and soda), the soda and alumina can be added as either alkoxides or hydroxides. Aluminum and sodium hydroxides contained in the waste sludge appear to contribute to the final glass composition in the same way as when they are added supplementarily. This property permits tailoring of the final waste glass composition, and it negates any need to remove sodium from the waste sludge. Experience has shown that zeolite added to the sludge is readily incorporated in the waste glass during melting. This capability provides a means for disposing of cesium contaminated zeolite ion-exchange resins.

The reactive mixing process necessarily results in the production of alcohol due to condensation reactions involving partially hydrolyzed metallic alkoxides. Since the reactive mixing of radioactive materials must be performed remotely, the alcohol could be either burned (perhaps in a waste heat boiler) inside of the canyon, or recycled to make the required alkoxides outside of the canyon. Fortunately, the steam/alcohol mixture is not explosive, only flammable, which should create no formidable safety problems. By using the same alcohol to control the rate of hydrolysis as is used to make the alkoxides, recycling is simplified and the economics of the process are removed.

Waste Glass Durability

As would be expected, compositions high in alumina appear to have excellent chemical durability as determined by soxhlet leach tests conducted at 100°C. These leach results show that the reference and simplified alkoxide processes give comparable results. Furthermore, over the range of glass compositions and cooling rates investigated the waste glasses made by the reactive mixing method are resistant to devitrification.

Waste Glass Fabrication

After the alcohol and water are removed at temperatures below 600°C, very little weight change occurs. Separation of the mixing and melting operations at this stage permits innovation in waste glass fabrication. The high melting rates and short residence times characteristic of the reactive mixing process will permit the design of smaller volume, higher throughput joule heated melters. On the other hand, the more viscous melts of the more chemically durable compositions can be prepared using either hot or cold wall in-can melting techniques. Cold wall techniques utilizing a remote heat source such as a laser would permit a variety of canister materials to be used, including low carbon steel, maleable cast iron, polycrystalline ceramics, glass and graphite.

REFERENCES

1. J. M. Pope and D. E. Harrison, "Containment of Nuclear Waste," Patent Application Case No. 49,146.

2. J. A. Kelley and R. M. Wallace, "Procedure for Determining Leachabilities of Radioactive Waste Forms," *Nucl. Tech.* 30: 47-51 (1976).

AN ESCA INVESTIGATION OF MOLYBDENUM CONTAINING SILICATE AND PHOSPHATE GLASSES

R. Nyholm
Institute of Physics
University of Uppsala
P.O. Box 530
S-751 21 Uppsala, Sweden

L. O. Werme
KBS (Nuclear Fuel Safety Project)
P.O. Box 5864
S-102 48 Stockholm, Sweden

INTRODUCTION

Investigations have shown that fission products containing borosilicate glasses sometimes show a separated, yellow, alkali molybdate phase. It has been found that additives to the glasses as well as changes in melting conditions influence this phase separation. Knowledge of state and structure of molybdenum in glasses is therefore important (see, e.g., ref. 1 and references therein). In this work simple silicate glasses containing MoO_3 have been investigated by ESCA (Electron Spectroscopy for Chemical Analysis) using as a reference a simple phosphate glass.

In most glass studies ESCA has only been applied to give information of the relative abundance of the elements in the surface layers.[2,3,4] However, in the last few years papers have appeared, showing how ESCA can apply to investigations of structural problems.[5,6,7] In those papers, the interest was focused on the oxygen 1s spectrum, which revealed a structure interpreted as originating from bridging and nonbridging oxygen atoms in the silicate network. In this paper, the primary interest is to study the Mo 3d spectrum in phase-separated and nonphase-separated glasses.

EXPERIMENTAL

The investigation was carried out on a Hewlett-Packard 5950A electron spectrometer, using mono-chromatized Al K α radiation (1487 eV) for excitation of the photoelectrons. The measurements were performed in a vacuum of approximately 100 nPa.

The samples were taken as glass chips from the interior of the glasses. In the initial stages of the study, various means were applied to minimize the surface contamination of the samples. However, it was established that the Mo 3d signal was unaffected by minor contaminations although the oxygen signal was found to be sensitive to pretreatments of the samples.

For this investigation, simple soda-lime glasses containing molybdenum were chosen. The compositions of the glasses, which for the lower MoO_3 content (2%) are identical to what was used in ref. 1, are given in Table 1. For the 2% MoO_3 glasses, both oxidizing and reducing (addition of tannic acid to the melt) conditions were employed. For the 4 and 8% glasses, only oxidizing conditions were used. All glasses, except the 8% silicate and the phosphate glasses, were produced both as nonphase-separated and phase-separated. The 8% Mo-silicate glass could only be produced phase-separated, and the 8% Mo-phosphate glass was only produced as nonphase-separated.

Table 1. Composition of the Glasses (in Weight %)

SiO_2	CaO	Na_2O	MoO_3
72.0	6.0	20.0	2.0
(73.0)	(6.5)	(19.7)	(0.8) (mol %)
70.5	5.9	19.6	4.0
69.0	5.8	19.2	8.0
P_2O_5	Al_2O_3	ZnO	MoO_3
71.8	11.0	9.2	8.0

For the 2% Mo-silicate glasses, the oxidizing conditions produced a noncolored, clear glass, while the reducing melts produced a dark brown glass in agreement with ref. 1. To check the effect of the reducing agent, samples were taken to EPR investigations. A resonance (g-value 1.905) attributed to Mo(III) was observed in both the oxidized and reduced glasses. A markedly higher intensity (comparable to what was reported in ref. 1) was observed in the reduced glass. However, a rough estimate based on the EPR signal intensity indicates that only 0.1% of the total Mo content is found as Mo(III) in the reduced glass.

RESULTS AND DISCUSSION

The recorded Mo-spectra are shown in Fig. 1. For the 2% Mo-silicate glasses, only the oxidized glass is represented. The reduced glass showed an identical Mo 3d spectrum, within the accuracy of the measurement. This can also be understood in view of the EPR-results discussed in the previous section.

The nonphase separated glasses (including the phosphate glass) is characterized by a double Mo 3d doublet (see Fig. 2). The major Mo $3d_{5/2}$ peak is situated at 232.5 eV, on the average, in the silicate glasses (see Table 2). (This coincides with the Mo $3d_{5/2}$ binding energy reported for MoO_3,[8] assuming a metallic Mo $3d_{5/2}$ binding energy of 227.94 eV.[9]) The additional Mo 3d peak was shifted 2.5 eV towards lower binding energies. This shift does not coincide with any reported shifts due to MoO_x (1.2 eV) and MoO_2 (3.4 eV).[8] However, there is some controversy as how to interpret photoelectron spectra from MoO_3 as well as mixed oxides (e.g., $SrMoO_3$) with formal Mo valence <6, due to the metallic properties of these oxides.[10] The peak generally ascribed to MoO_x (valence close to 5) can well be attributed to final state effects in the photoionization process, i.e., a satellite to the MoO_2 peak. Considering these arguments and comparing the found Mo 3d spectra from glasses with Mo-oxide spectra from the literature, the most probable interpretation of the doubled peaks in the glass spectra must be that they arise from Mo-atoms in two different chemical environments. This doubled peak is also present in the phosphate glass; the binding energy of the Mo $3d_{5/2}$ peak is shifted towards higher binding energy (233.7 eV), but with a second $3d_{5/2}$ peak shifted 2.5 eV towards lower binding energy. The shift in binding energy for the major $3d_{5/2}$ peak is most probably due to the phosphate environment. However, for the 8% Mo-silicate glass, as well as for the 2% and 4% Mo-silicate glasses, the 3d doublet at lower binding energy disappears when the glass is phase separated. In view of these findings, it seems reasonable to attribute the low binding energy 3d doublet to Mo-atoms as terminating atoms in the SiO_2 network and the major doublet to clusters of MoO_3 or alkali-molybdates. (MoO_3 and alkali molybdates can generally not be separated on the basis of binding energy measurements only.) Alternatively, the double peak structure is due to the formation of diffuse clusters of high MoO_3-content, where the low energy doublet corresponds to Mo-atoms at the diffuse boundary to the silicate structure. After a heat treatment, the clusters probably grow into clusters or droplets with more distinct boundaries in agreement with current theories for the decomposition of inhomogeneous systems.[11,12]

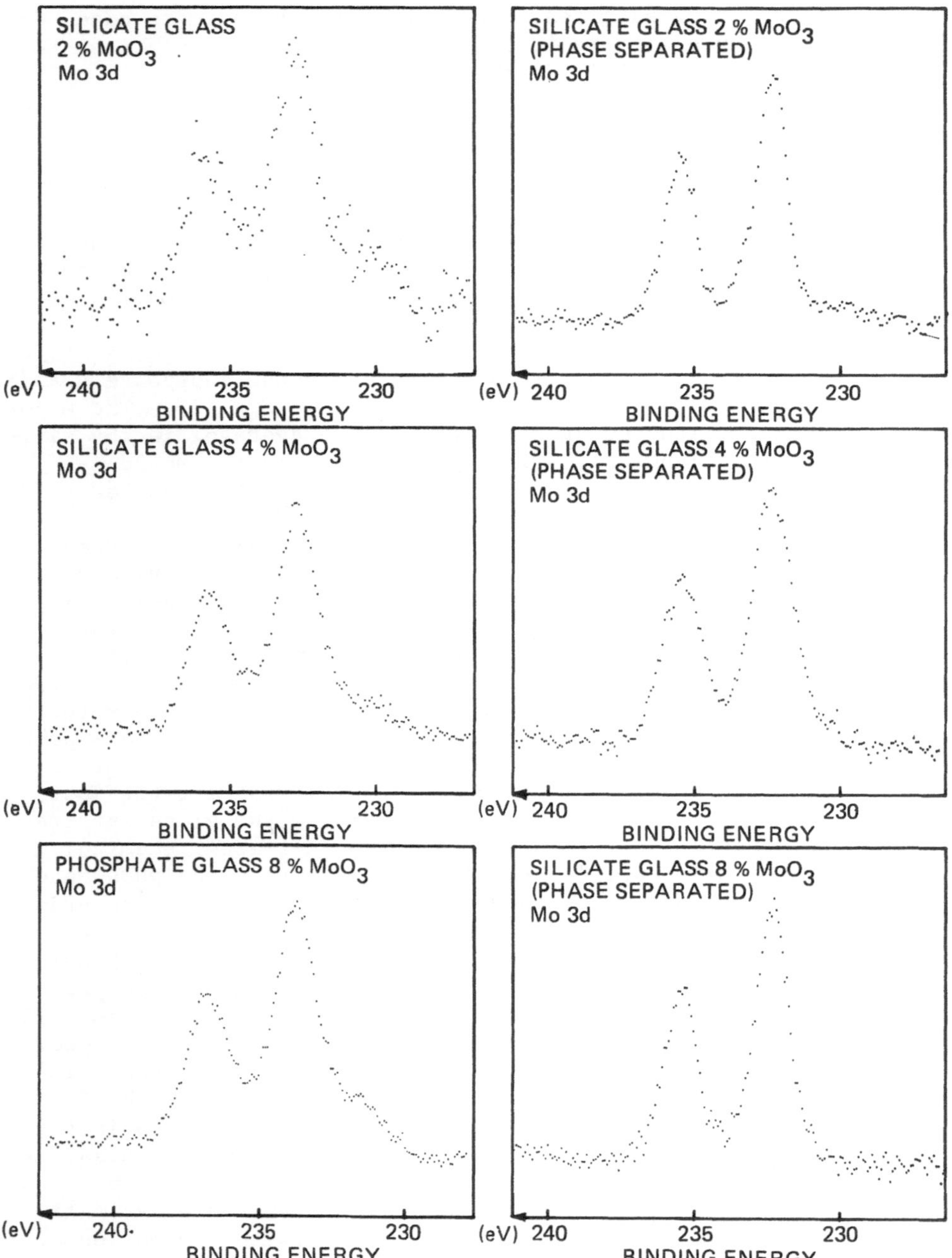

Fig. 1. Recorded Mo 3d spectra showing an extra 3d component for non phase separated glasses.

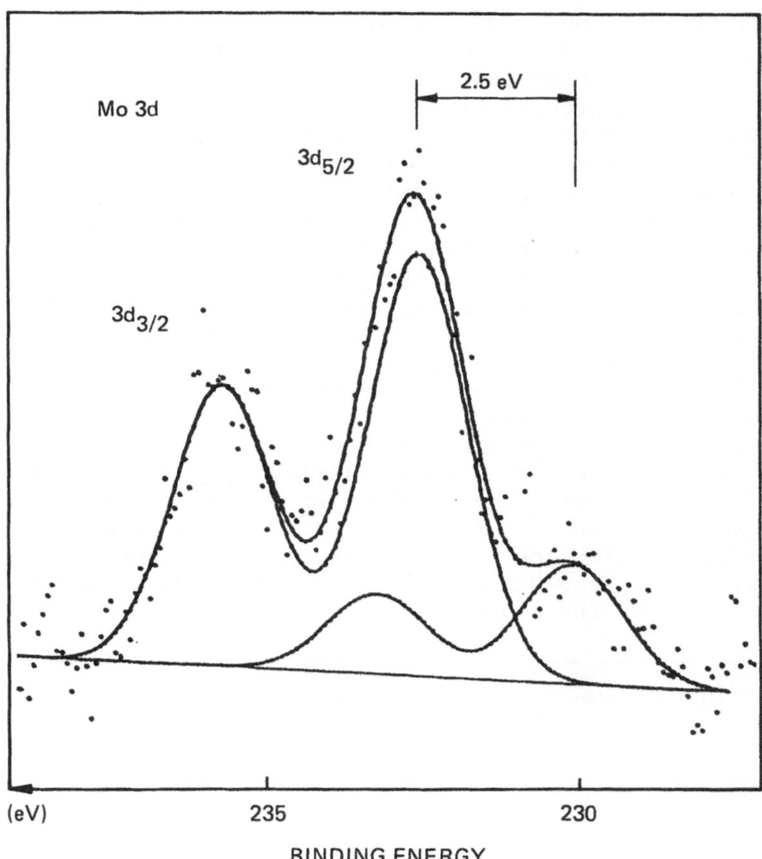

Fig. 2. Decomposition of the Mo 3d spectrum from 2% Mo-silicate glass (non phase separated)

Table 2. Binding Energies (eV) Observed in the Electron Spectra, Relative C 1s (285.0 eV)

Accuracy ± 0.3 eV

	Silicate glasses			Phosphate glass
	2% Ni	4% Mo	8% Mo	8% Mo
Si 2p	102.9	103.0		-
Ca $2P_{3/2}$	347.0			-
Na 1s	1071.9			-
O 1s	532.5	532.5	532.5	532.8
Mo $3d_{5/2}$	232.5	232.5	232.5	233.7
P 2p	-	-	-	135.2

Apart from the Mo-spectrum, only the oxygen 1s spectrum exhibited any structure. In the O 1s spectra (Fig. 3), a low energy component, shifted about 1.8 eV towards lower binding energies, was present. As was mentioned before, the intensity of this low energy peak varied with pretreatment of the sample and also during the course of the experiment. This was also true for the Na 1s line. The effect on Na has been observed before[2,3] and is probably due to the high mobility of the Na^+ ions in the silica network when subjected to X-radiation. Previous reports[5,6,7] on the low energy peak in the O 1s spectrum have attributed this peak to nonbridging oxygen atoms, the main peak being due to bridging oxygens in the silica network. In the work done, this assignment seems well established. However, the varying intensity of the low energy peak could indicate the presence of some impurities, but a contamination should give rise to a peak shifted 3 eV relative to the main peak.[7] No such peak was observed and in this investigation no particular efforts were made to clarify the origin of the intensity variations of the low energy oxygen 1s peak. For comparison, an O 1s spectrum from the phosphate glass is also shown (Fig. 4). As can be seen, this peak is also asymmetric, but on the high binding energy side. This was not investigated further since it was considered outside the scope of this study.

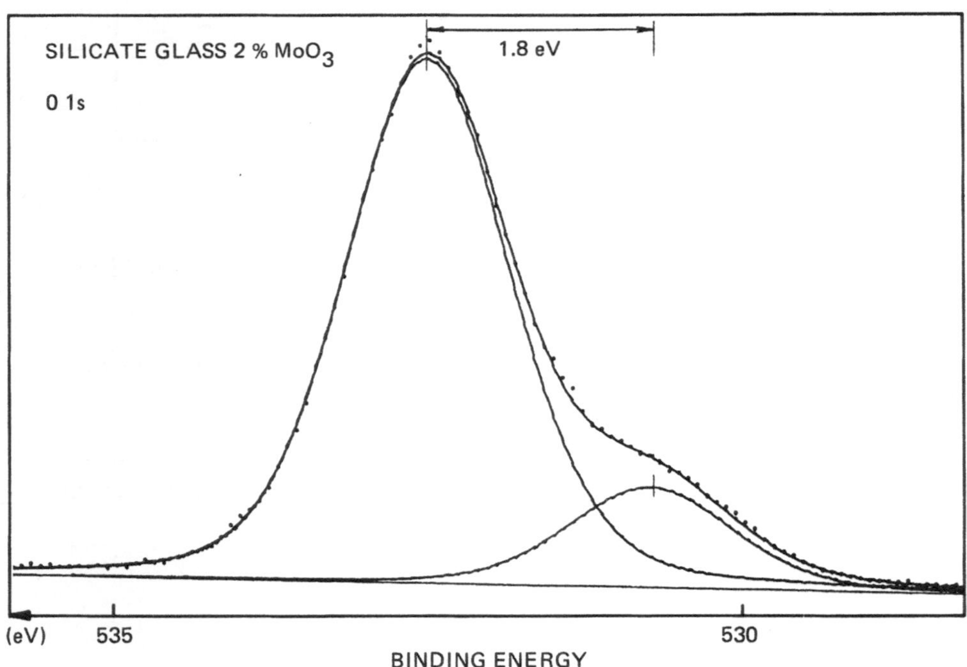

Fig. 3. O1s spectrum showing the bridging and non-bridging oxygen components.

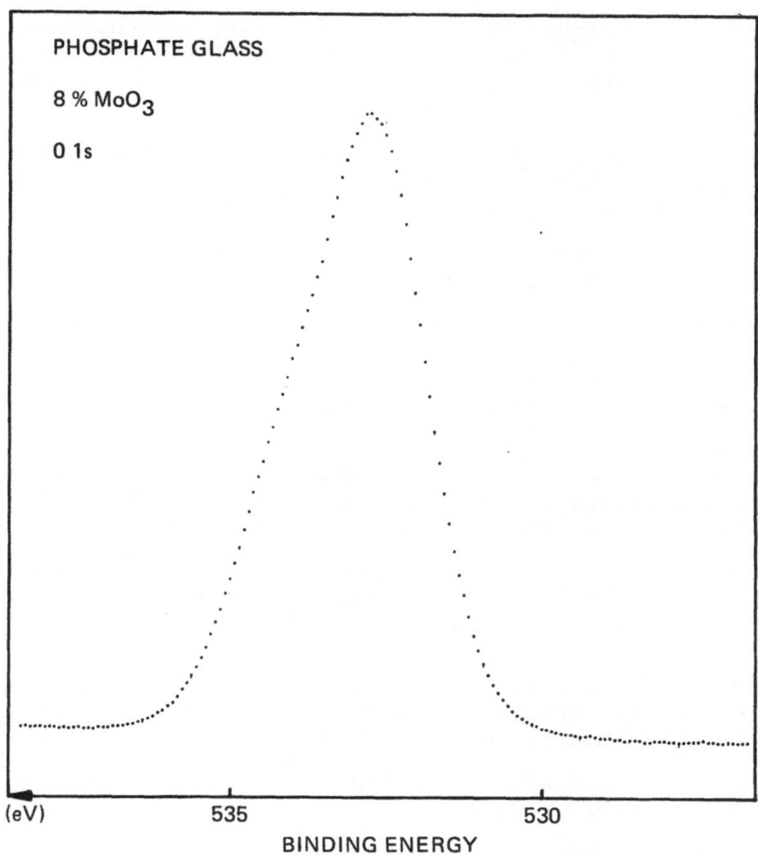

Fig. 4. O 1s spectrum from an 8% Mo-phosphate glass.

REFERENCES

1. B. Camara, W. Lutze, and J. Lux, "An Investigation of the Valency State of Molybdenum in Glasses Containing Fission Products," in: Scientific Basis for Nuclear Waste Management, Vol. 2, C. J. Northrup, ed., Plenum Press, New York (1980).

2. B. G. Dawkins, ESCA Studies on Leached Glass Forms, du Pont, Savannah River Plant, Report No. DP-MS-79-21.

3. J. Escard and D. Brion, "Possibilité d'Analyse Quantitative des Intensités en Spectroscopic Électronique: Application aux Verres," C.R. Acad. Sc., Paris, 276:945 (1973).

4. J.-C. Tournay, J.-H. Thomassin, P. Baillif, S. Scherrer, F. Champomier, F. Naudin, "Interaction Eau-Verre Comportement de Traces d'Aluminium," J. Non-Cryst. Solids 38&39:643 (1980).

5. R. Brückner, H.-U. Chun, and H. Goretzlki, "Photoelectron Spectroscopy (ESCA) on Alkali Silicate- and Soda Alumino-Silicate Glasses," Glastechn. Ber. 51:1 (1978).

6. Y. Kaneko, "Application of Photoelectron Spectroscopy of Glass," Yogyo-Kyokai-Shi (J. Ceram. Soc., Japan) 88:330 (1978).

7. J. S. Jen and M. R. Kalinowski, "An ESCA Study of the Bridging to Non-bridging Oxygen Ratio in Sodium Silicate Glass and the Correlations of Glass Density and Refractive Index," J. Non-Cryst. Solids 38&39:21 (1980).

8. C. Tenret-Noël, J. Verbist, et Y. Gobillon, "Etudes par Spectrométrie de Photoélectrons (ESCA) des étages d' Oxydation Plusiers Oxydes de Molybdéne et Dans l'Oxyde Mixte $CrMoO_4$," J. Microsc. Spectrosc. Electron. 1:255 (1976).

9. R. Nyholm and N. Mårtensson, "Core Level Binding Energies for the Elements Zr-Te (Z = 40-52)," J. Phys. C: Solid St. Phys., 13:L279 (1980).

10. N. Beatham, P. A. Cox, R. G. Egdell, and A. F. Orchard, "Core Photoelectron Spectra of Narrow-band Metallic Materials," Chem. Phys. Letters 69:479 (1980).

11. M. Hillert, "A Solid-Solution Model for Inhomogeneous Systems," Acta Met. 9:525 (1961).

12. J. W. Cahn, "On Spinodal Decomposition," Acta Met. 9:795 (1961).

STRUCTURAL AND REDOX PROPERTIES OF URANIUM IN CA-MG-AL-SILICATE GLASSES

Henry D. Schreiber, G. Bryan Balazs, Barbara J. Williams, and Stephen M. Andrews

Department of Chemistry
Virginia Military Institute
Lexington, VA 24450

INTRODUCTION

Despite the use of glass as a medium for the immobilization of uranium-rich nuclear wastes, much of the basic chemistry of uranium dissolved in glass still remains to be ascertained. In fact, only recently the presence of significant concentrations of U(V) ions, in addition to the presumed U(VI) and U(IV) species, in silicate glasses has been determined.[1,2] The valence state of uranium, as well as its structural site, within the glass-forming melt may control in part the bulk properties of the resulting glass.

The objectives of this experimental study were threefold: (1) to investigate the U(VI)-U(V)-U(IV) redox equilibria in Ca-Mg-Al-silicate glass as a function of base composition, melt temperature, imposed oxygen fugacity, and total uranium content, (2) to relate the optical spectra of the uranium-containing glasses to the structural sites occupied by individual uranium redox species, and (3) to compare the uranium redox equilibria to the redox equilibria of other multivalent elements in identical glasses.

EXPERIMENTAL PROCEDURES

Two Ca-Mg-Al-silicate compositions, labeled FAS (wt%: 5.6 CaO, 30.3 MgO, 9.9 Al_2O_3 & 54.2 SiO_2) and FAD (wt%: 19.2 CaO, 26.9 MgO, 2.3 Al_2O_3 & 51.6 SiO_2)[3], were employed as synthetic analogs of nuclear waste immobilization glasses. Homogeneous glass powders of these two base compositions were prepared from high-purity component oxides by repeated fusions, quenchings, and grindings. Nominal amounts of U_3O_8 were then intimately mixed in known proportions to the base compositions. Since both FAS and FAD melt at the relatively high temperature of 1500°C, the chosen compositions are somewhat idealized for nuclear waste immobilization.

Individual samples were prepared in high-temperature laboratory furnaces that were adapted to provide precise control of the sample atmosphere and temperature.[3] About 250 mg of glass powder in a small Pt crucible were suspended in the furnace. Sample temperatures were monitored by a thermocouple immediately adjacent to the sample capsule. Oxygen fugacities in the sample chamber were regulated by the flow of CO:CO_2 mixtures. In all cases, the samples were equilibrated under the desired conditions for about 24 hours. That this time was sufficient for uranium to attain redox equilibrium was shown by time studies.

The cylindrical charge of glass was then mounted in epoxy in order to section a slab that was ground and polished on both sides. The glass slab was typically about 1 mm thick and about 4 mm in diameter. The visible/near-IR absorption spectra (300 nm to 3000 nm) of the samples were recorded on a Beckman 5240 spectrophotometer with the glasses mounted in a 3 mm aperture. The total uranium contents of the glasses as determined by the electron microprobe on glass chips compared favorably with the nominal concentrations.

The remaining portions of the samples were then ground to fine powders and examined by indirect redox titrations.[3,5] These titrations measured the total reducing power of the sample. In the titration, about 100 mg of the powder were dissolved under an Ar atmosphere in a $HF/H_2SO_4/H_2O$ solution that contained excess Fe(III). The Fe(III) reacted stoichiometrically with the U(V) and U(IV) of the dissolved glass to form U(VI) and Fe(II) in aqueous solution. The resulting amount of Fe(II) was then determined with standardized Ce(IV) titrant using ferroin indicator.

RESULTS

The uranium redox equilibria in the silicate glasses can be given by the ionic equations:[2]

$$4U^{6+}_{(melt)} + 2O^{2-}_{(melt)} \rightleftharpoons 4U^{5+}_{(melt)} + O_{2(gas)} \quad (1)$$

$$4U^{5+}_{(melt)} + 2O^{2-}_{(melt)} \rightleftharpoons 4U^{4+}_{(melt)} + O_{2(gas)} \quad (2)$$

Rearrangement of the equilibrium constant expression for these equations yields,

$$-\log f_{O_2} = 4 \log R + b \quad (3)$$

where R = [U(V)]/[U(VI)] or [U(IV)]/[U(V)] and b is a function only of temperature and bulk composition. Thus, plots of $-\log f_{O_2}$ versus log R should yield straight lines of slope 4 for systems of constant temperature and compositions.[3] Two parallel straight lines for the U(VI)-U(V) and U(V)-U(IV) equilibria were then determined by successive approximations with the experimental data and equation 3 as constraints. The data were inconsistent with a simple U(VI)-U(IV) equilibrium. No evidence for the existence of U(III) in any glasses was found. Figure 1 presents the results for the U(VI)-U(V)-U(IV) redox systems in the glasses containing about 1 wt% uranium. The dependence of the equilibria on total uranium content is shown in figure 2.

The colors of the glasses that contain 1 wt% uranium change systematically from yellow to blue-green as the conditions of synthesis change from oxidizing to reducing. The optical spectra of the glasses were then matched to the uranium redox determinations in order to assign the absorption peaks to specific uranium redox states as shown in figure 3. The absorbances of the assigned peaks correlate well with the concentration of the absorbing species of uranium in a Beer's Law plot.

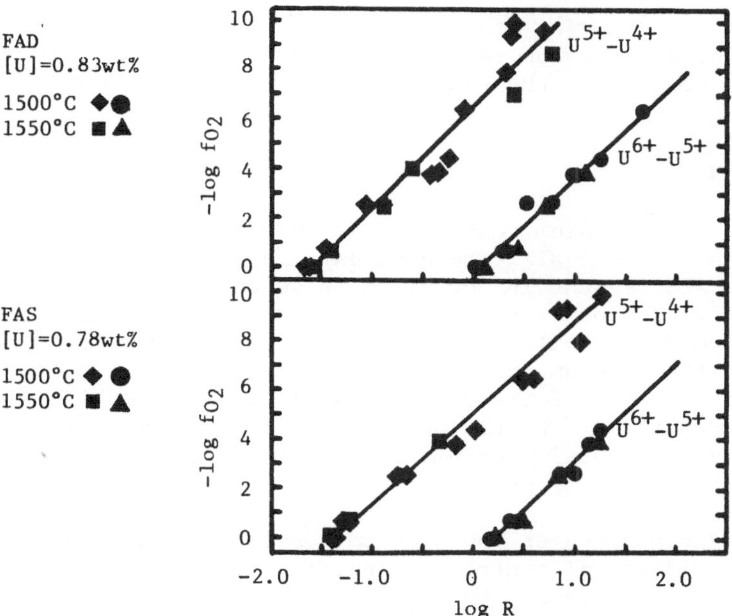

Fig. 1. Uranium redox equilibria. R is ratio of reduced to oxidized species; f_{O_2} is in atmospheres.

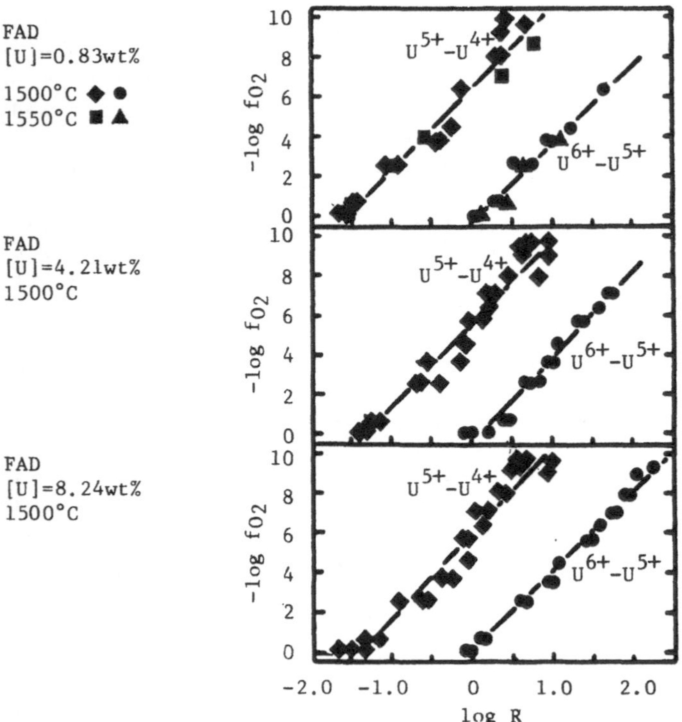

Fig. 2. Uranium redox equilibria as a function of total uranium content in the glasses. f_{O_2} is in atmospheres.

DISCUSSION

All three uranium redox species, U(VI), U(V), and U(IV), coexist in Ca-Mg-Al-silicate glasses. The proportion of each redox species present in the glass is dependent on the synthesis conditions. Composition FAS is consistently more reducing than composition FAD[3], while the imposed oxygen fugacity has a dramatic effect on the uranium redox states. Over a range of 50°C, no appreciable thermal effects on the equilibria were established for these compositions. Little or no variation in the uranium redox equilibria with total uranium content (from 0.8 to 8 wt%) was observed.

Upon comparison of the uranium results to those of other redox couples at identical synthesis conditions[5], a series of relative reduction potentials of the multivalent elements in Ca-Mg-Al-silicate glasses can be developed.[4] At a common temperature, oxygen fugacity, base composition, and total element concentration (1 wt%), the series is defined as Ce(IV)→Ce(III) > U(VI)→U(V) > Fe(III)→Fe(II) > U(V)→U(IV) > Cr(III)→Cr(II) > Eu(III)→Eu(II) >> Ti(IV)→Ti(III). This series is also displayed graphically in figure 4.

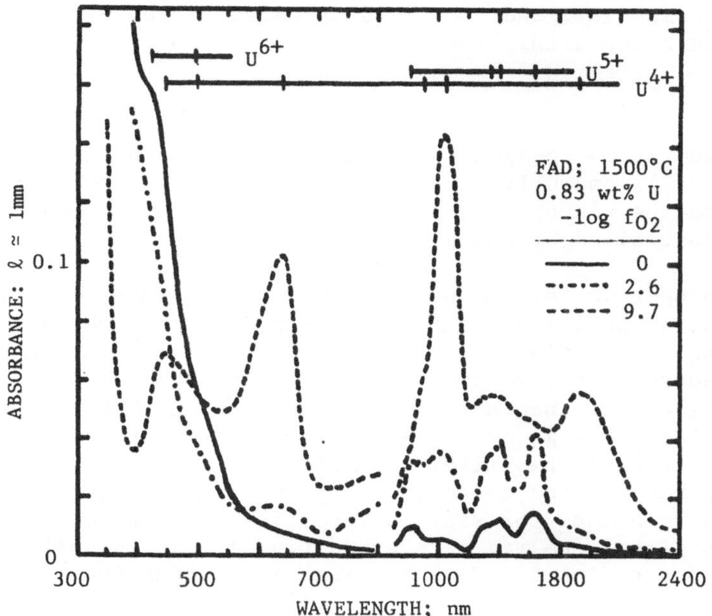

Fig. 3. Optical spectra of U-containing glasses.

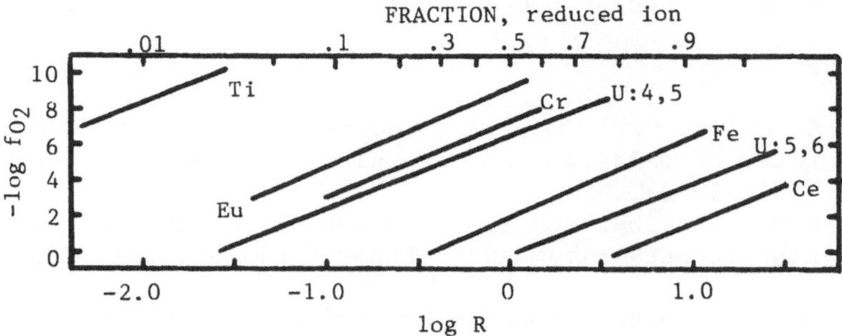

Fig. 4. Redox equilibria of redox couples in glass composition FAD at 1500°C containing about 1 wt % of the multivalent element.

Of some consequence is the fact that the uranium redox equilibria bracket that of the major redox component iron.

After the assignment of spectral absorbances to specific uranium redox states, a structural interpretation of the uranium ions in the glass could be obtained.[1] Consequently, this resulted in the identification of U(VI) in the glasses as the UO_2^{2+} (uranyl) entities, the presence of U(V) as either UO_2^{+} complexes or as 6-

coordinated species, and the probable existence of U(IV) in 8-coordinated sites within the glass structure.

SUMMARY

Uranium in Ca-Mg-Al-silicate glasses has been shown to exist as U(VI), U(V), and U(IV) ions, each with their unique structural site within the glass. The proportion of uranium in each redox state is controlled by the synthesis conditions.

The properties of the nuclear wasteform glasses are controlled by the bulk composition, equipment, temperature, time, atmosphere, and cooling rate during the processing of the glasses. However, these parameters in turn are ultimately explained in part by the redox chemistry of the melts. A fundamental understanding of the basic chemistry may go a long way in deciphering the wasteform properties — thermal stability, radiation damage, bulk properties, physical strength, and corrosion. By constraining uranium and other redox elements to exist in specific redox/structural states within the glass, one may obtain additional glass stability or other favorable properties for nuclear waste immobilization.

ACKNOWLEDGMENTS

Portions of the experimental research presented in this paper were supported by NASA (NSG-7355), the Research Corporation, and NSF. We are most grateful to C. W. Hartless for her technical assistance in the preparation of this manuscript.

REFERENCES

1. G. Calas, Experimental study of the behavior of uranium in magmas: states of oxidation and coordination, Geochim. Cosmochim. Acta 43: 1521 (1979).
2. H.D. Schreiber and S.M. Andrews, The redox states of uranium in synthetic basaltic magmas, Lunar & Planet. Sci. XI: 1000 (1980).
3. H.D. Schreiber, Redox states of Ti, Zr, Hf, Cr, and Eu in basaltic magmas: An experimental study, Proc. 8th Lunar Sci. Conf. 2: 1785 (1977).
4. H.D. Schreiber, T. Thanyasiri, J.J. Lach, and R.A. Legere, Redox equilibria of Ti, Cr, and Eu in silicate melts: Reduction potentials and mutual interactions, Phys. Chem. Glasses 19: 126 (1978).
5. H.D. Schreiber, H.V. Lauer, and T. Thanyasiri, Oxidation reduction equilibria of iron and cerium in silicate glasses: Individual redox potentials and mutual interactions, J. Non-Cryst. Sol. 38: 785 (1980).

STABLE PRODUCT LOW LEACH GLASSES

Suresh N. Karkhanis, Peter J. Melling, William S. Fyfe, G. Michael Bancroft

Departments of Chemistry, Geology and Centre for Chemical Physics
University of Western Ontario
London, Ontario, Canada N6A 5B7

ABSTRACT

Glasses can be tailored to produce leach solutions that will precipitate stable minerals with suitable sites for radioactive waste ions and good ion retention properties. A series of such SPLEG (Stable Product Low Leach Glass) glasses have been prepared and their leaching and hydrothermal devitrification behaviour studied. The glasses we have produced are based on tourmaline, montmorillonite and calcium phosphate and their hydrothermal devitrification products are apatites and clay minerals or zeolites.

INTRODUCTION

Glasses have attracted considerable interest as possible hosts for the long term containment of medium and high level nuclear-fuel wastes. The choice of an ideal glass for fixation of the fission products would be influenced by the following factors:

(a) the glass should be kinetically stable (i.e. the rate of dissolution in ground water should be very low);
(b) the fabrication temperature should be as low as possible to ameliorate problems with volatile, active, fission-products.

Techniques exist [1] to reduce volatilization so that glass fabrication temperatures above those now considered satisfactory could well prove acceptable. Criterion (b) could then be relaxed somewhat to accommodate more leach resistant glasses. Because of the high disorder of glasses with respect to the corresponding crystalline phases, all glasses should have higher solubilities than their crystalline equivalents (hence the SYNROC concept). Further all glasses should eventually (on a geological timescale) crystallize, particularly in the presence of warm ground water. Many studies from experimental petrology highlight this important point [2]. Hence we consider it desirable to add the following criteria to the choice of glass composition:

(a) That the glass crystallize to products with large arrays of lattice sites (for cations and anions) with good ion retention properties.
(b) That the product phases are stable under the pressure, temperature and chemical conditions of the repository.
(c) Waste ions are incorporated in the product phases.

With this in mind, we are developing a series of SPLEG (Stable Product Low Leach Glass) waste forms based on the concept that glasses can be tailored to devitrify in the presence of ground waters to give thermodynamically stable minerals which will easily accept nuclear waste ions in their structure. The glasses we chose to test this concept are based on a combination of the naturally occurring stable minerals, montmorillionite, tourmaline and apatite. These minerals were chosen not only for their natural stability, but also because they are commonly found in nature with a wide variety of impurity ions accommodated in their structure. The preliminary work reported in this paper indicates that the SPLEG concept can be realized, and this should provide an additional safeguard over and above those currently contemplated for a nuclear-fuel waste repository.

EXPERIMENTAL

Glasses were prepared from mixed dry powder batch materials consisting of either analytical grade chemicals or powdered natural materials. The composition of the natural material glass NATSPLEG-1 is given in Table 1. All of the other glasses reported here are based on that composition.

Melting temperatures are defined by the ability to pour a 1 ml sample from a 5 ml melt in a 20 ml capacity metal crucible into a metal mold, without fibres forming between the sample and melt. This definition is conservative in that it gives melting temperatures which are higher than those which would be required on a larger scale.

Samples were polished discs 2 mm thick and 10 mm diameter which were hydrothermally leached and devitrified in stainless steel bombs of 15 ml enclosed volume containing 10 ml of distilled water. We are aware that this volume of solution will give rapid saturation.

Table 1. Composition of NATSPLEG-1

(a) Major components (mol %) (X-ray fluorescence)

SiO_2	41.5	B_2O_3	4.0	Fe_2O_3	3.0	CaO	19.3
Al_2O_3	14.8	P_2O_5	5.5	MgO	9.8	Na_2O	1.1

(b) Impurity ions (ppm) (spark-source mass-spectrometric)

F	500	S	15	Cl	50
Sc	100	Zn	40	Ga	2
Ge	1	As	0.2	Se	0.03
Br	0.2	Rb	0.2	Sr	330
Ba	7	La	1	Ce	0.5
Pr	0.3	Nd	0.5	Sm	1
Eu	0.3	Hf	5	V	400
Cr	50	Ni	15	Cu	50
Y	1	Zr	60	Nb	15
Mo	1	Cd	0.1	Sn	5
I	1	Cs	0.1	Gd	3
Tb	0.5	Dy	3	Ho	0.5
Er	4	Tm	0.6	Yb	2
Pb	3				

Devitrified products were identified by X-ray powder diffraction (XRD) and scanning electron microscopy (SEM). Analysis of the leach solutions of SYNSPLEG-3 for Na, Ca, Cu, Si and Al was done using atomic absorption spectrometry (AA).

RESULTS AND DISCUSSION

Four glasses were tested to determine their leach rates and crystalline products. NATSPLEG-1 had a melting temperature of about 1400°C and a weight loss leach rate of 4.3×10^{-8} (Kg m^{-2} s^{-1}) for an experiment conducted at 300°C for 1 week. This compares well with that of naturally stable rhyolitic obsidian 1.52 $\times 10^{-7}$ (Kg m^{-2} s^{-1}) [3]. From this, SYNSPLEG-1 was modified to have a more realistic sodium content. This produced little change in the melting temperature. SYNSPLEG-2 was developed from SYNSPLEG-1 and has an increased boron content. This glass melts at 1350°C and has a weight-loss leach rate at 300°C for 1 week of 5.9×10^{-7} (Kg m^{-2} s^{-1}). Substitution of Mg by Cu in the composition of SYNSPLEG-2 produces a glass with a melting temperature of about 1280°C (SYNSPLEG-3) which can be prepared without pouring at 1150°C. The optical basicity of a glass [4] is a measure of the Lewis basicity of a glass as a solvent, in the same way that pH measures the basicity of water and has possible predictive value in choosing glass components. The change in optical basicity suggests there will be an increase in the leach rate. However, at only a 3% change in basicity the effect should be small.

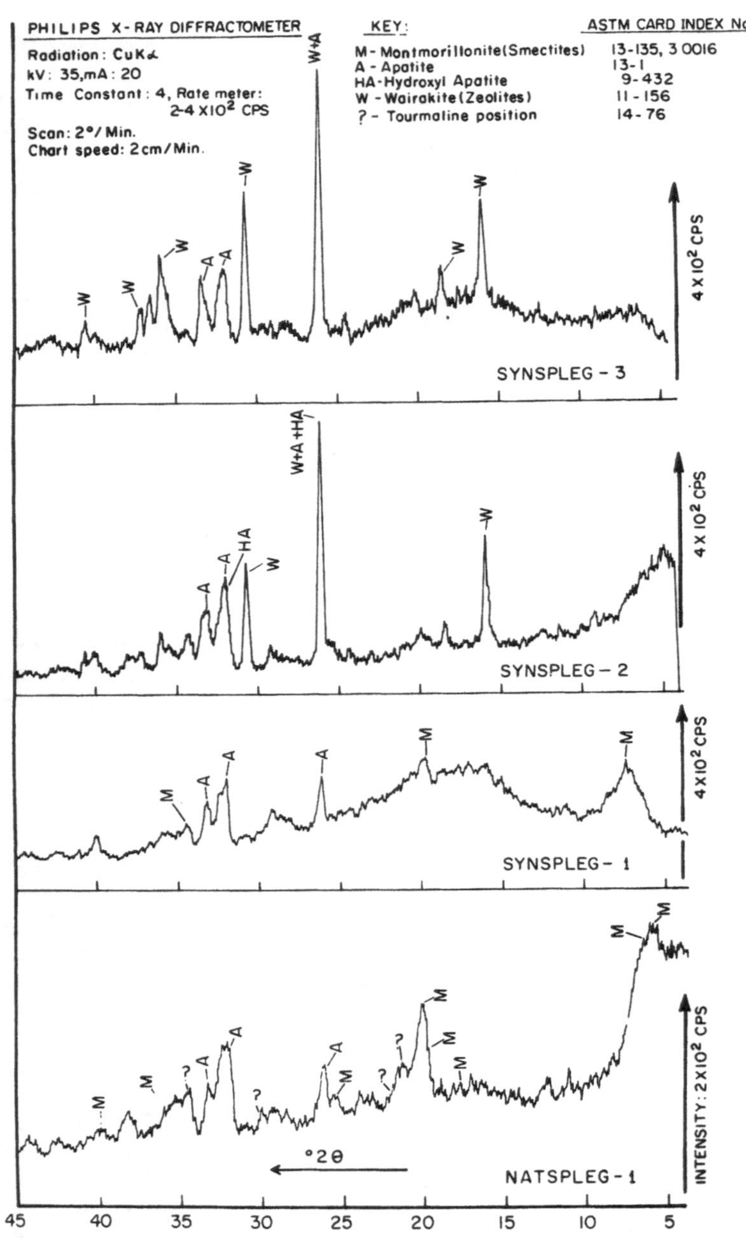

Fig. 1. X-ray diffraction patterns for products from glass, leaching experiments with distilled water at 300°C/1 week.

XRD patterns of the reaction products are given in Figure 1. For NATSPLEG, the products are apatite, montmorillonite and tentatively tourmaline. For SYNSPLEG-1 the products are the same. SYNSPLEG-2 gives hydroxyapatite and wairakite. Wairakite is a zeolite mineral and hydroxyapatite is a commonly occuring stable form of calcium phosphate. Annealed samples of SYNSPLEG-3 give similar products to SYNSPLEG-2. We realise, however, that there may be problems with the ion-exchange properties of zeolites.

Figure 2 shows SEM micrographs of two forms of calcium phosphate observed on the surface of SYNSPLEG-3. Figure 2(a) shows well formed hexagonal crystals of apatite and Figure 2(b) shows spherulites similar to those observed by Abe et al. [5,6] and identified by them as $\beta Ca(PO_3)_2$ with a trace of the γ form. The fine precipitate surrounding them is most likely wairakite (Figure 3).

From the data in Table 2, it is clear that the glass water interaction is not a simple process and there are many competing precipitation and preferential leaching processes [7].

Space does not permit a full discussion of these results. However, Ca and Cu are being precipitated as they are corroded and the 10% increase in volume on going from glass to products ($\rho_{glass} = 2.73$, $\rho_{apatite} = 3.2$, $\rho_{wairakite} = 2.25$) will result in a "self-sealing" layer on the surface of the glass which will act as a barrier to further reaction.

CONCLUSIONS

The work presented here shows that it is possible to tailor glasses to devitrify in the presence of water to give minerals [8] which have good potential for the fixation of nuclear waste. The ability to tailor glass to give particular mineral products raises the possibility of producing a glass which is in near equilibrium with the backfill materials and/or host rock of a waste repository. This should then reduce leach rates, a factor which would greatly enhance the security of a waste repository.

Further work, however, is required and has been initiated:

(a) to study the long term low temperature leaching properties of the glass and the resulting alteration products,
(b) to study glass-backfill-host rock interactions, and
(c) to investigate the behaviour of waste ions during the crystallization reactions.

(a)

(b)

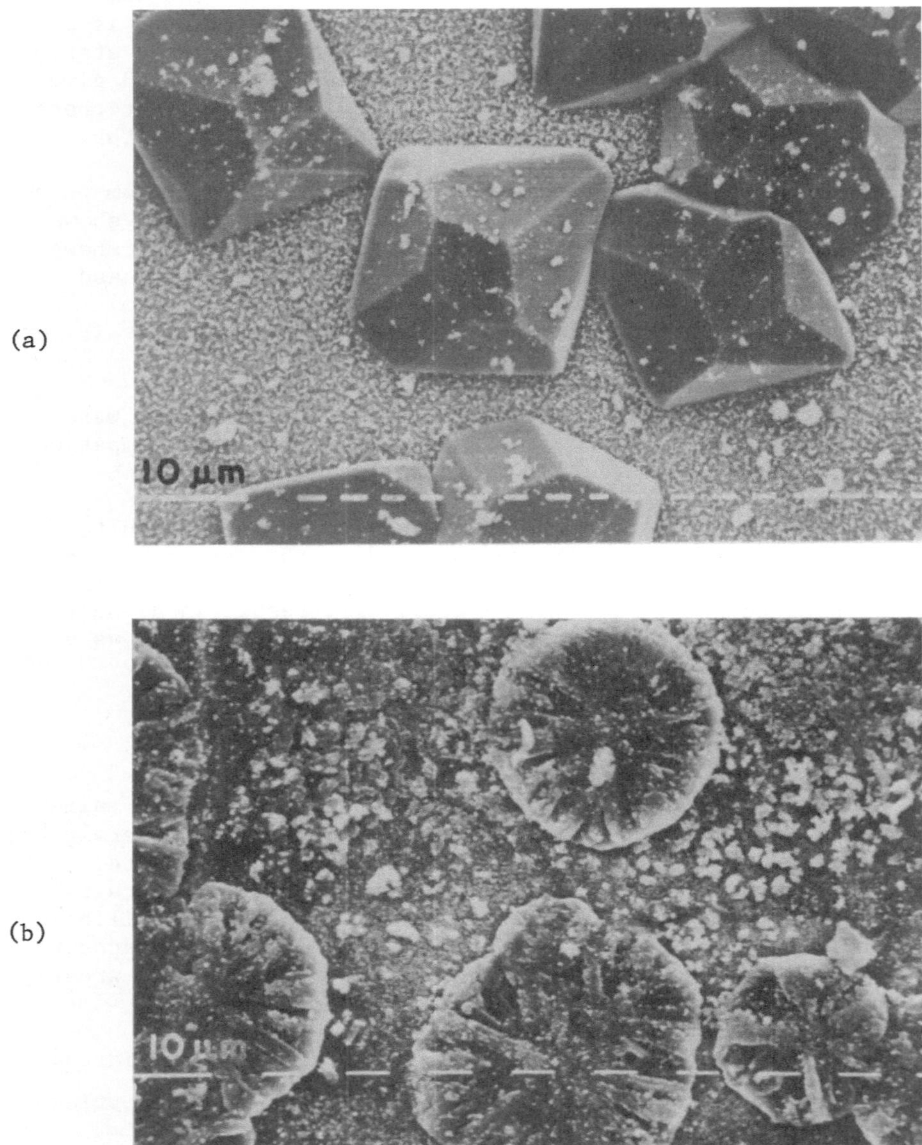

Fig. 2. Calcium phosphate crystals on the surface of SYNSPLEG-3 (300°C, 1 week)

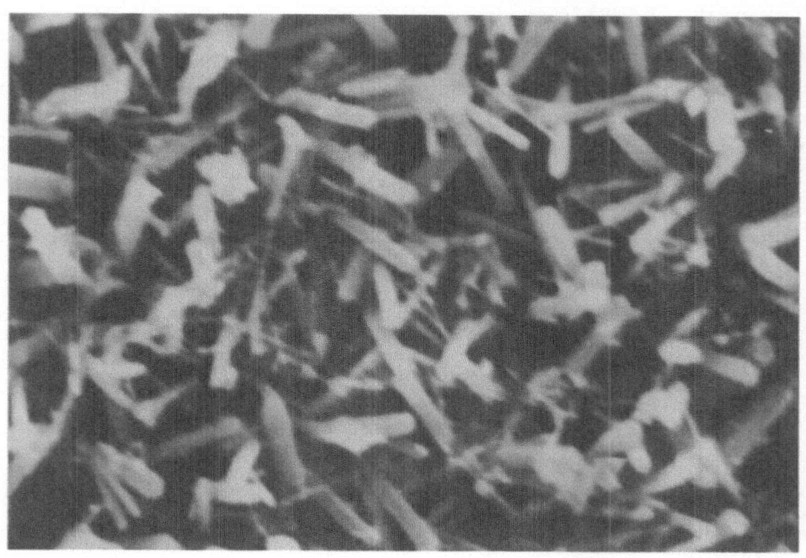

Fig. 3. Fine precipitate on the surface of SYNSPLEG-3

Table 2. Leach Rates in kg m^{-2} s^{-1} of Glass Based on Congruent Leaching for SYNSPLEG-3

	Ca	Na	Cu	Al	Si
100°C					
6 hrs	7.4×10^{-9}	1.6×10^{-7}	5.2×10^{-9}	—	—
1 day	3.0×10^{-9}	2.2×10^{-8}	1.3×10^{-9}	—	—
3 days	7.9×10^{-10}	1.7×10^{-8}	—	—	—
7 days	8.4×10^{-11}	3.2×10^{-8}	7.5×10^{-10}	—	—
200°C					
6 hrs	2.0×10^{-7}	9.8×10^{-7}	7.3×10^{-7}	2.1×10^{-7}	—
1 day	9.8×10^{-9}	3.2×10^{-7}	9.1×10^{-8}	7.9×10^{-8}	—
3 days	1.4×10^{-9}	1.9×10^{-7}	5.1×10^{-8}	3.5×10^{-8}	1.6×10^{-8}
7 days	8.4×10^{-11}	8.7×10^{-8}	1.3×10^{-9}	1.5×10^{-8}	8.4×10^{-9}
300°C					
6 hrs	—	1.4×10^{-6}	5.9×10^{-6}	1.6×10^{-6}	—
1 day	—	6.7×10^{-7}	2.7×10^{-8}	1.6×10^{-7}	1.2×10^{-7}
3 days	—	2.7×10^{-7}	2.2×10^{-8}	—	5.9×10^{-8}
7 days	6.7×10^{-7}	1.1×10^{-7}	1.4×10^{-8}	—	2.1×10^{-8}

ACKNOWLEDGEMENTS

We would like to thank John Forth for making the discs, B. Krønberg for the spark source data, R. Ringman for drafting assistance and A. Noon for photographic assistance. We also wish to acknowledge the financial contribution of Atomic Energy of Canada Ltd. and the comments of K. Harvey, P. Hayward and J. Tait.

REFERENCES

1. S. Weisenburger and K. Weiss, "Ruthenium Volatility Behaviour During HLLW-Vitrification in a Liquid Fed Ceramic Melter", pp. 51-68 in Scientific Basis for Nuclear Waste Management, vol. 1, ed. G. J. McCarthy, Plenum Press (1979).

2. A.D. Edgar "Experimental Petrology Basic Principles and Techniques", Clarendon Press (Oxford, 1973).

3. S.N. Karkhanis, G.M. Bancroft, W.S. Fyfe, and J.D. Brown "Leaching Behaviour of Rhyolite Glass", Nature, 284 (1980), 435-437.

4. J.A. Duffy and M.D. Ingram, "An Interpretation of Glass Chemistry in Terms of the Optical Basicity Concept", J. Noncryst. Solids, 21 (1976) 373-410.

5. Y. Abe, T. Arahori and A. Naruse, "Crystallization of $Ca(PO_3)_2$ glass below the glass transition temperature", J. Am. Ceram. Soc., 59 (1976) 487-490.

6. Y. Abe, K. Mori and A. Naruse "Role of OH radical in the Crystallization of Calcium-metaphosphate Glass" 14.13 - 15.19 "10th Int. Congress on Glass" (Kyoto, 1974).

7. R.W. Douglas and T.M. El-Shamy, J. Am. Ceram. Soc., 1:50 (1967).

8. R. Roy, E.R. Vance, N.H. MacMillan, and W. B. White, "Matrix Encapsulated Waste Forms: Theory and Application to Idealized Systems, Hydrated Radio and Encapsulant Phases", this volume (see index).

TITANATE WASTE FORMS FOR HIGH LEVEL WASTE - AN EVALUATION OF MATERIALS AND PROCESSES*

R. G. Dosch, A. W. Lynch, T. J. Headley and P. F. Hlava

Sandia National Laboratories**
Albuquerque, New Mexico 87185

INTRODUCTION

Recent reviews[1,2] outline the history, variety of approaches, and current status of the application of ceramic materials to radwaste immobilization. Common to most, and in particular to high level waste streams envisioned in fuel reprocessing, are problems associated with the transfer of processing methodology to a remote operation. In initial titanate studies done in our laboratory, a process was used in which nuclides in liquid waste were fixed via ion exchange or other sorption mechanisms on a sodium titanate column bed which was subsequently converted to a ceramic monolith by hot pressing.[3] Atomic scale distribution of waste constituents within the matrix (titanate) was realized and process flexibility allowed for some nuclide separation, i.e., I and Tc were not retained on the column and conditions were easily established such that Cs and Sr could also be segregated. This potential complemented the concept of waste form development for specific nuclides and was particularly attractive because this group of fission products includes those which: 1) are most difficult to immobilize in a ceramic which must otherwise accommodate all waste constituents, 2) contribute significantly to decay heat, and 3) have the highest potential for migration in a geologic setting. However, potential remote processing problems were foreseen, the most negative of which was column plugging by suspended solids which were introduced into the waste stream. Also, the ceramic product was an assemblage of at least 12 crystalline phases within a rutile matrix, only half of which could be identified thus preventing the

*This work was supported by the U.S. Department of Energy under contract DE-AC04-76-DP00789.

**A. U.S. Department of Energy facility.

use of thermodynamics and geological evidence in assessing long term stability and compatibility with host rock formations.

The current titanate work is directed toward process simplicity as well as enhanced waste form stability. Reference process steps include a batch contact of calcium titanate powder with liquid waste, volatilization of water and nitrates (700-775°C), and hot pressing (1250-1300°C at 1000 psi) to produce ceramic monoliths. Microstructural analyses of the ceramic show the waste constituents are contained in four types of crystalline phases within a rutile matrix which comprises 40-50% by volume of the material. These phases include oxides exhibiting structure types related to perovskite, zirconolite, hollandite, and metallic phases. The addition of boron (0.2-1.0% by weight) to the waste decreases hot pressing temperature (925-975°C), provides the option of atmospheric pressure sintering (950-975°C), and results in an additional phase, CaB_2O_4, in the waste form. The synthetic materials produced by the reference process were shown to be highly resistant to aqueous attack under hydrothermal (250°C) conditions.

Processing chemistry used in liquid waste stabilization has also been applied to solid Defense Waste (sludge). Consecutive additions of aqueous solutions of tetramethylammonium titanate and calcium nitrate (these materials react to form a calcium titanate precipitate) to a water slurry of the sludge results in a homogeneous mixture of sludge and calcium titanate. Techniques (hot pressing, atmospheric pressure sintering) used in consolidating these materials and the microstructures and properties of the processed ceramics are briefly discussed.

LIQUID WASTE PRECURSOR MATERIALS

The precursor material (referred to as calcium titanate powder) used in the reference process is prepared using the chemistry outlined empirically in Equations (1-3).

$$2Ti(OC_3H_7)_4 + NaOH \xrightarrow{MeOH} NaTi_2O_5H \text{ (solution)} \xrightarrow[\text{water}]{\text{acetone}} NaTi_2O_5H \text{ (powder)} \quad (1)$$

$$2NaTi_2O_5H + Ca(NO_3)_2 \longrightarrow Ca(Ti_2O_5H)_2 + 2NaNO_3 \quad (2)$$

$$Ca(Ti_2O_5H)_2 + Ca(NO_3)_2 + 2NaOH \longrightarrow Ca_2(Ti_2O_5H)_2 + 2NaNO_3 \quad (3)$$

The reaction between tetraisopropyl titanate (E. I. du Pont de Nemours and Co. - trade name TYZOR TPT) and sodium hydroxide involves the addition of a methanol solution of the hydroxide (10-15% by weight) to the alkoxide. The reaction product, a clear solution, is added to an acetone-water mixture resulting in hydrolysis and precipitation of sodium titanate. A finely divided powder is obtained by drying the precipitate at ambient temperature in vacuum. The acetone-water ratio used in hydrolysis defines the surface area of the powder, where a ratio of 10 produces a powder with a surface area in the 200-300 m^2/g range. Conversion to calcium titanate is done by equilibrating sodium titanate powder in 1 M $Ca(NO_3)_2$ solution of sufficient volume to provide a two-fold or greater excess of calcium based on the stoichiometry given in Eq. (2). Sodium hydroxide addition (Eq. 3) gives a product with a Ca:Ti mole ratio of 0.5 (this material is used in the reference process). The titanate powder is apparently involved in this reaction (as opposed to simple coprecipitation of calcium hydroxide) as final solution pH is significantly less (∿9.5) than that of a saturated solution of calcium hydroxide. The converted powder is filtered, washed with water, and dried at ambient temperature. Sodium content is typically less than 0.1% by weight.

This method of precursor preparation is applicable to niobates, tantalates, and zirconates (or mixtures thereof) as well as titanates. SYNROC "C" precursor materials have also been prepared by this method, leaving only liquid waste and temperature processing required to produce fully dense, SYNROC monoliths.[4]

TITANATE WASTE CERAMIC REFERENCE PROCESS

The initial process step is the contact of the precursor (Eq. 3) with an acid waste solution. The simulant used in this work represents the liquid waste from reprocessing one tonne of spent uranium fuel (30,000 MWD/tonne U) contained in 568 liters with a nitric acid concentration of 1-1.5 M. Only uranium and elements with no stable isotopes were omitted. A precursor to waste liquid ratio is used such that the final ceramic contains 25% by weight of waste constituent oxides. Under this constraint, the precursor-liquid mixture has the consistency of a thick slurry or paste. The mixture is dried and calcined at 700-775°C to produce a powder. The off-gas composition should be similar to that resulting from direct calcination of liquid waste since the only additional contribution of the precursor material would be water.

During the precursor-liquid contact, the bulk of the waste constituents are "fixed" on the powder by ion exchange, hydrolysis due to locally high pH on particulate surfaces, or other unknown mechanisms. The liquid phase in the slurry/paste contains calcium nitrate along with ∿30% of the Cs and small amounts of Sr. Drying and calcining results in a relatively homogeneous powder which receives no further processing prior to hot pressing.

Hot pressing was done in an argon atmosphere using carbon dies, a constant pressure of 1000 psi, and a heating rate of 50°C/min. Samples were heated at this rate until no further axial shrinkage (ram deflection) was observed and held for 10-15 minutes at the temperature where shrinkage stopped. Under these conditions, densification of the reference material began at 920°C and was complete at 1275°C.

The addition of boron (0.2-1.0%), either as elemental boron (amorphous) or the oxide, lowers processing temperature. Using hot pressing conditions given previously, maximum reference material densities (4.5 g/cc) are achieved at 925°C. Boron (0.75-1%) also promotes densification during atmospheric pressure sintering, where densities up to 98% of comparable hot-pressed materials are achieved during 1-hour soaks at 950-975°C. Samples are cold pressed at 4000 psi prior to sintering. The Fe-FeO equilibrium was used to control oxygen fugacity during sintering.

MICROSTRUCTURE AND LEACHING PROPERTIES

Electron microprobe and scanning transmission electron microscope (STEM) methods were used in microstructural analyses of the reference materials. Rutile comprised 40-50% of the ceramic and acted as a matrix phase. Rutile grains as large as 10 μm were observed, with some containing small amounts of Zr in solution. Waste constituents were distributed into four additional phases identified by electron diffraction as having the following structure types:

1. Zirconolite [$Ca(Ar,U)Ti_2O_7$] - Predominantly Ti with U, Ca, Zr, Gd, Nd, and Fe.
2. Perovskite ($CaTiO_3$) - Predominantly Ti and Ca with Nd, Gd, U, Sr, and Fe.
3. Hollandite (similar to $BaAl_2Ti_6O_{16}$) - a phase containing Ti along with small amounts of Cs, Ba, Rb, Zr, Mo, Fe, Cr, K, and Ca.
4. Metallic solid solutions - varying chemistries with major components generally from a group including Mo, Ru, Rh, with lesser amounts of Rh, Pd, Ni, and Fe.

A STEM photomicrograph of a selected area containing four of the five phases found in the material is shown in Fig. 1a. High resolution microscopy techniques are being used for further structure verification. Figure 1b is a lattice image from an area of a crystallite identified by electron diffraction as having a hollandite structure type. Lattice spacing obtained from the photograph, taken along the [001] direction, confirm the structure type identification. Qualitative interpretation suggests homogeneity both in the crystalline lattice and channels within the lattice. This technique is being used for comparison of crystalline phases in waste ceramics with single phase materials prepared with compositions representative of the nominal stoichiometry of mineral analogues.

It is interesting to note that the waste phase assemblage in the reference material closely resembles that of a SYNROC composition proposed by Ringwood[5] for commercial waste. While the

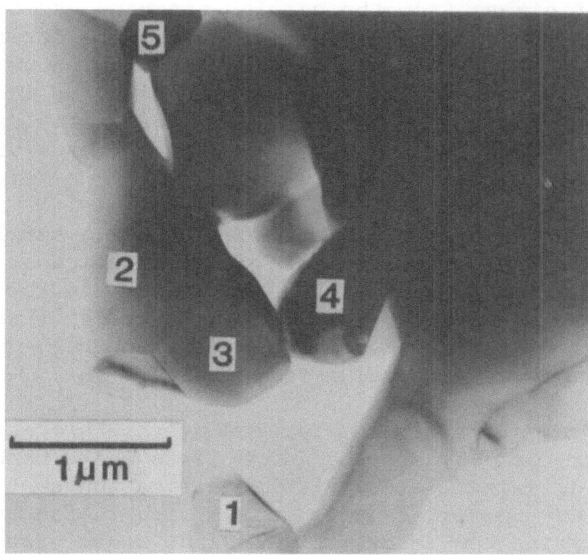

Figure 1a. Calcium titanate-based reference waste. STEM photomicrograph-phases shown are rutile (Ti) - #1,4; zirconolite (Ti, Ca, Nd, U, Gd, Zr, Fe) - #3; perovskite (Ti, Ca, Nd, Gd, U, Sr, Fe) - #2; and metallic (Mo, Ru, Pd, Rh, Ni) - #5.

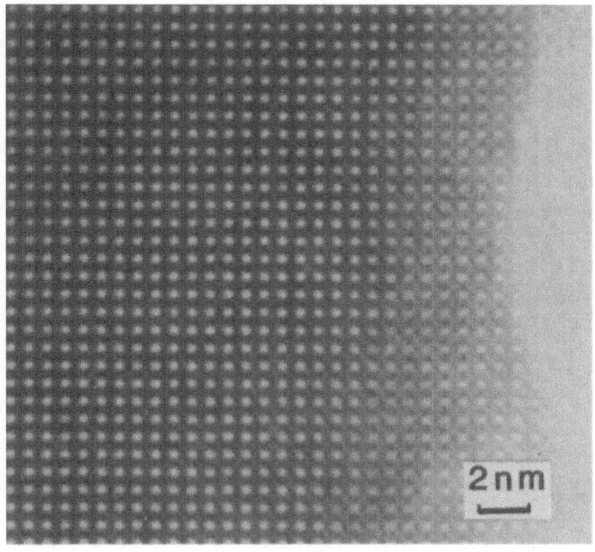

Figure 1b. Lattice image of hollandite phase in the [001] direction.

perovskite can originate from the calcium titanate additive used in this process, the other waste containing phases are determined by the waste constituents. Thus, it appears that the phase assemblage in the calcium titanate-based ceramic wastes would be independent of minor deviations in waste composition and of waste loading factors, although the relative amounts of these phases within the rutile matrix could vary.

An additional phase, CaB_2O_4, was identified in samples in which boron had been used as a sintering aid. Substitution of waste constituents into this phase could not be detected, and the titanate waste phase assemblage did not appear to be altered by the use of boron. The samples did have a finer grain structure, most likely the result of the lower processing temperature.

For our purposes, materials prepared by this process were compared in short-term (1-2 weeks) leach tests where monoliths were in contact with deionized water in fused silica tubes at 250°C. Sample water volume to surface area ratios were in the range of 15-30 ml/cm^2. During the tests, the leachate became saturated with silica as evidenced by container etching and precipitation of both particulate and colloidal silica upon cooling. Reference material leach rates for two consecutive one-week tests in units of $\mu g/cm^2 \cdot day$ were Cs: 8.7-3.2, Mo: 5.4-0.6, Sr: 0.24-0.3, Ca: 5-6, and Ba: 0.4-0.6 where the two values represent rates during the first and second week of the test, respectively. Attack on the matrix could not be visually observed and Ti, U, or Gd were not detected in leachates by inductive coupled plasma (ICP) analyses. Based on detection limits, leach rates for these elements were less than 0.2, 0.6, and 0.4 $\mu g/cm^2 \cdot day$, respectively.

Leach rates for reference material containing B, both hot pressed and sintered, were comparable to those given above. Boron leach rates, however, were in the range of 40-70 $\mu g/cm^2 \cdot day$ and after leaching, the surfaces were coated with a thin, adherent film of a white material having a calcium silicate composition. Electron microprobe examination of the interface indicated that the silica did not penetrate into the ceramic matrix.

SOLID DEFENSE WASTE (SLUDGE)

Processes developed for liquid waste solidification were applied to sludge wastes. The purpose was the development of a simple process leading to a homogeneous distribution of sludge within a titanate matrix. This was accomplished by in situ formation of calcium titanate precursors in aqueous slurries of sludge simulants (high Fe, high Al, and composite sludges representing Savannah River waste[6]). Aqueous solutions of tetramethylammonium titanate and calcium nitrate were added to the slurry, resulting in precipitation of a compound with an empirical stoichiometry

$Ca(Ti_2O_5H)_2$ which forms rutile and $CaTiO_3$ upon heating. The slurry was washed and dried producing a homogeneous powder suitable for hot pressing ($\leq$1150°C using 1% B as a sintering aid). Calcium titanate additions of 50% by weight were necessary to produce dense ceramics (4-4.5 g/cc for high Al and high Fe sludges, respectively) with open porosities of 0.1% or less.

Detailed microstructural characterization of the 50% sludge - 50% calcium titanate ceramic materials have not been completed. Initial scanning electron microscope and electron microprobe analyses indicated that the calcium titanate addition comprises a continuous matrix in the materials. Other phases with compositions depending on which simulant was used are dispersed within the matrix phase. In material containing the iron-rich sludge, areas containing Fe-Ni and Fe-Ti as major constituents are observed which possibly represent spinel and ilmenite compositions. Areas containing a mixture of Al, Ca, and Ti as major constituents were observed in the samples prepared from alumina-rich sludges. All the major phases in both materials appeared to contain varying amounts of the minor constituents of the sludge simulants.

One minor constituent, SiO_2, comprising about 0.4% by weight and in the form of 300-400-μm-sized particles used to simulate the addition of sand (used in supernate clarification) to sludge wastes had a significant effect on the properties of the final waste form. During hot pressing, a silica-alumina rich phase forms at the interface between the silica particles and the bulk of the material. The formation of this phase appears to have significant impact on the leaching properties; in general, the leach rates of these materials were relatively high. One-week soxhlet tests with monolithic samples resulted in weight losses ranging from 0.4 to 1.6% by weight. The high leach rates were associated in part with the alumino-silicate phase in the reaction zone surrounding the silica particles. This phase acts as a sink for other sludge constituents — Cs for example — and is highly vulnerable to aqueous attack in the soxhlet tests.

SUMMARY

Titanate-based ceramics containing 25% by weight of waste oxides were prepared by a straightforward process involving direct batch contact of an acidic, high-level liquid waste with a precursor powder, calcining to volatilize water and nitrates, and hot pressing (1250-1275°) to produce a fully dense ceramic. The ceramic has a rutile matrix containing waste constituents distributed in three crystalline oxide phases — perovskite, zirconolite, hollandite — and in metallic phases present in minor amounts as metallic blebs. Addition of boron lowered processing temperatures for both hot pressing and atmospheric pressure sintering to the 925-975°C range. Leach rates were less than 10 $\mu g/day \cdot cm^2$ for waste constituents and matrix attack was negligible at 250°C.

Titanate-matrix ceramics were prepared from solid waste sludges by in situ precipitation of calcium titanate in slurried sludge followed with consolidation by hot pressing. Fully dense ceramics were obtained; however, relatively high leach rates were measured for the materials. They were attributed to an aluminosilicate phase, formed during hot pressing, which was vulnerable to aqueous attack.

REFERENCES

1. Alternative Waste Form Peer Review Panel, L. L. Hench, Chr., "The Evaluation and Review of Alternative Waste Forms for Immobilization of High Level Radioactive Wastes," Report No. 2, DOE/TIC - 11219 (June 1980).
2. R. Roy, "Solid Phase Immobilization of Radioactive Waste," PSU-004, March 1980.
3. R. G. Dosch, "Chapter 8 - Ceramic Forms for Nuclear Waste," in: Radioactive Waste in Geologic Storage, S. Fried, ed., ACS Symposium Series 100, American Chemical Society, Washington,D.C., (1978).
4. R. G. Dosch and A. W. Lynch "Solution Chemistry Techniques in SYNROC Preparation," SAND80-2375 to be published (1980).
5. A. E. Ringwood, "Safe Disposal of High-Level Radioactive Wastes," Publ. No. 1438, Research School of Earth Sciences, Australian National University, Canberra (1980)
6. J. A. Stone, S. T. Goforth, and P. K. Smith, "Preliminary Evaluation of Alternative Forms for Immobilization of Savannah River High Level Waste," DP-1545 (1979).

THE STABILITY OF PEROVSKITE AND SPHENE IN THE PRESENCE OF BACKFILL AND REPOSITORY MATERIALS: A GENERAL APPROACH

H. W. Nesbitt, G. M. Bancroft, S. N. Karkhanis and W. S. Fyfe

Centre for Chemical Physics and Geology Department
University of Western Ontario
London, Ontario N6A 5B7, Canada

INTRODUCTION

A variety of glasses[1,2] and minerals[2,3,4] has been suggested as immobilizing waste forms for reprocessed nuclear fuel waste. The kinetic reactivity of these and other materials has been determined by measuring their leach rates, usually in distilled water. We have recently emphasized[5] that the thermodynamics of mineral and glass reactions should also be studied closely. Under equilibrium conditions, the mineral or glass will suffer no weight loss; and consequently, it should be advantageous to choose only thermodynamically stable materials to host reprocessed nuclear fuel waste. However, it must be emphasized that the thermodynamic stability of a phase can be determined only in relation to a specific chemical environment. Thus, the chemical environment must be carefully defined, because a phase which is thermodynamically stable in one environment may be very unstable in another. For example, we have already suggested[5] that perovskite is thermodynamically stable in igneous rocks very low in SiO_2 and high in CaO, but it is unstable thermodynamically in gabbros, granites, and in most natural waters.

In this paper, we show that the thermodynamic stability and kinetic reactivity of waste forms will be influenced — and even controlled — by the nature of the backfill and repository materials. We emphasize that it is necessary to consider the nuclear waste host, backfill and repository rocks as one chemical system, so that the whole system is unreactive. To illustrate this philosophy, we give an example of our approach using sphene ($CaTiSiO_5$) as the nuclear waste host (without radwaste components), clays as backfill, and granite as the repository system. Clays and granites have generally been considered in nuclear waste programs, and we show that sphene is thermodynamically stable relative to perovskite in waters equilibrated with this system. Sphene has other desirable properties as a waste host which will be discussed.

THE EXAMPLE

Environmental, economic and engineering considerations may greatly restrict the repository rock types available for the storage of nuclear wastes. We arbitrarily chose a granite as a repository rock. Any other rock type or material could have been chosen, for our approach is applicable regardless of the nature of the repository or vault material. Our granite is found to contain quartz, plagioclase, potassium feldspar, and minor amounts of biotite, ilmenite and sphene. Backfill and waste host materials must now be chosen which are chemically compatible with the granite. Clays are commonly considered for backfill because of their sorptive capacity, their impermeability and swelling properties when wetted.[6,7] A choice has to be made as to which, if any, clay minerals are compatible with the mineralogy of the granite. The stability fields at 25°C of the feldspars, chlorite (as a proxy for biotite), kaolinite and Na, K, Mg and Ca clays are illustrated in Fig. 1, along with the compositional ranges of waters equilibrated with the various minerals.[8-11] As illustrated, Na-bentonite can equilibrate with albite ($NaAlSi_3O_8$), illite with microcline ($KAlSi_3O_8$), and Mg-bentonite with chlorite. Backfill materials containing bentonites and illite are therefore compatible with the abundant minerals in the granite. The anorthite ($CaAl_2Si_2O_8$) stability field lies at the top of Fig. 1D, but plagioclases in granites contain only 5 to 10% anorthite, and the plagioclase stability field is larger than that of anorthite. If anything then, Ca feldspar will provide an even larger $[Ca^{++}]$ in equilibrated waters.

Having resolved the backfill-granite compatibility problem in this simplified case, the compatibility of the host materials with the granite and backfill must be considered. Minerals in the granite such as sphene and ilmenite are obvious first choices to host the waste products. Sphene especially has a number of attractive properties as a host. Sphene is known to contain a large variety of elements of widely varying charge and size;[12] the alkalis, alkaline earths, transition metals, rare earths and actinides. It has a relatively low melting point of 1382°C.[12] Of greater importance, sphene can be stabilized in natural waters equilibrated with our choice of granite repository, and clay backfill which have $\log[Ca^{++}]/[H^+]^2$ of 12-14, and $\log[SiO_2]_{aq}$ of -3.1 (Figs. 1 and 2). Sphene stability can be ensured by adding calcite (or perhaps a calcium zeolite) to the backfill to

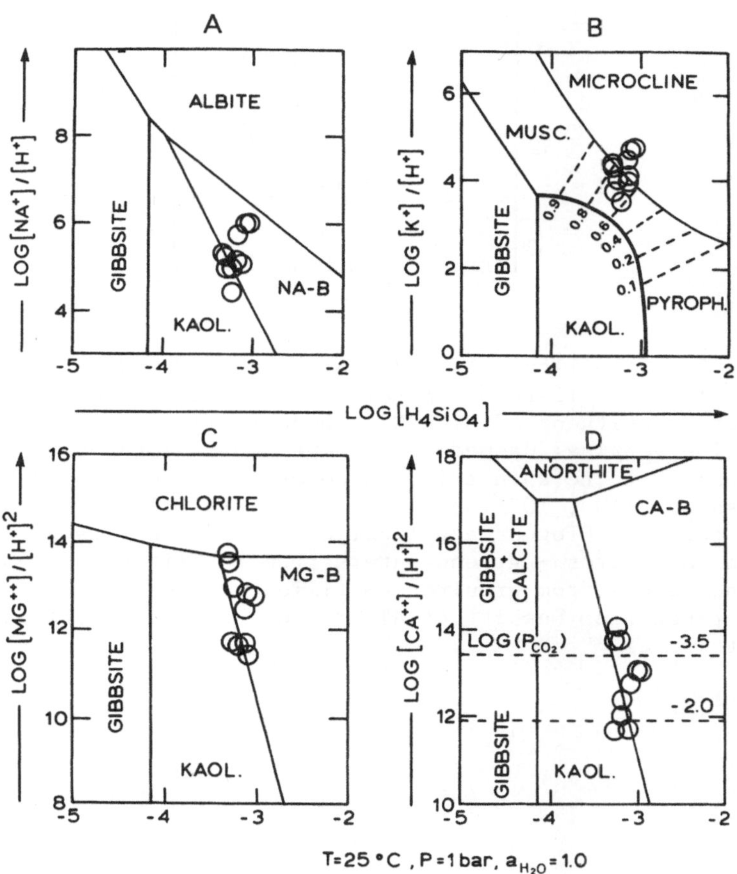

Fig. 1. Illustrates clay and other silicate equilibrium assemblages at 25°C and 1 bar. The dashed curves numbering 0.1 to 0.9 represent illites of composition $K_nAl_2(Si_{4-n}Al_n)O_{10}(OH)_2$ (where n has the value denoted on the dashed curves). Data from Nesbitt.[9] The circles represent analyses from dilute inflow waters to Saline Valley.[11] These waters are very similar in composition to those from other common natural waters[10] (see Fig. 2 and compare with Fig. 1D). Kaol = kaolinite; B = bentonite; Musc = muscovite; pyroph. = pyrophyllite.

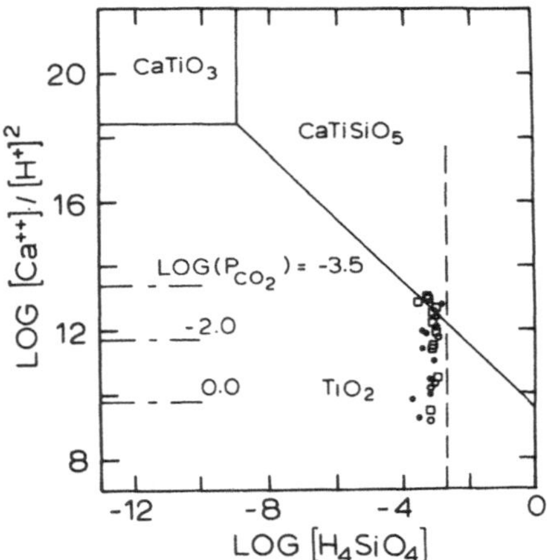

Fig. 2. Stability relations in the system $CaO-TiO_2-SiO_2-CO_2-H_2O-HCl$. The titaniferous minerals are perovskite ($CaTiO_3$) which is a major constituent of SYNROC, sphene ($CaTiSiO_5$), and rutile (TiO_2). The vertical dashed curve represents amorphous silica saturation. Calcite saturation in solutions of differing CO_2 partial pressures (10^0, 10^{-2} and $10^{-3.5}$ bars) is illustrated by the dash-dot curves. Ground waters collected from basalts and gabbros are shown as squares — ground waters from rhyolites as open circles and ground waters from granites as filled circles. Thermochemical data was taken from Nesbitt et al.,[5] and the ground water data from White et al.[10]

ensure the high $[Ca^{++}]/[H^{+}]^2$ values, while the feldspars in crushed granite ensure the high SiO_2 (aq) values in all solutions migrating through the backfill. Additional assurance of sphene stability could be achieved by including rutile in the backfill. Figure 2 indicates that sphene would actually grow in this environment. Our whole system — repository rock, clays with additives and the sphene based host — is chemically compatible and internally thermodynamically stable.

RESULTS AND DISCUSSION

The above example should not be considered a solution to the disposal of nuclear waste, although it may well have practical significance. The example does emphasize our philosophy by which nuclear waste disposal systems can be designed to store wastes in thermodynamically stable minerals. The utility of our approach is apparent from Fig. 2: sphene rather than perovskite emerges as a likely host candidate. Sphene can be stabilized thermodynamically in many natural waters, and in our granite-clay system. Perovskite cannot be stabilized in either. Hydrothermal experiments[13] have recently shown that sphene is more stable than previously determined. Independently, P. J. Hayward[14] has also recognized sphene as a useful host candidate, and is doing research on sphene-based synthetic waste forms.

Leach tests on sphene under equilibrium conditions are not yet complete, but our preliminary leach results in distilled water show that the leach rates for an Ontario sphene [27.60% CaO, 42.20% TiO_2, 30.20% SiO_2 by XRF] are lower than that reported earlier for a perovskite single crystal from Magnet Cove Arkansas, but comparable to, or higher than, leach rates for perovskite grains from Western Australia (Table 1). Our results are in semiquantitative agreement with those obtained on the same sample by P. J. Hayward.[14] Both sphene and perovskite leach rates are comparable to those obtained on high stability glasses.[15,16]

It is not surprising that the perovskite and sphene leach rates are comparable in the distilled water tests. In both cases the ΔG° for the dissolution reactions are very negative and for:

$$CaTiO_3 + 2H^+ \rightleftharpoons Ca^{++}_{(aq)} + TiO_2 + H_2O \quad (1)$$

$$CaTiSiO_5 + 2H^+ + H_2O \rightleftharpoons Ca^{++}_{(aq)} + H_4SiO_{4(aq)} + TiO_2 \quad (2)$$

$\Delta G^\circ_1 = -104$ kJ and $\Delta G^\circ_2 = -53$ kJ at 25°C.[17] The large negative ΔG° values will provide a large driving force for both reactions in distilled water. However, under equilibrium conditions ($\log[Ca^{++}]/[H^+]^2 \sim 13$ and $\log[H_4SiO_4] \sim -3$), $\Delta G = 0$ for reaction (2), and sphene should suffer no weight loss. For $\log[Ca^{++}]/[H^+]^2 > 13$ and $\log[H_4SiO_4] \sim -3$ (i.e., in the sphene stability field), ΔG is positive and sphene should grow. At present, we are doing leach tests on sphene using calcite plus quartz to produce $\log[Ca^{++}]/[H^+]^2 > 13$ and $\log[H_4SiO_4] \sim -3$, along with perovskite. Under these conditions perovskite should convert to sphene. After one month's leaching at 100°C, we indeed obtain a weight increase in the sphene.

Table 1. Leach Rates of Sphene, Perovskite and Glasses in kg m^{-2} sec^{-1} [a]

	25°C	100°C
$CaTiSiO_5$	2.6×10^{-10}	2.8×10^{-10}
$CaTiO_3$ (crystal)		2.2×10^{-9}
$CaTiO_3$ (grains)	$<1 \times 10^{-11}$ [b]	1.3×10^{-10}
Rhyolite	1.0×10^{-10}	4.1×10^{-9}
Borosilicates[c]	$>6 \times 10^{-12}$	$>1.2 \times 10^{-9}$

[a] kg m^{-2} sec^{-1} = 1.16×10^{-4} g cm^{-2} day^{-1}. All results are given for the static leaching of disks in the first week unless otherwise noted. The glass leach rates are determined from weight losses, while the titanate leach rates are determined by the amount of Ca^{++} in solution assuming incongruent dissolution. The weight losses and solution analyses for the sphene and perovskite crystals are consistent with reactions (1) and (2) occurring. All leach rates drop off dramatically after longer periods of leaching.[5,16]

[b] This sample had been leached before at higher temperatures so that this value will be comparatively low.

[c] Ref. 15. 30-day leach test. These glasses give substantially lower leach rates than previously reported for other borosilicates.

Two other points are important when considering the overall disposal system. First, we must assess the influence of higher temperature near the radwaste container on the mineral stability fields, and this will be the topic of a future paper. Second, choice of backfill materials for a particular host rock can be aided by examining the minor alteration products associated with microcracks in the surrounding rock. Since the flow rate would have been very low through such cracks, the minor alteration product (e.g., for granite these would probably include calcite and complex clays) would be in local equilibrium with the granites and solutions. The backfill phases should thus be similar to these minor alteration products in the rock near the repository.

We began by choosing a repository rock, and having done so, deduced which materials would be suitable for backfill and host materials. The procedure can be reversed; the host material can be selected, the backfill and repository materials compatible with it then determined. A more sophisticated approach may be taken using computer programs such as EQUILAS[18] to calculate the most stable mineral assemblage for a given chemical system. Hopefully, the approach and computer programs such as EQUILAS will find wide usage in the nuclear waste field for the entire repository system including radwaste.

ACKNOWLEDGEMENTS

We are grateful to Atomic Energy of Canada for financial support; and to P. J. Hayward of AECL for supply of the sphene, analysis and helpful discussion.

REFERENCES

1. J. E. Mendel, The Storage and Disposal of Radioactive Waste as Glass in Canisters, Battelle Pacific Northwest Lab. Rep. PNL-2764 December (1978), 71 pages.

2. T. D. Chikalla and J. E. Mendel, eds., Ceramics in Nuclear Management, Proceedings of the International Symposium Ceramics in Nuclear Waste Management, Cincinnati, CONF-790420, US Department of Energy, Technical Information Center, Oak Ridge, 1979, 377 pages.

3. G. J. McCarthy, Nuclear Waste Ceramics: Materials, Considerations, Process Simulation and Product Characterization. Nuclear Technology 32, 92-105 (1977); see also ref. 2, pp. 274-276.

4. A. E. Ringwood, S. E. Kesson, N. G. Ware, W. Hibberson and A. Major, The Synroc Process. A Geochemical Approach to Nuclear Waste, Geochemical Journal 13, 141-165 (1979); see also ref. 2, pp. 174-179.

5. H. W. Nesbitt, G. M. Bancroft, W. S. Fyfe, S. N. Karkhanis, A. Nishijima and S. Shin, On the Thermodynamic Stability and Kinetics of Dissolution of Perovskite, Nature, in press.

6. Kärn-Bränsle-Säkerhet, Sweden, Handling and Final Storage of Unreprocessed Spent Nuclear Fuel, 235 pages (1978).

7. W. S. Fyfe and Z. Haq, Nuclear Waste Disposal: Geochemical and Other Aspects, Geological Survey of Canada Paper 79-10, 55-59 (1979).

8. R. M. Garrels and C. Christ, Minerals, Solutions and Equilibria, McGraw-Hill, 450 pages (1965).

9. H. W. Nesbitt, Estimation of the Thermodynamic Properties of Na, Ca and Mg Beidellites, Canadian Mineralogist 15, 22-30 (1977).

10. D. E. White, J. D. Hem and G. A. Waring, Chemical Composition of Subsurface Waters, U.S. Geol. Survey Prof. Paper 440 F (1963), 67 pages.

11. L. A. Hardie, The Origin of the Recent Non-marine Evaporite Deposit of Saline Valley, Inyo County, California, Geochim. Cosmochim. Acta 32, 1279-1301 (1968).

12. W. A. Deer, R. A. Howie and J. Zussman, Rock Forming Minerals, Vol. 1, Ortho and Ring Silicates, Longmans, pp. 69-76 (1965).

13. J. A. Hunt and D. M. Kerrick, The Stability of Sphene; Experimental and Geologic Implications, Geochim. Cosmochim. Acta 41, 279-288 (1977).

14. P. J. Hayward, personal communication, AECL, Pinawa, Canada.

15. K. B. Harvey, personal communication, AECL, Pinawa, Canada.

16. S. N. Karkhanis, G. M. Bancroft, W. S. Fyfe and J. D. Brown, Leaching Behaviour of Rhyolite Glass, Nature 284, 435-437 (1980).

17. R. A. Robie, B. S. Hemingway, and J. R. Fisher, Thermodynamic Properties of Minerals and Related Substances, U.S.G.S. Bull. 1452, 456 pages (1978).

18. H. C. Helgeson, personal communication.

TAILORED CERAMIC NUCLEAR WASTE FORMS:

PREPARATION AND CHARACTERIZATION*

A. B. Harker, C. M. Jantzen, P. E. D. Morgan, and D. R. Clarke

Rockwell International Science Center
1049 Camino Dos Rios
Thousand Oaks, CA 91360

INTRODUCTION

The concept of immobilizing the radioactive elements of nuclear waste in an assemblage of mineral phases was originally introduced by Hatch.[1] Since that time, a number of other mineralogic-ceramic assemblages have been developed. Implicit in this concept is the idea of using additives to alter the waste composition to tailor it chemically to produce crystalline phases after consolidation which closely approximate natural mineral assemblages of proven stability over geologic time scales.

For consolidation of the defense nuclear waste stored at the Savannah River Plant, alumina tailoring was chosen. The philosophy involved in the tailoring approach for the alumina-based assemblage takes advantage of the existing chemical composition of the nuclear waste as much as possible, to minimize the required tailoring and produce a solid ceramic form with high waste loading and large volume reduction, while reducing the required processing steps. The tailoring was designed to produce specific host phases for the waste radionuclides such that the bulk of the ceramic form would be made up of highly insoluble phases, containing no radioactive material, and providing microstructural isolation of the radiophases to improve the leach resistance of the form. The amount of inert phases is determined by the elements in the waste, the tailoring elements, the waste loading, and the consolidation technique. The processing steps are being designed to produce a fully dense strong ceramic with low surface area, microstructural isolation of the radiophases, and no continuous glassy phases.

*This effort supported by U.S. Department of Energy through Rockwell Energy Systems Group under Prime Contract No. DE-AC09-79-ET41900, "Development of Tailored Ceramic for Geologic Storage of Nuclear Wastes."

TAILORING

The tailoring concept uses the high Al and Fe content of SRP waste with the addition of sufficient rare earth or titania plus alumina to incorporate most radionuclides into a magnetoplumbite phase, with uraninite being the uranium host.[2] The non-radioactive elements remaining form primarily spinel phases. In addition to these three primary phases a fourth phase is normally present whose identity depends upon which tailoring agent is in excess as shown in Table I.

Tailoring of the SRL "composite" composition's waste with alumina and rare earth oxide, so that each alkali ion is balanced by one trivalent ion, as for $Cs_{0.5}La_{0.5}(Fe,Al)_{12}O_{19}$*, produced the four phase assemblage, corundum-spinel-magnetoplumbite-uraninite.[3] Magnetoplumbite is the major phase and the c/a ratio of the hexagonal unit cell, a distinctive indicator of the type of magnetoplumbite present, suggested an $XY_{12}O_{19}$ magnetoplumbite formulation (c/a = 3.922)**. The x-ray spectra recorded from the four phases during STEM analysis indicate the elemental distribution among the phases shown in Table I. A "composite" composition sample with ion sieve added was tailored specifically for the enhancement of the magnetoplumbite phase. The x-ray spectrum recorded from the tailored magnetoplumbite shows that it contained Al, Fe, Ni, La, Na, Sr, Cs, Si, Ce and K, the latter element introduced by the presence of the ion sieve.

The magnetoplumbites are accommodating structures because they are composed of spinel blocks which have the usual IV and VI coordinated sites for cations while interspinel layers have unusual V fold sites for small cations. The interspinel layers also accommodate large cations of ~ 1.15 - 1.84 Å radius, replacing oxygen in XII fold sites in the anion close-packed structure.[4] The naturally occurring mineral, $CaAl_{12}O_{19}$, is known to be geologically stable and phase-compatible with both corundum and spinel.[5]

The range of crystalline solid solubility in the CaO-Nd_2O_3-Al_2O_3 ternary system was chosen as a model for alumina-rare earth

---

* $La_{0.5}Na_{0.5}Al_{12}O_{19}$ was not previously known but was made phase-pure (c/a = 3.948) contrasting with β-$NaAl_{11}O_{17}$(c/a ~4.02) and $LaAl_{11}O_{19}$(c/a = 3.957); $La_{0.5}Na_{0.5}Fe_{12}O_{19}$ is known (P.D.F. 15-605); poorly crystallized $Cs_{0.5}La_{0.5}Al_{12}O_{19}$ was made phase pure at 1165°C (c/a ≈ 3.95).

** c/a ratios for the structurally related β-alumina type structures are ~ 4.0 - 4.1 while true magnetoplumbites fall in the range of 3.92 - 3.98.[6]

TABLE 1

CURRENT HIGH Al TAILORED CERAMIC MINERALOGY AND WASTE ELEMENT DISTRIBUTION

	Nominal Composition	Elements Detected
Principal Phases		
Spinel	(Mg,Fe,Ni,Mn) $(Al,Fe)_2O_4$	Al, Mn, Fe, Ni, Ti, Trace Mo, Si
Uraninite	UO_2	U, Trace Zr, Th
Magnetoplumbite	$XY_{12}O_{19}$ where $X = Ca,Sr,Ba,RE_{0.5}+$ $Na_{0.5}$, etc. $Y = Al,Fe,Ti,Si$	Al, Mn, Fe, Ni, La, Na, Ca, Ti, Trace Sr, Cs, Ba, Si, Mo, Ce, K
Phases Present with Excess Tailoring		
Corundum (excess Al_2O_3)	Al_2O_3	Al, Trace Fe, Ti
Perovskite (excess REO)	$RE(Al,Fe)O_3$	La, Al, Fe, Trace Ca, Na, Ce
Pseudobrookite (excess TiO_2)	$(Fe,Ti,Al)_2TiO_5$	Fe, Ti, Al

tailoring. A partial solid solution exists between the calcium magnetoplumbite, $CaAl_{12}O_{19}$ and a high temperature neodymium aluminate phase, up to $Ca_{0.95}Nd_{0.05}Al_{11.95}O_{18.95}$, at 1300°C, the temperature of the ternary section comparable to the hot-pressing temperature.

In the CaO-Nd_2O_3-Fe_2O_3 ternary system, chosen as a model for iron-bearing magnetoplumbites in an oxidizing environment, neither $CaFe_{12}O_{19}$ nor $NdFe_{12}O_{19}$* magnetoplumbite structures form at 1300°C and Nd^{3+} does not stabilize $CaFe_{12}O_{19}$ as La^{3+} does. However, $CaAl_6Fe_6O_{19}$ solid solutions are known to form, and it is expected that extensive magnetoplumbite solid solutions would exist in the CaO-Nd_2O_3-Fe_2O_3-Al_2O_3 system.

When iron is reduced to the ferrous state it substitutes crystal chemically for Mg^{+2}. Magnetoplumbites of the complex type $Sr_{0.5}Nd_{0.5}Mg_{0.5}Al_{11.5}O_{19}$, $Nd_{0.75}Na_{0.25}Mg_{0.5}Al_{11.5}O_{19}$ and $NdMgAl_{11}O_{19}$ have been synthesized in this laboratory to determine the expected solid solubility of Fe^{+2} magnetoplumbite anologues. Since spinel, $MgAl_2O_4$, is compatible with $LaAlO_3$ and $NdAlO_3$ perovskites, it is assumed that similar relations will hold in the reduced pseudoternary $(Mg,Fe,Ni)O$-RE_2O_3-Al_2O_3 system, and therefore, a schematic quaternary tailoring diagram can be drawn (Fig. 1a). The range of solid solution of the magnetoplumbite[2] illustrates that the phase compatibility regions are quadrilaterals rather

*$LaFe^{+2}Fe^{+3}{}_{11}O_{19}$ is known (P.D.F. 15-605).

than planar triangles through the quaternary tetrahedrom (Fig. 1a,b). The enhanced size of the compatibility quadrilaterals, due to extensive solid solution of several of the component phases, allows a wide variability of waste compositions to be accepted by a predetermined phase assemblage at a given waste loading.

Using these formulations with the various SRP compositions without aluminum dissolving[3] allows waste loadings of 30-50% for composite compositions and up to 90% waste loadings for high alumina compositions. With this tailoring formulation the necessity of adding alumina reduces the waste loading to below 30% for the extreme high iron compositions; however, work is currently in progress on tailoring modifications to increase the waste loadings for high iron cases.

PREPARATION

The powder preparation and consolidation procedures were designed to produce a dense ceramic, with a small (<1 μm) uniform grain size containing no continuous intergranular amorphous phases. The small grain size helps prevent the occurrence of microcracking during fabrication due to differential or anisotropic thermal contraction or phase transformation. Spray calcining of the simulated waste powder produces ceramic precursors which are substantially x-ray amorphous so that crystallization to the final phase assemblage occurs during the consolidation step (reactive hot pressing or hot isostatic pressing) allowing advantage to be taken of the transient, extremely fine, grain structures occurring during the phase transformation. This in turn minimizes the temperatures* that must be used to obtain high density so that volatility effects, especially those of Cs, are minimized.[2] It is envisaged in any practical application, cold isostatic pressing of the pre-calcined, e.g., spray dry, roasted sludge, would be followed by encapsulation of the pellet for subsequent hot isostatic pressing at ∿50,000 psi below 1200°C.

In the consolidation process it is necessary to achieve the correct oxidation states for several elements, e.g., U^{4+} and the correct Fe^{2+}/Fe^{3+} ratio. In rare earth oxide-alumina-titania tailored assemblages reducing conditions during consolidation favor spinel and perovskite with lesser amounts of magnetoplumbite while

*Laboratory consolidation by reactive hot pressing in the range 1200-1300°C at 4000 psi is used for high Al tailored ceramics surrounded by Ni or Al_2O_3. Before reactive hot pressing the powders are cold isostatically pressed to ∿50,000 psi. Titania additions to High Al Tailored Ceramics are used primarily as a sintering aid and lower hot pressing temperatures 100-150°C. Tailoring for high Fe compositions only necessitates hot pressing at ∿1050°C.

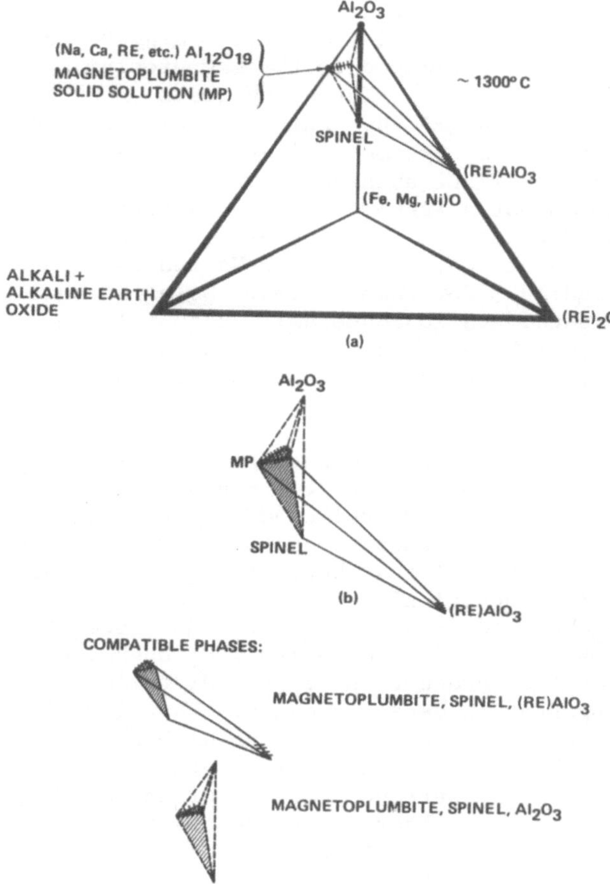

Figure 1.

a) Tailoring diagram showing magnetoplumbite solid solution and the compatibility quadrilaterals for the phase assemblages produced during hot-pressing: Al_2O_3-magnetoplumbite-spinel and spinel-magnetoplumbite-(RE)AlO_3 perovskite.

b) Distorted view of the compatibility quadrilaterals showing the three-dimensional solid solution of magnetoplumbite (hatched region) and the magnetoplumbite-spinel plane (shaded) that the two compatiibility quadrilaterals share in common.

more oxidizing environments (i.e., Ni-NiO buffer) favor magnetoplumbite over spinel and perovskite. At the oxygen partial pressure of the Ni-NiO buffer uraninite is the stable uranium host but more oxidizing environments produce defect uranate phases.

CHARACTERIZATION

The microstructure produced in the rare earth oxide-alumina tailored ceramic form is shown in Fig. 2. The bright field transmission micrograph shows the lath-like magnetoplumbite grains adjacent to a uraninite region. The dark field micrograph demonstrates the absence of continuous intergranular glass and shows that only thin strips ($<$ 2 nm wide) of amorphous material exist between isolated grains in the ceramic.

Solution analysis of static dissolution tests on these forms at 90°C in DI water show that sodium and cesium are the most readily leached elements from these structures. Lattice parameter data indicates that these elements are replaced in the magnetoplumbite interspinel layers by hydronium ions.[7] The leach tests on both multiphase and pure phase $Na_{0.5}La_{0.5}Al_{12}O_{19}$ samples show that sodium leach rates of $\sim 10^{-5}$ gr/cm^2-day (geometric surface area and normalized for elemental content) are obtainable in these ceramics, with cesium retention being about a factor of 2 to 3 better. Similar tests on both MCC-76-68 commercial waste glass and SRL borosilicate glass from simulated defense wastes show Na and Cs leach rates to be between 10^{-4} and 10^{-3} for these forms. Solution analysis for Sr and U in the leached tailored ceramics show release rates of $< 10^{-6}$ gr/cm^2-day, with the structural elements, Al, Fe, Ti, etc., all being below the detection limits of atomic absorption ($< 10^{-7}$ gr/cm^2-day).

XRD analysis of the surfaces of the leached samples give evidence of passivation layer formation. After two months exposure to static deionized water, a La_2O_3-Al_2O_3 tailored sample showed surface precipitates or recrystallization products. X-ray dispersive analysis of the surface crystallites showed the surface scale to be composed primarily of elemental Al. The presence of this element combined with the observed x-ray diffractograms showed the leached surface to be primarily γ-AlOOH (boehmite).[8] Boehmite has also been found as a surface product on TiO_2-Al_2O_3 tailored samples when TiO_2 is not present in excess amounts. When TiO_2 is present in excess amounts an iron-titanium-aluminum oxyhydroxide, hoegbomite, is observed.[8] The simple γ-AlOOH recrystallization product would be expected from the surface reaction of the alumina or spinel with the deionized water. There is evidence from the 14-day leach tests that some intermediate alumina-hydrate, tohdite ($5Al_2O_3 \cdot H_2O$), may form as a precursor to γ-AlOOH formation.

The origin of these passivation layers appears to be from dissolution of the non-radiophases, spinels and corundum. The pure phase magnetoplumbite leach studies showed no recrystallization or passivation layer formation, and XRD patterns of leached tailored ceramic surfaces have shown the magnetoplumbite reflections to be enhanced with respect to those of spinel and corundum. The formation of the passivation layers and the dissolution resistance of the tailored ceramic form is shown in the micrograph of a leached, ion beam-thinned foil (Fig. 3). This rare-earth oxide-alumina tailored multiphase ceramic of crystalline grains less than 10 nm thick was exposed to DI water at 90°C for 24 hours. As can be seen, the magnetoplumbite grains still show complete crystallinity while fibers can be seen forming on the surface. Similar tests were conducted on a thinned section of a ceramic sample over-tailored with titania, containing pockets of amorphous

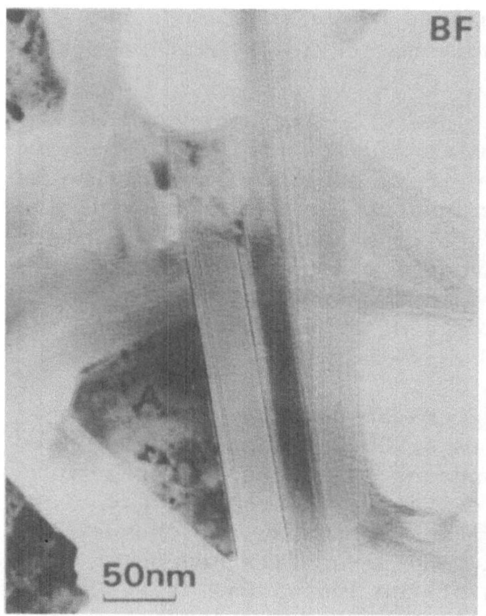

Fig. 2 Bright field (BF) and dark field (DF) transmission electron micrographs of the microstructures of an La_2O_3-Al_2O_3 Tailored Ceramic. The light regions arrowed are thin (~ 1 nm) films of glass. The long lath-like grains are the magnetoplumbite phase and the region A is uraninite.

Fig. 3 La_2O_3-Al_2O_3 Tailored Ceramic after leaching the TEM foil for 24 hr. in deionized water at 90°C. Fibers of boehmite (γ-AlOOH) formed but the lath-shaped magnetoplumbite grains appear unaffected. Even the 2.2 nm c-axis-spacing lattice planes of the magnetoplumbite grain arrowed can still be seen.

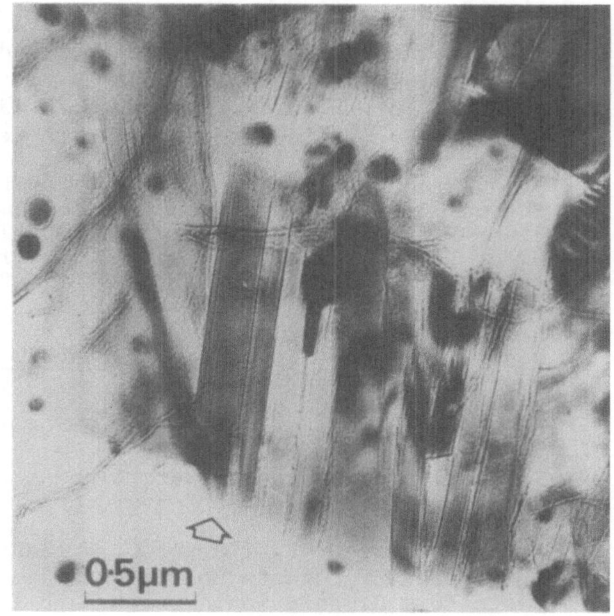

material. After 24 hours of exposure to DI water at 90°C the amorphous material was totally dissolved, leaving behind only the crystalline matrix.

The ease of dissolution of the intergranular amorphous material in an incorrectly tailored ceramic, and its affinity for radionuclides such as cesium, demonstrate its undesirability in a solid nuclear waste form. Providing an excess of the tailoring additives required for producing the host crystalline phases for the glass-forming elements, and using highly reactive powders can, however, produce a fine grain multiphase ceramic from nuclear waste composition without any continuous intergranular amorphous phase.

CONCLUSION

Simple tailoring and processing can be used to accommodate all of the radionuclides in SRP waste at 30-90% waste loading into magnetoplumbite and uraninite radiophases by taking advantage of the natural elemental content of the waste. The remaining waste elements and tailoring additives form inert encapsulant phases such as spinel, corundum, pseudobrookite, and perovskite depending on the type of tailoring. Examination of the individual phases before and after leaching has demonstrated that continuous intergranular amorphous phases enhance leachability and, therefore, highly reactive powders have been used to produce fine grain ceramics and limit the intergranular amorphous material. The magnetoplumbite has proven to be leach resistant and to accept a wide range of elements in solid solution.

REFERENCES

1. L. P. Hatch, "Ultimate Disposal of Radioactive Wastes, Am. Scientist 41:410 (1953).
2. P. E. D. Morgan, D. R. Clarke, C. M. Jantzen and A. B. Harker, High Alumina Tailored Nuclear Waste Ceramics, Submitted J. Am. Ceram. Soc. (1980).
3. J. A. Stone, S. T. Goforth, Jr., and P. K. Smith, Preliminary Evaluation of Alternative Forms for Immobilization of Savannah River Plant High-Level Waste, SRL-DP-1545:98 (1979).
4. W. D. Townes, J. H. Fang and A. J. Perrota, The Crystal Structure and Refinement of Ferromagnetic Barium Ferrite, $BaFe_{12}O_{19}$, Z. Krist. 125: 437 (1967).
5. H. Curien, C. Cuillemin, J. Orcel and M. Sternberg, La Hibonite, Nouvelle Espece Minerale, C. R. Acad. Sci. Paris 242:2845(1956)
6. J. M. P. J. Verstegen and A. L. W. Stevels, The Relation Between Crystal Structure and Luminescence in β-alumina and Magnetoplumbite Phases, J. Lumin. 9:406 (1974).
7. M. W. Brieter, G. C. Farrington, W. L. Roth and J. L. Duffy, Production of Hydronium Beta Alumina from Sodium Beta Alumina and Characterization of Conversion Products, Mat. Res. Bull., 12:895 (1977).
8. A. B. Harker, D. R. Clarke, C. M. Jantzen, P. E. D. Morgan, The Effect of Interfacial Material in Tailored Ceramic Nuclear Waste Form Dissolution, Proceedings International Symposium on Symposium on Surfaces and Interfaces, Berkeley (1980).

RECENT PROGRESS ON SYNROC DEVELOPMENT

A. E. Ringwood,* K. D. Reeve,† J. D. Tewhey‡

*Australian National University, Canberra, Australia
†Australian Atomic Energy Commission, Sydney, Australia
‡Lawrence Livermore Laboratory, Livermore, Calif. 94550

ABSTRACT

The bulk leachability of a sample of SYNROC, based on cesium extraction by deionized water at 85°C, fell from an initial value of 0.6 g/m^2.day on the first day to 7 x 10^{-3} g/m^2.day after 25 days. This latter leach-rate is about 300 times less than that for PNL76-68 borosilicate glass under similar conditions. A modest extrapolation of existing data suggests that SYNROC would be less leachable than glass by a factor exceeding 1000 after 55 days. Preliminary leach results for other elements are also reported. They indicate that the leachability of SYNROC, fabricated according to our current specifications, would be very much smaller than that of borosilicate glasses under realistic geologic conditions. Detailed studies of the effects of radiation damage on the crystal structures and radionuclide retention of SYNROC minerals under geologic conditions have been completed and confirm that the ability of SYNROC to immobilize HLW safely for millions of years is not adversely affected.

INTRODUCTION

SYNROC is a titanate ceramic composed of three main minerals - zirconolite $CaZrTi_2O_7$, "hollandite" $BaAl_2Ti_6O_{16}$, and perovskite $CaTiO_3$. These minerals have the capacity to accept into their crystal lattices nearly all of the elements present in high-level wastes (HLW). Similar minerals occur in nature where they have survived in a wide range of geochemical-geological environments for periods up to 2000 million years. It is this evidence of long-term geologic stability, combined with experimental observations showing that these minerals are extremely resistant to the attack of hydrothermal solutions, which suggests that SYNROC would provide a superior method of immobilizing HLW[1]. Research programs at our respective laboratories are currently aimed at optimizing the preparation and properties of SYNROC.

The SYNROC formulation currently under study is given in Table 1, together with compositions of constituent minerals. The intimately mixed powders are calcined at 800°C in an $Ar-H_2-H_2O$ atmosphere near the Fe-FeO oxygen fugacity buffer and then hot-pressed to theoretical density at 1100-1150°C.

LEACHABILITY OF SYNROC AND BOROSILICATE GLASSES

The basic factors governing the leachability of silicate glasses have been well-documented by Hench[2]. HLW borosilicate glasses typically have low silica and high alkali contents. When subjected to attack by aqueous solutions, alkali metals and B_2O_3 tend to be selectively leached from the surface layers which become enriched in silica. Because of the relatively low leachability of silica, these outer layers protect the underlying glass and strongly influence the bulk leachability of the glass. Further leaching occurs primarily via diffusion of cations through the surface layer*. It follows that the leachability of borosilicate glasses will be strongly influenced by factors controlling the solubility of silica in groundwaters.

Table 1. Compositions of SYNROC and its Constituent Minerals

	Hollandite 40%	Zirconolite 35%	Perovskite 25%	Bulk Composition
TiO_2	71.0	50.3	57.8	60.5
ZrO_2	0.2	30.5	0.2	10.8
Al_2O_3	12.9	2.5	1.2	6.3
CaO	0.4	16.8	40.6	16.2
BaO	16.	-	-	6.4
Total	100.5	100.1	99.8	

Currently preferred SYNROC formulation (wt.%): SYNROC additives (above) = 80, HLW calcine = 10, extra TiO_2 = 9, extra Al_2O_3 = 1.

*With more intense leaching, as the silica-rich skin is continually dissolved, a porous outer layer of insoluble oxides is formed, typically including P, Ca, Zr, Sr and the rare earths. This skin is often fragile and tends to spall off. It therefore does not usually provide a major barrier to further attack by aqueous solutions[3].

The silica in the surface layer is essentially in an amorphous state. On the other hand, the free silica present in most natural rocks consists of crystalline quartz. The groundwaters in most geological environments are therefore saturated with quartz. However, the solubility of quartz in groundwater is much less than that of amorphous silica. The difference ranges from a factor of 8 at 75°C to 2.5 at 300°C. It follows that groundwaters obtaining access to borosilicate glasses will nearly always be undersaturated with respect to the amorphous silica-rich surface layers.[3]

Many of the standard methods for evaluating the leachability of borosilicate glasses have been based upon procedures which ignore this key factor. Thus, the popular Soxhlet test is based upon closed-system leaching under isothermal conditions and is often carried out in pyrex glassware. The leaching solution therefore rapidly becomes saturated with amorphous silica derived both from the borosilicate glass sample and the pyrex container, so that the amorphous silica layer on the surface of the sample becomes insoluble. Usually, the leach rate is observed to drop rapidly with time, falling to quite low values. Results of this type have created a widespread impression that glass is a satisfactory wasteform of low intrinsic leachability. In other testing procedures widely employed, particularly at elevated pressures and temperatures, the glass is leached isothermally under closed system conditions in a relatively small volume of water. The water rapidly becomes saturated with silica (and other components) dissolved from the glass and the leach rate again falls dramatically. Tests of this type provide a false sense of security.

In order to correspond to realistic geologic conditions, the wasteform should be leached in aqueous solutions which are undersaturated with respect to amorphous silica and other components. The single-pass leaching system developed by Coles et al.[4] is one of the very few methods which meet these requirements. The IAEA procedure[5] in which the solution is changed frequently and analysed, whilst not as realistic as single-pass leaching, may represent a reasonable compromise. At present, single-pass leaching has not been carried out above 75°C. It is most desirable that the technique be extended to cover leaching under pressure and temperatures to 300°C.

Coles and coworkers[6,7] have carried out an important series of single-pass leaching experiments on a typical borosilicate glass (PNL 76-68). Over a period of 300 days, the leach rates for most elements did not decrease substantially with time (Fig. 1). This behaviour apparently arises because the leaching rates are largely controlled by steady-state dissolution of the siliceous skin by SiO_2-undersaturated solutions. In this respect, the experiments provide a realistic model for leaching in typical natural environments.

The leaching behaviour of titanate minerals is unique. Our observations show that when they are in contact with aqueous solutions, both under natural weathering conditions and at high pressures and temperatures, univalent and divalent cations (e.g. Cs^+, Sr^{2+}) tend to be leached from a near-surface layer leaving a thin skin enriched in TiO_2. Further leaching occurs by diffusion of cations through the titania-rich skin, combined with dissolution of the outer layer which is, however, extremely slow. A primary reason for the great resistance of many titanate minerals to leaching and weathering is the extreme insolubility of titanium dioxide in aqueous solutions. Its solubility in groundwaters is about 1000 times less than that of amorphous silica[3]. Moreover, unlike silica, titania is not known to occur in an amorphous form in nature. The principal weathering product of titanate minerals, leucoxene, is essentially finely crystalline rutile. Leucoxene and rutile are widely distributed in rocks so that groundwaters would normally be saturated with TiO_2 (unlike the situation for amorphous SiO_2). This factor, combined with the very low solubility of TiO_2 in groundwater, accounts for the very resistant nature of many titanates.

LEACHING OF SYNROC

Currently we are studying the comparative leachabilities of SYNROC, its constituent minerals, and borosilicate glasses. Preliminary results are discussed below. Because cesium is one of the most mobile elements, its leachability has been monitored for evaluation purposes. As the preparation of SYNROC has improved, the Cs leach rate has steadily decreased. Optimum performance has not yet been achieved and further improvements are anticipated.

Figure 1 shows that the leach rate of SYNROC falls dramatically with time. This is believed to be caused by two factors:

(a) Owing to lack of complete equilibration during heat-treatment a significant amount of cesium may occur not in the hollandite lattice but at grain boundaries and in metastable minor phases which are more readily leachable. Thus, the high initial rate is caused by selective leaching of this loosely-bound cesium. This problem can be minimized by maintaining a high degree of chemical homogeneity in the feedstock and careful control of heat-treatment and redox conditions.

(b) After the initial high leach rate, the leachability decreases proportionally to the square root of time. We interpret this to reflect selective leaching of surficial monovalent and divalent cations, leaving growing surface layers enriched in insoluble TiO_2 (and ZrO_2). Cesium ions must diffuse through this insoluble protective layer to be available for leaching. The square root time dependence is a consequence of this latter process[2].

The borosilicate glass (Fig. 1) showed leachabilities close to

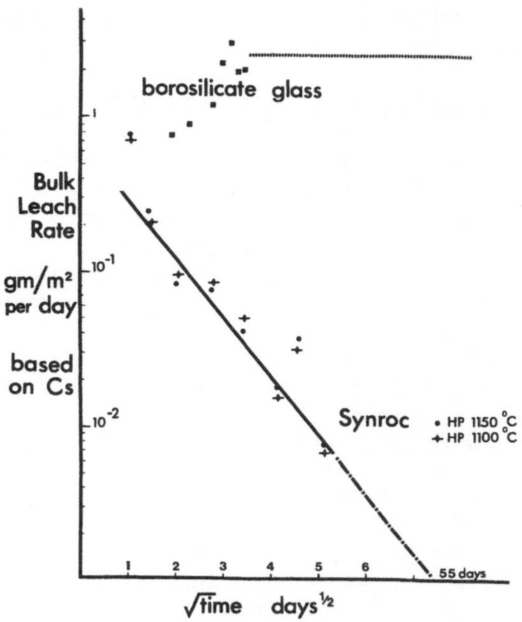

Fig. 1. Bulk leach rates (based upon Cs-loss) versus time for two samples of SYNROC (0. % Cs) hot-pressed at 1100°C and 1150°C respectively. Sample was in the form of 110-275 micron grains and the area quoted is calculated geometric surface area. Leachant was deionized water at 85°C, and was changed at frequent intervals corresponding to each analysis point. Also shown are similar measurements carried out on a grain fraction in the same size range of borosilicate glass PNL 76-68. The top horizontal line is based on ref.(7) showing that the bulk leach rate (Cs-based) for drilled cores of PNL 76-68 remains constant between 9 and 55 days during single-pass leaching in deionized water at 75°C.

1 g/m^2.day in 3-day leaches up to the eighth day. After that, the leaching water was changed daily and leach-rate increased to 2 g/m^2.day. This increase is probably a result of SiO_2 saturation reached in the longer runs. The shorter runs correspond more closely to the conditions achieved in single-pass leaching[4,6,7]. The latter method applied to cores drilled from the same glass showed that Cs leachability was constant between the 9th and 52nd days[7].

The drastic fall in leach rate with time is likely to be a key characteristic of SYNROC. After 25 days of leaching, the Cs leachability of SYNROC is about 300 times smaller than glass. With a relatively small extrapolation of the data to 55 days, the leachability of Cs from SYNROC would be more than 1000 times smaller than that for glass under conditions of single-pass leaching.

Analogous behaviour is shown by the leaching of a synthetic Cs-bearing hollandite under similar conditions to those described in Fig. 1. After 44 days, the bulk (Cs) leachability had fallen by more than two orders of magnitude to 4×10^{-3} g/m^2.day, whilst the bulk leachability based on barium was about an order of magnitude smaller than that based on Cs. The same authors had also shown that the leachability of Sr and Ca are intermediate between those of Cs and Ba, whilst the leachability of Zr is about 2.5 orders of magnitude smaller. Ti and Al concentrations in leach solutions were below the limit of detection. Leachabilities of actinides and rare earths are expected to be closer to those of Ti, Zr and Al.

The above results are regarded as being highly encouraging for SYNROC samples that are carefully prepared according to recommended procedures. They demonstrate that after 20 days exposure to circulating water at $85^{\circ}C$, the leachability of alkalis and alkaline earths from SYNROC is more than a hundred times smaller than from borosilicate glasses. A modest extrapolation of existing data suggests that these elements would be about 1000 times less leachable in SYNROC as compared to glass beyond 50 days. Additional data[3,6] show that the leachabilities of trivalent and quadrivalent elements are much smaller, probably by 1 or 2 orders of magnitude, than those of alkalis and alkaline earths.

RADIATION DAMAGE

A key technical question is the effect of radiation damage (dominantly from alpha recoils) on the structure of minerals proposed for waste immobilization. Sinclair and Ringwood[8] have investigated the topic by studying the crystal structures of a collection of naturally-occurring zirconolites and perovskites of differing ages which contain varying amounts of U and Th. These minerals have accordingly received varying cumulative radiation doses, ranging from 5×10^{17} to 10^{20} α/gm. These doses can be directly related to the doses which would be received by the same minerals occurring in SYNROC containing 10% of HLW, over a given interval of time (the "SYNROC age").

They found that natural zirconolite receiving an α-dose equivalent to a SYNROC age of 1000 years remained fully crystalline but expanded in volume by 2 percent (1.3×10^{18} α/gm). With more intense irradiation (8×10^{19} α/gm) (equivalent to a SYNROC age of 4×10^5 years) natural zirconolite transformed to a cubic defect-fluorite-type structure with a volume increase less than 3 percent. It appears that the main effect of the radiation dose was to cause a degree of disorder of the cations, leaving the fluorite-type anion lattice essentially intact. This behaviour persisted up to SYNROC ages exceeding 10^6 years. With even heavier irradiation, zirconolites ceased to diffract X-rays, i.e., they became metamict according to conventional terminology. However, electron diffraction and Mossbauer studies of these samples showed that they possessed a considerable degree of short-range order and contained numerous small crystalline domains. Moreover, the expansion in volume from the original state was less than 3 percent. The results showed that even with this enormous degree of radiation damage, the zirconolite crystals, although metamict, were still essentially in the crystalline state and in no way resembled the highly disordered structure of a true glass.

Similar studies of natural perovskites possessing SYNROC ages up to 20,000 years showed that this degree of irradiation caused volume increases up to 1.8%. The X-ray powder patterns of the perovskites were essentially unchanged. Comparative studies showed that the perovskite lattice is even more resistant to α-irradiation than zirconolite.

Oversby and Ringwood[9] analysed a collection of natural zirconolites and perovskites for uranium and lead by isotope dilution, and for Pb isotopic composition. These studies confirmed that zirconolites possess a remarkable capacity to immobilize U, Th and their decay products, even though they have suffered extreme degrees of radiation damage. For example, Sri Lanka zirconolites which have experienced α-doses exceeding 8×10^{19} α/gm have nevertheless remained as closed systems to U, Th and Pb for 550m.y.[9]. Limited data on perovskites also indicate that this mineral will be sufficiently stable under radiation to be a satisfactory host for HLW elements.

Results of irradiation over slow geological timescales have been complemented by an experimental study of the effects of fast neutron irradiation on SYNROC over a period of one month[10]. The volume expansion of SYNROC minerals which had experienced similar degrees of radiation damage was found to be the same on both long and short timescales.

The leaching behaviour of zirconolites and perovskites which have suffered varying amounts of radiation damage are currently under study. At $200^{\circ}C$, bulk leach rates (based on·U loss) were 3.5×10^{-4} g/m^2.day for Kaiserstuhl zirconolite (1.3×10^{18} α/g) and 1.8×10^{-3} g/m^2.day for Sri Lanka zirconolite (8×10^{19} α/g). These leach rates are very low in comparison to borosilicate glass at this temperature and demonstrate that extreme radiation damage does not lead to unacceptably high leachability.

Whilst the above results are most encouraging for SYNROC, a recent study of the effects of radiation damage in borosilicate glass by Dran et al.[11] is highly disquieting. They found that although borosilicate glass containing HLW would leach relatively slowly for the first 2000 years, after this period, the α-recoil damage paths start overlapping and the leach rates increase drastically by more than a factor of 20. They estimate that after this stage is reached, the remaining lifetime of a borosilicate glass cylinder 0.5 meter in diameter would only be about 100 years! The implications of this work are extremely serious, and emphasize the need for continued development of ceramic wasteforms. Recent results[12] show that crystalline nonsilicate minerals such as perovskite are much more resistant to these effects.

REFERENCES

1. A. E. Ringwood, S. Kesson, N. Ware, W. Hibberson and A. Major, "The SYNROC Process: A Geochemical Approach to Nuclear Waste Immobilization," Geochem. J. 13: 141 (1979).
2. L. L. Hench, D. Clark and E. Yen-Bower, "Corrosion of Glasses and Glass Ceramics," Nucl. Chem. Waste Manage. 1: 59 (1980).
3. A. E. Ringwood, V. Oversby and S. Kesson, "Comparative Leach Testing of SYNROC, SYNROC Minerals and Borosilicate Glasses," manuscript in preparation (1980).
4. D. G. Coles, H. C. Weed and J. S. Schweiger, "Single-pass Leaching of Nuclear Melt Glass by Groundwater," in Radioactive Waste in Geologic Storage, S. Fried, Ed., ACS Symposium Series, 100: 93 (1978).
5. E. D. Hespe, "Leach Testing of Immobilized Radioactive Waste Solids, A Proposal for a Standard Method," At. Energy Rev. 9: 195 (1971).
6. D. G. Coles and F. Bazan, "Continuous-flow Leaching Studies of Crushed and Cored SYNROC," submitted for publication, J. Nucl. Technol. (1980).
7. D. G. Coles and coworkers, unpublished results (personal communication with D. G. Coles).
8. W. Sinclair and A. E. Ringwood, "Effects of Nuclear Radiation on the Crystal Structures of Zirconolite and Perovskite," Geochem. J., in press.
9. V. M. Oversby and A. E. Ringwood, "Lead Isotopic Studies of Zirconolite and Perovskite and Their Implications for Long Range SYNROC Stability," J. Waste Manage., in press.
10. K. D. Reeve and J. Woolfrey, "Accelerated Irradiation Testing of SYNROC Using Fast Neutrons," Aust. Ceram. Soc. J. 16: 10 (1980).
11. J. C. Dran, M. Maurette and J. Petit, "Radioactive Waste Storage Materials: Their α-Recoil Aging," Science 290: 1518.
12. J. C. Dran, M. Maurette, J. Petit and B. Vassent, "Radiation Damage Effects on the Leach Resistance of Glasses and Minerals: Implications for Radioactive Waste Storage," this volume (see index).

MATRIX-ENCAPSULATED WASTE FORMS: APPLICATION TO IDEALIZED SYSTEMS, COMMERCIAL AND SRP/INEL WASTES, HYDRATED RADIOPHASES AND ENCAPSULANT PHASES

Rustum Roy, E.R. Vance, G.J. McCarthy* and W.B. White

Materials Research Laboratory, The Pennsylvania State University, University Park, PA 16802

INTRODUCTION

This paper describes the encapsulation strategy as applied to microscopic-scale encapsulation in ceramics composed of micron-sized grains of possibly more leachable radiophases intimately surrounded by micron-sized grains of more insoluble phases.

We define "radiophase" explicitly as any phase (in the phase rule sense) which incorporates radionuclides into its structure. Figure 1 draws the distinction between two strategies for radionuclide immobilization in oxide phases (whether glassy or crystalline). In the strategy indicated on the left, the radiophases must each be very insoluble. This is the basis of most present glass and ceramic waste form thinking. On the right is shown the encapsulated waste form theory where the radiophase is physically surrounded by inert phases containing no radionuclides. The advantage of the latter strategy is immediately obvious; one can encapsulate a more soluble radiophase by one or more of the many nearly totally inert ceramic phases available.

In the following we address the following questions:

a) Under what conditions of radiophase fractional content (commercial or defense wastes, etc.) is the encapsulation strategy recommended. What are viable encapsulants?

---

*Now at Departments of Chemistry and Geology, North Dakota State University, Fargo, ND 58105.

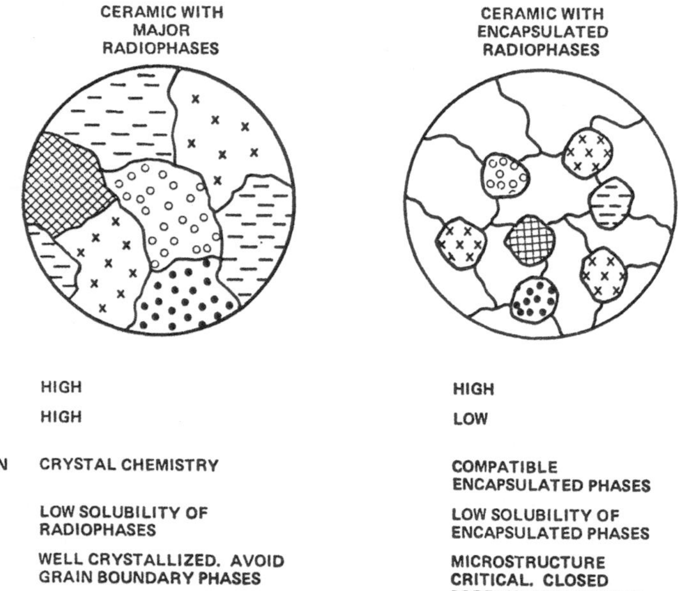

	CERAMIC WITH MAJOR RADIOPHASES	CERAMIC WITH ENCAPSULATED RADIOPHASES
WASTE LOADING	HIGH	HIGH
RADIONUCLIDE LOADING	HIGH	LOW
CHEMICAL DESIGN CONSTRAINTS	CRYSTAL CHEMISTRY	COMPATIBLE ENCAPSULATED PHASES
PROTECTION	LOW SOLUBILITY OF RADIOPHASES	LOW SOLUBILITY OF ENCAPSULATED PHASES
PROCESSING	WELL CRYSTALLIZED. AVOID GRAIN BOUNDARY PHASES	MICROSTRUCTURE CRITICAL. CLOSED PORE VOLUME NEEDED

Fig. 1. Two strategies for radionuclide immobilization in oxide phases (idealized situations shown for simplicity).

b) How do the properties of encapsulated waste forms compare to others for commercial? And how <u>can</u> they be compared?

c) How would the processing compare in complexity to making glasses or other ceramics?

We also point out for the first time thermodynamic advantages of hydrated phases over virtually all anhydrous candidate radiophases of use for the system of commercial radionuclides. The encapsulation concept is not new. At some level (micro) it was evident in the work of Evans and MacDonald[1], and their clay adsorption. It was explicitly used at a more macroscopic level in the Arrance patent[2]. The Belgian PAMELA[3] process relies on it at the 1 cm level, and is an example of a highly developed encapsulated waste form. The advantages of the hydrated waste forms have not previously been explicitly enunciated, although the works of Schulz and Kupfer[4] and Barney[5], on the so-called aqueous silicate process have utilized such phases.

EARLY EXPERIMENTS WITH ENCAPSULATION OF COMMERCIAL WASTES

In 1971-72 this approach of encapsulation of commercial

Table I. Results of Hot-Pressing Using Ceramic Matrices with PW-4 Mix.

Matrix	X-ray Phase Assemblage	Degree of Consolidation
Sand	S-α-Quartz + M-X	Excellent
Mullite	S-PW-4m + M-Mullite + W-X	Poor
Alumina	S-α-Al_2O_3 + S-PW-4m	Poor
Ludox*	S-Cristobalite + M-X	Excellent
Silicic Acid	S-Cristobalite + M-X	Average
Mullite Gel	S-PW-4m + M-X	Good
1:1 Gel**	S-PW-4m	Good
Alumina Gel	S-PW-4m	Poor

*Colloidal SiO_2 sol in ammonia (duPont). **1 $Al_2O_3 \cdot 1SiO_2$.
For x-ray phase assemblage, S = strong, M = medium, W = weak.
X = unknown phase(s) characteristic of reaction between matrix and PW-4 mixture.

waste* calcine in a series of matrices was explored with preliminary experiments[6]. Table I summarizes this work, in which both low-melting glasses and a variety of obvious ceramics such as SiO_2, Al_2O_3, etc., were used.

PW-4 type calcine was prepared by our "supercalcine" solution mixing technique[7]. Vacuum hot-pressing at temperatures up to 1000°C produced poor consolidation. First attempts at encapsulation utilized 1:1 mixtures of Corning codes 7720 and 7740 frits and PW-4 calcine. The hardest, densest pellets resulted after pressing at 780° and 825°C, respectively. No measurable weight loss of the pellets occurred after 48 hour exposure to 40°C distilled water. This early study was conducted before our leach-testing laboratories were established; weight loss was the only indication of leaching resistance.

Other encapsulants were also evaluated: α-quartz[8], alumina, mullite, amorphous SiO_2, alumina gel and (Al_2O_3 + SiO_2) gel. The hot pressing routine was 1000°C/10 min/2500 psi. With 1:1 mixtures of encapsulant and PW-4 calcine, good consolidation was obtained with α-quartz and amorphous silica matrices, but not with the alumina or mullite. In the latter cases, somewhat better products were obtained with gels than with crystalline powders. For the

---

*At that time designed as a candidate waste form for the Retrievable Surface Storage Facility where high loading and minimum volume were desirable.

α-quartz[8] and the amorphous silica, it was clear from x-ray diffraction that some reaction between the waste calcine and the encapsulated phases had occurred; much of the amorphous silica had crystallized to form cristobalite. This seemed to be a pre-condition for consolidation. However, there was some reaction between the PW-4 and the mullite, yet consolidation was poor. Thus an additional liquid-phase sintering mechanism would seem to be operative in the silica encapsulation. No weight losses were observed in the aforementioned leach test with silica encapsulation, but losses were observed in the other materials--in general, the poorer the consolidation, the greater the weight loss in the leach test.

Waste loading was clearly too high in these early tests. In a follow-up study[9] supported by the Atomic Energy Commission through Battelle, Pacific Northwest Laboratories, the same simulated waste was used along with the promising α-quartz matrix. Waste loading was dropped to ∿25 wt % and a small amount (∿7 wt %) of a lead borosilicate glass frit was added to provide controlled liquid phase sintering. The processing conditions were 1200°C/10 min/3000 psi. In the standard Soxhlet test, the products had leach resistances comparable to those of borosilicate waste glasses[9].

THE ENCAPSULATED WASTE FORM: MECHANISM OF PROTECTION

Figure 1 shows that the (conceptual) encapsulated waste form provides protection against radiophase dissolution by surrounding all radiophases with inert matrices. Ideally there should be zero <u>open</u> porosity. Of course, if the concentration of radiophase were high enough one could obtain a connected series of radiophase crystals leading to the surface. Taking the simplest case of equally-sized cubes, standard percolation theory[10] shows that it would require some $\gtrsim$20 vol.% radiophase to reach the connected radiophase condition. Thus if one keeps encapsulant and radiophases approximately the same size all wastes with low concentration of radiophases should be very effectively handled by this strategy.

DEFENSE WASTES: EXPERIMENTAL DATA

Unlike commercial-reprocessed waste, which contains $\geq$50 wt % of fission-products and actinides, high-level defense wastes contain only ∿0.1 wt % of fission products. Clearly then all defense wastes become significant candidates for this strategy. No data exist on the phase composition and microscopic nature of, say, the Savannah River sludges. However, earlier work[11] would suggest that gels of the trivalent elements Al, Fe, aged in high ionic concentration at modest temperatures, would have transformed to a γ-series oxyhydroxide of the boehmite-lepidocrocite family. Such a phase could not incorporate ions such as Cs^{+}, Sr^{2+} and Pu^{4+}, etc. Most of these ions would be either in solution, or present as other radiophases. Upon heating at moderate temperatures (even lower hydrothermally) one could therefore create a large (99:1) preponderance of inert encapsulant phases (Al_2O_3 and/or AlO·OH). Compositional design would therefore involve two objectives: tailoring additives to react with Cs, Sr, etc., to produce the most insoluble radiophases, and adding further to the stable encapsulating phases.

Savannah River Simulations

Preliminary experiments in ceramic production from simulations of Savannah River waste sludges, type compositions of which are widely published[12], showed spinel and hematite/corundum to dominate the phase assemblages. Each of these should be leach-resistant since they occur as resistate minerals. Throughout the experimentation we utilize the sol-gel technique developed by Roy et al. in 1952[13] and employed routinely in these laboratories for preparing both ceramics and glasses. Such mixing via solution allows one to coat any radiophase particles at an atomic level. However, even with tailoring additions of Al_2O_3 and SiO_2 (at the solution stage) and Si metal to act as an O_2 buffer, the Fe-rich versions of the sludge yielded the relatively soluble uranates as one of the radiophases, and it was difficult to crystallize the Na into nepheline. Little difference in the phase assemblages of the materials was observed whether they were sintered in Ar or air, or were vacuum hot-pressed in graphite dies. Firing temperatures ranged to ∿1000°C for Fe-rich material, to ∿1300°C for Al-rich preparations. However, the use of Fe metal as an O_2 buffer allowed the leach-resistant[14] UO_{2+x} to form in both argon-sintered and vacuum hot-pressed preparations. The phase assemblages of the various simulations are shown in Table II. No particularly undesirable inert phases were observed, apart from in the air-fired materials. Hydrothermal leaching experiments were carried out for seven days at 300°C/300 bars in deionized water. X-ray diffraction showed all phases except the nepheline to be unaltered. No disintegration of monoliths took place, though alteration of the surfaces was observed--the nepheline was largely transformed into analcime. Solution analyses showed the predominant leached species to be Na, with virtually no solution of Al or Fe. If the nepheline → analcime transformation also takes place at more realistic (∿ 100°C) leaching temperatures, an Na-removal step in the processing sequence may be advantageous. However, the nepheline may be dilute enough to not form a connected network and it may not carry much Cs^+. Soxhlet and Paige tests are under way.

We have found that the Cs-loaded chabazite, a possible agent for clean-up of the supernate, allows near-quantitative retention of Cs on heating to 1100°C, so heating the zeolite with the sludge in calcination or final firing in an open system would not lead to volatility problems from this source. The retention of Cs after the zeolite has reacted with the sludge is another matter. In any

Table II. Phases in Simulated SRP Wastes (Second-Generation).

Starting Material	Firing Conditions		Phase Assemblage
High-Iron* (+4% Al_2O_3 + 14% SiO_2 + 2% Fe)	Air	1000°C	Nepheline, Na/Ca uranate, plagioclase, spinel, hematite
	Argon	1000°C	Nepheline, UO_{2+x}, spinel, fcc metal
	Hot-press	1000°C	Nepheline, UO_{2+x}, spinel, fcc metal
High-Alumina* (+5% SiO_2 + 2% Fe)	Air	1200°C	Spinel, corundum, amorphous
	Argon	1200°C	Plagioclase, corundum, spinel, UO_{2+x}
	Hot-press	1200°C	Nepheline, corundum, spinel, UO_{2+x}, plagioclase
Composite* (+4% Al_2O_3 + 14% SiO_2 +2% Fe)	Air	1050°C	Spinel, plagioclase, hematite, corundum, nepheline, Na/Ca uranate
	Argon	1050°C	Two spinels, UO_{2+x}, nepheline, FeO
	Hot-press	1050°C	Two spinels, UO_{2+x}, nepheline, FeO

Percentages are in mole %. A few weak x-ray reflections remain unidentified. *See ref. 12; no Al removal.

event, the ceramic approach would very likely utilize closed-system firing such as hot isostatic pressing. It also seems likely that pollucite, a good candidate for ^{137}Cs fixation[15,16] would be formed in our simulations, from compatibility studies on air-fired preparations, but this needs further investigation.

INEL Simulations

The Al_2O_3-rich waste stream has a similar chemical make-up to that of SRP sludge so we would anticipate similar behaviors. The zirconia calcines are quite different, however. Pellets fired at 800°C with the aid of binders (Berreth, private communication; see also ref. 17) were obtained by courtesy of INEL and x-ray diffraction showed they consisted mainly of CaF_2, monoclinic ZrO_2 and fluorapatite. This agrees with INEL experience (Valentine, private communication). Firing in air at temperatures $\geq$1000°C caused nearly complete F_2 loss, which is probably undesirable even though some of the F_2 is lost in vitrification of zirconia calcine[18].

Vacuum hot-pressing at 1100°C or firing at temperatures up to 1300°C in welded Pt capsules produced negligible modification of the phase assemblage; the hot-pressed material was reasonably consolidated and Soxhlet testing showed a leach rate of $\sim 5 \times 10^{-5}$ g/cm^2·day for the first three days. These results are comparable to those on the vitrified material[18]. Analysis of the leachates is under way, as is hydrothermal testing.

FUNDAMENTAL ADVANTAGES OF HYDRATED RADIOPHASES AND ENCAPSULANTS

Finally we introduce here a fundamental consideration regarding the design of materials to last for millenia in the near-surface terrestrial environment. Details will be published elsewhere. The most likely environments for radwaste disposal are the seabed and either a silicate or halide host rock. One must assume that in any kind of failure we will generate an environment which will be characterized in the near field by modest temperatures (100-300°C) and the presence of water. If we are interested in oxide materials of maximum thermodynamic stability in these environments it is evident from the thousands of extant phase diagrams that the vast majority of solid phases in equilibrium with an aqueous phase under these conditions are hydrated (or more accurately hydroxylated). The exceptions are the tetravalent oxides SiO_2, TiO_2, ZrO_2, ThO_2. But except for the [fluorite] host for U^{4+} and the rare earths, these phases cannot serve as radiophases, but could make excellent encapsulants. All other radiophase assemblages would be less soluble if one or more phases were hydrated. Boehmite (AlO·OH) is even less soluble in H_2O (near neutral pH) than corundum (Al_2O_3). Such empirical data are widely available. See for example the detailed solubility data by the NBS group[19], for the system Al_2O_3-CaO-H_2O, and the dozens of phase diagrams in systems involving two or more common oxides and H_2O produced in the fifties and sixties by the Geophysical Laboratory in Washington and The Pennsylvania State University.[20]

What this redirection of emphasis indicates is that those interested in tailormaking radiophases (and encapsulant phases) should be looking at other candidate phases. Among these especially the amphiboles and layer structures are prime candidates, and an enormous literature already exists on the substitution of various ions in synthetic versions of such phases. Because of the XII, VI and IV-fold sites in micas and clays, as compared to only the IV-sites and the large cages in the zeolites, the former will be the most important. These structures also offer the enormous advantage of utilizing existing precursor templates to adsorb specific ions to more easily prepare the desired phase, even where metastable. As for all radiophases, the effects of the appropriate levels of radiation on the phases chosen will need to be studied.

Hints already exist in the pleochloric halos in mica that crystal structural damage is not particularly different from that in anhydrous phases and may indeed be less severe.

The advent of the technology of hydrothermal ceramics for refractories in refineries, and based on the relatively crude yet widespread experience with hydrothermally reacted "autoclaved" cement blocks signals the technological feasibility of such approaches. The Oak Ridge group[21] have already demonstrated this for the case of the autoclaved concretes.

CONCLUSION

The encapsulation approach should be valid, almost axiomatic, for defense waste. However, there are still problems to be investigated experimentally. These are (a) because of the dilution, it is difficult to confirm the geometry of the radionuclide-bearing phases relative to that of the matrix: one almost has to use the inverse approach by making leach measurements, (b) deciding between using the highly reactive oxyhydroxide sludges themselves or sintered calcine to be coated, (c) verification of the insolubility of the encapsulant phases in a variety of groundwaters, and (d) the production of ceramics of near-zero porosity, using hot-isostatic pressing, or incorporation in either silicate or phosphate cements.

REFERENCES

1. E.J. Evans and J.F. MacDonald, Hot Pressed Silicate Materials for the Fixation of Fission Products. CRER-795 (1959).
2. F.C. Arrance, "Method for Disposal of Radioactive Waste and Resultant Product," U.S. Patent No. 3093593 (1963).
3. W. Heimerl, Solidification of HLW Solutions with the PAMELA Process, in: "Ceramics in Nuclear Waste Management," T.D. Chikalla and J.E. Mendel, Eds., Dept of Energy Publication CONF-790420 (1979), p. 97.
4. W.W. Schulz and M.J. Kupfer, Solidification and Storage of Hanford's High-Level Radioactive Liquid Wastes, in: "High-Level Radioactive Waste Management," M.H. Campbell, Ed., American Ceramic Society, Washington, DC (1976), p. 54.
5. G.S. Barney, Fixation of Radioactive Waste by Hydrothermal Reaction with Clays, ibid., p. 108.
6. G.J. McCarthy and M. Lovette, Use of Hot-Pressing to Fix Solid Radioactive Wastes in Glass, Bull. Amer. Ceram. Soc. 51:655 (1972).
7. G.J. McCarthy, High Level Waste Ceramics: Materials Consolidation, Process Simulation, and Product Characterization, Nucl. Tech. 32:92 (1977).
8. G.J. McCarthy, Quartz Matrix Isolation of Radioactive Wastes, J. Mat. Sci. 8:1358 (1973).
9. G.J. McCarthy and M.T. Davidson, Ceramic Nuclear Waste Forms:

II. A Ceramic-Waste Composite Prepared by Hot Pressing, Bull. Amer. Ceram. Soc. 55:190 (1976).
10. J.W. Essam, Percolation Theory, Repts. Prog. Phys. 43:883 (1980).
11. G.W. Oosterhout, Morphology of Synthetic Submicroscopic Crystals of α- and γ-FeOOH and of γ-Fe_2O_3 Prepared from FeOOH, Acta Cryst. 13:932 (1960).
12. J.A. Stone, S.T. Goforth, Jr., and P.K. Smith, Preliminary Evaluation of Alternative Forms for Immobilization of Savannah River Plant High-Level Waste, DP-1545, E.I. DuPont de Nemours and Co., Savannah River Laboratory, Aiken, SC (1979).
13. R. Roy, Aids in Hydrothermal Experimentation: II. Methods of Making Mixtures for Both 'Dry' and 'Wet' Phase Equilibrium Studies, J. Amer. Ceram. Soc. 39:145 (1956).
14. D.E. Grandstaff, A Kinetic Study of the Dissolution of Uraninite, Econ. Geol. 71:1493 (1976).
15. S. Komarneni, G.J. McCarthy and S.A. Gallagher, Cation Exchange Behavior of Synthetic Cesium Aluminosilicates, Inorg. Nucl. Chem. Lett. 14:173 (1978).
16. D.M. Strachan and W.W. Schulz, Characterization of Pollucite as a Material for the Long-Term Storage of Cesium-137, Bull. Amer. Ceram. Soc. 58:865 (1979).
17. K.M. Lamb. S.J. Priebe, H.S. Cole and B.D. Taki, A Pelleted Waste Form for High-Level ICPP Wastes, in: "Ceramics in Nuclear Waste Management," T.D. Chikalla and J.E. Mendel, Eds., Dept. of Energy Publication CONF-790420 (1979), p. 224.
18. J.R. Berreth, D. Gombert, II, and H.J. Cole, Vitrification of ICPP High-Level Zirconia Calcine, in: "Ceramics in Nuclear Waste Management," T.D. Chikalla and J.E. Mendel, Eds., Dept of Energy Publication CONF-790420 (1979), p. 183.
19. L.S. Wells, W.F. Clarke, and H.F. McMurdie, J. Res. NBS 30:403 (1943).
20. E.M. Levin, C.R. Robbins and H.F. McMurdie, "Phase Diagrams for Ceramists," M.K. Reser, Ed., Amer. Ceram. Soc., Columbus, OH (1964).
21. H.O. Weeren, J.G. Moore and E.W. McDaniel, Waste Disposal by Shale Fracturing at ORNL, in: "Scientific Basis for Nuclear Waste Management," G.J. McCarthy, Ed., Plenum, NY (1979), p. 257.

PHASE EQUILIBRIA LEACHING CHARACTERISTICS AND CERAMIC PROCESSING OF SYNROC D FORMULATIONS FOR U.S. DEFENSE WASTES

H. Newkirk, F. Ryerson, D. Coles, C. Hoenig,
R. Rozsa, C. Rossington, F. Bazan, and J. Tewhey

Lawrence Livermore National Laboratory
Livermore, California 94550

INTRODUCTION

High-level U.S. defense wastes consist of reprocessed wastes which have been neutralized with excess NaOH. The neutralization process has resulted in the precipitation of an insoluble sludge containing most of the radionuclides except cesium, which is present in a supernatant salt solution. Defense wastes have been generated by a variety of reactor and separations processes, therefore, they tend to be compositionally heterogeneous.

The SYNROC concept, as applied to commercial high-level nuclear waste, was introduced by Ringwood in 1978[1]. The SYNROC formulation for immobilization of commercial wastes, designated SYNROC C, consists of the three-phase assemblage, hollandite ($BaAl_2Ti_6O_{16}$), perovskite ($CaTiO_3$), and zirconolite ($CaZrTi_2O_7$). An alternative SYNROC formulation for the immobilization of high-level defense wastes, designated SYNROC D, was proposed by Ringwood in 1979[2,3]. The assemblage of coexisting phases in SYNROC D are perovskite, zirconolite, nepheline ($NaAlSiO_4$) and spinel ($R^{2+}O \cdot R_2^{3+}O_3$). Cesium from the supernate is to be immobilized in hollandite. In the current processing scheme, presynthesized granules of hollandite will be added to calcined SYNROC D powders prior to hot processing or sintering.

DISPOSITION OF DEFENSE WASTES COMPONENTS IN SYNROC D

The major components of Savannah River Plant (SRP) sludge calcines and their tank-to-tank limits of variation are given below[4]. The principal radwaste components in the sludge are fission products; ^{90}Sr,

* Work performed under the auspices of the U.S. Department of Energy by the Lawrence Livermore National Laboratory under contract number W-7405-eng-48.

^{147}Pm, ^{137}Cs, ^{151}Sm, ^{134}Ce, and ^{144}Ce. The principal alpha emitters are ^{238}Pu, ^{241}Pu, ^{239}Pu, ^{240}Pu, and ^{244}Cm although the gross alpha level is low, 0.5 mCi/gallon of sludge[5]. ^{137}Cs is the major radwaste component in the supernate.

Component Average Composition		Limits of Variation (wt%)
Fe_2O_3	5.8 - 57.7	39.4
Al_2O_3	5.3 - 83.4	30.9
MnO_2	3.9 - 10.9+	10.9
U_3O_8	1.4 - 13.4	3.6
CaO	0.4 - 3.9	2.9
NiO	0.9 - 9.9	4.9
SiO_2	0.4 - 0.9+	0.9
Na_2O	2.2 - 5.6+	5.6
Na_2SO_4	0.5 - 1.0+	1.0
		100.1

The disposition of inert and radwaste components of SRP wastes in SYNROC D formulations has been determined by means of optical microscopy, XRD, XRF, SEM, STEM, electron microprobe analysis and autoradiography. A summary of results are presented herein.

Sodium

Although sodium is present in SRP sludges in relatively small amounts, it is of some concern to SYNROC developers because it does not exhibit extensive solid solution in the titanate or spinel phases of SYNROC D. The important criteria for a suitable sodium host are that it be (1) thermodynamically compatible with the SYNROC D phase assemblage, (2) exhibit a resistance to leaching, and (3) require a minimum of additive components for its formation. Nepheline appears to fulfill these criteria. Nepheline contains 22 wt% sodium, therefore, a SYNROC D formulation consisting of 25% nepheline will immobilize the 5.6% sodium present in an average sludge. Nepheline stoichiometry requires the addition of silica at a 2:1 wt ratio to sodium and, in the case of alumina-deficient sludges, small additions of Al_2O_3. Experimental work at Australian National University (ANU) has shown that sodium can be effectively immobilized in rare earth or niobium perovskites ($Na_{.5}REE_{.5}TiO_3$ or $NaNbO_3$) if the sodium content of the sludge is ≤ 2 wt%[6].

Uranium and Other Actinides

Uranium replaces Zr in zirconolite, forming a $CaUTi_2O_7$ component. Uranium can also replace Ca in zirconolite or perovskite with charge balance achieved by substitution of Al^{3+} and/ or Fe^{2+} for Ti^{4+}. Actinide-doping experiments on SYNROC C formulations with ^{239}Pu, ^{241}Am, and ^{244}Cm have been done at Oak Ridge National Laboratory. Alpha autoradiography analysis has shown the actinides to be equally partitioned in zirconolite and perovskite. The other coexisting SYNROC phases were found to be actinide-free[7]. Rare earth elements are also partitioned equally between zirconolite and perovskite.

Strontium

In SYNROC C and SYNROC D formulations, Sr is strongly partitioned into perovskite with lesser amounts going into zirconolite. Experiments at ANU with a two-phase SYNROC formulation (perovskite absent) have shown that zirconolite can accept up to 1.4 wt% Sr into solid solution. SYNROC D formulations made from SRP sludges containing 10 wt% U_3O_8 will have high percentages of zirconolite and minor perovskite. Zirconolite will be the principal host for Sr in waste forms with high U content.

Cesium

The small amount of Cs that is present in SRP sludge is partitioned into a silicate phase. Cs has been observed in dilute solid solution in nepheline and as tiny, discrete blebs of pollucite ($CsAlSiO_6$) coexisting with nepheline. In the processing scheme currently being developed for SYNROC D, the Cs from the supernatant solution will be incorporated into hollandite. In current experiments, 15 wt% of presynthesized hollandite containing 3.4 wt% Cs is added in millimeter sized "chunks" to SYNROC D (Fig. 1a, b). The Cs concentration in the final waste form is 0.5 wt%. These Cs concentrations have not been "optimized" and are subject to change based on phase equilibria and production technology criteria.

Iron, Aluminum, Manganese, and Nickel

Fe, Al, Mn, and Ni are the principal inert components in SRP sludge, collectively comprising from 75 to 95 wt% of the sludge. Each of these components can exhibit order-of-magnitude variation from tank to tank. The inert components form spinel solid solutions which can comprise as much as 60% of the final waste form. Titanium is also present as an ulvospinel component ($2R^{2+}O \cdot TiO_2$) in solid solution. Heat treatment and processing of SYNROC D near the Ni-NiO buffer ($f_{O_2} = 10^{-9.5}$ at 1050°C) is within the magnetite ($FeO \cdot Fe_2O_3$) field, thereby facilitating the synthesis of spinels. As a consequence of extensive solid solution between end-members, spinel is a "forgiving" phase in SYNROC D. The principal role of the SYNROC additives in SYNROC D formulations is to synthesize the radwaste-containing phases. That being the case, the amounts and proportions of additives are similar for the wide variety of SRP sludge compositions. The major inert components in the sludge are relatively insensitive to the additive components (except TiO_2) and end up as a complex spinel solid solution that is compatible with the bulk chemistry of the sludge (Fig. 2).

Fig. 1a. SEM photomicrograph of SYNROC D microstructure. Phases present are zirconolite (white), perovskite (medium gray euhedral crystals), spinel (medium gray irregular crystals) and nepheline (black). Scale bar at upper right is 5 microns.

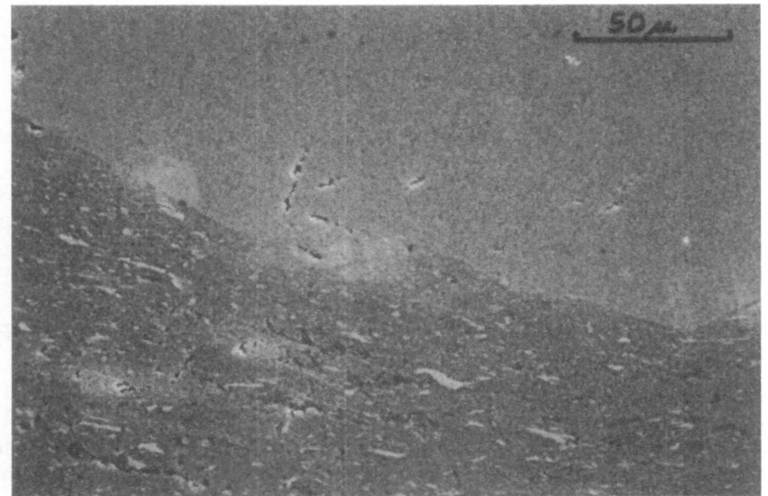

Fig. 1b. SEM photomicrograph of SYNROC D with granule of presynthesized hollandite (top, light gray). Matrix of SYNROC D consists of zirconolite, perovskite, spinel, nepheline and nickel metal. Scale bar is 50 microns.

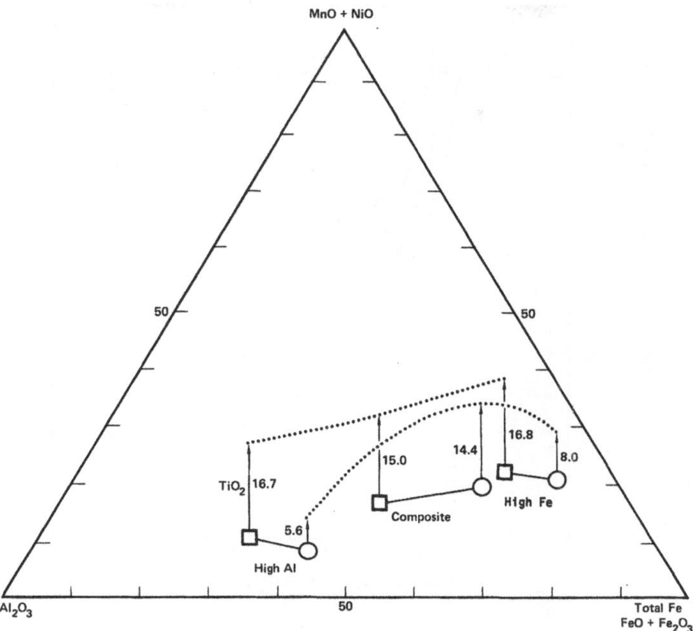

Fig. 2. Bulk chemistry of three SRP sludges plus SYNROC additives (squares) and the corresponding spinel compositions (circles) are plotted with respect to Al_2O_3 - (FeO + Fe_2O_3) - (MnO + NiO). Spinel compositions are iron-rich with respect to bulk compositions because aluminum goes into nepheline and zirconolite. TiO_2 content of spinels and sludge/SYNROC D calcine are indicated graphically (arrows) and numerically. Spinels have a wide compositional range and are considered to be a "forgiving" phase.

SYNROC Additives

The principal SYNROC additives in SYNROC D formulations are TiO_2, ZrO_2, SiO_2, CaO, and in the case of the extreme compositions, Fe_2O or Al_2O_3. The latter are added to assure the synthesis of spinel. A small amount of Ni powder is added as a "getter" for excess oxygen. The amount of additives is dependent on (1) sodium content of the sludge, (2) uranium content and the level of uranium loading in zirconolite, and (3) an arbitrary zirconolite/perovskite ratio. A fixed amount of TiO_2 is added for incorporation into the spinel-ulvospinel$_{ss}$ (Fig. 3).

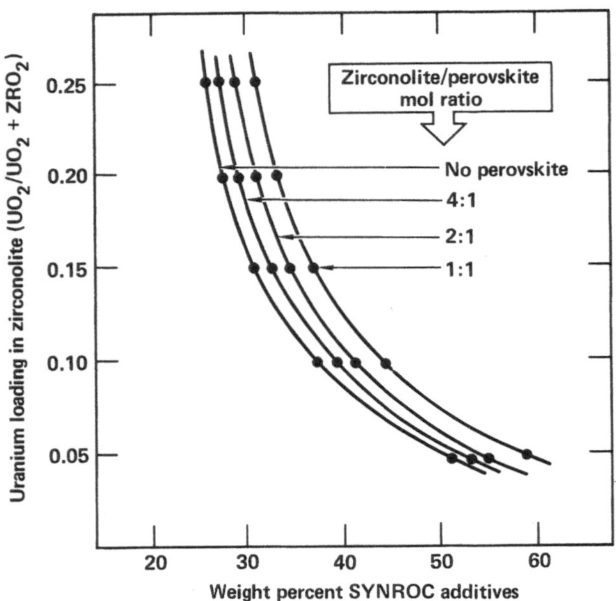

Fig. 3. Weight percent SYNROC D additives for SRP composite sludge plotted as a function of uranium loading in zirconolite and the desired zirconolite/perovskite ratio. Additives are TiO_2 (Z,P, spinel), CaO (Z,P), ZrO_2 (Z), and SiO_2 (nepheline).

LEACHING OF SYNROC D

Leaching studies of SYNROC D have been done by means of static, high temperature experiments and continuous-flow experiments (MCC-4)[8]. The data reported in Table 1 are from high-temperature experiments (distilled water, powdered sample, 150°C, one day). The elements reported are the only ones observed in the leachate. Analysis was done by means of XRF.

Table 1. Leaching Results of SYNROC D and Cs-Bearing Hollandite

Leach rate = g SYNROC or hollandite/$m^2 \cdot$ day

	Ca	Sr	U	Cs
SYNROC D (Z,P,N,S)	$<8 \times 10^{-3}$	7×10^{-4}	$<1 \times 10^{-4}$	
Cs-Bearing Hollandite				7.8×10^{-3}

CERAMIC PROCESSING OF SYNROC D

The flow sheet in Figure 4 depicts the current experimental methods that are being employed at LLNL to produce SYNROC D samples containing presynthesized Cs-bearing hollandite. The starting material for SYNROC D (high Fe, high Al and composite compositions) is simulated sludge obtained in 55 gallon quantities from Southwestern Chemical Corporation. Hot pressing temperatures for SYNROC D are 1000-1150°C. Hot pressing temperatures for hollandite are 1200-1400°C.

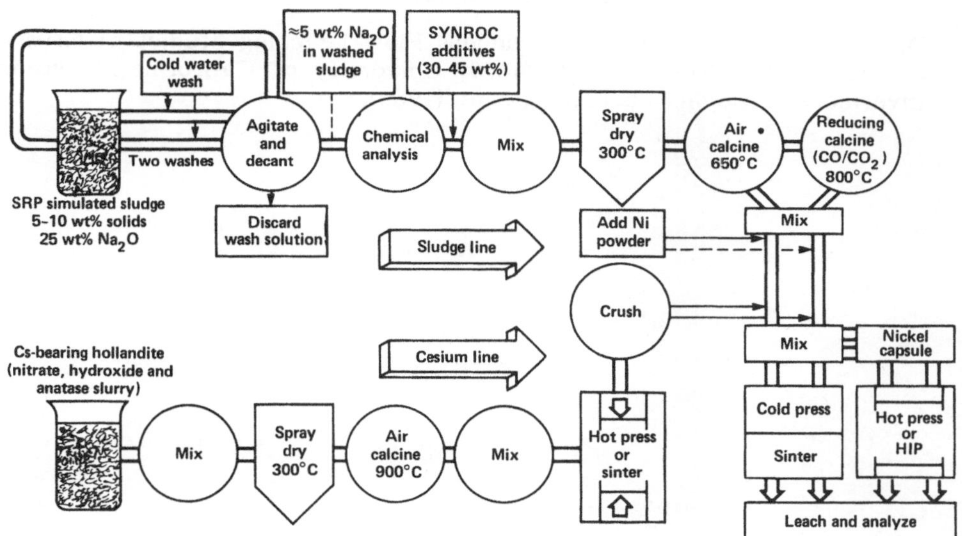

Fig. 4. Flow Diagram for Laboratory Scale Production of SYNROC D

REFERENCES

1. A. E. Ringwood, Safe Disposal of High-Level Nuclear Reactor Wastes: A New Strategy, Australian National University Press, 64 p., (1978).

2. A. E. Ringwood, S. E. Kesson, N. G. Ware, W. O. Hibberson, and A. Major, The SYNROC Process: A Geochemical Approach to Nuclear Waste Immobilization, Geochem. Journ., 13, 141-165, (1979).

3. A. E. Ringwood, S. E. Kesson, and N. G. Ware, Immobilizaton of U.S. Defense Nuclear Waste Using the SYNROC Process, in: Scientific Basis for Nuclear Waste Management, 2, C. Northrup, ed (1980).

4. J. A. Stone, S. T. Goforth, Jr., and P. K. Smith, Preliminary Evaluation of Alternative Forms for Immobilization of Savannah River Plant High-Level Waste, DP-1545, DuPont Savannah River Laboratory, 98p., (1979).

5. J. A. Stone, J. A. Kelley, and T. S. McMillan, Sampling and Analysis of SRP High-Level Waste Sludges, DP-1399, DuPont Savannah River Laboratory, 50p., (1976).

6. A. E. Ringwood, Personal Communication (1980).

7. P. Angelini, D. P. Stinton, R. W. Carpenter, J. S. Vavruska and W. J. Lackey, Phase Identification and Partitioning of Elements in Sol-Gel-Derived SYNROC, Amer. Cer. Soc. Bull., 59, 394, (1980).

8. D. G. Coles and F. Bazan, Continuous-Flow Leaching Studies of Crushed and Cored SYNROC, UCRL-84679, University of California, Lawrence Livermore National Laboratory, 29p., (1980).

SOL-GEL TECHNOLOGY APPLIED TO CRYSTALLINE CERAMIC NUCLEAR WASTE FORMS*

P. Angelini, W. D. Bond, A. J. Caputo, J. E. Mack,
W. J. Lackey, D. A. Lee, and D. P. Stinton

Oak Ridge National Laboratory
Post Office Box X
Oak Ridge, Tennessee 37830

INTRODUCTION

The sol-gel process is being developed for the solidification and isolation of high-level nuclear fuel waste. Three gelation methods are being developed for producing alternative waste forms.[1] These include (1) internal gelation for producing spheres of up to 1 mm diam suitable for coating, (2) external gelation, and (3) water extraction methods for producing material suitable for alternate ceramic processing. This report addresses the internal gelation process and the characteristics of materials produced by that method. Gel derived materials of the Synroc[2] compositions containing up to 70 wt % simulated Savannah River Plant (SRP)[3] waste have been produced.

GELATION

A generic flowsheet for the preparation of gel spheres by the internal gelation method is shown in Fig. 1. Feed solutions containing cations of the desired waste form additives and the waste components are used to prepare a feed "broth" that contains all the necessary ingredients to form gel spheres when aqueous droplets of "broth" are dispersed into a heated immiscible organic liquid. Gelation occurs by the release of ammonia from hexamethylenetetramine (HMTA) in the feed broth (Eq. 1).

*Research sponsored by the Office of Waste Operations and Technology, U.S. Department of Energy, under contract W-7405-eng-26 with Union Carbide Corporation.

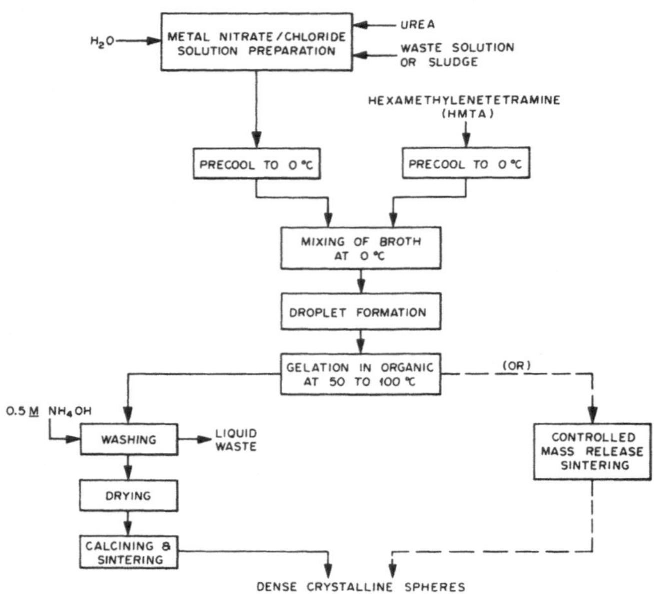

Fig. 1. Generic flowsheet for preparing spheres by the internal gelation process.

$$(CH_2)_6N_4 + 6H_2O \leftrightarrow 4NH_3 + 6CH_2O \quad . \qquad (1)$$

Urea is also present in the feed broth and is necessary to complex the metal cations and avoid premature gelation of feed broth by HMTA at 0°C. The added urea also neutralizes excess acid in the metal salt solutions and reduces the acidity from pH values below 0 to about 2. Cations of the additives in the Synroc-B and Synroc-D compositions are usually derived by dissolving the appropriate amounts of the respective salts (Zr, Al, Ca, Ba). Titanium is added as $TiCl_3$ or TiO_2 sol and silicon as SiO_2 sol. The simulated waste has been introduced either as a solution or as a slurry.

Droplets of the feed broth are formed by metering it to a vibrating nozzle submerged beneath the immiscible organic liquid surface. Droplet size is governed by the feed flow rate and the imposed vibrational frequency. The droplets gel in 5 to 15 s of free fall in a heated column of organic liquid. Gelation depends on the composition of feed solutions and the temperature of heated organic medium. Kinetic studies[4] of NH_3 formation by Eq. (1) show that the reaction is rapidly accelerated by increasing temperature, and it proceeds with an activation energy of about 100 kJ/mol. The reaction is second order, depending on both the HMTA and hydrogen ion concentrations.

The immiscible organic liquid used throughout this work was a solution of 33 vol % trichloroethylene and 67 vol % 2-ethyl-1-hexanol. About 0.5 vol % of a surfactant was added to this solution to prevent the droplets from adhering to the forming column wall. We fabricated an internal gelation system capable of producing 100 g batches of alternative waste form microspheres. The heated 1.3-m column is maintained at 45 to 50°C, which is sufficient to affect gelation of the droplets. The spheres are collected and then washed with isopropyl alcohol to remove the adhering organic layer and finally with 0.5 M NH_4OH to remove NH_4NO_3, urea, and any unreacted HMTA and additive anions such as NO_3^{1-} and Cl^{1-}. It has been found part of the soluble $Ca(OH)_2$ and $Ba(OH)_2$ is leached from the gelled spheres upon washing. Materials have been produced with Cs, Sr, Nd, Mo, and Ru dopants to study the side waste streams and the effects of drying and sintering. Both washed and unwashed gel spheres have been successfully produced.

DRYING

Drying tests investigated batches of Synroc-B (1% simulated waste) material in either the washed or unwashed state. The Synroc-B (1% simulated waste) is referred to as Synroc-B for the remainder of this report. A brief summary of these results is as follows: (1) washed spheres dried without humidity at 100 and 300°C resulted in sintered densities of 89 and 95% of theoretical, respectively; particles dried with humidity resulted in sintered densities from 94 to 97%; (2) among the different batches, the sintered density varied only 2 to 5%, (3) unwashed spheres dried in air without humdity resulted in sintered densities of 45 to 54% and produced cracked spheres (the interest in unwashed spheres is due to cesium retention), and (4) drying at different humidity levels varied the sintered density of particles by 2 to 3%. These effects will be investigated further with a new, programmable dryer.

Synroc-B spheres have been formed with dopant amounts of Cs, Ru, Nd, Sr, and Mo. Initial measurements indicated that essentially all the cesium added (0.5%) was removed during sphere washing, while the other elements were retained. Thus, the drying of unwashed spheres was initiated in order to retain the cesium. Air drying of unwashed material resulted in cracked spheres. Vacuum drying produced essentially crack-free spheres [Fig. 2(a)]. Washed, air dried [Fig. 2(b)], and sintered spheres were crack-free. The cesium content of sintered spheres appears to depend upon the sintering conditions and perhaps to a lesser extent upon the drying conditions. Spheres unwashed, dried, and sintered at 1025°C had two small and very broad unidentified x-ray diffraction peaks. At 1225°C, there were broad $CaZrTi_2O_7$ (zirconolite) peaks and another small and broad unidentified peak.

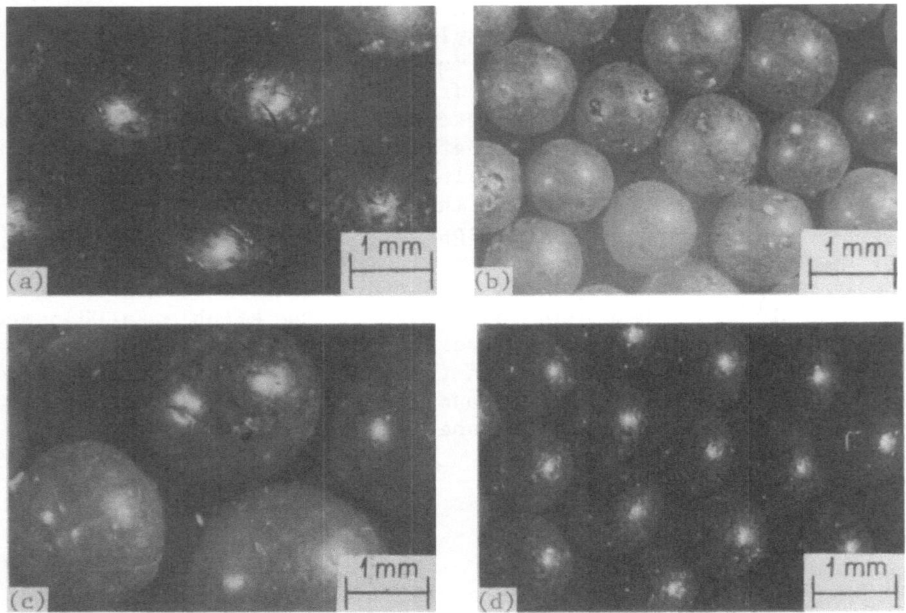

Fig. 2. As-dried surface macrograph of internal gelation derived waste forms. (a) Synroc-B (1% simulated waste) unwashed gel, vacuum dried; (b) Synroc-B (1% simulated waste) washed gel, air dried; (c) Synroc-D (70% simulated SRP waste) unwashed gel, nitrogen dried; (d) Synroc-D (70% simulated SRP waste) washed gel, air dried.

The phase identification of washed spheres is presented in the sintering section of this paper.

Drying tests were also performed on gel derived waste forms of (1) Synroc-B with about 10% simulated SRP waste and (2) Synroc-D (70% simulated SRP waste, 30% additives). The unwashed spheres of Synroc-B with 10% SRP waste showed only a small amount of cracking when dried under vacuum, and again washed spheres showed very little cracking but will probably have a low cesium content. Unwashed spheres of Synroc-D dried in nitrogen [Fig. 2(c)] and washed spheres dried in air [Fig. 2(d)] produced essentially crack-free spheres. Sintering, phase identification, and coating of gel-derived Synroc-D remain to be performed.

THERMAL ANALYSIS

Thermal analysis studies have been made on internal gelation Synroc-B microspheres. Washed and unwashed gel microspheres dried and not dried have been examined by thermogravimetric analysis (TGA), differential scanning calorimetry (DSC), differential thermal analysis (DTA), and evolved gas analysis (EGA) using mass spectrometry (MS). Water was a major component evolved — even from dried (200°C) microspheres. Thermograms from TGA and EGA-MS showed evidence for the release of different kinds of water — outer sphere or surface water, inner sphere or coordination water, and water formed from organic combustion and from hydroxide dehydration. Unwashed wet gel microspheres evolved several gases near 300°C from decomposition of urea and HMTA. Ammonia was released at two peak temperatures (160 and 220°C) — the first from HMTA hydrolysis and the second from urea decomposition. The thermal analysis of washed wet gel microspheres showed mainly the release of water at 100°C. A small water peak at about 350°C was probably due to hydrocarbon combustion. A CO_2 release at 600°C was due to calcium carbonate decomposition. Dried washed gel samples also released significant amounts of water. Water and other masses were released at higher temperatures than from the respective undried material. Activation energies were 36 kJ/mol for surface water, 179 kJ/mol for CO from urea, and 293 kJ/mol for CO_2 from organics combustion.

SINTERING

The sintering behavior of washed and dried Synroc-B gel-derived microspheres has been investigated. Results in this section considers only this type of material. Dilatometry results showed that sintering of the Synroc microspheres occurred primarily between 550 and 800°C. Shrinkage of the microspheres continued from 800 to 1000°C but at a much slower rate. The density obtained when sintering in Ar—4% H_2 or in air at 825°C was about 3.94 Mg/m^3 (94% T.D.). The maximum density (95% T.D.) was obtained near 1000°C. At higher temperatures, the microspheres did not sinter to such high densities. For example, microspheres sintered at 1125°C had densities of about 3.88 Mg/m^3 (92% T.D.), and microspheres sintered at 1225°C had densities of about 3.75 Mg/m^3 (89% T.D.). The rate used to heat the microspheres from room temperature to the sintering temperature (50 to 200°C/h) did not influence the final sintered density. The densification at all temperatures was complete in less than 4 h.

The formation of phases in internal-gelation-derived Synroc-B has been observed after sintering at various temperatures. Phase identification of batches heated to 825 and 875°C in Ar—4% H_2 showed very broad peaks of TiO_2 and $CaZrTi_2O_7$. Formation of the phases hollandite and perovskite had not begun at these temperatures. The broad x-ray diffraction peaks indicate an extremely fine grain size. Metallography of samples fired at these temperatures showed no visible grains. Sintering at 925 and

975°C in Ar-4% H_2 revealed that TiO_2 and $CaZrTi_2O_7$ were becoming more crystalline and $BaAl_2Ti_6O_{16}$ (hollandite) was beginning to form at 975°C. Metallography showed very fine grains. By 1025°C, distinct peaks of $CaZrTi_2O_7$ and $BaAl_2Ti_6O_{16}$ were present and TiO_2 had disappeared. Also present at this temperature were about 10 small peaks that could not be identified with confidence because the major peaks for the unidentified compounds were apparently obscured by coincident lines from other phases. Phase identification of microspheres sintered in Ar—4% H_2 at higher temperatures revealed the same phases. A trace of $CaTiO_3$ (perovskite) had formed by 1225°C. The shortage of $CaTiO_3$ from this batch is not understood because it has been observed by x-ray diffraction in other sol-gel batches.

Slightly different results were obtained when batches of Synroc-B were sintered in air. The same phases, TiO_2 and $CaZrTi_2O_7$, were present at temperatures up to 975°C. Again a trace of hollandite formed at 975°C. At 1025 and 1075°C TiO_2 remained the major phase along with $CaZrTi_2O_7$. There were also minor amounts of alumina and hollandite. A careful examination of the hollandite phase revealed that the composition was actually $BaAl_2Ti_5O_{14}$ instead of $BaAl_2Ti_6O_{16}$. The x-ray diffraction patterns of these phases are very similar. When batches of Synroc-B microspheres were sintered in Ar—4% H_2 at 1025 and 1075°C, the formation of the Synroc phases was complete and no TiO_2 remained.

LEACHING

Leach testing of sol-gel derived crystalline waste forms has only recently begun. Primary emphasis has been on developing the sol-gel processing techniques for producing crystalline ceramics that are known or are expected to have excellent leach resistance. Leach apparatuses are being assembled to perform tests specified by the Material Characterization Center.[5] Test results will be used in conjunction with verification of phase formation to characterize the sol-gel-derived waste form.

COATING OF GEL-DERIVED WASTE FORMS

Pyrocarbon coatings with densities less than 1 Mg/m^3 were deposited at temperatures as low as 1000°C from acetylene. The purpose of these deliberately low-density buffer coatings was to prevent cracking of the dense pyrocarbon outer layer. Cracking could otherwise result because of a thermal expansion mismatch between the Synroc kernel and the dense pyrocarbon layer. Many pyrocarbon coatings with densities of about 2.0 Mg/m^3 were applied at temperatures as low as 1100°C with acetylene and 1200°C with propylene [Fig. 3(a)]. Preliminary results indicate that very

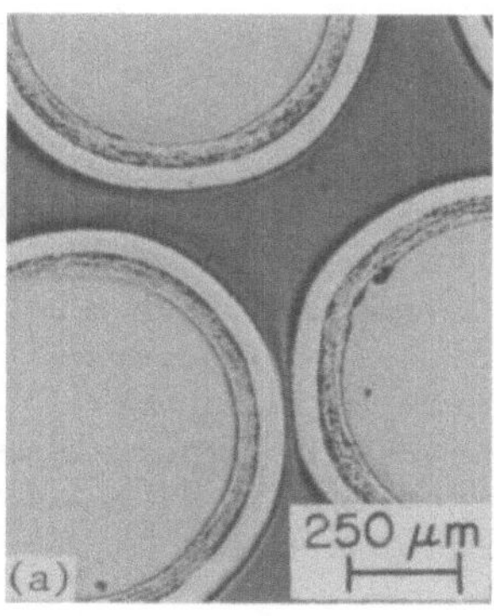

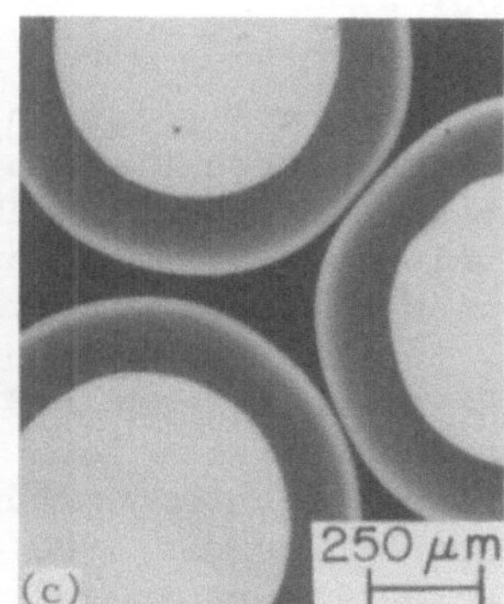

Fig. 3. Pyrolytic-carbon-coated and silicon-carbide-coated gel-derived Synroc spheres. (a) micrograph of two-layer pyrolytic carbon coating; (b) radiograph showing SiC outermost layer deposited at 1400°C; (c) radiograph, SiC outermost layer deposited at 900°C.

high-density pyrocarbon[6] was deposited at the temperatures mentioned. The results of chlorine leach at 1000°C indicated that less than one particle in 10^4 was cracked or defective. From previous experience, we expect that defective fractions of less than one particle in 10^6 can be obtained. Temperatures of 1100 to 1200°C could still be too high for some applications. Efforts are in progress to deposit coatings at lower temperatures.

Many pyrocarbon coatings have also been deposited in an attempt to contain molten Synroc during the 1400 to 1500°C SiC coating process. This has proven to be quite difficult because most pyrocarbon coatings crack when heated to 1400°C because of the large volume increase that occurs when the kernels melt. Initially 75% of the coatings cracked when the particles were heated to 1400°C. Modifications in the buffer coating procedure to produce thicker and less dense coatings have reduced the cracking to 3 to 5% of the batch. We expect to be able to eliminate all cracking in the near future.

Silicon carbide coatings have been applied to several pyrocarbon coated batches. These coatings were applied at 1400 [Fig. 3(b)] and 1500°C. It was encouraging to note that even though the kernels melted and penetrated the buffer they did not reach or react with the SiC coatings. These SiC coatings were very dense (98% T.D.) and are expected to be very leach resistant. Silicon carbide has also been applied at 700 to 1000°C [Fig. 3(c)] on pyrocarbon-coated particles. The objective of the experiment was to obtain a high-density SiC layer at the lowest possible temperature. The coatings are currently being evaluated.

SUMMARY

Internal gelation has been used to produce ceramic spheres of various alternative nuclear waste compositions. A gelation system capable of producing 100-g batches has been assembled and used for development. Waste forms containing up to 70 wt % simulated Savannah River Plant waste have been produced. Dopants such as Cs, Sr, Nd, Ru, and Mo were used in some experiments to observe side waste streams and sintering effects. Synroc microspheres were coated with both low-density carbon, high-density impermeable carbon, high-temperature dense SiC, and SiC deposited at temperatures near 900°C. Other gelation methods and other alternative waste forms are being developed.

ACKNOWLEDGMENTS

The authors would like to thank C. E. DeVore, R. Hickey, R. L. White, W. H. Elliott, and J. C. McLaughlin for their technical assistance; F. L. Layton and D. A. Costanzo of the Analytical Chemistry Division for their assistance; H. E. McCoy, L. C. Williams, W. J. Lackey, R. E. Blanco, and R. G. Donnelly for reviewing the manuscript; S. Peterson for technical editing; and Rhonda Castleberry for preparing the manuscript for publication.

REFERENCES

1. W. J. Lackey, P. Angelini, F. L. Layton, D. P. Stinton and J. S. Vavruska, Sol-Gel Technology Applied to Glass and Crystalline Ceramics, in "Waste Management '80," Vol. 2, sponsored by the University of Arizona College of Engineering and the Department of Energy, pp. 391–417 (1980).
2. A. E. Ringwood, S. E. Kesson, N. G. Ware, W. O. Hibberson, and A. Major, The Synroc Process. A Geochemical Approach to Nuclear Waste Immobilization, Geochem. J. 13(4): 141–65 (August 1979).
3. J. A. Stone, S. T. Goforth, Jr., and R. K. Smith, Preliminary Evaluation of Alternative Forms for Immobilization of Savannah River Plant High-Level Waste, DP-1545 (December 1979).
4. H. Tada, Decomposition Reaction of Hexamine by Acid, J. Am. Ceram. Soc. 82: 255–63 (1960).
5. Pacific Northwest Laboratory, Material Characterization Center Workshop on Leaching of Radioactive Waste Forms — Summary Report, PNL-3318 (April 1980).
6. D. P. Stinton and W. J. Lackey, Influence of Process Variables on Permeability and Anisotropy of Biso-Coated HTGR Fuel Particles, ORNL/TM-6087 (November 1977).

THE CHARACTERIZATION OF NUCLEAR WASTE FORMS BY EPR SPECTROSCOPY

L. A. Boatner, M. M. Abraham, and M. Rappaz

Solid State Division
Oak Ridge National Laboratory*
Oak Ridge, TN

INTRODUCTION

Electron paramagnetic resonance (EPR) spectroscopy is one of the most powerful, site-specific techniques currently available for the study of certain types of impurities and defects in solids.[1] The technique is site-specific since the EPR spectrum can reflect the local environment (e.g., the crystalline electric-field symmetry and strength) of a paramagnetic ion or defect on a microscopic basis. In an approach to the containment and isolation of nuclear waste in which the primary waste form is a crystalline substance, the basic concept is to incorporate radioactive ions in an inert crystal lattice. Ideally, in such an incorporation, the radioactive ions will occupy substitutional sites in the lattice that are normally occupied by ions that constitute the inert crystalline host material. EPR spectroscopy yields the maximum amount of information when applied to exactly this type of solid-state situation (i.e., to paramagnetic impurities diluted in a crystalline substance). Accordingly, this technique can be extremely valuable in characterizing both ceramic waste forms and crystalline systems that are analogs of resistant natural minerals. Renewed interest in both of these types of alternative waste forms has recently been stimulated by findings relative to the stability of borosilicate glasses under certain hydrothermal conditions. It should be noted, however, that EPR spectroscopy has also been extensively applied to investigations of impurities and defects in glasses and, accordingly, its application to the problem of characterizing radioactive waste forms is not restricted to crystalline materials.

*Operated by Union Carbide Corporation for the U.S. Department of Energy under contract W-7405-eng-26.

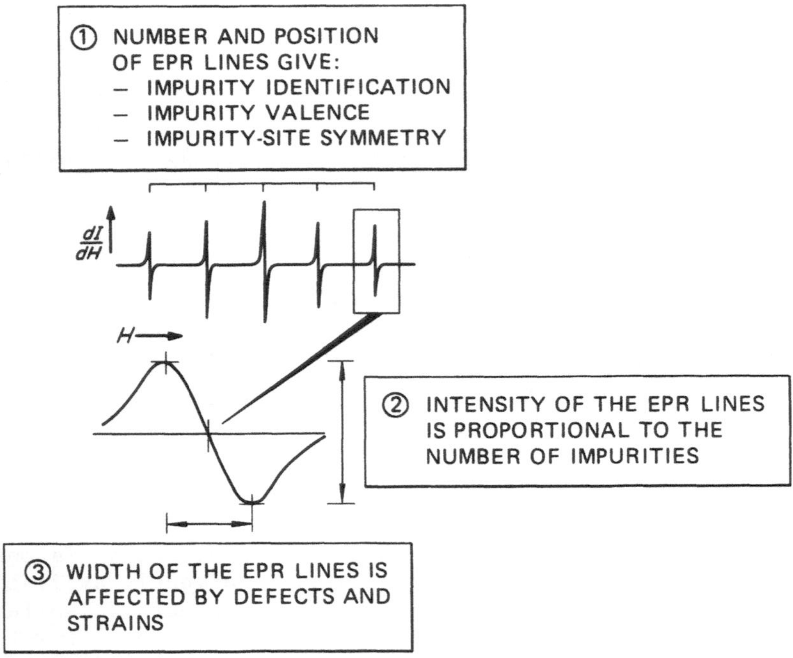

Fig. 1. The type of information provided by EPR.

Basically, EPR spectroscopy consists of the observation of magnetic dipole transitions (usually $\Delta M_S = \pm 1$) between electronic energy levels that are split by various perturbations. Most frequently, EPR deals with the electronic ground state of the paramagnetic impurity or center and an external magnetic field is usually applied in order to lift the ground state degeneracy. Transitions between the resulting magnetic-field-split Zeeman levels are then induced by photons whose frequency lies in the microwave region (e.g., 9 to 35 GHz for magnetic fields ranging between 3 and 12 kG). Since it is difficult to vary the microwave frequency in an EPR experiment, the energy difference corresponding to the various transitions is matched by varying the amplitude of the external magnetic field. Consequently, an EPR spectrum consists of one or more absorption lines whose number and magnetic-field positions will reflect: i) the type of impurity being studied, ii) the valence state of this impurity, and iii) the local crystal field symmetry at the impurity site. The type of information that can be obtained by means of EPR spectroscopy is summarized in Fig. 1. In addition to the information noted above, EPR can also be used to obtain quantitative analytical results as well as information concerning imperfections in the host material (e.g., radiation-induced defects and strains). In the following discussion, the types of information that can be obtained using EPR will be treated in more detail and specific examples will be presented.

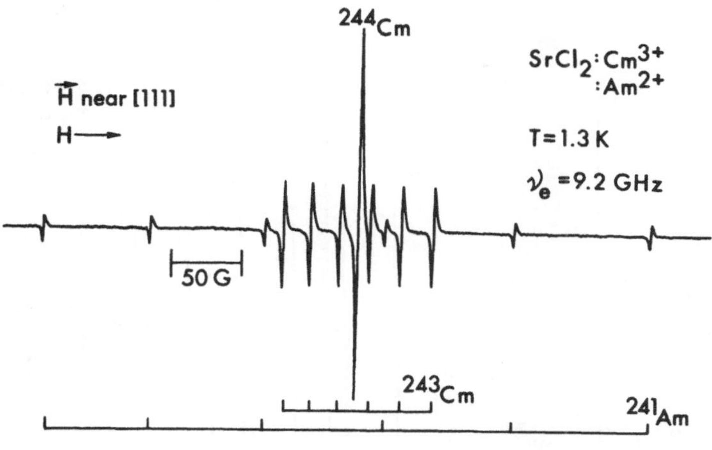

Fig. 2 EPR spectra of ^{243}Cm, ^{244}Cm, and ^{241}Am in a single crystal of $SrCl_2$.

IMPURITY IDENTIFICATION, VALENCE, AND SITE SYMMETRY

In its simplest, most direct application to the characterization of crystalline (or amorphous) waste forms. EPR specroscopy can be used as a purely qualitative analytical tool for the identification of paramagnetic impurities or defects in solids. (A paramagnetic impurity or defect arises when the net orbital and spin angular momentum of the electronic system of such an entity is nonzero.) Nuclei that have an odd number of protons or neutrons (or both) are characterized by a nonzero nuclear spin I and an associated nuclear magnetic moment $\mu = g_N \beta_N I$ where g_N and β_N are respectively the nuclear g-value and nuclear magneton. In general coupling between this nuclear magnetic moment and the "electronic" magnetic moment of the paramagnetic ion results in a so-called "hyperfine" splitting or structure in the EPR spectrum. This hyperfine structure can often be used to make a positive identification of paramagnetic impurities. As an example, the EPR spectrum of the actinides ^{243}Cm, ^{244}Cm, and ^{241}Am in the fluorite structure host $SrCl_2$ is shown in Fig. 2.[2] For ^{243}Cm, the nuclear spin I = 5/2 so that 2I + 1 = 6 hyperfine lines are observed; for ^{241}Cm, I = 0, so 2I + 1 = 1 and only one transition is observed (i.e., there is no hyperfine structure for the even-even nucleus); and for ^{241}Am, I = 5/2 so that 2I + 1 = 6 hyperfine lines are present in the spectrum of this isotope. The number of hyperfine lines, coupled with other spectroscopic information, can frequently be used to make a positive identification of impurities. In the case of naturally-abundant impurities, the intensity ratio between hyperfine lines of the various isotopes often provides a unique identification.

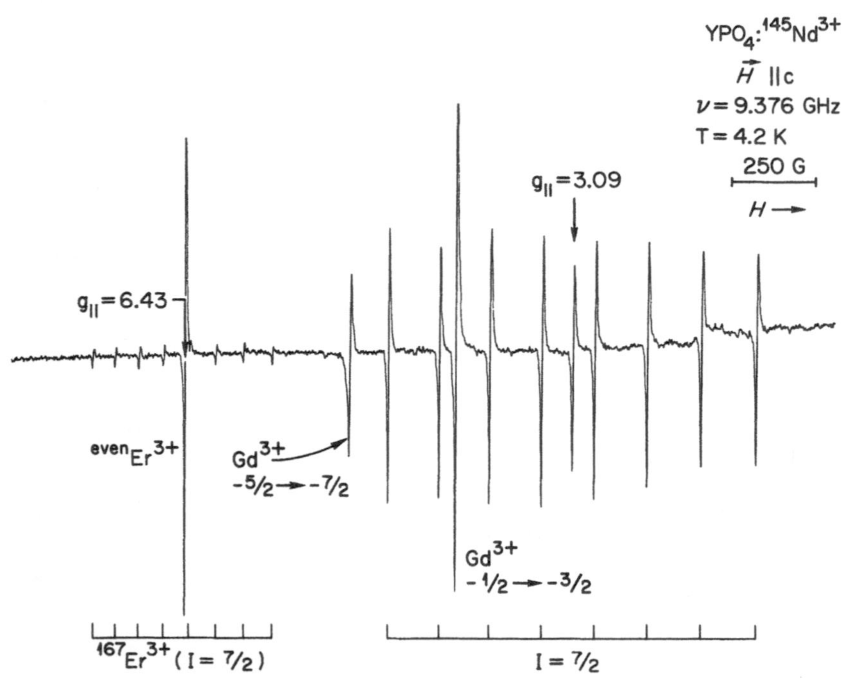

Fig. 3. EPR spectrum obtained using isotopically enriched ^{143}Nd. Transitions due to Er^{3+} and Gd^{3+} are also present.

EPR spectroscopy can be used not only in determining that a given paramagnetic impurity is present but can provide an identification of its valence state. Such an identification can be made from the number of observed EPR transitions coupled with their position as a function of the strength of the applied magnetic field and its orientation relative to the axes of a host crystal. The utility of this approach for the identification of the valence states of impurities in solids can be enhanced by the ability to dope intentionally samples with isotopically enriched elements.

An example of an EPR spectrum in which the spectral complexity has been reduced by employing an enriched isotope is illustrated in Fig. 3. Here a single crystal of YPO_4 (xenotime) has been deliberately doped with the enriched isotope ^{145}Nd. The characteristic 8-line hyperfine pattern coupled with the measured spectroscopic parameters can be used to identify the responsible ion as $^{145}Nd^{3+}$. The spectrum of Nd^{3+} is of particular interest since its electronic properties (it is characterized by a $4f^3$ electronics configuration) are analogous to those of trivalent uranium which has a $5f^3$ configuration. By investigating the specroscopic properties of Nd^{3+} it is possible to identify and investigate impurities of ^{238}U and avoid the

Table 1. PW-4b Composition and Valence States

	MOL %	VALENCE	DETECTED BY EPR		MOL %	VALENCE	DETECTED BY EPR
La	26.4	+3	+2	Pd	4.1	+2, +4	+3
Ce		+3, +4	+3	Sr	3.5	+2	
Pr		+3, +4	+3	Ba	3.5	+2	
Nd		+3	+3, +4	Rb	1.3	+1	
Pm		+3	+3	U	1.4	+3 → +6	+3, +4, +5
Sm		+2, +3	+3	Th		+4	
Eu		+2, +3	+2	Np	0.2	+3 → +6	+4, +6
Gd		+3	+3	Pu		+3 → +6	+3, +6
Tb		+3, +4	+3, +4	Am		+3 → +6	+2, +4
Dy		+3	+3	Cm		+3	+3
Ho		+3	+2, +3	Fe	6.4	+2, +3	+1, +2, +3
Er		+3	+3	Na	1.0	+1	
Tm		+2, +3	+2	PO_4	3.2		
Yb		+2, +3	+3	Tc	9.0	+7	+4
Lu		+3		Rh		+2, +3, +4	+2
Zr	13.2	+4		Te		−2, +4, +6	
Mo	12.2	+6 → +2	+3, +5	In		+3	+2
Ru	7.6	+2, +3, +4 +6, +8	+3	Ni		+2, +3	+1, +2, +3
Cs	7.0	+1		Cr		+2, +3, +6	+3, +5 (+4) +1 (+2)

necessity of employing the radioactive uranium isotopes ^{238}U or ^{235}U. It should be noted that EPR transitions due to Gd^{3+} and Er^{3+} (naturally abundant) are also present in the spectrum shown in Fig. 3.

From Table 1, it is apparent that EPR spectra have been observed for most of the elements that are present in the composition of PW-4b nuclear waste. The second column of Table 1 lists the chemical valence states usually found for these elements in accordance with the periodic table. The third column lists the valence states for which an EPR spectrum has been reported in the literature. For special cases such as La^{2+}, unusual valences induced by the crystal host and/or irradiation have been observed by EPR.

Lanthanide orthophosphate compounds (i.e., analogues of the mineral MONAZITE) are currently being investigated as a potential primary waste form (or a waste-form phase) for the isolation of actinide and other radioactive isotopes.[3,4] Flux-grown single crystals of these substances have been intentionally doped with rare-earth, actinide, and iron group impurities and EPR spectroscopy (supplemented by optical and Mössbauer studies) was used in identifying the valence of various impurity ions in the solid state. These investigations have shown that it is possible to incorporate divalent, trivalent, tetravalent, and pentavalent ions from the actinide, lanthanide, and iron groups in the lanthanide orthophosphates.

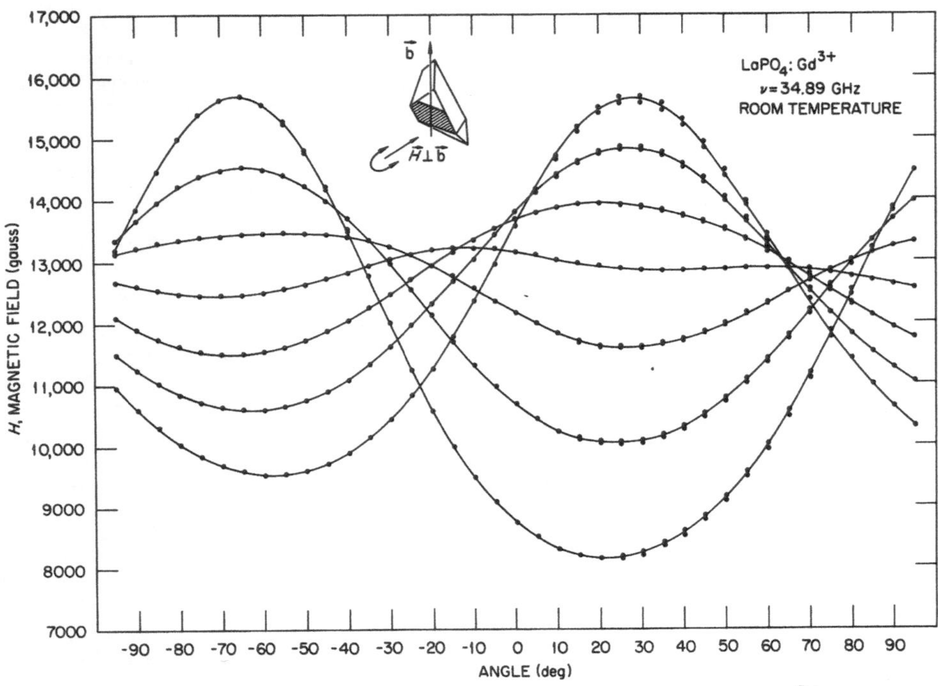

Fig. 4. Angular variation of the EPR transitions of Gd^{3+} in $LaPO_4$ illustrating the spectral symmetry properties.

It should be noted that the valence state of a paramagnetic impurity can also frequently be determined using a polycrystalline sample (e.g., a powder or solid ceramic body), and it is often not necessary to use a macroscopic single crystal specimen.

When a free ion, whose electronic wavefunctions are of the form $|LSJM_J\rangle$, is placed in a host lattice it experiences a crystalline electric field due to the surrounding ions of the lattice. This electric field generally removes the 2 J+1 degeneracy of the free-ion energy levels. The number of multiplets resulting from this interaction depends on the symmetry of the crystalline electric field and, therefore, the symmetry properties of an EPR spectrum can reflect those of the surrounding crystal field. An example of this type of dependence is shown in Fig. 4 where the angular variation of the spectrum of Gd^{3+} in $LaPO_4$ is plotted.[5] For the case of the lanthanide orthophosphates, EPR spectoroscopic studies of Gd^{3+} were used to show that each unit cell of $LaPO_4$ contains two magnetically inequivalent, but otherwise equivalent, lanthanum sites. The amount by which the free-ion energy levels are split is, of course, related to the strength of the crystalline electric field and this field strength is also often manifested in the properties of an EPR spectrum. A dependence of the paramagnetic resonance spectrum on the local crystal field symmetry and strength can be used to obtain information

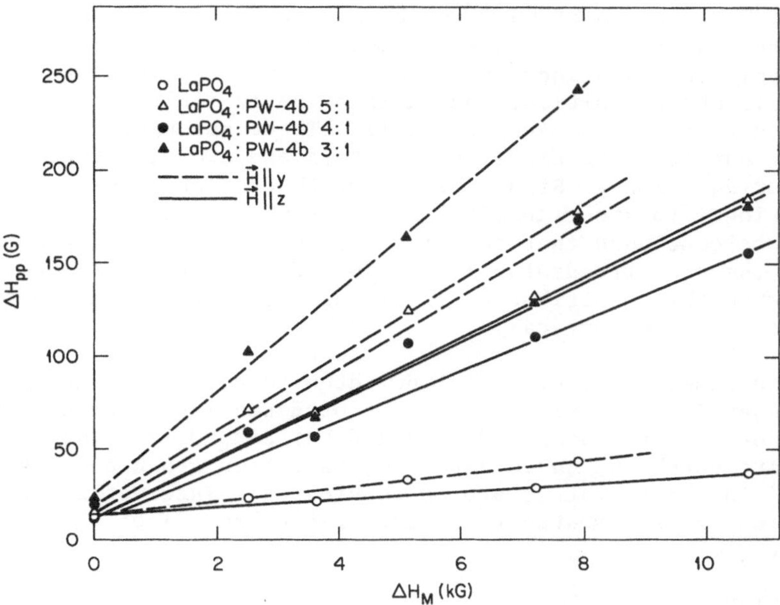

Fig. 5. Average peak-to-peak linewidths ΔH_{pp} vs magnetic-field separations ΔH_M for the symmetric transitions $M \leftrightarrow M-1$ and $-M \leftrightarrow -M+1$ (M-1/2,3/2,5/2,7/2) of Gd^{3+} in single crystals of $LaPO_4$ doped with various amounts of PW-4b simulated waste.

pertaining to the local crystal structure and to detect the possible formation of complex associated defects arising, for example, from deviations from stoichiometry or radiation effects.[6,7]

QUANTITATIVE RESULTS AND DEFECT STUDIES

EPR spectroscopy can provide useful information via the line intensity which is proportional to the number of paramagnetic impurities. Therefore, some indication of the nature of the phase diagram can be obtained for mixed host-impurity systems. Quantitative analytical results obtained using EPR are, however, often not as accurate as those available from other techniques.

Information that EPR spectroscopy can also provide in order to characterize nuclear waste forms can be obtained from the linewidths of EPR transitions (Fig. 1). The EPR linewidths are sensitive to defects and strains, and paramagnetic impurities such as Gd^{3+} can be used to probe the "quality" of the host crystal structure. This technique offers some advantages over x-ray diffraction techniques, since any broadening of the x-ray diffraction peaks is screened by the intrinsic linewidths resulting from the instrument characteristics and the crystallite size. Single crystals of $LaPO_4$ have been grown with the addition of various amounts of PW-4b simulated waste, and the EPR spectra of Gd^{3+} in these samples have been compared with that obtained for a pure $LaPO_4$ single crystal. Figure 5 shows quantitatively the relationship between the average linewidth of two symmetric EPR transitions and their magnetic field separation for pure and PW-4b doped $LaPO_4$ crystals. These measurements were carried out for four different crystals and two different magnetic field orientations. Since the linewidth of each transition is proportional to its magnetic field separation from the central line, it can be concluded that the line broadening is "inhomogeneous" (i.e., the line broadening is due to a distribution of line positions with this distribution resulting from a distribution of impurities in the crystal).

The sensitivity of Gd^{3+} linewidths to defects or strains is also under investigation as a means of detecting metamictization phenomena in orthophosphates. Mixed $Ln_{1-x}An_xPO_4$ crystals, where Ln is a rare earth and An an actinide, have been grown and EPR spectra of Gd^{3+} in these systems are of considerable interest for detecting radiation damage created by α-particles or recoil of the daughter nuclei.

CONCLUSION

Although the examples of the application of EPR to waste form characterization described here have emphasized studies of lanthanide orthophosphates, the same techniques can be applied to perovskite, hollandite, zirconolite, etc. single crystals or powders or to amorphous materials.

REFERENCES

1. A. Abragam and B. Bleaney, "Electron Paramagnetic Resonance of Transition Ions," Oxford University Press, New York (1970).
2. L. A. Boatner and M. M. Abraham, Electron Paramagnetic Resonance from Actinide Elements, Rep. Prog. Phys. 41:87 (1978).
3. L. A. Boatner, G. W. Beall, M. M. Abraham, C. B. Finch, P. G. Huray, and M. Rappaz, Monazite and Other Lanthanide Orthophosphates as Alternate Actinide Waste Forms, in: "Scientific Basis for Nuclear Waste Management," Vol. II, C. J. Northrup, ed., Plenum Press, New York (1980).
4. L. A. Boatner, G. W. Beall, M. M. Abraham, C. B. Finch, R. J. Floran, P. G. Huray, and M. Rappaz, in: "The Management of Alpha-Contaminated Wastes," IAEA-SM-246/73, International Atomic Energy Agency, Vienna, Austria (in press).
5. M. Rappaz, M. M. Abraham, L. A. Boatner, and J. O. Ramey, EPR Spectroscopic Characterization of Gd^{3+} in the Monazite-Type Rare-Earth Orthophosphates: $LaPO_4$, $CePO_4$, $PrPO_4$, $NdPO_4$, $AmPO_4$, and $EuPO_4$, Phys. Rev. (in press).

VALENCE STATES OF ACTINIDES IN SYNTHETIC MONAZITES

K. L. Kelly,[†*] G. W. Beall,[†*] J. P. Young,[††] and L. A. Boatner[+]

Oak Ridge National Laboratory[**]
Oak Ridge, Tennessee 37830

ABSTRACT

The valence states of various actinides doped into the lanthanide orthophosphates ($LnPO_4$) have been investigated. Actinide-doped $LnPO_4$ single crystals were grown by means of a flux technique and the actinide valence states were determined by optical absorption spectrophotometry. Radiation damage effects were also studied in these systems. The lanthanide orthophosphates are found in nature in the form of the mineral monazite and synthetic analogs of this substance represent a promising primary containment medium for the isolation of high-level nuclear wastes.

INTRODUCTION

Orthophosphates formed by elements in the first half of the lanthanide transition series (i.e., La through Gd) are primary constituents of the mineral monazite. The long-term stability of this material under a variety of geological conditions has been established by radiometric measurements of the $^{207}Pb/^{206}Pb$ ratio[1] and other geological dating techniques[2] which indicate that monazite crystals range in age from 5×10^8 to 2×10^9 yr. Monazite ores are a natural source

†Chemistry Division, ORNL; ††Analytical Chemistry Division, ORNL;
+Solid State Division, ORNL.
*Current address: Radian Corporation, 8500 Shoal Creek Blvd., Austin, TX 78766.

**Operated by Union Carbide Corporation for the U.S. Department of Energy under contract W-7405-eng-26.

of both thorium and uranium. Concentrations of 15 wt% UO_2[3] and 14 wt% ThO_2[4] are not uncommon and, in some cases, even higher thorium concentrations have been reported. Due to the incorporation of these naturally-occurring actinides, monazite crystals have been subjected to long-term α-particle and α-recoil events. The inherent resistance of monazite to heavy-particle radiation effects is demonstrated by the fact that this mineral does not undergo significant metamictization and is always found in the crystalline state.[5] The three factors noted above, i.e., long-term geological stability, a high capacity for natural actinide impurities, and resistance to radiation effects, combine to make analogs of monazite an attractive alternative to borosilicate glass for the primary containment of high-level radioactive wastes.[6,7] Accordingly, a series of investigations was initiated to determine the chemical properties of various actinides doped into the lanthanide orthophosphates. These investigations included an identification of the actinide valence states and studies of changes in the host crystal produced by radiation damage and subsequent annealing.

EXPERIMENTAL

Single crystals of every lanthanide orthophosphate, with the exception of $PmPO_4$, have been grown from a $Pb_2P_2O_7$ flux using the technique described by Feigelson.[8] Following the crystal growth process, the crystals were either mechanically separated from the solidified flux or the flux was dissolved by boiling in concentrated nitric acid in order to free the entrained samples. An examination of undoped single crystals of $LaPO_4$ showed that this material was transparent over the entire spectral region of interest and, accordingly, $LaPO_4$ was selected as a host for the incorporation of actinide impurities. Single crystals of $LaPO_4$ containing either ^{238}U, ^{242}Pu, ^{237}Np, ^{241}Am, or ^{246}Cm were prepared from starting mixtures which contained 10.0 wt% UO_2, PuO_2, or NpO_2, 0.5 wt% Am_2O_3, or 0.7 wt% Cm_2O_3 relative to La_2O_3. The actual concentrations of the actinides in the doped crystals were determined by means of analytical techniques and these results are given in Table 1. It is believed that the actinide ions occupy random substitutional lanthanum sites in the $LaPO_4$ crystals.

Table 1. Concentrations of Actinides in $LaPO_4$ Crystals

Atom	Radiochemical	Spark Source Mass Spectroscopy
Np	1.5 wt%	3.0 wt%
Pu	4.3	6.0
Am	0.19	0.20
Cm	0.25	0.1
U	-	1.7

In order to obtain optical absorption spectra, the actinide-doped orthophosphate crystals were placed in a quartz capillary that was then mounted in a microscope spectrophotometer.[9] Since this is a single-beam instrument, a previously recorded background spectrum was stored in a Tektronix 4051 computer and was subtracted from the spectrum of each sample in order to obtain the true absorption. Spectral scans were initiated at a wavelength of 1100 nm in every case, but the cutoff wavelength was determined by the particular characteristics of each sample and it varied from 500 to 300 nm.

RESULTS AND DISCUSSION

Single crystals of pure $CePO_4$ that were grown using the technique noted above were black and appeared to contain Ce(IV) as well as Ce(III). Attempts to estimate the relative proportions of each valence state present in the crystals were, however, not successful. The samples were crushed, dissolved in boiling concentrated sulfuric acid, and titrated with a standardized Fe(II) solution in order to determine the amount of Ce(IV). A second aliquot of the solution was then oxidized to obtain the amount of cerium present as Ce(III). Complete dissolution of the $CePO_4$ crystals proved to be extremely difficult, however, and rapid reprecipitation always occurred. These factors eliminated any possibility of performing an accurate, direct wet-chemical analysis. Additionally, optical absorption spectrophotometry was not considered as an experimental method for this determination because the absorptivity of Ce(IV) is much greater than that of Ce(III) and the anticipated absorption peaks overlap. Accordingly, the presence of Ce(IV) completely obscures the Ce(III) peaks and a spectrophotometric analysis of the ratio of these ions is not practical.

Crystals of $LaPO_4$ doped with $\sim$ 1.7 wt% ^{238}U were characterized by a bright green color, and the optical absorption spectrum (see Fig. 1) shows that uranium is primarily in the tetravalent state.[10] (The dotted curve in Fig. 1 represents the spectrum of U^{4+} in solution.) Several small peaks were observed at 630, 790, and 900 nm which indicate that a small amount ($\sim$ 5%) of U^{3+} is also probably present. Since ^{238}U has an extremely long half-life (10^9 yr), radiation damage is not a factor and, therefore, the $La(^{238}U)PO_4$ crystals were not annealed prior to the spectrophotometric measurements.

Plutonium-doped single crystals of $LaPO_4$:5.0 wt% Pu exhibited a dark purple coloration. The optical absorption spectrum shown in Fig. 2 indicates that plutonium (like uranium) is also incorporated in $LaPO_4$ in the tetravalent state.[11] The dotted curve in Fig. 2 represents the the spectrum of Pu(IV) in aqueous $HClO_4$. There is no indication of the presence of Pu^{3+} in the $La(Pu)PO_4$ spectrum shown in Fig. 2. This crystal was annealed for 12 h with no noticeable change in its appearance; however, an absorption spectrum of the post-annealed crystal did exhibit some subtleties which differed

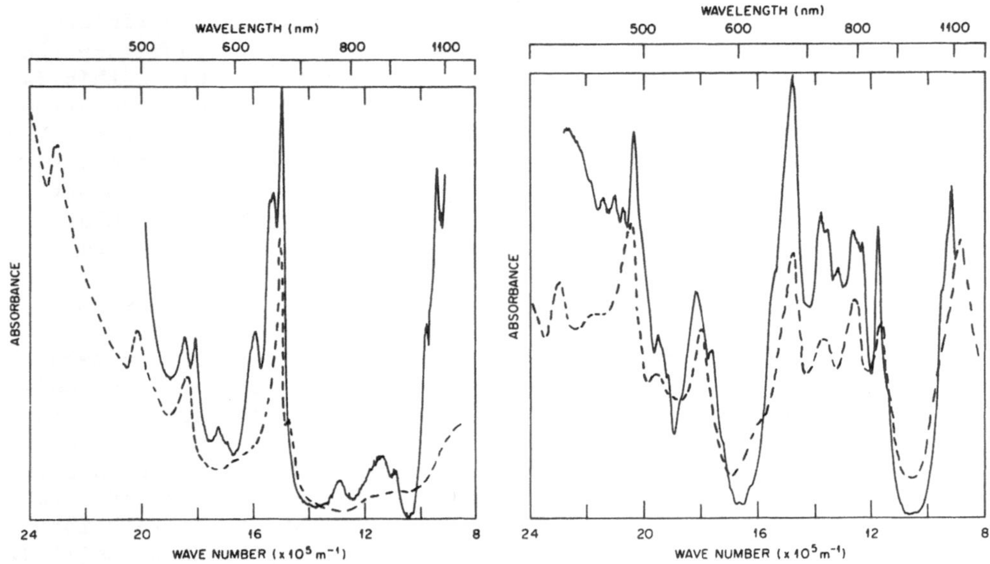

Fig. 1. Solid curve: spectrum of $La(U_{1.7\ wt\%})PO_4$; dashed curve: solution spectrum of U^{4+} in $HClO_4$.

Fig. 2. Solid curve: spectrum of $La(Pu_{5.0\ wt\%})PO_4$; dashed curve: solution spectrum of Pu^{4+} in $HClO_4$.

from the original spectrum. These differences are not clearly understood at this time, but the absorption peaks in the spectrum of the annealed crystal still correlate with those in the spectrum of tetravalent plutonium. Accordingly, fundamental changes in the chemical nature of the plutonium dopant did not occur with annealing at moderate (∿ 500°C) temperatures.

The absorption spectra of $LaPO_4$ crystals doped with 0.2 wt% ^{241}Am show that americium is incorporated in the trivalent state (Fig. 3).[12] The americium-doped $LaPO_4$ crystals were characterized by an amber color when removed from the Pt crystal growth crucible. This coloration could have been due either to the americium ions themselves or to radiation-induced color centers in the $LaPO_4$ host. When the $La(^{241}Am)PO_4$ crystals were annealed at ∿ 500°C for 16 h they decolorized indicating that the amber color is due to radiation damage and is not associated with optical absorption arising from americium ions. The amber color returned to the crystals within a short time following their removal from the annealing furnace. The series of spectra shown in Fig. 4 was obtained as a function of time following the anneal. The bottom spectrum in Fig. 4 was recorded 2 h after the anneal and the middle and top spectra were recorded 22 h and 192 h, respectively, following the anneal. The increasing broad band absorption evident in the visible region in these three spectra is

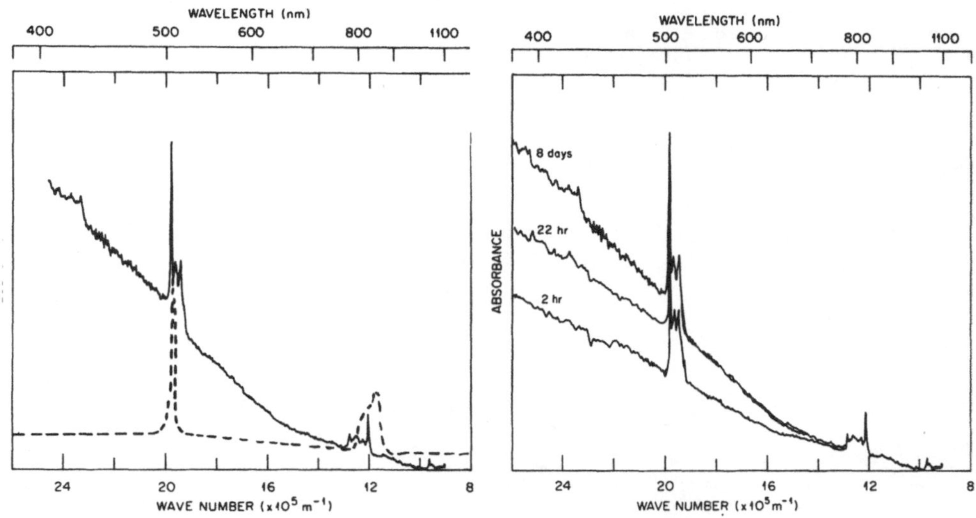

Fig. 3. Solid curve: spectrum of $La(Am_{0.2\ wt\%})PO_4$; dashed curve; absorption spectrum of Am^{3+} in fayolite.

Fig. 4. $La(Am_{0.2\ wt\%})PO_4$ spectra for crystals annealed at 500 °C. Time periods are delays after the anneal.

associated with the return of the amber color to the crystal. A qualitative determination of the type of radiation responsible for the damage which resulted in the amber coloration was made by means of a simple experiment. A pure $LaPO_4$ single crystal was placed in juxtaposition with an americium-doped specimen. This allowed radiation to impinge on the undoped crystal but did not subject the pure sample to a α-particle or α-recoil damage. The undoped specimen remained colorless following a 16 h exposure to these conditions. This result suggests that radiation damage in the americium-doped crystal is primarily due to α-particles and nuclear recoil.

Single crystals of $LaPO_4$ doped with ∿ 0.2 wt% ^{246}Cm are also characterized by an amber color. The optical spectra obtained for the $La(^{246}Cm)PO_4$ specimens did not exhibit any absorption peaks that could definitely be assigned to curium (see Fig. 5). The two small absorption peaks shown in Fig. 5 are due to a residual americium impurity that was present in the curium dopant. Curium is expected to be in the 3+ valence state, and the absence of any observable Cm(III) absorption peaks can be explained by the poor spectral sensitivity of this ion coupled with its low concentration in the crystal. Tetravalent curium, however, should have been detected since its spectral sensitivity is much higher. The present observations are not definitive but are completely consistent with the presence of Cm(III) in

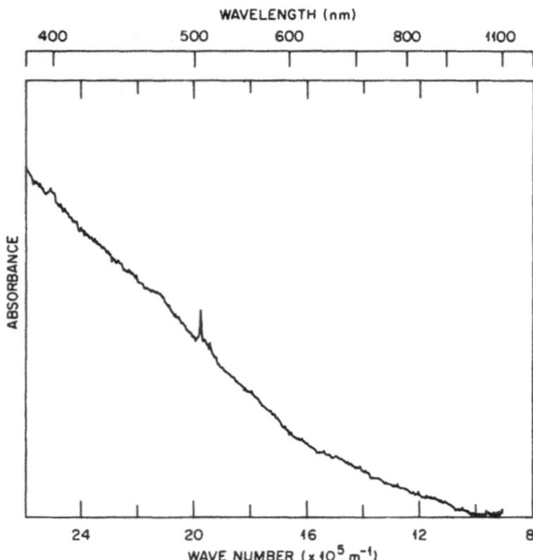

Fig. 5. Absorption spectrum of $La(Cm_{0.2\ wt\%})PO_4$.

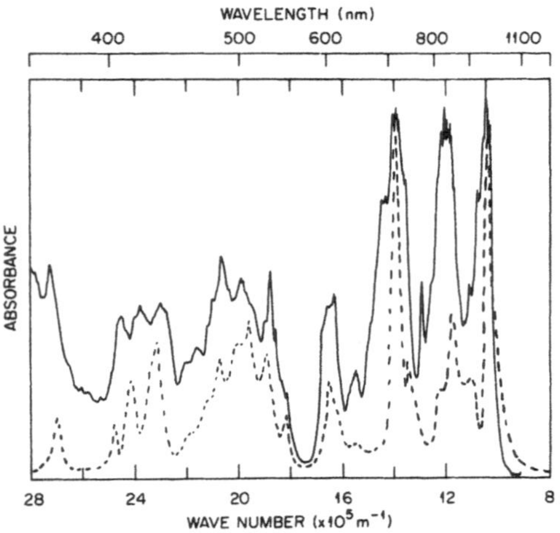

Fig. 6. Solid curve: spectrum of $La(Np_{2.0\ wt\%})PO_4$; dashed curve; solution spectrum of Np^{4+} in $HClO_4$.

the crystals. Following a 16 h anneal, the Cm-doped $LaPO_4$ samples became colorless. The amber color began to return shortly after the specimens were removed from the annealing furnace as it had in the case of the ^{241}Am-doped samples. Attempts to observe absorption peaks from Cm(III) were made using colorless, freshly annealed $La(Cm)PO_4$ crystals but these attempts were not successful. A conclusive identification of Cm(III) will require additional experimentation and possibly the application of other spectroscopic techniques.

The absorption spectrum of $La(Np)PO_4$ ($\sim$ 2 wt% Np) is shown in Fig. 6. These crystals are characterized by a bright green color. A comparison of the $La(Np)PO_4$ spectrum (solid line) with that of Np(IV) in an aqueous solution of $HClO_4$ (i.e., the dashed curve in Fig. 6) shows that tetravalent neptunium is present in the orthophosphate single crystal. No changes in the crystal color as a function of time were observed in the case of ^{237}Np-doped $LaPO_4$. Investigations of the ^{237}Np daughter produced following the decay of ^{241}Am in $LaPO_4$ have been carried out by means of Mössbauer spectroscopy. These results indicate that the ^{237}Np valence state can be either 5+ or 3+.

SUMMARY

The purposes of the present optical spectrophotometry and annealing studies of actinide-doped lanthanide orthophosphates were to dentify the valence states of actinide impurities as incorporated in host $LnPO_4$ lattice and to determine if radiation-induced coloraion of these crystals could be observed. Radiation-produced color enters apparently were formed in the cases of La(^{241}Am)PO_4 and a(^{246}Cm)PO_4 as evidenced by the presence of an amber color in the rystal that did not exhibit any f-f spectral character but that ould be completely removed by annealing the crystals at a moderate emperature. The amber color returned within several hours after the nneal.

The valence states of the actinides doped into single crystals f $LaPO_4$ were found to be U(IV), U(III), Pu(IV), Np(IV), Am(III), nd most probably Cm(III). Since it is known that trivalent and etravalent ions are generally less mobile in geologic media than entavalent or hexavalent ions, the present determination that the ctinides are in these "less mobile" valence states in $LaPO_4$ is a ignificant factor in reinforcing the viability of employing anaogs of monazite as a primary containment form for high-level transranic wastes. In order to fully understand the solid state chemcal properties of mixed lanthanide-actinide phosphate systems and he implications of these properties for long-term storage under a ide variety of environmental conditions, more intensive experimenation and analysis will be required.

EFERENCES

1. Paul Pasteels, Eclogae Geol. Helv. 63:31 (1970).
2. O. H. Leonardos, Jr., Econ. Geol. 69:1126 (1974).
3. C. M. Gramoccioli and T. V. Segalstad, Am. Miner. 63:757 (1978).
4. T. Kato, Min. Journ. (Japan) 2:224 (1958).
5. R. C. Ewing, Am. Miner. 60:428 (1975).
6. L. A. Boatner, G. W. Beall, M. M. Abraham, C. B. Finch, P. G. Huray, and M. Rappaz, "Scientific Basis for Nuclear Waste Management," Vol. II, C. Northrup, ed., Plenum Press (1980), p. 289.
7. L. A. Boatner, G. W. Beall, M. M. Abraham, C. B. Finch, R. J. Floran, P. G. Huray, and M. Rappaz, to be published in the Proceedings of the International Symposium on the Management of Alpha-Contaminated Wastes, sponsored by the International Atomic Energy Agency and the Commission of the European Communities, Vienna, Austria, 2-6 June (1980).
8. R. S. Feigelson, J. Am. Ceram. Soc. 47:257 (1964).
9. J. P. Young, R. G. Haire, R. L. Fellows, and J. R. Peterson, J. Radioanalyt. Chem. 43:479 (1978).
0. D. M. Gruen and R. L. McBeth, J. Inorg. Nucl. Chem. 9:290 (1959).
1. D. A. Costanzo (unpublished results).
2. C. B. Finch and G. W. Clark, J. Phys. Chem. Solids 34:922 (1973).

CRYSTAL CHEMISTRY AND PHASE RELATIONS IN THE SYNTHETIC MINERALS OF CERAMIC WASTE FORMS. II. STUDIES OF URANIUM-CONTAINING MONAZITES

D.D. Davis, E.R. Vance and G.J. McCarthy*

Materials Research Laboratory
The Pennsylvania State University
University Park, PA 16802

ABSTRACT

$Ca_{0.5}U_{0.5}PO_4$ can be synthesized in a neutral atmosphere or a closed system at ∿1200°C by ceramic techniques and it forms complete solid solutions with $Ca_{0.5}Th_{0.5}PO_4$ and $NdPO_4$. Uranium appears to occur in monazite essentially in only the tetravalent state. The possible role of some other ions in monazite was also studied.

INTRODUCTION

Monazite (a light rare earth-phosphate) is a prime candidate for the immobilization of actinides and rare earths in ceramics derived from nuclear wastes.[1-3] In part I of this series[3] the compositional flexibility of the monazite structure was reported: $NdPO_4$ and $Ca_{0.5}Th_{0.5}PO_4$ formed by firing in air at ∿1200°C were found to form a complete solid solution and this solid solution would incorporate up to 20 mole % of $Ca_{0.5}U_{0.5}PO_4$ under these conditions. The present paper extends this work and also examines monazite as a host for actinides of different valence, both by themselves and in the presence of other radwaste ions, corrosion products, and process chemicals.

*Departments of Chemistry and Geology
North Dakota State University
Fargo, ND 58105

EXPERIMENTAL

All preparations were made up from mixed solutions of metal nitrates and $(NH_4)H_2PO_4$. The mixtures were dried, calcined at ∿500°C, ground, pressed into pellets using ∿$5x10^4$ psi and fired at ∿1200°C, either in open systems with different atmospheres, or in welded Pt capsules. X-ray diffraction data were recorded with a standard diffractometer using a graphite monochromator and CuKα radiation.

RESULTS AND DISCUSSION

Uranium in Monazite

From a chemical point of view, U is probably the main actinide in both reprocessing and defense waste, so it will tend to dominate the "impurity" chemistry of monazite phases in nuclear waste ceramics to which phosphate has been added. Previously[3] it was found that $Ca_{0.5}U_{0.5}PO_4$ (monazite-structured) could not be formed phase-pure. However, some monazite was observed by the solution-mixing technique followed by firing in air at ∿1200°C. In the present work phase-pure material was produced by firing stoichiometric mixtures at ∿1200°C in Ar, N_2, or a closed system. Firing at ≤1050°C under these conditions was insufficient. Vacuum hot-pressing in graphite dies at 1200-1400°C was not successful in producing the desired phase, but subsequent firing in Ar at 1200°C produced phase-pure material. These results are to be contrasted with those of Muto et al.,[4] who found that ningyoite (hydrated $Ca_{0.5}U_{0.5}PO_4$ having an orthorhombic structure) could be converted to the monazite structure by heating in Ar at 900°C, but not at 950°C. At 950°C it decomposed to uraninite plus residual monazite.

In the present work, heating phase-pure $Ca_{0.5}U_{0.5}PO_4$ in air above ∿700°C produced decomposition of some of the monazite into the same unknown phase that was observed together with the monazite in stoichiometric calcines fired in air at 1200°C. Subsequent firing in Ar at ∿1100°C, however, increased the proportion of monazite, with the remainder consisting of uraninite. A possible explanation of these results is that, in air, the oxygen partial pressure is too high to allow all the U to exist as U^{4+}. Decomposition into a U^{6+}-rich phase might allow a low-melting calcium phosphate to form with some loss of the phospate by volatilization and subsequent heating in the neutral atmosphere could allow the U^{6+} to reform U^{4+} with an increase in the amount of monazite.

Heating $Ca_{0.5}U_{0.5}PO_4$ in a strongly reducing atmosphere (H_2 or 5% H_2/N_2) at ∿1200°C also caused decomposition into uraninite plus residual monazite. This result, coupled with the above data on the response of $Ca_{0.5}U_{0.5}PO_4$ to heating in air, shows that $Ca_{0.5}U_{0.5}PO_4$

is stable only over a relatively small range of oxygen partial pressures. For preparations heated in Ar at 1200°C, a continuous solid solution of $Ca_{0.5}U_{0.5}PO_4$ in $NdPO_4$ and $Ca_{0.5}Th_{0.5}PO_4$ was observed.

No alteration of $Ca_{0.5}U_{0.5}PO_4$ powder was observed by x-ray diffraction after three weeks' contact with deionized water at 200°C and 300 bars.

The possibility of U adopting a valence other than +4 in monazite was examined. (Tetravalent U is compensated by divalent Ca in $Ca_{0.5}U_{0.5}PO_4$.) No evidence of any lattice parameter change relative to pure $NdPO_4$ was observed in $(Nd_{1-x}, U_x)PO_4$ ($x \leq 0.4$) preparations fired at 1200°C in air, N_2 or 5% H_2/N_2, so U^{3+} does not seem to be a favored form of U in monazite. To examine the possibility of U^{+} formation, we studied preparations of $(Nd_{1-3x}, U_x, Ca_{2x})PO_4$ and $(Nd_{1-2x}, U_x, (Na/K/Rb)_x)PO_4$ stoichiometries fired at 1000-1200°C and in which $x \leq 0.2$. The monazite lattice parameter showed no significant deviation from predictions made on the basis that the U enters the lattice in the tetravalent state, with half the Ca or alkali forming a second phase or being volatilized as Ca phosphate or an alkaline oxide. These results, however, do not preclude the possibility that U^{3+} or U^{5+} could enter the lattice in quite dilute amounts. As an illustration of this point, no evidence of U^{5+} solubility in zircon was observed by x-ray diffraction in $(Zr_{1-2x}, Y_x, U_x)SiO$ preparations, though small amounts ($\approx$0.1 wt %) of U^{5+} can indeed exist.[5]

Uranium-Containing Monazites for Defense Waste Solidification

The high level waste sludges at Hanford and Savannah River contain substantial amounts of U but only traces of rare earths. In order to crystallize this U in the monazite structure type, the solid solution phase will have to be compatible at crystallization temperatures with the major spinel and corundum structure phases. The required compatibility studies are still in progress but to date, none of these experiments has indicated incompatibility of monazite with spinels formed in air-fired simulated Savannah River crystalline ceramics or the simple oxides of Fe, Al, Ti or Cr. It has been found that $NdPO_4$ is not compatible with CaF_2, a major component of some INEL calcines. These phases reacted to form a fluorapatite and $(Ca,Nd)F_{2+x}$ in an 1100°C sealed capsule experiment.

CONCLUSIONS

Our earlier observation that $Ca_{0.5}U_{0.5}PO_4$ cannot be prepared by calcination and firing in air that has been confirmed with numerous additional experiments. However, this phase can be prepared under inert or reducing atmospheres or in sealed systems where oxygen content is fixed. The possibility that charge compensating substitutions from other ions in waste could stabilize U^{3+} or U^{5+} was explored. It was found that the principal valence state of uranium was U^{4+} in analogy to $Ca_{0.5}Th_{0.5}PO_4$. So far, it appears that the U in Hanford and SRP sludges could be incorporated into a monazite structure phase if it were desirable to do so, but that monazite structure phosphates are not compatible with the high CaF_2 calcines at INEL.

ACKNOWLEDGEMENT

This research was supported by the Department of Energy through a subcontract from the Rockwell Energy Systems Group.

REFERENCES

1. G.J. McCarthy, High-Level Waste Ceramics: Materials Considerations, Process Simulation and Product Characterization, Nucl. Tech. 32:92 (1977).
2. G.J. McCarthy, W.B. White and D.E. Pfoertsch, Synthesis of Nuclear Waste Monazites, Ideal Actinide Hosts for Geologic Disposal, Mat. Res. Bull. 13:1239 (1978).
3. G.J. McCarthy, J.G. Pepin and D.D. Davis, Crystal Chemistry and Phase Relations in the Synthetic Minerals of Ceramic Waste Forms. I. Fluorite and Monazite Structure Phases, in: "Scientific Basis for Nuclear Waste Management," Vol. 2, C.J. Northrup, ed., Plenum, New York (1980).
4. T. Muto, R. Meyrowitz, A.M. Pommer and T. Murano, Ningyoite, a New Uranous Phosphate Mineral from Japan, Amer. Mineral. 44:633 (1959).
5. E.R. Vance and D.J. Mackey, Optical Study of U^{5+} in Zircon, J. Phys. C: (Solid State Phys) 7:1898 (1974).

CHARACTERIZATION OF IRON-ENRICHED SYNTHETIC BASALT FOR TRANSURANIC CONTAINMENT

J. E. Flinn, S. P. Henslee, P. V. Kelsey,
R. L. Tallman, and J. M. Welch

Materials Technology Division, EG&G Idaho Inc.
P. O. Box 1625, Idaho Falls, ID 83415

INTRODUCTION

Considerable quantities of low-level and transuranic (TRU) wastes are being temporarily stored at Idaho National Engineering Laboratory (INEL). Estimates show that stored and buried wastes at INEL will approach 4 x 10^5 m^3 by 2000 AD. Main combustible components of INEL waste forms are paper, wood, and plastics, and main noncombustibles are concrete, metals, sludge, and soil. Two prime benefits of solidification are: (a) substantial volume reduction and (b) stabilized waste forms suitable for either short- or long-term storage.

Slagging pyrolytic incineration (SPI) is being considered to convert stored and buried wastes into a slag.[1] In this process, combustibles are burned and noncombustibles, including metals, are oxidized into a molten slag, with further processing conducted in a heated tundish, i.e., an electromelter, where the molten slag is allowed to homogenize (within a reasonable time period) and then cast into large, cylindrical metal containers. Analyses of INEL waste slags[2] show them similar in composition and appearance to natural basalts, but rich in iron.

Most of the characterization assessments of synthetic slag wastes have been run on small melts, which are highly vitreous and were not representative of castings that normally result from a combined SPI-tundish process. Penberthy Electromelt of Seattle, Washington, produced ∿90-kg slag castings with compositions approximating average INEL waste as well as average wastes containing various amounts of soil from the INEL site.

In total, four melts and subsequent castings were made. This paper summarizes the evaluation of these castings.

MELTING AND CASTING

Large-scale castings were made for four compositions (representing the major oxide components associated with average waste with various soil additions[2]) using an electromelt process. The melt batches were premixed from industrial grade powders and soil, where applicable. The charge was usually completed in one day, and the melter held at 1300 to 1480°C overnight, with pouring usually accomplished the next day. The canisters used to contain the molten slag through subsequent cooling were galvanized steel and/or mild steel, 38.1 cm in diameter x 43.2 cm high x 0.18 cm thick.

Before pouring, the bottoms of three canisters used in the initial tests were filled ∿2.5-cm deep with sand for insulating purposes. Furthermore, two small holes were drilled into the bottom of each canister to vent air and steam from the sand layer. The fourth casting was cooled via a 7.6-cm thick steel plate placed under the canister for heat extraction. All canisters effectively contained their melts.

GENERAL STRUCTURAL FEATURES

The overall appearance of the interiors of castings was rock-like. In general, the texture or grain size near canister walls was finer than that near the center of the castings. The overall grain size for castings is considerably finer than that observed for natural basalt in Idaho. This is expected since cooling of these castings was considerably faster (less than 24 hrs from pouring temperature to room temperature) than the cooling of most natural basalts.

Each of the castings displayed a friable crust and a center cavity. The crust probably was caused by a faster cooling rate on the surface than in the interior since it was directly exposed to air. This exposure would also promote thermal shock. The porous cavity represented approximately 10 percent of the total casting volume.

Plugs from various locations within the castings were removed by core drilling. The plugs were sectioned into discs, then analyzed and tested. By chemical analysis, each casting was homogeneous. The nominal compositions are shown in Table 1. The most notable difference for the series is the A-30 casting, with exceptionally high iron content and ferrous-to-ferric iron ratio.

TABLE 1. Nominal Chemical Composition of A-Series Electromelt Slags

Slag Type	Oxide Compounds (wt%)								
	SiO_2	Al_2O_3	Fe_2O_3	FeO	TiO_2	CaO	MgO	K_2O	Na_2O
A-0	40.4	10.2	15.6	14.8	0.07	8.6	3.4	1.8	5.0
A-20	42.9	10.6	9.8	15.8	0.30	10.8	3.3	1.9	4.4
A-30	47.8	11.0	2.5	19.6	0.24	9.6	2.9	2.1	4.2
A-40	51.3	11.0	6.1	11.7	0.05	9.8	2.4	1.7	6.0

PHASE AND MICROSTRUCTURE

Crystalline phases present in the A-series slags are augite, magnetite, and probably hematite (in some areas) as determined by X-ray diffraction microscopic examination and EDS analysis.

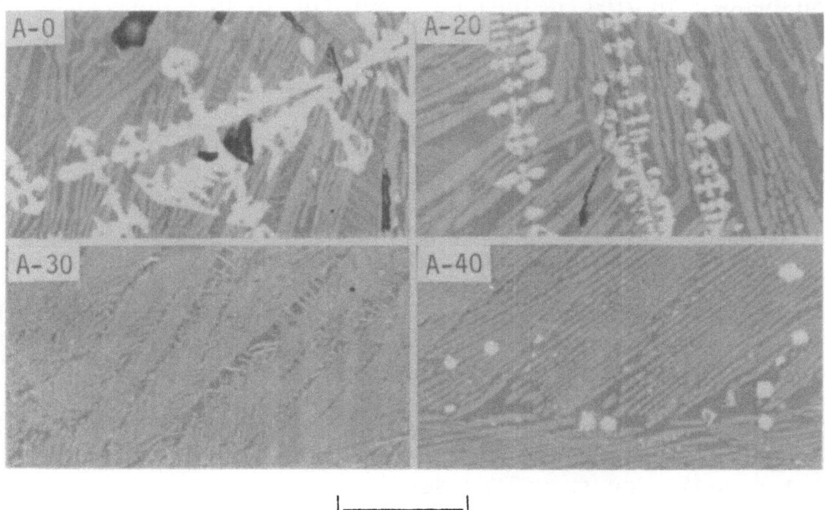

Figure 1. Photomicrographs of A-series slags. All samples were taken from the upper middle sections of the castings and their chemical compositions are shown in Table 1.

The microstructure of the A-0 casting, shown in Figure 1, is feathered acicular augite (light gray regions) with large spinel dendrites (white regions within the matrix). The darker gray area (matrix) is glass.

The microstructure of A-20 specimens (Figure 1), is similar to that of the A-0 casting except that there is less spinel and the augite crystals are slightly larger. The microstructures of the bulk of the A-30 specimens (Figure 1) are dominated by two phases, fine-acicular augite and glass; very little spinel was observed. The microstructure of A-40 specimens shows a marked decrease in the amount of spinel present (Figure 1). However, the A-30 specimen taken near the air-exposed surface showed typical levels of spinel, augite, and glass when compared to A-0 and A-40 structures.

Results show that a) number and size of the spinel grains decrease with decreasing total iron and/or Fe_2O_3 contents and b) the degree of devitrification increases and grain size decreases with increasing ratios of silica to iron oxides.

LEACHING

Rectangular-shaped monoliths (2.3 to 2.9 g) for leaching assessments were sectioned from plug cores. The monoliths were abraded and polished, cleaned, and weighed on a microbalance, then suspended in 70°C deionized water for 14 d. The volume of leachant was 40 ml/cm^2 of specimen surface. Results were determined from weight loss measurements, and from leachant analyses by emission spectroscopy.

The results are shown in Table 2. The agreement between the two measurements is very good. The leach rates range from 30×10^{-6} g/cm^2·d for A-0 to 8×10^{-6} g/cm^2·d for A-40 slags. These values are ten to thirty times higher than those reported for more vitreous melt specimens.[2]

The normalized compositions of the leachates were about the same for all leach specimens from the four castings. The leached composition was 47 wt% SiO_2, 29 wt% Na_2O, 18 wt% Al_2O_3, 5 wt% K_2O, 0.6 wt% CaO, 0.3 wt% Fe_2O_3, and 0.3 wt% MgO. This composition seems to be a derivative from the glass regions of the slags with very little contribution from either the augite or spinel phases.

Leach rates (Table 2) increased with increasing iron and decreasing silica content, and with increasing glass-phase content. Microcracking increased with iron content, increasing the surface area available for leaching. A high pH from the dissolution of alkali in a crack can promote the dissolution of

TABLE 2. LEACHING RESULTS FOR ELECTROMELT SLAG SPECIMENS (14 day Static Test at 70°C)

Slag Type	Specimen No.	Surface Area (cm^2)	Weight Loss (mg)	Leach Rate ($\mu g/cm^2 \cdot d$)	Total Oxide[a] Leach Rate ($\mu g/cm^2 \cdot d$)
A-20	1	5.730	1.500	18.7	19.8
	2	5.768	1.719	21.3	22.0
	3	5.582	1.825	22.6	26.4
A-40	4	5.969	0.650	7.8	10.8
	5	5.381	0.749	9.9	13.1
	6	5.618	0.704	9.0	11.3
A-30	7	5.204	1.089	14.9	17.8
	8	5.732	1.360	16.9	20.2
	9	5.215	1.086	14.9	18.8
	10	5.599	0.968	12.3	15.8
A-0	11	5.553	2.764	35.6	38.0
	12	5.652	2.616	33.1	37.2
	13	5.730	1.042	13.0	17.7

a. From leachant analyses, converted to oxides.

Al_2O_3 and SiO_2. Dissolution could be further enhanced at crack tips by the presence of high tensile stresses.

MECHANICAL BEHAVIOR

Mechanical property measurements were obtained on the slag specimens; these included splitting tensile strengths (ASTM 496-71), fracture toughness, and diamond pyramid microhardness. In addition to the hardness values, the extent and nature of the microcracks from six indentation load levels were determined.

The splitting tensile strengths and fracture toughness values determined on the plug samples removed from the castings are shown in Table 3. A-20 appears to be slightly stronger than either A-0 or A-30. This behavior can be explained in part by analysis of the failed specimens. The fracture surfaces produced from the tests for A-0 and A-30 are similar in surface roughness and general appearance, while the failed surface of the A-20 was less coarse. The implication is that the A-20 contained a higher density of micro-cracks and thus required more energy to propagate those micro-cracks. This assumption is reinforced by the size of the particles produced by the splitting tensile test. The A-20 produced a noticeably finer particle, generating more surface area than either A-0 or A-30, and thus requiring more energy.

Fracture toughness values (Table 3) for the slags were lower than for the borosilicate glass. This can be explained again by the relative amount of micro-cracking exhibited by the various samples. The borosilicate glass showed no visible signs of micro-cracking while all three slags were extensively micro-cracked. The presence of micro-cracks could be considered advantageous with respect to fracture toughness through multiple crack paths etc., but this is not necessarily the case since the toughness values for the slags do not correlate to the amount of micro-cracking. All else being equal, the borosilicate glass should exhibit greater fracture toughness values since there was no micro-cracking. The effect of micro-cracking should be noticeable in the crack propagation rate where multiple crack paths and crack branching play an important role and would subsequently decrease the rate.

Microhardness of spinel phases was identical to that of the remaining matrix. Measured micro-crack lengths produced by indentations (Figure 2), showed no apparent correlation with phases present, i.e., a preference for the micro-cracks to follow phase boundaries rather than crossing phases was not seen.

TABLE 3. SPLITTING TENSILE STRENGTH AND FRACTURE TOUGHNESS RESULTS

	Splitting Tensile, MPa		Fracture Toughness, MPa$\sqrt{m}$	
Slag Type	Strength	Std. Deviation	K_{IC}^{SR}	Std. Deviation
A-0	13.23	3.92	0.92	0.13
A-20	16.02	5.56	0.95	0.06
A-30	12.65	3.53	1.13	0.20
Borosilicate Glass	43.3	14.39	1.41	0.09

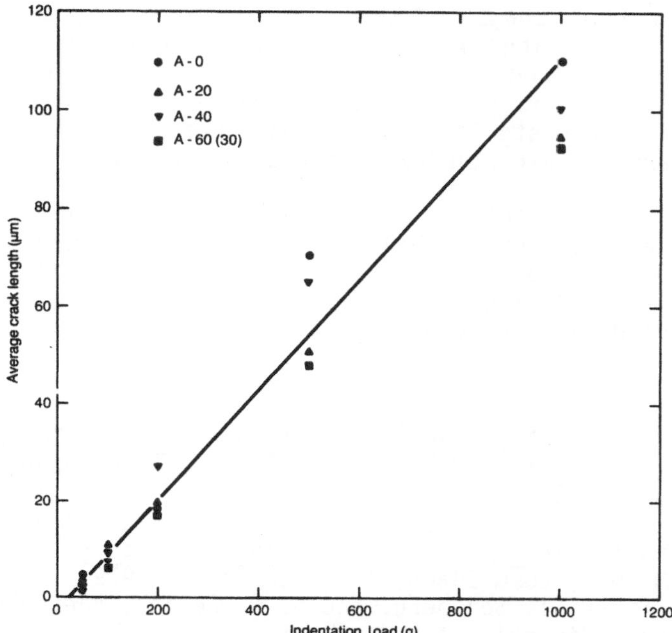

Figure 2. Indentor load effect on average crack length.

The results and discussion on the durability observations, i.e., leaching and mechanical properties, strongly suggest that controlled cooling of the melts to promote more devitrification should be explored. This approach would reduce the amount of glass phase as well as minimize micro-cracking.

SUMMARY AND FUTURE CONSIDERATIONS

The electromelt process and the resulting iron-rich castings offer great promise for rendering nuclear waste into a stable form. The process offers great flexibility with regard to both compositional variation of the incoming waste and the high rates at which the waste can be introduced and cast. The cast product, a fine-grained basalt-like material, shows excellent homogeneity with little or no reaction to the steel containment. The preliminary mechanical and chemical durability data show the form to have adequate containment properties for TRU waste. However, work presently underway to improve these properties through additives and controlled cooling cycles has greatly enhanced the durability of the waste form. Furthermore, recent evidence indicates that divalent iron (Fe^{2+}) included in the crystalline phases of granites and basalts imparts a resistance to leaching[3] of uranium and other actinide ions[4,5]. The mechanism for this resistance is believed to be the Fe^{2+}/Fe^{3+} buffer system which prevents oxidation of actinides in the waste form and inhibits reactions with carbonate and sulfate complexes in the ground water.

The flexibility of the electromelt process, combined with the potential superior durability of the iron-enriched synthetic basalt, indicate that this waste form system should be considered for all types (high level and low level) of nuclear waste consolidation and containment.

REFERENCES

1. N. D. Cox et al., Figure-of-Merit Analysis for TRU Waste Processing Facility at INEL, TREE-1293, EG&G Idaho, Inc., October 1978.
2. J. E. Flinn et al., Summary of FY-1979 Material Support Studies for SPI - Migration and Immobilization, EGG-FM-5041, EG&G Idaho, Inc., September 1979.
3. J. A. Speer, T. Soldberg, and S. W. Becker, "Petrography of U-Bearing Minerals of the Liberty Hill Pluton, South Carolina: Phase Assemblages and Migration of U in Granitoid Rocks," unpublished manuscript.
4. G. W. Beall, G. D. O'Kelley and B. Allard, An Autoradiographic Study of Actinide Sorption of Climax Stock Granite, ORNL-5617, June 1980.
5. G. W. Bird, "Geochemistry of Radioactive Waste Disposal," Geoscience Canada 6: 199-204 (1980).

ACKNOWLEDGEMENTS

This work was supported by the U. S. Department of Energy, Assistant Secretary for Nuclear Energy, Office of Waste Management under DOE Contract Number DE-AC07-76ID01570.

TREATMENT OF CLADDING HULLS BY THE HIPOW PROCESS*

Hans T. Larker and Ragnar Tegman

High-Pressure Laboratory ASEA AB

S-915 00 Robertsfors, Sweden

ABSTRACT

The conditions for densifying and bonding Zircaloy cladding hulls from spent LWR fuel to blocks by the HIPOW (hot isostatic pressing of waste) process have been studied. Fully dense and mechanically strong blocks of Zircaloy can be made without additives at temperatures around 1000°C. A volume reduction of about seven times and surface area reduction of more than 300 times, compared to typical loose-filled cladding hulls remaining after the chop-leach operations in a reprocessing plant, can be obtained. A study of a possible process for industrial scale has been made. Handling under water can prevent any fire hazard in the preparation sequence. The use of a special hermetically sealed double-wall metal container encasing the hulls during the densification in the hot isostatic press virtually eliminates the problem of lasting contamination of this equipment, thus greatly simplifying service and maintenance. One hot isostatic press can serve a reprocessing line with an LWR fuel capacity of 800 tons/year. Fines (residues) from fuel dissolution and alpha-contaminated ashes from incincerated organic materials in the plant may also be incorporated in the Zircaloy blocks. Tritium can quantitatively be contained in these blocks.

INTRODUCTION

The HIPOW process has been presented earlier in applications for making fully dense blocks of crystalline, high-melting radioactive materials of mineral structure.[1,2] Starting material could

*Research sponsored by SKBF/Project KBS.

be, for example, supercalcine, titanate ion exchanger, or SYNROC powders, but it could also be cladding hulls or residues from reprocessing of spent fuel and alpha-contaminated ashes obtained from incineration of contaminated organic material in a reprocessing plant (Fig. 1). The characteristic for this process is that the processing at high temperature takes place in a hermetically sealed metal container. The hot pressing is carried out isostatically by surrounding the container with argon gas at high pressure.[3] Because of the very efficient application of pressure, fully dense material, even in the form of large blocks, can be made at a comparatively low temperature — 50 to 70% of the melting point of the material.

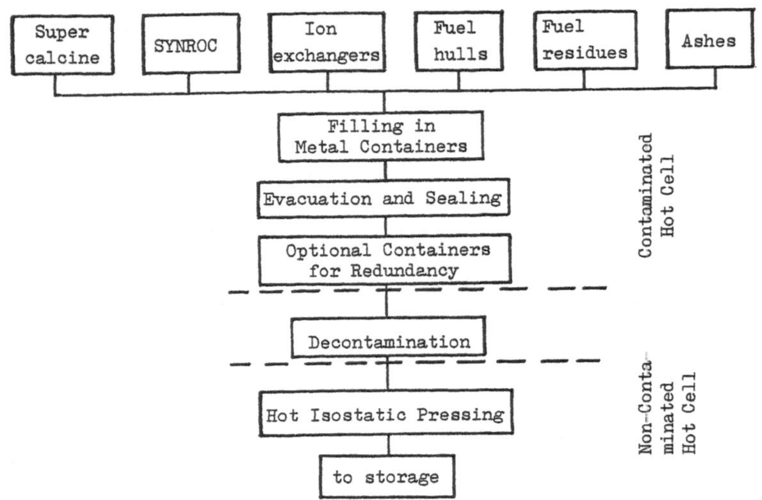

Fig. 1. Formation of fully dense blocks by the HIPOW process from different radioactive waste products including the principle for prevention of lasting contamination of the hot isostatic press.

APPLICATION OF THE HIPOW PROCESS FOR CLADDING HULLS

Zircaloy cladding hulls, remaining after the chop-leach operation in a reprocessing plant, are a type of non-high-level radioactive waste. The hulls contain long-lived alpha emitters confined within a few microns of the hull surfaces. The type and amount of these materials require the same efficient prevention from migration into the biosphere as is the case for HLW after the initial thousands of years. The material is bulky, about 0.4 m^3/ton spent fuel, and great care must be taken to prevent fire while the surface to volume ratio is high because zirconium after catching fire can burn even in water.

Suggested methods of embedding cladding hulls in concrete, bitumen, or low-melting metal alloys give products of high to moderate volume and questionable long-term corrosion resistance. Melting in the form of zirconium eutectics (e.g., with copper and/or nickel) gives a fairly low volume, but segregation of the radionuclides may be a problem and may contribute to unsatisfactory long-term corrosion resistance.

A good process for cladding hulls should preferably give a product with

1. very high corrosion resistance in the repository,
2. minimum gross volume and small surface-to-weight ratio, and
3. dimensions to fit handling devices in the repository used for other types of waste.

Guidelines for the process might be

1. simple and few active processing steps (even at the expense of complicated nonactive steps),
2. very low ignition risk, and
3. low probability of lasting contamination of the high-temperature equipment.

The costs should, of course, be kept as low as possible, but comparisons between competing processes should be based on the costs of the total system including the final repository costs.

Zircaloy has excellent resistance to water corrosion,[4,5] However, most of the activity of the cladding hulls is confined within a few microns of the surfaces and is not protected efficiently from long-term attack by the groundwater. If the hulls could be densified to solid blocks with the activity distributed throughout the entire volume, a waste form with very good long-term fixation of the radioactive elements could be obtained. Because of the high melting point of Zircaloy (above 1800°C), direct melting is very difficult. Furthermore, there would be a tendency for the surface oxides and the fission products to enrich the slag collected at the surface of the ingot, and much of the protective function of the Zircaloy metal would be lost. Most of the volatile constituents would have to be collected from the off-gases, and the high-temperature equipment would be heavily contaminated.

The HIPOW process offers a possibility to bond the Zircaloy cladding hulls to fully dense blocks of the desired size by a high temperature-high pressure treatment in the solid state while enclosed in a hermetically sealed sheet metal containment.

LABORATORY-SCALE EXPERIMENTS

Procedure

Zircaloy-2 tubes (OD 12.25 mm, wall 0.8 mm), black oxidized and treated as for BWR fuel manufacture (oxide layer thickness, ∿1 μm) were cut in 25-mm-long lengths and washed in water. Most of the tubes were then flattened one by one between punches using a press force of about 85 kN. Cylindrical containers made of low carbon steel or unalloyed titanium were used to contain the materials for hot isostatic pressing. They had an ID of 80-85 mm, an inner height of 80 mm, and a wall thickness of 3 or 5 mm. Three different methods of filling the containers with tube pieces (simulated cladding hulls) were used.

In the first method untreated simulated hulls were filled at random. This gave a fill density of 0.9 kg/dm^3 or a fill factor of only 0.14. As it is known from long experience that a container filled with a loose material of such a low fill factor will deform and probably leak during the hot isostatic pressing, titanium powder was poured in to fill the spaces between the hulls. A fill factor of at least 0.6 was achieved easily. This is usually high enough to give a shape-stable compaction.

In the second filling procedure, flattened hulls were poured into the container giving a fill density of 2.4 kg/dm^3 or a fill factor of about 0.36. Titanium powder was then added during repeated tapping of the container. To avoid segregation of the powder and the hulls, the latter were fixed by a wire mesh at the top before starting to fill the powder. A fill factor of >0.6 was obtained.

In the third method flattened hulls were filled into the containers but no further material was added. The fill factor of flattened hulls was, however, found to be only about 0.4 even if a container that is large relative to the size of the flattened hulls is used. This is too low to obtain a controlled shape of the container with its content during the hot isostatic pressing. A punch was therefore used to axially compress the hulls in three consecutive filling steps. The pile of hulls partly sprang back when the force on the punch was released, but by securing the top lid of the container to the wall during the last pressing, a final fill density of 4.2 kg/dm^3 was obtained after pressing with a pressure of 100 MN/m^2.

In all three cases after filling the containers, lids with evacuation pipes were welded to the containers. The evacuation pipes were then attached to a vacuum pump for 12-24 h while the filled containers were heated to a temperature of 150°C to outgas the interior. The evacuation pipes were then crimped and the containers hermetically sealed in preparation for hot isostatic pressing.

The processing parameters selected were 150 MPa argon gas pressure and 3 h sustain time at a temperature which was varied between 900-1250°C in 50°C steps. The heating rate was 400°C/h and the cooling rate was 300°C/h.

Results

The containers filled according to all three methods behaved satisfactorily during the hot isostatic pressing. Even the containers of 3-mm-thick titanium sheet were shown to be thick enough to prevent penetration by the sharp corners of the flattened Zircaloy hulls during the compaction. However, a strong reaction occurred between iron containers and Zircaloy at a temperature of 1100°C or higher, and a 3- to 4-mm-thick intermediate layer of low strength was formed. No such problems were observed with titanium containers even at a 1250°C processing temperature.

The minimum temperature required to obtain strong and fully dense bodies of hulls (6.54 kg/dm^3; i.e., ∿100% TD) was 950°C. The density was the same for all the higher temperatures, but at 900°C a density of only 4.7 kg/dm^3 was achieved. The material to which titanium powder is added can be processed under the same conditions as pure Zircaloy.

Test bars were cut from the blocks to measure the strength. Because the material was difficult to machine, bend testing was selected and test bars 5 x 5 x 40 mm were made. A bend strength in 3-point bending of 700 MN/m was obtained for temperatures above 950°C, and an elongation of 2 to 6% was observed. There was a tendency for the fracture to follow the joints between hulls at 950°C but not at the higher temperatures.

The microstructure of the Zircaloy material without additions is shown in Fig. 2 for 950, 1000, 1050 and 1150°C hot isostatic pressing temperatures. At the higher temperatures considerable grain growth occurs; this becomes more pronounced as the temperature increases.

Good material is obtained with all the above-mentioned filling techniques, but the alternative procedure using hulls without additives is the most interesting for many reasons. The total weight and volume will be 1.5 to 5 times higher in the other two alternatives. The cost for added material is also a negative factor, and material with titanium added might not be as good from a corrosion viewpoint as pure Zircaloy. Of course, an addition of Zircaloy powder is possible but would be costly.

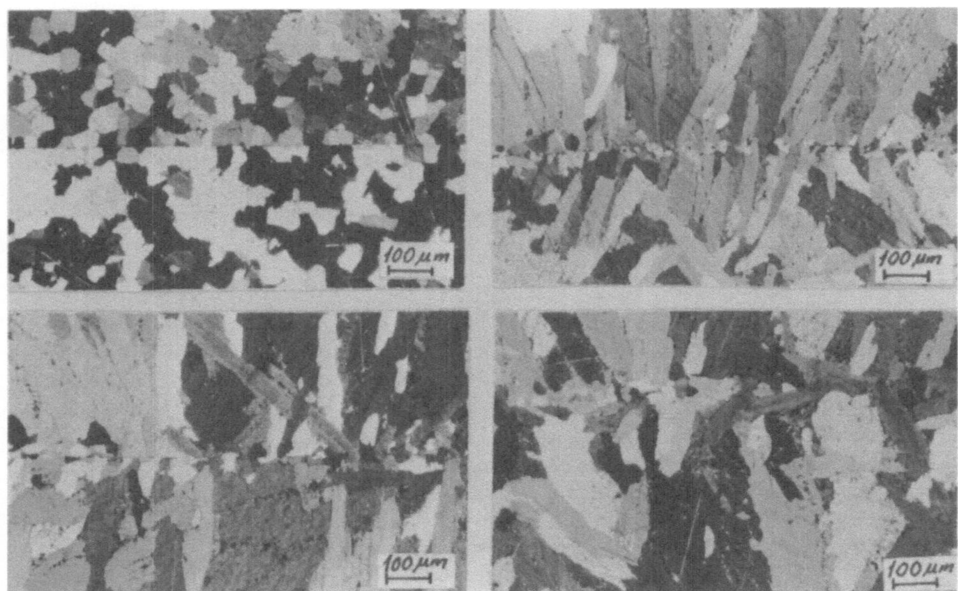

Fig. 2. Microstructure and joint of densified Zircaloy hulls. HIPOW processed at 150 MPa and 3 h at 950°C (top left), 1000°C (top right), 1050°C (bottom left), and 1150°C (bottom right).

OUTLINE OF THE HIPOW PROCESS FOR CLADDING HULLS IN TECHNICAL SCALE

The laboratory-scale experiments with simulated fuel hulls show that fully dense and well-bonded bodies of 100% hulls can be made at temperatures around 1000°C. However, a precompaction of the hulls in the container by a punch was used, and this technique would complicate considerably the handling of real, radioactive hulls. Filling titanium or Zircaloy powder in the voids of the pile of hulls requires material additions of at least 50% of the hull volume. The volume of the finished product will increase accordingly. This is even more pronounced if hulls that are not flattened previously are used. In this case the volume of the added material will be 4 to 5 times that of the cladding hulls. A process without adding foreign material would give both the minimum volume and probably the best long-term leach resistance.

A new type of gas-impermeable metal container to contain hulls during the hot isostatic pressing can solve the filling problem (Fig. 3). The design of this container is such that it readily will compress axially on application of an external gas pressure until support from the compacted hulls is obtained. When the hot isostatic pressing proceeds at elevated temperature, a three-dimensional (isostatic) compaction follows until the Zircaloy block is fully densified. The appearance of such a compacted container filled with flattened hulls is seen to the right in Fig. 3. The container is made with double walls. Since they are sealed separately by welding, the probability that a leak will develop through both of the containers during the hot isostatic pressing can be kept very low. The high-temperature equipment and the hot isostatic press can be maintained therefore in a nonradioactive environment, which considerably facilitates maintenance.

Alternatively, the handling of the hulls before they are sealed into the container can be made under water or in a protective gas to avoid any fire hazard. In the process outlined below, handling under water for the first processing steps is used because the hulls normally come from storage in a water basin.

The size of the finished blocks is such that the same equipment can be used in transport and final repository as will be used for lead-titanium-clad HLW glass containers in the Swedish KBS I concept.[6] About 1850 kg of hulls from about 6 tons of fuel is contained in such a block with 550 to 600 mm diameter and about 1600 mm total height, including a "COGEMA type gripping head."

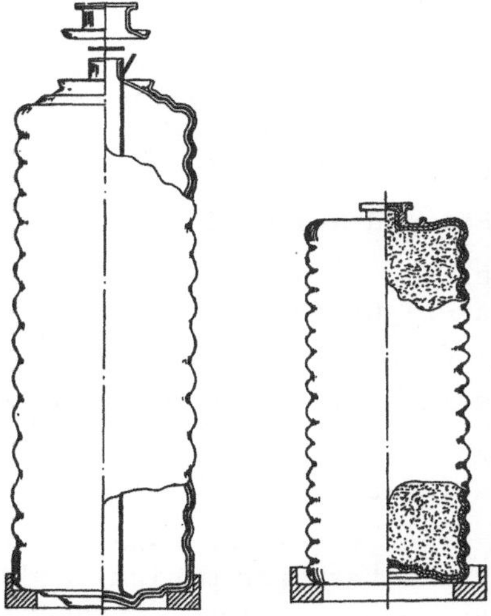

Fig. 3. Axially resilient, double-wall container for the HIPOW process. Prefabricated container with lids and drainage/evacuation tube (left) and filled with Zircaloy hulls and fully densified (right).

The basic steps of the process are the following:

1. Baskets with hulls are collected from storage under water and the material fed (under water) through a coarse sieve which removes large objects like end pieces. These could be filled into separate containers and processed to blocks along a similar route as the cladding hulls.
2. The hulls pass a flattening device. It may consist of large-diameter rolls which flatten the hulls,[7] or equipment with reciprocating jaws.
3. The flattened hulls and other small objects fall by gravity into a prefabricated double wall container of approximately 700 mm in diameter and 2000-mm cylindrical height. It has an inner container of 5-mm-thick titanium and a 5-mm stainless steel outer container. The containers are prepared for separate closures (Fig. 3). The containers have corrugated walls and ends allowing for axial compression when outer gas pressure is first applied.
4. When the container is full, the inner lid is put into position.
5. The container is lifted out of the water basin and is rinsed with pure water to remove the crud and other external contaminants.
6. The water in the container is sucked out through a Zircaloy-2 draining tube (e.g., diam 10/7 mm) ending at the lowest part of the container. The tube was put into position in the prefabricated container.
7. The inner lid is welded tight.
8. The draining tube is attached to a vacuum line and is evacuated to approximately 0.1 mbar, while it is heated to 150°C to remove the remaining water in the container.
9. The tube is then crimped and sealed.
10. The stainless steel cap is placed over the inner container closure and welded gas-tight to the outer container.
11. The container, which is now hermetically sealed and ready for hot isostatic pressing, may be decontaminated if necessary.
12. It is moved through air locks to the cell and inserted into the hot isostatic pressing unit.
13. The hot isostatic press is closed, and the argon gas pressure (which is the pressure transfer medium) and the temperature are increased according to a preselected schedule. The maximum pressure is set at 150 MPa and the temperature at 1050°C. This relatively high furnace temperature reduces the time for heating the center to a minimum of 950 to 1000°C. A sustain time of 3 h after the center has reached minimum 950°C is used.

14. Temperature and pressure of the hot isostatic press are decreased, and the compacted block of Zircaloy is taken out of the press. The original sheet metal container of titanium and stainless steel now appears as layers metallurgically bonded to the Zircaloy block, and the thickness of each has grown to 6 to 8 mm.

15. The cooled block is transported out of the cell to intermediate storage and is ready for transport to final storage.

It seems suitable to have one such HIPOW line for each reprocessing line with an annual capacity of 800 tons spent LWR fuel. The entire hot isostatic processing, including the heating and the cooling period, can then be made inside the pressure vessel of the hot isostatic press. This is of great advantage in order to maintain the tritium quantitatively in the solid Zircaloy, even after this high-temperature treatment as is discussed below.

A processing cost estimate from hulls in the water basin storage area to finished, fully dense blocks indicates that the costs will be less than 0.01¢/kWh even with the relatively long processing times involved in this case.

TRITIUM HANDLING

The Zircaloy cladding of spent fuel from light water reactors normally contains from 100 to 300 ppm total hydrogen. The tritium content is about 3000 times less or 0.03 to 0.1 ppm with a resulting activity of 0.3 to 1 Ci/kg Zircaloy. Above about 400°C the hydrogen can diffuse rapidly in the metal and into the surrounding gas phase. An inherent characteristic of hot isostatic pressing, which is very valuable in this case, is that the total gas volume surrounding the container with Zircaloy during the treatment at elevated temperature is limited and closed. The walls of the pressure vessel, which are kept at low temperature, cannot be penetrated by hydrogen or tritium. The maximum amount of tritium that can be found in the high-pressure gas in the enclosed space at equilibrium and maximum processing temperature is about 0.1% of the total. Upon cooling most of the tritium in the gas will diffuse back into the Zircaloy, and the equilibrium pressure at the temperature when this process ceases to operate is probably several orders of magnitude lower. This means that 99.9 to 99.999% of the tritium can be confined in the Zircaloy block after the process.

CONCLUSION

The suggested process gives a product that has the minimum possible volume and should satisfy very high requirements for long-term isolation from the biosphere. The special double-wall metal container allows for axial precompression upon application of external gas pressure. It can be used to produce compacted blocks of controlled shape even of low-density materials like hulls as well as other types of radioactive materials (e.g., calcine). Thus the handling in the radioactive environment becomes simple even if the container manufacture is rather complex (but fully inactive). The hot isostatic pressing operation is straightforward and in a type of equipment which is usually remotely operated even in normal industrial production (e.g., tool steel production).

An option that seems very attractive is to place small Zircaloy-2 containers filled with fuel residues (fines) or ashes near the center of the container before filling it with flattened hulls. They would be compacted to fully dense "islands" in the big Zircaloy block.

REFERENCES

1. H. T. Larker, pp. 207-10 in Scientific Basis for Nuclear Waste Management, vol. 1, ed. McCarthy, Plenum Press, New York (1979).
2. H. T. Larker, pp. 169-73 in Ceramics for Nuclear Waste Management, ed. T. D. Chikalla and J. E. Mendel, NTIS, Springfield, Va. (1979).
3. H. T. Larker, "Hot Isostatic Pressing - Characteristics and Prospects in Industrial Use," pp. 329-37 in High Pressure Science and Technology, ed. B. Vodar and Ph. Marteau, Pergamon Press (1980).
4. K. F. Flynn, R. E. Barletta, L. J. Jardine, and M. J. Steindler, p. 154 in Scientific Basis for Nuclear Waste Management, vol. I, ed. McCarthy, Plenum Press, New York (1979).
5. W. H. Smyrl, Behavior of Candidate Canister Materials in Deep Ocean Environments, SAND 77-0243, Sandia Laboratories (1977).
6. Handling of Spent Nuclear Fuel and Final Storage of Vitrified High Level Reprocessing Waste, KBS (Nuclear Fuel Safety Project), Box 5864, S-102 48 Stockholm, Sweden (1977).
7. G. Spenk, H. Frotscher, H. Gräbner, and H. Kapulla, "Conditioning and storage of spent fuel cladding hulls by rolling and embedding in concrete," IAEA, International Symposium on the Management of Alpha-Contaminated Wastes, Vienna, Austria, June 2-6, 1980, IAEA-SM-246/17.

BITUMINIZATION OF LOW-LEVEL LIQUID WASTE

Teruo Tokubuchi and Shigeki Kitajima*
Hiroshi Kuribayashi, Takuro Yagi and Tomoyoshi Kagawa**

*Nuclear Power Dept., Kyushu Electric Power Co.
1-82, Watanabe-douri 2-chome, Chuou-ku, Fukuoka, Japan
**Nuclear Project Div., JGC Corporation
14-1, Bessho 1-chome, Minami-ku, Yokohama, Japan

PREFACE

Solidification of radioactive wastes with bitumen has a number of preferable features compared with solidification with cement. Bituminization technology has been developed mainly in European countries such as West Germany, France and Belgium and some of the processes are in practical use at present in various countries of the world.

In Japan, the first bituminization plant "DRUM MIXER" was constructed for the Japan Atomic Energy Research Institute (JAERI) by JGC Corporation in 1973. However, this plant was intended for the treatment of radioactive wastes from laboratories. Regarding wastes generated at commercial nuclear power plants, Kyushu Electric Power Co. has just started the operation of a facility for the bituminization of liquid waste from its PWRs by using the DRUM MIXER process. In Japan, besides these two DRUM MIXER plants presently in operation, three are under construction and two are on the drawing boards. The general features of the bituminization process will be reviewed and the results of fundamental research work and actual operation of the DRUM MIXER will be discussed.

1. DRUM MIXER

In bituminization, there are continuous processes using extruders or thin film evaporators in addition to batch processes. JGC is working on a continuous process using an extruder and a batch process using a mixing evaporator called a DRUM MIXER. The DRUM MIXER is easy to operate and is widely applicable to the treatment of many kinds of wastes, because of the simplicity and flexiblility of the batch process.

Bitumen (or asphalt) is fed into the DRUM MIXER and heated to temperatures ranging from 150° to 200°, preferably 170° to 180°C. Heating is effected by a heat transfer medium which flows through the jacket and the rotor of the DRUM MIXER.

Figs. 1 and 2 show the schematic process flow and constructional outline of the DRUM MIXER.

Liquid waste is fed into the preheated liquid bitumen in the DRUM MIXER by a pump. The feed rate is so controlled that the temperature of the bitumen is kept constant. If the temperature of bitumen in the DRUM MIXER is higher than the preset temperature, the control valve opens wider to let a larger amount of the liquid waste flow into the DRUM MIXER and, consequently, to lower the temperature of bitumen within the DRUM MIXER. In the opposite case, the operation proceeds in the reverse direction.

Water in the liquid waste evaporates while the waste is being uniformly mixed with the heated bitumen by the mixer which revolves at the rate of 30 to 150 rpm within the DRUM MIXER.

As a result, non-volatile salts in the liquid are dispersed in the form of fine dry particles into the bitumen and coated by it. This is why a stable solid product can be obtained, after solidification through cooling, of which the leaching rate of radioactive salts in water is extremely low.

After passing through a condenser, the water vapor generated from the liquid waste in the DRUM MIXER is condensed and stored in a condensate receiving tank, and after oily matter in the condensate has been removed, it may be released. The radioactivity of the condensate is normally very low, the decontamination factor (DF) being about 10^3. The DRUM MIXER is evacuated so that the pressure therein is kept slightly lower than atmospheric pressure during the mixing operation.

After a given amount of the liquid waste has been added, the feeding pump is stopped and the flow of the heating medium is switched from the heater to a cooler to prevent over-heating of the bitumen in the DRUM MIXER. When the temperature of the bitumen has lowered to a proper level, the content is poured into a drum by gravity flow.

The dimensions of the DRUM MIXER required for producing one drum product (200 l) within an operating time of four to six hours per batch are approximately 800 W x 1,800 L x 800 H in mm.

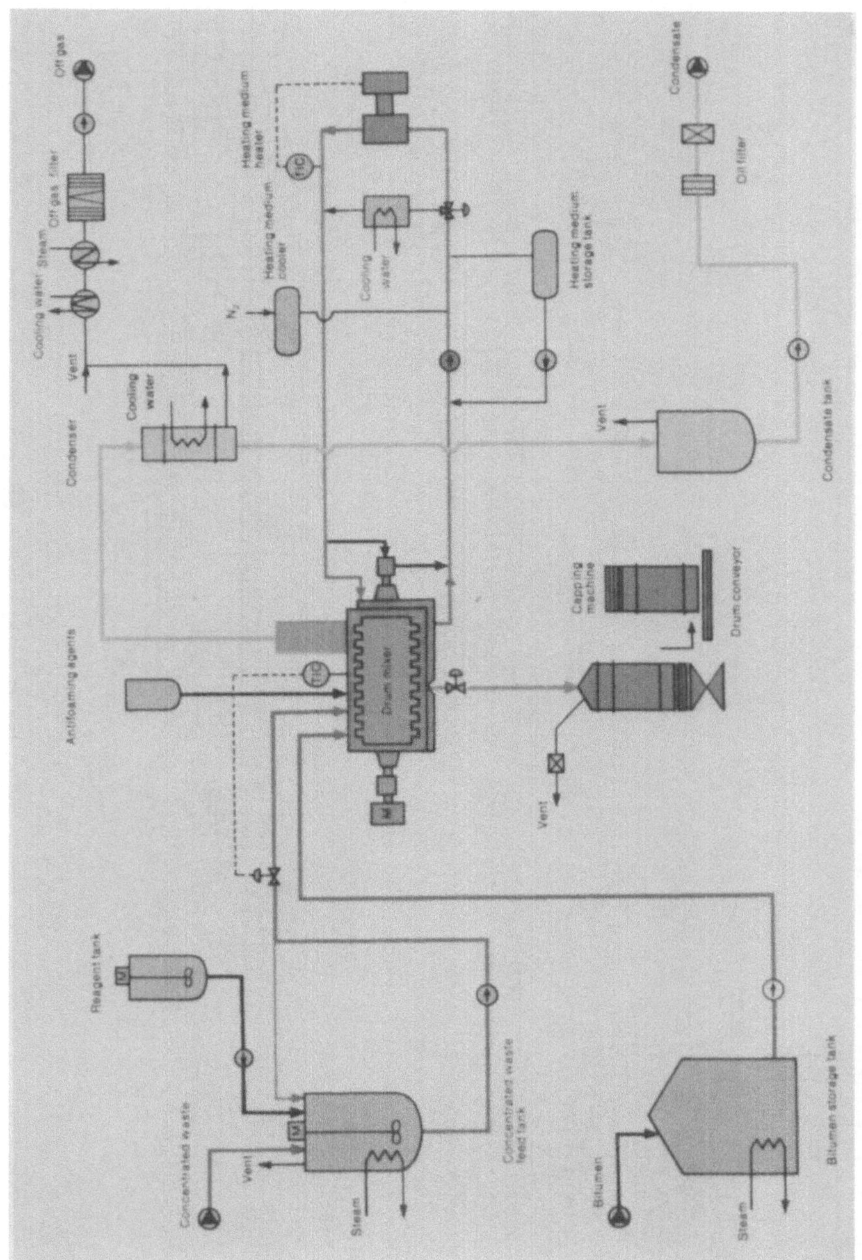

Fig. 1. Bituminization plant "drum mixer" schematic flow diagram.

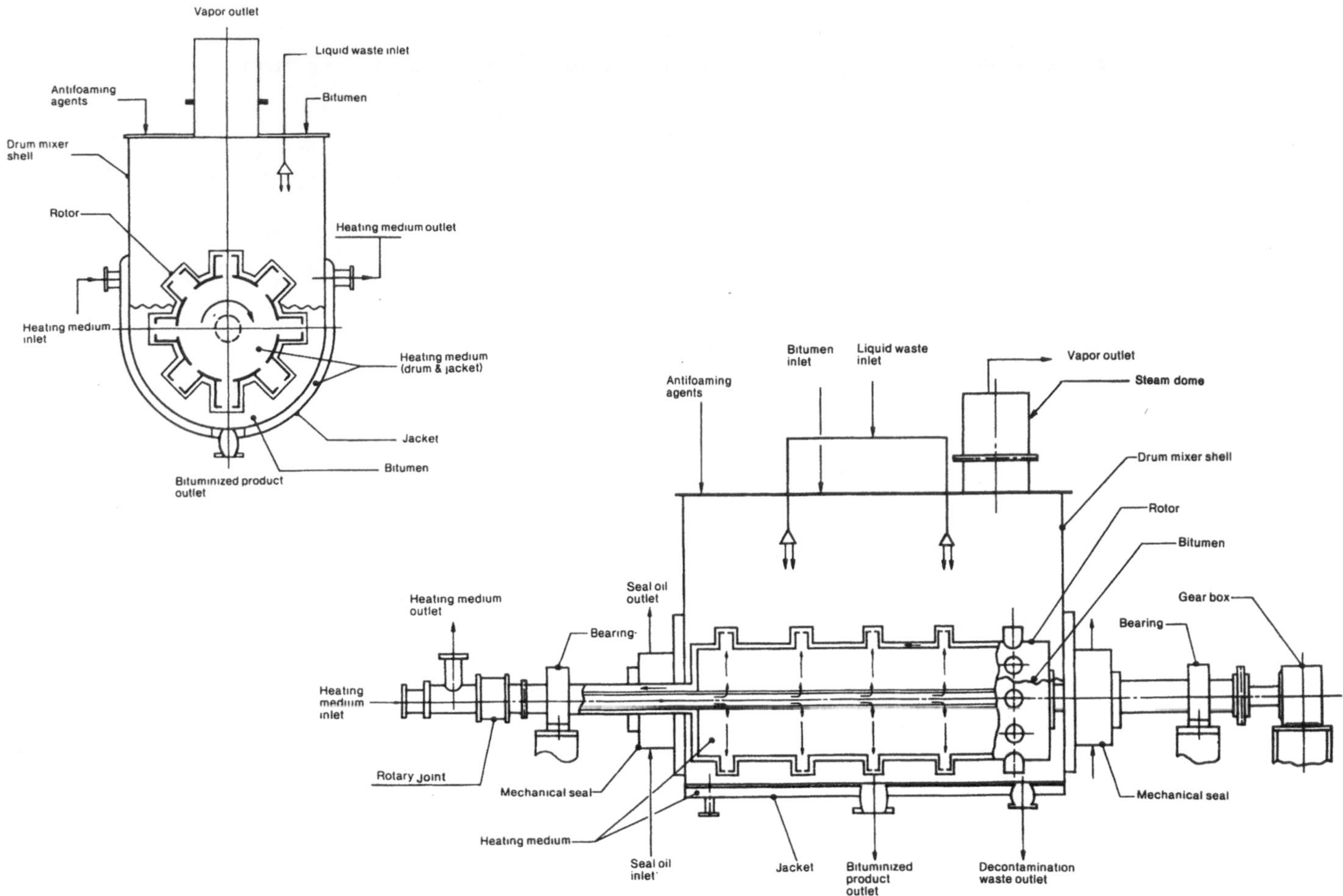

Fig. 2. Constructional outline of drum mixer.

2. WASTE

Wastes which can be treated by the DRUM MIXER are concentrated liquid waste containing salts such as sodium borate, sodium sulfate, etc., spent ion-exchange resins (bead, powder) and filter sludge.

Kyushu's plant was designed to treat 140 l/hr of liquid waste containing sodium borate, with its concentration presumed to be 21,000 ppm in terms of boron. At present, it is not set up to treat spent ion-exchange resins, but has been used to treat equipment and other drains which are less active and contain less salts, in order to reduce the activity release to the environment.

3. OPERATION RESULTS

Actual (active) operation started in the middle of April 1980 after extensive test (inactive) operations. The results of both operations are summarized as follows:

(1) Evaporation Capacity

Compared with the design capacity of 140 l/hr, the actual capacity was found to be as high as 200 l/hr. This decreased to 150 l/hr after 20 successive batch operations. This capacity is expected to be maintained in further operations.

The total volume of liquid waste treated from April to September 1980 at Genkai amounted to 75 m^3 as shown below:

Drain	Total Volume (m^3)	Activity (μCi/ml)	Non-volatile Content (wt%)
Equipment*	23	10^{-5}	10.0
Equip.* + Laundry	52	10^{-3}	2.0

*After neutralization

(2) Overall Volume Reductivity

Up to now, at Genkai, 75 m^3 of liquid waste has been bituminized into 41 drums, each with 200 l capacity. Overall volume reduction ratio, calculated at 9.1, seems rather high. This may be accounted for by the fact that the equipment/laundry drains had not been preliminarily concentrated to a sufficient extent due to the early stage of the hot test operation.

(3) Mechanical Performance

It has been confirmed that the pumps and related instruments for transporting liquid waste, bitumen, and heating medium are provided with the required precision and reliability. Furthermore, it has been found that the thermal elongation of the DRUM MIXER and rotating mixer is within allowable limits, and that the durability and tightness of mechanical seals and the rotary joint (for supplying the heating medium to the rotating mixer) are adequate.

(4) Decontamination of the DRUM MIXER

For maintenance reasons, it is necessary for workers to have access to the DRUM MIXER without exposure to radiation. In this regard, it is possible to decontaminate the DRUM MIXER by washing it with a mixture of hot water and non-flammable organic solvent.

After a month's operation, decontamination by using solvent and water was tried and the dose rates before and after decontamination were measured and found to be 2.0 mrem/hr and less than 0.1 mrem/hr at the outer surface of the DRUM MIXER (including thermal insulation), respectively.

(5) Condensate Water

Oily matter, probably originating from bitumen, is observable in the condensed water. It is possible, however, to reduce this oily matter to less than 1 ppm by using an oil separator and active carbon adsorbent bed. The decontamination factor for the condensate water, which is normally 10^3 for sodium and boron, decreases to 10^2 if foaming occurs during operation. For this reason, the use of an anti-foaming agent is recommended.

(6) Typical Operational Data

Shown in Table 1 are typical operational data. Moreover, a record chart showing the temperature inside the DRUM MIXER and that of the heating medium is shown in Fig. 3.

Table 1. Typical Operational Data

Bitumen charge per batch	(kg)	130
Liquid waste feed rate	(l/hr)	150
Mixing ratio (non-volatile waste/bitumen)	(wt/wt)	40/60
Mixing temperature	(°C)	180
Heating medium temperature (inlet ave.)	(°C)	242
Heating medium flow rate (rotor)	(m^3/hr)	25
(jacket)	(m^3/hr)	15
Rotation rate of mixing rotor	(rpm)	60
Average duration per batch (equipment drain)	(hr)	6
(equipment and laundry drain)	(hr)	40
Volume reductivity (equipment drain)	(-)	4.3
(equipment and laundry drain)	(-)	18.6

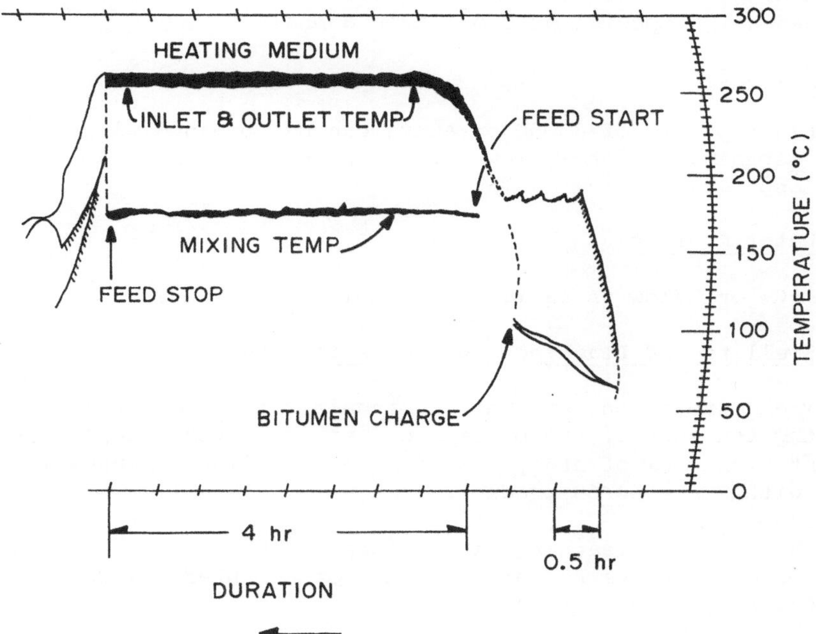

Fig. 3. Typical temperature record chart.

4. BITUMINIZED PRODUCT PROPERTIES

(1) Specific Gravity

Regarding boric acid with a Na/B ratio = 1.0, namely equimolally neutralized liquid waste, a solid with a specific gravity of more than 1.3 is obtainable at a salt to bitumen mixing ratio of 40/60.

(2) Softening Point

The softening point of the raw bitumen has a decisive effect on that of the product. Straight 40/60 asphalt gives a product with a softening point of 50° to 60°C. Higher softening points can be obtained using special or blown type asphalt.

(3) Flash Point

Flash points of 330°C or above can be obtained with straight 40/60 asphalt. Slightly lower flash points are obtainable with blown asphalt.

(4) Water Content

1.0% or below is easily attainable.

(5) Swelling and Leaching Tendencies in Water

The correlation between the bituminization conditions and the swelling tendency of the bituminized solid is very complex and is presently an item of study. So far, no swelling tendency is observable with S/B = 40/60 product.

The observed leaching rate values of 10^{-3} to 10^{-6} $cm^3/(cm^2.day)$ are considered to be better than that of cemented solids by one to two decimal places.

STUDIES ON SINTERED TITANATES AND ZEOLITES AS HOSTS FOR MEDIUM LEVEL RADIOACTIVE WASTE

S. Forberg, T. Westermark and L. Fälth

Dept. of Nuclear Chemistry, The Royal Institute of Technology, S-100 44 Stockholm, Sweden

Inorganic Chemistry 2, Institute of Technology, S-220 07 Lund, Sweden

INTRODUCTION

In Sweden considerable interest has been paid in recent years to potentially safer alternatives to bitumen or concrete fixation of MLW from nuclear reactors, e.g., organic ion exchangers containing activated corrosion products and fission products from failing fuel element encapsulations. A system for transferring long-lived activities from organic to inorganic ion exchangers, i.e. titanates and zeolites, was demonstrated on a bench scale.[1,2] In comparison with currently practised techniques, this system might result in smaller volumes and more stable final products.

The stability depends of course on the conditioning of the inorganic ion exchangers. The purpose of this work was to study and compare the quality of the products obtained by hot pressing and hot isostatic pressing[3] in order to guide the choice of full-scale equipment.

CHOICE OF MATERIALS AND TESTS

The only authentic materials available for sintering hitherto are the inorganic ion exchangers from bench scale experiments[1,2] with one batch of spent organic ion exchangers. To cover the span of probable concentrations in the titanates, the Swedish nuclear power stations were asked for the maximum content of each single corrosion product in organic exchangers from BWR as well as from PWR. From the answers a list of upper limits for all elements from chromium to zinc was erected as a guide for the experiments. Occasionally, much lower concentrations may be encountered. As a lower limit, one-fourth of the upper limit for each element was chosen. Relative variations of single elements were not actualized at this stage.

In both cases some sodium was expected to remain with the corrosion products in the titanate. The effect of sodium upon the stability was not obvious. Having the practical possibility to displace the sodium selectively with either an acid or an ammonium salt, these alternatives were found worthwhile to study.

Besides mechanical strength and absence of pores, the leaching rate in water at the repository, probably a rock cave at about 40 m depth, is the vital point of these products. The near absence of α emitters and the very low concentration of transmutating nuclides render processes like, e.g., swelling or metamictization of minor importance. The heating effect developed is correspondingly low and the temperature to be considered is well below 20°C.

For the leaching tests, radioactive isotopes were needed, e.g., strontium for the titanates and cesium for the zeolites. For additional information, tracers for cobalt, sodium and cesium were introduced in the titanates. For the first screening tests of the many similar materials, 3-day Soxhlet leach tests were considered adequate and chosen for their simplicity, leaving more thorough and realistic tests to be done later. For scientific and understanding purposes, the characterization of internal structures is of foremost importance. X-ray diffraction with a Guinier-Hägg powder camera was the available method.

EXPERIMENTAL DETAILS AND RESULTS

Preparatory studies

Before sintering the labeled materials, the crystallization and recrystallization as a function of temperature were followed for two zeolites and for sodium titanate loaded with ∿5 wt % of corrosion products. In two samples of titanate, the remainder of the sodium was displaced with stoichiometric amounts of acid or ammonium salt respectively. The acid also displaced some of the corrosion products. Later displacements of sodium were performed solely with an ammonium salt.

The crystalline phases obtained after heating to successively higher temperatures are presented in Table 1. Due to low concentrations, the fate of individual elements could not be followed. To study possible elemental behavior, a series of titanates was saturated with important single elements. The crystal structures formed are shown in Table 2.

TABLE 1

Crystalline phases identified by means of a Guinier-Hägg powder camera after heating for two hours in open platinum crucibles. The abundances were judged from the intensity of diffraction lines.

A = anatase, R = rutile, X = unidentified

Starting material	(Surplus) counter-ion	Temperature (°C)	Type of structure (wt %)
Synthetic	Na^+	∿ 900	amorphous
mordenite	Cs^+	∿1000	amorphous
		∿1200	$CsAlSi_5O_{12}$ (new phase)
Zeolite P	Na^+	∿ 900	nepheline
	Cs^+	∿ 600	plagioclase
		∿1100	pollucite + quartz
Titanate	Na^+	∿ 400	amorphous
with ∿ 5 %		∿ 700	A 30, R 20, $NaFeTi_3O_8$ 50
corrosion		800-1200	R 50, $NaFeTi_3O_8$ 50, X
products	H^+	500- 600	A
		700- 800	R 85, $Sr_{0.2}Cu_{0.8}Ti_{1.1}O_3$ 15
		900-1200	R 85, perovskite 15, X
	NH_4^+	∿ 500	A
		600- 700	A 50, R 50
		900-1200	R 85, perovskite 15, X

Preparation of starting materials

After the transfer process, the zeolites are expected to remain essentially in the sodium form, the cesium constituting minor concentrations. Hence, the zeolite P studied was prepared simply by adding a carrier-free cesium tracer to an aqueous slurry of the zeolite. The mordenite came from authentic transfer experiments.

The titanates were prepared by exchanging partially, in chromatographic columns, the sodium with a mixture of corrosion products according to the stipulated upper and lower limits. For practical reasons, chromium was added in the chromic form, although chromate also occurs in the organic ion exchangers. All other corrosion products were introduced in the form Me^{2+}. The concentrations of

TABLE 2

Appearance and disappearance of crystalline phases during heating of single-element titanates to successive, higher temperatures for two hours in open platinum crucibles (Guinier-Hägg powder camera). A = Anatas, R = rutile. CPI = corrosion product ion at the exchange. The numerals at 1200 °C indicate relative abundances.

CPI	Mn^{2+}	Fe^{2+}	Co^{2+}	Cu^{2+}	Zn^{2+}
Temperature					
1200°C	R 7 $MnTiO_3$ 3	R 7 Fe_2TiO_5 3	R 7 $CoTi_2O_5$ 3 No $CoTiO_3$	R 6 Cu_2TiO_3 3 Cu_3TiO_5 1	R 6 Zn_2TiO_4 4
1000°C				Cu_2TiO_3 No CuO	
900°C	$MnTiO_3$ No Mn_2O_3				
800°C		No A			No A
700°C	No A	R Fe_2TiO_5	$CoTiO_3$ No A	No A	R
600°C	R Mn_2O_3	A	R	CuO	$ZnTiO_3$
500°C	A		A	R	A
400°C				A	

corrosion products were analysed by means of X-ray fluorescence by Tekniska Roentgencentralen, Stockholm. The results are shown in Table 3.

At either level of corrosion products the excess of sodium in one column was displaced with a stoichiometric amount of ammonium. In all cases, the salt solutions were displaced from the columns with pure water before emptying and drying. After drying at 600°C, where the individual titanate beads break into splinters, each material was thoroughly mixed in order to smooth the fractionation inevitably occurring in the columns.

Pressing

The hot pressing was performed in graphitic cylinders under nitrogen at atmospheric pressure. The materials were axially compressed with 500 N cm^{-2} while heating to 1070°C for 2 hours.

TABLE 3

Concentration in g/kg of corrosion products in titanates after hot isostatic pressing, according to XRF analyses. The loading level and the surplus counter-ion are indicated. Tracers: ^{22}Na, ^{57}Co, ^{85}Sr, ^{134}Cs.

Level	High		Low		0
Counter-ion	Na^+	NH_4^+	Na^+	NH_4^+	Na^+
Cr	6	5	1	1	
Mn	7	8	2	3	
Fe	18	33	20	19	
Co[a] . Fe, unsintered				4	2
Ni	1	1	<0.5	<0.5	
Cu	15	16	2	2	
Zn	12	11	2	3	

[a]Co was designed to yield 1/5 of the concentration of Ni.

For hot isostatic pressing the dried materials were packed in steel cylinders and pressed with a piston at room temperature to the limit where the cylinder began to expand. After coating the well-filled cylinders with "Fiberfrax", a ceramic fibrous felt of an aluminium silicate, lids with central steel tubes were welded on. These containers were placed in a furnace at 600°C for 4 hours while evacuating through the piping. After leach testing, the tubes were closed, by squeezing, cut, and welded. As an extra measure against contamination of the press, each sample was provided with a similar, tightly fitting outer container.

The hot isostatic pressing was performed by ASEA's High Pressure Laboratory at Robertsfors, Sweden. The temperature, 1200°C, was kept for 2 hours at an argon pressure of 150 MPa. Most of the containers failed during the pressing, probably due to welding deficiencies. At proper functioning, the steel containers are heavily compressed and deformed. Many containers were hardly deformed, revealing leakage at an early stage of the pressing cycle.

Internal structures

The hot pressed as well as the hot isostatically pressed materials were examined by means of a Guinier-Hägg powder camera. Exactly the same crystallinic phases appeared in the two cases. An attempt was made to quantify the abundances from optical densities of diffraction lines. No significant difference was found

TABLE 4

Crystalline structures and abundances, judged from visual inspection of diffraction lines from Guinier-Hägg powder patterns, after pressing at 1070°C and isostatic pressing at 1200°C. Low percentages may be uncertain within a factor of 2. Starting materials, loading levels and counter-ions according to Table 3.

R = rutile, C = crichtonite e.g. $Fe^{2+}_{16}Fe^{3+}_{14}Ti_{66}O_{169}$.

Starting material	Type of structure
Mordenite, authentic, Na^+	Amorphous
Zeolite P, Na^+	Nepheline
Titanate, High, Na^+	R 60, $NaFeTi_3O_8$ 30, C 10
Titanate, High, NH_4^+	R 90, C 5, Ilmenite 5
Titanate, Low, Na^+	$Na_2Ti_6O_{13}$ 60, R 20, $NaFeTi_3O_8$ 20
Titanate, Low, NH_4^+	R 90, C 5, Ilmenite 5
Titanate, authentic, Na^+	$Na_2Ti_6O_{13}$ 70, $NaFeTi_3O_8$ 30

between results from the two sintering processes. The findings are shown in Table 4.

Leaching tests

For the purpose of screening, the Soxhlet leach test was the method of choice. Since there is no accepted international standard, the conditions were chosen from a standard proposal with one exception: Too small pieces of material were available for monolithic leaching; therefore, the materials were crushed and sieved, and ∿1 g of the fractions 250-500 μm were placed in open steel cylinders with bottoms of stainless steel nets, and leached for 3 days. A few grains of the same fractions were studied topographically by means of SEM; somewhat larger amounts were ground for diffraction studies.

After leaching, the flask containing the 300 ml of leachate was boiled for one day after adding carriers and nitric acid to decontaminate the glass walls. Cobalt and strontium were precipitated as carbonates, the centrifugate was evaporated to dryness, and the tracers were measured by means of a Ge(Li) detector. The gamma line of ^{85}Sr from the solid material could not be separated from the annihilation radiation from ^{22}Na. Therefore, strontium was analyzed by means of the $K_{\alpha}X$ radiation. The specific surface of leached samples was measured by the BET method, with krypton. In most cases the surface was very well defined. The surfaces and the leaching rates are given in Table 5.

TABLE 5

Leaching rates, μg $cm^{-2}d^{-1}$, obtained from Soxhlet tests on ∿ 1 g of crushed samples, 250-500 μm. The materials, classified as in Table 3 and 4, were sintered by hot isostatic pressing or by hot pressing, indicated by HP. For Cs the leached percentage is also given. The surfaces were measured by the BET method.

Material		Tracer					Surface
		Na	Co	Sr	Cs	%	cm^2/g
Mordenite Auth.	HP				2.5	0.13	170
Zeolite P Na^+	HP				3.0	1.3	1460
Titanate High	Na^+		2.3	7.1			1450
Titanate High	NH_4^+	51	40	61	320	35	370
Titanate Low	Na^+	3	4.6	0.6	48	22	1530
do.	HP	1.5	0.3	2.3	16	2.9	590
Titanate Low	NH_4^+	56	90	14	160	2.9	60
Titanate, Auth.	Na^+		0.04		25	34	4490
do.	NH_4^+		2.7				1730

DISCUSSION

Due to the leakage of argon, the hot isostatic pressing did not appear to have an advantage. The extremely large surfaces of authentic titanates shown in Table 5 must be ascribed to this phenomenon. The merits of the two sintering methods could not be objectively compared from the results obtained. Therefore, more experiments are needed and planned. These are studies on the conditioning of the titanates previous to the sintering, on the welding technique and performance, and at a higher sintering temperature, 1300°C, known to yield very good results.[4]

The upper and lower limits of the concentration of iron in the titanates were designed to equal those of zinc. Apparently the mean content of the iron increased about 2 wt % during the isostatic pressing. In addition, the concentration increased much more in a layer in contact with the steel container. The effect upon the mean leach rate of each sample might be considerable and could vary with the material and the tracer. It has to be considered in future studies.

The leached fraction of cesium from some titanates are too large to be linear with the time during the tests, 3 days. The thinnest grains may have been nearly exhausted. Fortunately, this is scarcely the situation for the other tracers. According to the transfer process,[1,2] the leach rate for cesium is of practical interest only for the zeolites.

Although well defined, the largest surfaces may not have been effectively accessible for leaching and the corresponding leach rates must be treated with some care.

Formally, a leach rate of 3 $\mu g\ d^{-1}\ cm^{-2}$ for the zeolites corresponds to the leaching out of cesium from a surface layer of 0.15 cm during 300 years. The same leach rate of strontium from titanates corresponds to a depth of 0.08 cm for 300 years. Bearing in mind that the temperature of the deposit will be at least 80°C lower, and the water flow much lower, the examples mentioned here might be quite satisfying. After a more successful sintering, the leach rates may be still better and a leaching test at more realistic conditions might be worthwhile. Before then a report in Swedish will appear with more details than the present study.

ACKNOWLEDGMENT

We are grateful for all assistance offered by members of each department and of Studsvik Energiteknik AB, and to the staff of the sponsor, the National Council for Radioactive Waste, for valuable discussions.

REFERENCES

1. R. Arnek and S. Forberg, "A System for the Transfer of Long-Lived Radioactive Nuclides from Spent Resins to Zeolites-Titanates," Report PRAV 3.19, National Council for Radioactive Waste, Box 5864, S-102 40 Stockholm, Sweden (1979).

2. S. Forberg, T. Westermark, R. Arnek, I. Grenthe, L. Fälth and S. Andersson, "Fixation of Medium Level Wastes in Titanates and Zeolites," Symposium on the Scientific Basis for Nuclear Waste Management, Boston, Mass., Nov. 26-30 (1979).

3. H. T. Larker, Hot Isostatic Pressing for the Consolidation and Containment of Radioactive Waste, in: "Scientific Basis for Nuclear Waste Management," G. J. McCarthy, ed., Plenum Press, N.Y. (1979).

4. S. Forberg, T. Westermark, H. Larker and B. Widell, Synthetic Rutile Microincapsulation, in: ibid.

DIFFUSION OF CESIUM IN CONCRETE

K. Andersson, B. Torstenfelt, and B. Allard

Department of Nuclear Chemistry
Chalmers University of Technology
S-412 96 GÖTEBORG, SWEDEN

INTRODUCTION

Concrete has been suggested as a possible encapsulation or overpack material for long-term storage of radioactive waste, particularly in conjunction with the deposition of low and medium level waste [1]. The low and medium level waste might contain a wide range of long-lived radionuclides, such as ^{14}C, ^{59}Ni, ^{90}Sr, ^{99}Tc, ^{129}I, ^{135}Cs, ^{137}Cs, actinides, etc, which due to their long half-lifes would be potentially hazardous if released into the groundwater.

A study of the sorption and diffusion of cesium and iodide in slag cement paste and concrete in different aqueous environments is described in this paper.

MASS TRANSFER IN A CONCRETE BARRIER

Concrete may be described as a porous solid with water filled pores. For a mass transport of radionuclides in this medium diffusion in the pores would be the dominating mechanism. The overall diffusivity in a porous solid is largely related to the porosity, but also to properties such as tortuosity and constrictivity, as well as sorption reactions on the solid surfaces.

Diffusion due to a concentration gradient can in general be described by the following equation (one dimensional diffusion with constant diffusivity)[2]:

$$\frac{dc}{dt} = D \frac{d^2c}{dx^2} \tag{1}$$

where c = concentration in liquid ($moles/m^3$), t = time (s), D = diffusivity (m^2/s), and x = length (m).

If there is a reversible sorption reaction on the solid, the diffusion can be described by :

$$\frac{dc}{dt} = D/(K_d \rho_p + 1) \frac{d^2c}{dx^2} \tag{2}$$

where K_d = distribution coefficient (m^3/kg), and ρ_p = density of the solid (kg/m^3). Thus, the diffusivity will decrease by a factor ($K_d \rho_p$+1) by the sorption. From measurements of sorption and mass transfer rates the diffusivity can be calculated.

PARAMETERS OF IMPORTANCE FOR THE MASS TRANSPORT IN CONCRETE

The properties of the solid concrete and the pore water in the concrete changes dramatically during the initial hydration process of a freshly moulded concrete [3]. This process would require a minimum of one year. After this period of time the further degradation processes leading to volume reduction and structure changes are much slower (thousands of years)[4]. Mass transport studies in concrete must therefore be performed on samples aged at least one year. (A detailed description of cement chemistry, etc is given in ref. 3.)

Solid Phase

The solid phase consists of a cement matrix and a rock ballast. The properties of both these materials and the effects of changes in any of them must be known. The structure of the cement matrix changes with time due to crystallization, leaching, and chemical reactions with species in water or air. The sorption capacity of the ballast may also affect the mass transfer rate.

Liquid Phase

The composition of the liquid phase influences the mass transfer rate in the pores in various ways. The leaching and degradation of the cement matrix will be dependent on pH of the solution as well as on the presence of certain aggresive agents such as Mg^{2+}, HCO_3^-/CO_3^{2-}, SO_4^{2-}, etc. Agents forming complexes with the transported species would also be of importance for the mass transfer rate.

EXPERIMENTAL

Choice of Nuclides

The sorption of iodide and cesium and the diffusion of cesium were studied using ^{131}I (half-life 8.04d) and ^{134}Cs (half-life 2.06y).

The nuclide concentrations in both the batch and the diffusion experiments were about 10^{-8}M. No inactive carrier was used.

Solid and Liquid Phases

The solid phase was either a concrete made from slag cement or the pure cement paste without ballast. Samples aged 1-2 years, (provided by the CBI, The Swedish Cement and Concrete Research Institute), were used in the experiments. All samples had been stored in the absence of air, either in saturated $Ca(OH)_2$-solution or covered with an asphalt layer. The compositions of the test samples are given in Table 1.

The aqueous phases were artificial pore water (two different compositions used), artificial sea water (typical of Swedish West Coast), artificial Baltic water (average of Swedish East Coast), artificial groundwater, and artificial groundwater with added carbonate, sulphate, and magnesium, respectively. The water compositions are given in Table 2.

All the experiments, both sample preparation, and batch and diffusion measurements, were performed in an inert atmosphere (N_2) to avoid carbonatization of the solid.

Table 1. Test Samples

No.	Composition [a]	Water/cement ratio	Age [b] (months)
C1	Paste, A	0.35	17
C2	Paste, B	0.35	12
C3	Paste, A	0.40	25
M1	Mortar[c], A	0.40	25
M2	Mortar[c], A	0.60	25

[a] A: 47.6% CaO, 29.3% SiO_2, 7.9% Al_2O_3, 1.3% Fe_2O_3, 4.9% MgO, 3.8% SO_3, 1.0% K_2O, 0.7% S^{2-}.

B: 47.3% CaO, 28.5% SiO_2, 7.7% Al_2O_3, 2.0% Fe_2O_3, 4.7% MgO, 2.2% SO_3, 1.1% K_2O, 0.7% S^{2-}.

[b] Time between preparation and start of sorption/diffusion experiments. C1 and C2 were covered with asphalt during storage and C3, M1, and M2 were stored in sat. $Ca(OH)_2$-sol.

[c] Cement:aggregate = 1:4 (sand, 4 mm)

Table 2. Water Compositions

Ion	SW		BW		GW		PW1		PW2	
	mg/l	mM	mg/l	mM	mg/l	mM	mg/l	mM	mg/l	mM
Na^+	10560	459	4980	216	65	2.84	10000	435	7000	304
K^+	380	9.71	179	4.58	3.9	0.100	22000	563	12000	307
Mg^{2+}	1270	52.3	599	24.7	4.3	0.177				
Ca^{2+}	398	9.93	188	4.68	18	0.448	10	0.250	10	0.250
Sr^{2+}	13.3	0.151	6.3	0.071						
Al^{3+}							75	2.78	75	2.76
F^-	1.4	0.071	0.66	0.033						
Cl^-	18980	535	8950	252	70	1.97				
Br^-	64.5	0.807	30.4	0.381						
HCO_3^-	140	2.29	140	2.29	123	2.01				
SO_4^{2-}	2650	27.6	1250	13.0	9.6	0.100	400	4.17	400	4.17
BO_3^{3-}	24.7	0.420	11.6	0.198						
SiO_2					12	0.206	20	0.343	15	0.258
pH	8		8		8.2		13.9		13.2	

SW = Artificial sea water I = 1.2
BW = Artificial Baltic water I = 0.56
GW = Artificial groundwater I = 0.0084
PW = Artificial pore water

Batch and Diffusion Measurements

The sorption of I and Cs in the different waters was measured on crushed cement paste and concrete by a batch technique[5]. Distribution coefficients (mol/kg solid per mol/m^3 liquid) were determined after a contact time of 1 week.

The diffusivity in a solid can be determined either under steady state or non-steady state conditions. The steady state diffusion can be determined by studying the mass transport from one solution to another, separated by a thin slice of the solid. Non-steady state diffusion can be studied by analyzing the concentration profile in a solid after a certain contact time with a solution containing the diffusing species (preferably with a geometry allowing diffusion in only one direction). In this study, only non-steady state diffusion experiments were performed.

Test samples with a diameter of 2 cm were coated with an epoxy resin on all sides but one of the planar end surfaces, and fitted

Table 3. Distribution Coefficients (K_d, m^3/kg)

Nuclide	Solid	GW	$GW+HCO_3^-$ [a]	$GW+SO_4^{2-}$ [b]	$GW+Mg^{2+}$ [c]	BW	SW	PW1	PW2
^{131}I	C1	0.007	0.003	0.013	0.007	0.005	0	0.019	
	C2	0.005				0.007	0	0.016	
	C3	0.003				0.003	0		0.017
	M1	0	0			0.005	0.001		0.001
	M2	0		0.004	0.002	0	0		0.003
^{134}Cs	C1	0	0.005	0.004	0	0.001	0	0	
	C2	0				0.002	0	0	
	C3	0				0.001	0		0.007
	M1	0.240	0.850			0.020	0.020		0
	M2	0.240		0.330	0.335	0.020	0.006		0

[a] Tot. HCO_3^- : 623 ppm.

[b] Tot. SO_4^{2-} : 50 ppm.

[c] Tot. Mg^{2+} : 20 ppm.

into metal tube holders. These holders were made to fit into a specially built grinding device, used for studying concentration profiles in solids (see Fig. 1). The samples were equilibrated with pore water and then put into a pore water solution containing the radionuclide. The nuclide concentration in the water was kept constant during the experiment. After a contact time of 2-3 months the concentration of the radionuclide as a function of penetration depth in the solid was analysed. Typical concentration profiles are shown in Fig. 2. A detailed description of the experimental procedure is given elsewhere[6].

RESULTS AND DISCUSSION

Batch Measurements

The measured distribution coefficients for cesium and iodide are given in Table 3.

The sorption of cesium on concrete shows a similar qualitative ionic strength dependence as has been observed for the pure ballast material, i.e. an increased sorption with decreasing ionic strength[7]. This is, however, not the case for pure cement paste, which shows practically no sorption of cesium at all, indicating that the sorption in the concrete might be largely due to the ballast. Also the sorption in pore water systems is generally low. The sorption in groundwater increases by addition of sulphate and magnesium, and a still larger increase is induced by carbonatization.

Table 4. Diffusivities (D) of Cesium Calculated from Concentration Profiles.

No.	$D \times 10^{14}$ (m^2/s)	
	Sample 1	Sample 2
C1	6.4	7.0
C2	-	2.0
C3	2.3	1.7
M1	3.5	2.3
M2	3.7	8.1

The highest sorption of iodide is found for cement paste in pore water. No sorption was found in sea water. The only added ion that seems to have any enhancing effect on the iodide sorption is sulphate.

Diffusivity Measurements

The concentration profile in the solid may be expected to follow the equation[2]:

$$\frac{c}{c_o} = \operatorname{erfc}\left(\frac{x}{2\sqrt{Dt}}\right) \quad (3)$$

which is the solution of eqn (1), with the prevailing initial and boundary conditions in the experiments. The diffusivity was calculated by curve fitting the experimental data to the equation, by the method of least squares (C f Fig. 2).

The efforts to measure the diffusion of ^{131}I were not successful, due to the low diffusivity and short half-life (8.04 d) of the isotope.

The average diffusivity for Cs calculated from the experiments (see Table 4) is $4x10^{-14} m^2/s$. Due to the low sorption of I, a diffusivity of about 10^{-14} m^2/s would be expected for I as well. As the sorption may not decrease the diffusivity by more than a factor of 4 in any of the samples studied, the main resistance to mass transfer must be due to geometric factors.

CONCLUSIONS

The low diffusivity observed for cesium in slag cement implies that the mass transport rate of cesium through intact slag cement paste and concrete would be very low. Assuming a diffusivity of 10^{-1}

m^2/s it would take 3900 years before the concentration of cesium on one side of a 0.4 m thick concrete wall would reach 1 % of the concentration on the other side. To reach 10 % of the concentration it would take 9600 years, and for 50 % over 50 000 years would be required.

Fig. 1. Grinding device for the study of concentration profiles in solids.

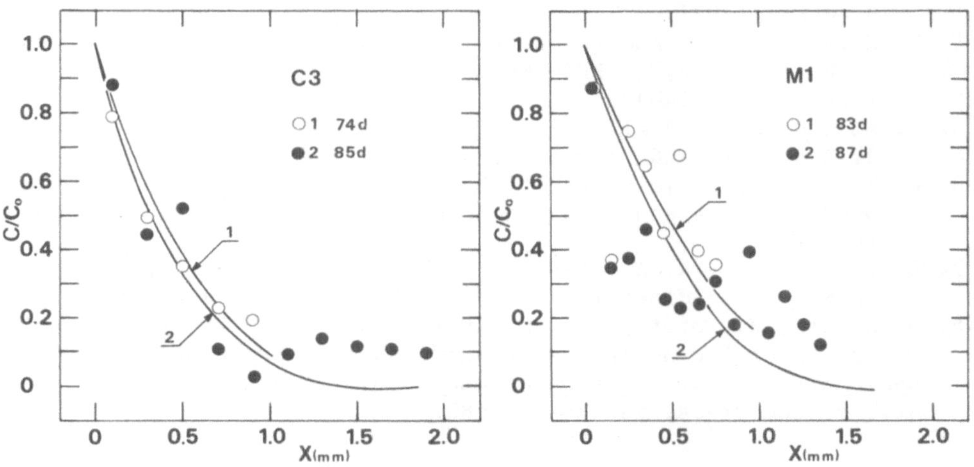

Fig. 2. Measured concentration profiles (dots) and equation (3) fitted to data (curve).

The choice of ballast would have a large influence on the sorption and the mass transfer rate of cesium in concrete (an increase in retention, i.e. water velocity/nuclide velocity, by a factor of at least 10 seems feasible).

The addition of suitable "getters" with high sorption of cesium, iodide or any other radionuclide might improve the already good properties of concrete as an encapsulation material for low and medium level radioactive waste furthermore[8].

Thus, an intact concrete encapsulation or overpack in a waste repository constitutes an effective barrier in preventing the release of ^{137}Cs (half-life 30 y) into the groundwater.

ACKNOWLEDGEMENT

This work was supported by the Swedish Research Council for Radioactive Waste (PRAV). The skillful laboratory work of Ms Wanda May and Ms Lena Eliasson is also gratefully acknowledged.

REFERENCES

1. N. Rydell, C. Thegerström, M. Cederström, "ALMA - a Study of a Repository for Low and Medium Level Waste in a Rock Cavern", paper presented at IAEA Int. Symp. on the Underground Disposal of Radioactive Wastes, Helsinki, 2-6 July 1979.
2. J. Crank, "The Mathematics of Diffusion", Oxford University Press, Oxford 1956.
3. S. G. Bergström, G. Fagerlund, L. Rombén,"Evaluation of Properties and Function of Concrete in Connection with Final Storage of Nuclear Fuel Waste in Rock", KBS-TR-12, Kärnbränslesäkerhet, Stockholm 1977.
4. I. Neretnieks, K. Andersson, L. Henstam, "Leakage of Ni-59 from a Rock Repository", KBS-TR-101, Kärnbränslesäkerhet, Stockholm 1978 (in Swedish).
5. B. Allard, G. W. Beall, "Sorption of Americium on Geologic Media", J. Env. Sci. and Health, 6, 507, (1979).
6. K. Andersson, B. Torstenfelt, B. Allard, "Sorption and Diffusion Studies of Cs and I in Concrete and Cement Paste", PRAV Report, in press.
7. B. Torstenfelt, K. Andersson, B. Allard, "Sorption of Sr and Cs on Rocks and Minerals. Part I", PRAV Report, in press.
8. B. Allard, B. Torstenfelt, K. Andersson, and J. Rydberg, "Possible Retention of Iodine in the Ground", pp. 673-80 in Scientific Basis for Nuclear Waste Management, vol. 2, ed. C.J.M. Northrup, Jr., Plenum Press, New York (1980.

PRECIPITATION OF RADIOSTRONTIUM IN SOIL*

Brian P. Spalding

Environmental Sciences Division
Oak Ridge National Laboratory
Oak Ridge, Tenne

INTRODUCTION

Radiostrontium exists in soil largely as an exchangeable cation but under certain natural or induced conditions can exist in a precipitated phase such as $Ca(Sr)CO_3$ or $Ca(Sr)HPO_4$ (Francis 1978). It is the close chemical similarity of Sr and Ca, the dominant dissolved and mobile cation in most shallow groundwaters, which leads to the mobility of radiostrontium in soil. Buried low-level radioactive waste can be leached laterally by groundwater and result in contamination of surrounding soils as has been observed in the solid waste disposal areas at the Oak Ridge National Laboratory (Webster 1979). It is desirable in the management of these contaminated soils to design a method to fix or immobilize the radiostrontium to minimize its spread and prevent its discharge into surface waters. Recently, a technique for improving the adsorption of radiostrontium by soil with alkali metal hydroxides was described (Spalding 1980). The mechanism of this enhanced calcium (hence, radiostrontium) adsorption by raising soil pH is depicted in Fig. 1. The added alkalinity neutralizes soil exchangeable acidity and ionizes surface hydroxyl groups producing an increase in the number of cation exchange sites. These new exchange sites then have the alkali metal, e.g., Na^+, as their associated cation; this condition creates considerable exchange selectivity for radiostrontium/calcium over this added Na but, since radiostrontium is still held only as an exchangeable cation, it remains potentially mobile by exchange with dissolved Ca in groundwater. It is the purpose of the present study to learn whether precipitated Ca-Sr phases can be formed in soil in competition with

*Research sponsored by U.S. Department of Energy under contract W-7405-eng-26 with Union Carbide Corp.

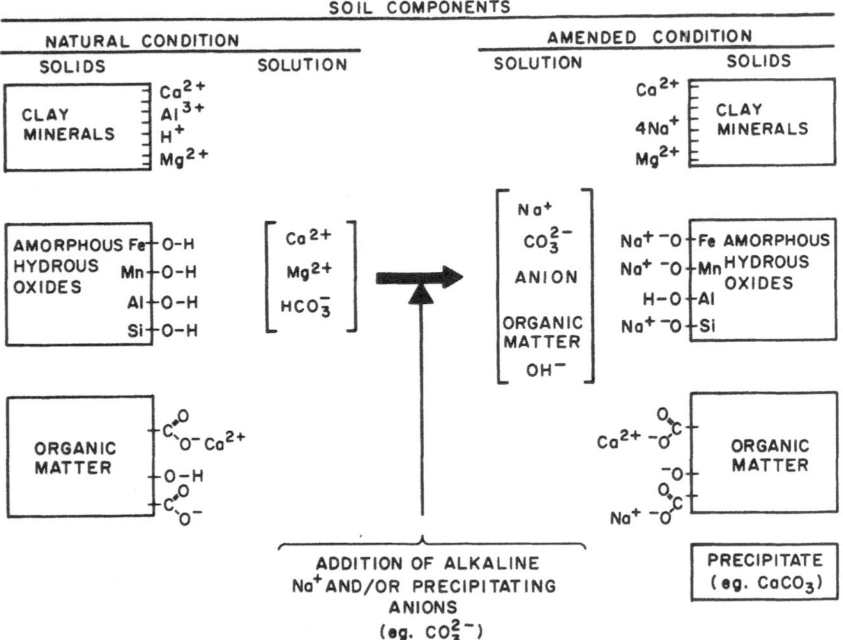

Fig. 1. Changes in the adsorptive characteristics of soil amended with alkaline sodium salts.

such changes in exchange selectivity and capacity. Candidate insoluble Ca-Sr phases include the carbonate, oxalate, fluoride, phosphate, aluminate, and silicate. More important than the formation of these phases in soil is the knowledge of how isotopically/isomorphically exchangeable any precipitated activity would be with dissolved Ca in a leaching groundwater. These questions were addressed experimentally by treating soils with a fixed quantity of the particular precipitating anion solution and, subsequently, spiking this mixture with an increment of $CaCl_2$ containing either ^{45}Ca, ^{85}Sr, or ^{22}Na and observing the radioisotopic distribution. Two additional increments of $CaCl_2$ were sequentially added and the redistributions of radioisotopes observed. Such a batch equilibrium method was designed to simulate a chemically-modified soil coming into contact with a radiostrontium contaminated groundwater and the subsequent leaching of any nascent precipitated phase by additional groundwater.

METHODS AND MATERIALS

The twelve soils used in this study were collected from within the established low-level radioactive waste disposal areas of the Oak Ridge National Laboratory. A description of these soils, their preparation, and the methods used for their chemical characterization have been given previously (Spalding 1980). These soils showed a considerable range in indigenous exchangeable Ca^{+2}, Mg^{+2}, and acidity (Al^{+3} and H^{+}): 1.6-26.8, 0.8-7.0, and 3.1-20.6 meq/100 g, respectively. Organic matter contents also varied widely from 0.1 to 6.8% and indigenous pH from 4.0 to 7.3.

The distributions of each isotope (^{45}Ca, ^{85}Sr, and ^{22}Na) between solution and insoluble phase were determined individually in duplicate using a batch equilibrium method. Soils were chemically amended by adding 20 ml of 0.1 $\underline{N}$ Na solutions of OH^{-}, F^{-}, CO_3^{2-}, $C_2O_4^{2-}$, PO_4^{3-}, AlO_2^{-}, or SiO_3^{2-} to 5 g of air-dry soil in a 30-ml polypropylene screw-cap centrifuge tube and shaking lengthwise at 100 oscillations/min for 2 hr. Radioisotopic precipitation was then initiated by adding 1 ml of 0.0828 $\underline{N}$ $CaCl_2$ and shaking as above for 16 hr; this initial increment of $CaCl_2$ also contained 22, 26, or 163 kBq/ml of ^{85}Sr, ^{22}Na, or ^{45}Ca, respectively. The suspension was centrifuged at 35,000 RCF for 10 min and an aliquot removed for activity determination; a 5-ml aliquot was used for ^{85}Sr and ^{22}Na while a 0.5 ml aliquot was used for ^{45}Ca. After counting, the 5 ml aliquot was replaced in the centrifuge tube; the 0.5 ml aliquot in the ^{45}Ca determination was not replaced. Isotopic/isomorphic exchange of the precipitated activity was determined by adding a second increment of 1 ml of 0.349 $\underline{N}$ $CaCl_2$, without radioisotopes, and shaking for a second 16 hr. After centrifuging as above, an aliquot was removed for counting, again reassembled in the centrifuge tube, and a third increment of 1 ml of 1.85 $\underline{N}$ $CaCl_2$ added. The contents were again shaken for 16 hr, centrifuged, and an aliquot removed for counting. Precipitation and isomorphic exchange reactions in the absence of soil were also determined for each chemical at each $CaCl_2$ increment; these soil-free activity precipitations were, thus, observed at sequential Ca/anion equivalent ratios of 0.04, 0.22, and 1.14.

Counting of ^{22}Na and ^{85}Sr was performed with a well-type NaI (Tl) gamma scintillation detector wired to a multichannel analyzer. Two to ten minutes counting times were generally adequate for an uncertainty in the count rate of less than 1%. Calcium-45 was counted by liquid scintillation of the 0.5 ml aliquot with 5 ml of Aquasol (New England Nuclear) cocktail. Correction for chemical/color quenching was made for each sample using an internal standard by spiking each sample with a known amount of ^{45}Ca and recounting. Correction for the amount of Ca and Na removed from each sample with each 0.5 ml aliquot was made during the subsequent calculations.

The total amount of Ca and Na in the soil suspension was determined as the sum of the amounts added in each increment of $CaCl_2$, the amounts in the Na anion solutions, and the amounts originally present in the soil as measured by the exchangeable cations. Total Ca was computed and herein defined to include indigenous exchangeable Mg because of their nearly identical chemical behavior in soil (Russel 1973). From these total amounts and the observed radioisotopic distributions, the milliequivalents of adsorbed and precipitated Ca and Na were calculated. Since the Ca in the solid phase can be present as either or both an exchangeable cation and a precipitate, the amount of precipitate was calculated as the difference between the measured total adsorbed Ca and the calculated amount of exchangeable soil Ca. This amount of exchangeable Ca was, in turn, calculated as the difference between the calculated total exchangeable cations and the exchangeable Na as determined from the ^{22}Na distribution, assuming that Na was adsorbed only as an exchangeable cation. The total exchangeable cation content of a soil at the measured equilibrium pH following chemical treatment was calculated using the linear relation between pH and total exchangeable cations at a given ionic strength (Spalding 1980). Interpolations were calculated between the measured exchangeable cation contents at the high pH's of NaOH treatments and the low pH's of the controls for each soil.

The above calculations are valid for the first increment of $CaCl_2$ only; after the second and third $CaCl_2$ additions, it cannot be assumed that the previously precipitated ^{45}Ca and ^{85}Sr activities were in equilibrium with the newly-added $CaCl_2$ in solution. But it is this difference between the observed radionuclide distribution after the third $CaCl_2$ addition and the adsorption calculated from both the cation exchange capacity and the total amount of added anion available for precipitation which can be used as a measure of the amount of ^{45}Ca or ^{85}Sr which is not in isotopic/isomorphic equilibrium with the solution. Hence, this difference between the observed and calculated distributions is inversely related to the leachability of radiostrontium of each precipitated phase in soil.

To calculate the amount of added anion which remained available for Ca-Sr precipitation, assumptions about the reaction of each anion in soil must be made. In the case of Na_2CO_3, the soil exchangeable acidity consumes substantial amounts of the added carbonate yielding the bicarbonate which does not precipitate Ca. From the measured pH, the proportion of the inorganic carbon in the carbonate and bicarbonate forms was calculated (Stumm and Morgan 1970); the amount of CO_3^{2-} remaining after soil reaction was assumed to be the amount of $Ca(Sr)CO_3$ formed since there was excess Ca present. Similar calculations were made for Na_3PO_4, where HPO_4^{2-} and PO_4^{3-} form insoluble Ca phases while $H_2PO_4^-$ does not. In addition, a portion of the added PO_4^{3-} was assumed to form an insoluble precipitate ($AlPO_4$) with the exchangeable soil acidity which is composed mainly of Al^{3+} (Coleman and Thomas 1967); this additional sink leaves less phosphate available to form Ca phases. In the case of F^- and $C_2O_4^{2-}$, equilibrium pH's were always well above the highest pK_a for each anion indicating that there would be no loss in their availability due to protonation. Fluoride and oxalate, also, do not form particularly insoluble Al or Fe phases and alternate anion consumption by soil acidity was assumed not to occur. In the case of aluminate and silicate, very little precipitated phase formed in any soil and these isotopic disequilibrium calculations were not performed.

RESULTS AND DISCUSSION

In the absence of soil, all anions at low Ca/anion ratios except aluminate and hydroxide quantitatively precipitated either ^{45}Ca or ^{85}Sr activity (Table 1). However, when this precipitated activity was reequilibrated with additional $CaCl_2$ to Ca/anion ratios greater than 1, silicate and fluoride precipitates showed considerable isotopic/isomorphic exchange as evidenced by the decrease in the percent activity precipitated. Assuming complete precipitation of each anion and complete isotopic/isomorphic equilibrium, each value in the last column of Table 1 should be 86%. The fact that the percent activity precipitated by the carbonate, oxalate, and

Table 1. The isotopic exchange of ^{45}Ca and ^{85}Sr in precipitated Ca phases after sequential additions of $CaCl_2$

		Ca/anion equivalent ratio (% activity precipitated)		
Anion	Isotope	0.04	0.22	1.14
CO_3^{2-}	^{45}Ca	97.4	99.2	98.5
	^{85}Sr	95.5	99.6	99.4
$C_2O_4^{2-}$	^{45}Ca	99.1	99.7	99.8
	^{85}Sr	97.9	99.7	99.1
PO_4^{3-}	^{45}Ca	99.9	99.9	98.3
	^{85}Sr	99.8	99.9	98.5
F^-	^{45}Ca	95.6	99.2	87.1
	^{85}Sr	99.6	99.7	62.8
SiO_3^{2-}	^{45}Ca	97.6	99.5	64.0
	^{85}Sr	87.1	96.4	19.1
$Al(OH)_4^-$	^{45}Ca	10.8	85.3	79.8
	^{85}Sr	0.8	7.1	7.7

phosphate anions remained little changed in going from conditions of excess anion to excess cation indicated that the nascent precipitated activity was not in equilibrium with the solution. Thus, these phases, if they can be formed in soil, have potential for radiostrontium immobilization.

Table 1 also shows that the fluoride, silicate, and aluminate precipitates discriminated between ^{45}Ca and ^{85}Sr, being generally more selective for ^{45}Ca. With CO_3^{2-}, $C_2O_4^{2-}$, and PO_4^{3-}, selectivity between the nuclides could not be observed since both activities remained quantitatively precipitated at each Ca/anion ratio. However, in the presence of soil, the CO_3^{2-} treatment showed a definite selectivity for ^{45}Ca (Fig. 2). These results show selectivity for precipitation in identical experiments except for the radioisotope used. Two of the three soils depicted in Fig. 2 exhibited ^{85}Sr selectivity with silicate and aluminate addition. But the most general conclusion from these results is that little discrimination between ^{45}Ca and ^{85}Sr exists in these soils either in their natural or chemically-modified conditions. The results subsequently discussed were based on ^{85}Sr adsorption measurements only and, hence, will include whatever selectivity these additional soils may exhibit.

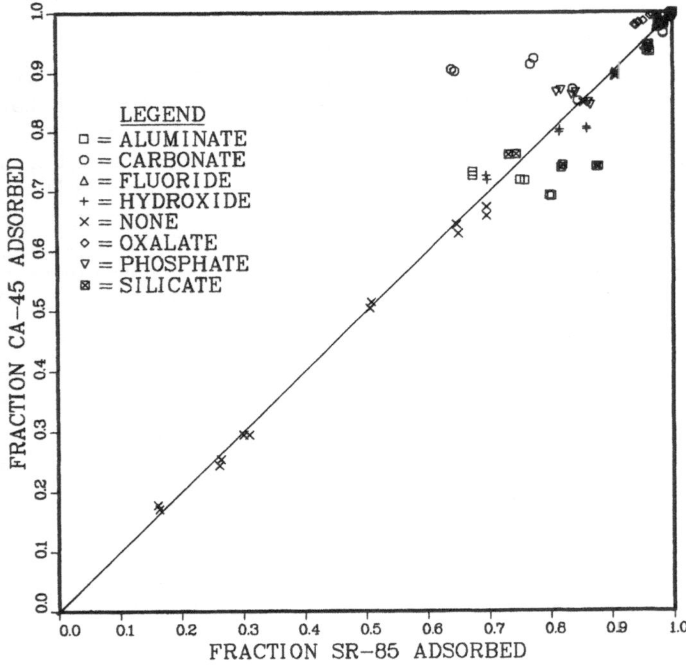

Fig. 2. Isotopic selectivity between ^{45}Ca and ^{85}Sr in chemically-amended and untreated soils.

Following the initial addition of $CaCl_2$, each chemical modification resulted in the precipitation of differing amounts of activity. When the calculated precipitated phase is expressed as a percent of the total Ca in the soil-anion-$CaCl_2$ system, the average precipitated phase for the nine soils decreased in the order: fluoride (73%), oxalate (55%), phosphate (26%), carbonate (24%), silicate (0.6%), and aluminate (0.4%). For the first four of these anions, the actual amount of precipitated phase was proportional to the indigenous exchangeable Ca and Mg content (Fig. 3). Oxalate and fluoride always formed more precipitate than the carbonate or phosphate most likely since a substantial portion of the latter anions was consumed in neutralizing soil acidity. The correlation coefficients between the amount of precipitated phase and the indigenous Ca + Mg were 0.80 (F^-), 0.96 (CO_3^{2-}), 0.93 ($C_2O_4^{2-}$), and 0.84 (PO_4^{3-}). The correlations between the amounts of precipitated phase and exchangeable soil acidity were equally significant, although negative, due to the high correlation between exchangeable acidity and Ca + Mg ($r = -0.81$, $n = 12$). The correlation between the amount of precipitated phase and the indigenous soil pH was almost equally high for each precipitate since pH is generally directly proportional to the sum of exchangeable bases in soil (Coleman and Thomas 1967). Thus a simple determination of soil pH has value in predicting the amount of radiostrontium which can be precipitated in a given soil.

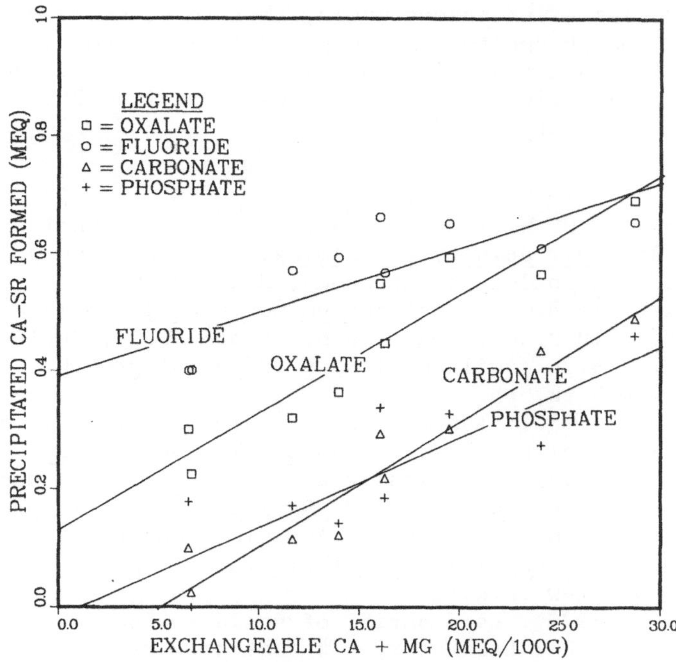

Fig. 3. Relationship between the amount of Ca (^{85}Sr) precipitated in soils and indigenous exchangeable Ca + Mg.

Regardless of the absolute amounts of precipitated phase, which has been formed in soil, it is more important to know just how well the activity in that phase is removed from isotopic/isomorphic equilibrium. Following the third $CaCl_2$ addition, the distribution of ^{85}Sr was calculated under the assumption of complete equilibrium as described in the methods section. The average difference between the calculated and observed values were: fluoride (12%), oxalate (12%), phosphate (10%), and carbonate (18%). Although these differences cannot be used to quantitatively compare the various anions for their degree of removal from equilibrium, they do establish that significant amounts of precipitated activity can be strongly removed from isotopic/isomorphic exchange with the solution or groundwater phase. The absolute comparison of these different precipitations can only be established with exhaustive leaching (both in terms of amounts of Ca and time) after treatment as has been described elsewhere (Spalding 1981). However, fluoride would appear to form a more exchangeable phase than the carbonate, oxalate, or phosphate based on the observations in Table 1. Oxalate, although it shows potential in these experiments, can ultimately lead to a more rapid movement of Sr in soil because it is readily metabolized by soil microorganisms with production of both acid and reducing conditions in the soil. Several oxalate-treated soil columns have turned gray, implying reducing conditions, several days following treatment and Sr breakthrough was accelerated (unpublished observations). The choice of precipitating anions would, therefore, seem to be limited to carbonate and phosphate. Although preliminary column leaching experiments indicate that carbonate is slightly more effective than phosphate in immobilizing radiostrontium (Spalding 1981), both chemicals have potential.

REFERENCES

N. T. Coleman and G. W. Thomas. 1967. The basic chemistry of soil acidity. In: Soil acidity and liming, R. W. Pearson and F. Adams, eds., American Society of Agronomy, Madison, Wis.

C. W. Francis. 1978. Radiostrontium movement in soils and uptake by plants. TID-27564. Nat. Tech. Inform. Ser., Springfield, Va.

E. W. Russel. 1973. Soil conditions and plant growth. 10th ed., Longman, New York, pp. 90-101.

B. P. Spalding. 1980. Adsorption of radiostrontium by soil treated with alkali metal hydroxides. Soil Sci. Soc. Am. J. 44:703-709.

B. P. Spalding. 1981. Chemical treatments of soil to decrease radiostrontium leachability. J. Environ. Qual. (in press).

W. Stumm and J. J. Morgan. 1970. Aquatic chemistry — an introduction emphasizing chemical equilibria in natural waters. John Wiley & Sons, New York.

D. A. Webster. 1979. Land burial of solid radioactive waste at Oak Ridge National Laboratory — a case history. In: Management of low-level radioactive waste, M. W. Carter, A. A. Moghissi, and B. Kahn, eds., Pergamon Press, New York.

LOADING AND LEAKAGE OF KRYPTON IMMOBILIZED IN ZEOLITES AND GLASS

A.B. Christensen, J.A. Del Debbio, D.A. Knecht,
and J.E. Tanner

Exxon Nuclear Idaho Company, Inc.
P.O. Box 2800
Idaho Falls, Idaho 83401

INTRODUCTION

Krypton-85 is formed in nuclear power reactors and remains trapped until the fuel is reprocessed. Federal regulations limit the release of Krypton-85 (^{85}Kr) to the environment, requiring recovery and storage of 85% of the ^{85}Kr produced in commercial light-water reactors after January 1, 1983.[1] One of the long-term storage options involves encapsulating ^{85}Kr in zeolites or glasses at high pressure and temperature.[2-5]

This paper presents experimental results for krypton encapsulation and leakage in sodalite, zeolite 5A, and Vycor "Thirsty" glass. The results show that all three materials are feasible for ^{85}Kr immobilization and long-term storage, although zeolite 5A and "Thirsty" Vycor are preferable due to lower leakage rates.

EXPERIMENTAL

The sodalite was specially prepared by W. R. Grace and has the formula $Na_6[(AlO_2)_6(SiO_2)_6]\cdot xH_2O$. The cubic unit cell diameter is 8.86 A and consists of adjoining beta cages separated by 2.2 A apertures. The beta cages can contain varying amounts of intercalated NaOH, which can be removed by leaching in water.

Zeolite 5A is available commercially, and was obtained from W. R. Grace. It has the formula $(0.8Ca+0.2Na)_{12}[(AlO_2)_{12}(SiO_2)_{12}]\cdot xH_2O$. The cubic unit cell has a 12.26 A diameter and is characterized by a 3 dimensional framework consisting of large (11.4

A) alpha cages and small (6.6 A) beta cages. The alpha cages are connected by 5A apertures and beta cages are connected to alpha cages by 2.2 A apertures.

The "Thirsty" Vycor samples used are the porous form of Vycor glass made by Corning Glass, and obtained from G. L. Tingey of Battelle North-West Laboratory. "Thirsty" Vycor is a glass leached to give a 96% SiO_2-4% BO_2 glass. The average pore diameter is 40 A with a 28% void fraction.

Krypton was encapsulated by zeolite or glass samples at temperature and pressure conditions shown in Figure 1. Typical activation conditions under ∿0.2 Torr vacuum to remove water are shown as dashed lines in Figure 1.

The apparatus consists of a pressurizing system described previously[3], and a 6 or 25ml Leco pressure capsule (State College, PA 16801) fitted with a cooling jacket and two independently controlled heaters to provide uniform temperature of ± 5°C along the capsule. Some of the thermocouples were calibrated at 650°C using a magnesium bath and showed less than 2°C error. Since the capsule seals were kept at 500°C, only the lower half of each capsule was used to encapsulate, while a steel plug filled the upper half to eliminate convection. The capsule was unloaded and loaded in a dry glove box to minimize water sorption.

Krypton loading was determined by rapidly heating the samples under vacuum and measuring the released gas composition using a CEC magnetic-section mass spectrometer.

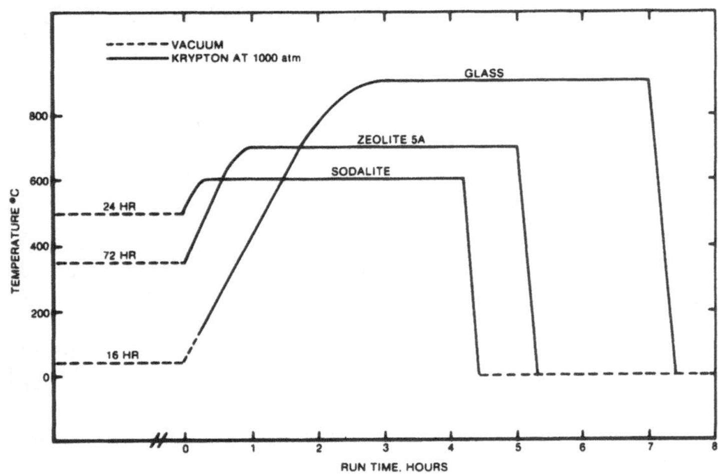

Figure 1: Pretreatment and encapsultion run conditions.

Short-term krypton leakage measurements were made by thermogravimetric analysis (TGA) up to 900°C with a scan rate of 20°C/min. This method was accurate only for gas leakage above 5%.

Long-term (1-100 days) leakage measurements were made by heating samples in evacuated sealed quartz tubes and periodically analyzing the released krypton using mass spectrometry.

RESULTS AND DISCUSSION

A summary of krypton encapsulation and leakage is shown in Table 1 and Figure 2 for sodalite, zeolite 5A, and "Thirsty" Vycor.

Zeolite 5A showed the largest loading per unit weight. Due to the higher density of glass rods over pilled zeolite 5A, the volumetric loading of glass is comparable to zeolite 5A. The least amount of krypton was released below 800°C from Vycor and the most from sodalite. Thus, based on loading and short-term leakage, zeolite 5A and Vycor appear the most promising storage materials. Detailed behaviour is described in the following:

While krypton loadings in sodalite were larger in samples with about half the intercalated NaOH removed, higher leakage rates were also measured (Run 2). When a larger ion like potassium was exchanged for the sodium ions, the leakage was not reduced, possibly due to the concurrent leaching of the NaOH (Run 3).

TABLE 1

SUMMARY OF ENCAPSULATION AND HIGH TEMPERATURE LEAKAGE DATA

UN #	MATERIAL	PRETREATMENT CONDITIONS		ENCAPSULATION CONDITIONS		Kr LOADING	TGA DATA
		Temperature (°C)	Time (hrs)	Pressure (atm)	Temperature (°C)	(STPcc/g)	%Kr Lost Prior to 800°C
1	Sodalite	575	16	1630	575	20	20
2	Leached Sodalite[a]	575	17	1020	600	27	60
3	K+Exch Sodalite	450	15	1020	600	20	100
4	Zeolite 5A	100	16	1020	550	62	26
		250	1				
5	Zeolite 5A	100	16	1020	600	53	6.5
		250	1				
6	Zeolite 5A	350	72	1020	700	60	F0.2
7	"Thirsty" Vycor Glass	200	9	1090	890	19	F0.2
		500	1				

Mole ratio of excess NaOH to sodalite = 0.28 compared with sodalite which is 0.5.

All samples were evacuated at 160 mm Hg except for zeolite 5A, Run #6 which was evacuated at 10mm Hg.

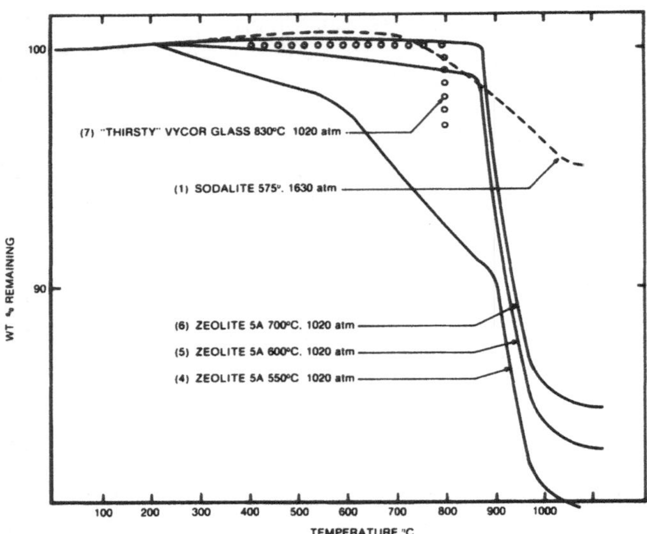

Figure 2: Thermogravimetric Curves Showing Weight Remaining of Krypton from Sodalite, Zeolite 5A, and "Thirsty" Vycor as a Function of Temperature, TGA Data.

X-ray analysis of sodalite has shown that structural changes occurring during pretreatment and encapsulation affect krypton loading. Figure 3 shows the correlation of peak splitting in the 211 reflection (2-4) and partial decomposition to carnegieite (5) during krypton loading.

As compared with unencapsulated material, Frame 2 shows a large peak splitting, a unit cell contraction and low krypton loading. Frames 3 and 4 show much less peak splitting, a unit cell expansion and greater krypton loadings. X-ray analysis of leached sodalite (removal of intercalated NaOH) has shown no peak splitting and a 2% increase in unit cell size with a krypton loading of about 30 cm^3/g. Frame #5 shows the presence of a large amount ($\sim$60%) of a decomposition product tentatively identified as a carnegieite phase,[6] which does not trap krypton. Tests have indicated that carnegieite forms during pretreatment or encapsulation near 600°C and 1600 atm. Carnegieite will not form at 550°C and 1000 atm.

Although the apertures interconnecting the zeolite 5A cages are larger than the krypton atomic diameter, krypton is trapped in zeolite 5A by a pore closure at high temperature and pressure, as shown by the loss of crystallinity in the powder x-ray pattern shown in Figure 4. The higher short-term leakage in Figure 2 for Run 4 compared to Runs 5 and 6 is explained by a retention

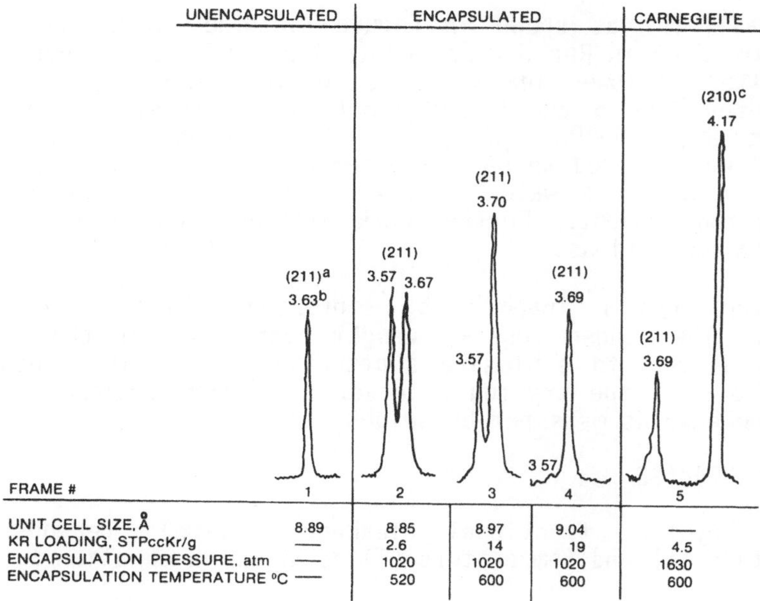

Figure 3: X-ray powder diffraction patterns of sodalite showing the correlation between peak splitting and partial decomposition with krypton loading

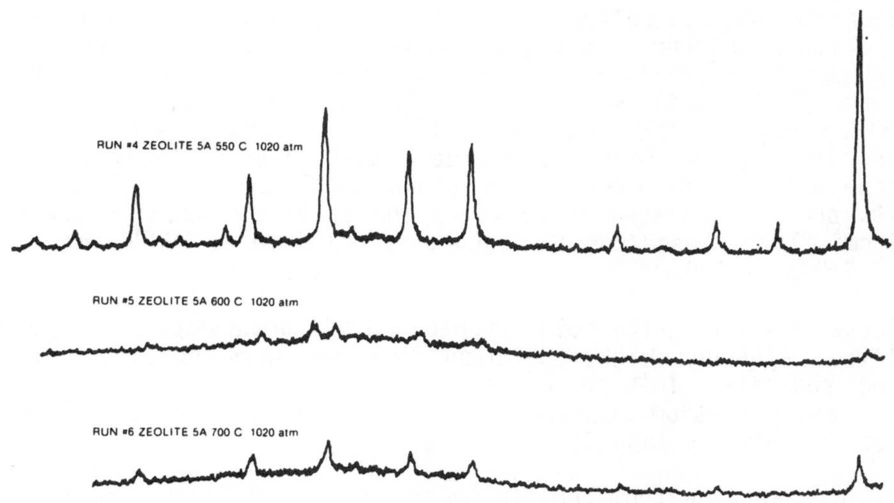

Figure 4: X-ray powder diffraction patterns of zeolite 5A, run 4-6 (see Table I).

of crystallinity as shown in Figure 4. The higher sintering temperature used in Run 6 compared to Run 5 (Table 1) apparently has resulted in lower leakage as shown in Figure 2. The x-ray patterns for Runs 5 and 6 are identical. Preliminary studies show that above 600°C and approximately 20 gaseous cm^3 STP/g of sorbed water, zeolite 5A can decompose to an anorthite form. However, in Run 6 a water content of 9 cm^3 STP/g yielded no decomposition product. Further work will be made to define decomposition conditions.

Similar krypton trapping by sintering "Thirsty" Vycor is observed. The sudden loss of weight near 800°C in the TGA in Figure 2 is due to explosive shattering of the glass powder, probably due to the krypton release. At lower temperatures no krypton release is measured by weight loss.

LONG-TERM LEAKAGE

The krypton fractional leakage ($Q_t/Q\alpha$) depends on storage time (t) and temperature (T) as shown in equation 1:

$$Q_t/Q\alpha = t^{\frac{1}{2}}\left(\frac{36 D_0}{\pi R_0^2}\right) \exp\,(-E/2RT) \qquad 1$$

This equation describes krypton leakage from sodalite but has not been tested for zeolite 5A and "Thirsty" Vycor. To measure temperature effects on leakage, results are plotted as log (% leakage/$t^{\frac{1}{2}}$), which decreases with leakage, against 1/T.

Sodalite leakage follows the $t^{\frac{1}{2}}$ relation, except for an initial period of higher leakage, probably due to some small cracks in the sodalite crystals.[3] Long-term tests were made at 150, 210, and 288°C, with Kr and water loadings of 15-25 and 8-50 cm^3 STP/g respectively and up to 81 days storage times. The krypton leakage was found to decrease with krypton loading and increase with water loading, as shown in Figure 5. If these results are extrapolated to 10 years and initial krypton loading of 22 cm^3 STP/g, less than 1% leakage is obtained at 190°C and 10 cm^3 STP/g of H_2O.

Tests were made with both leached and 50% potassium-exchanged sodalite. Although higher krypton loadings were obtained for leached sodalite, leakage tests over 20 days at 150°C showed leached and potassium-exchanged sodalites leaked 2 and 10 times as fast as ordinary sodalite, respectively.

Krypton leakage from zeolite 5A and "Thirsty" Vycor were measured for 30 and 68 days at 200°C. White zeolite 5A had an apparent initial leakage of 0.09% increasing to 0.099% in 68 days, "Thirsty" Vycor had a very low initial leakage (0.002%) rising to 0.004% in 68 days. Both are superior to sodalite which releases 2-3%

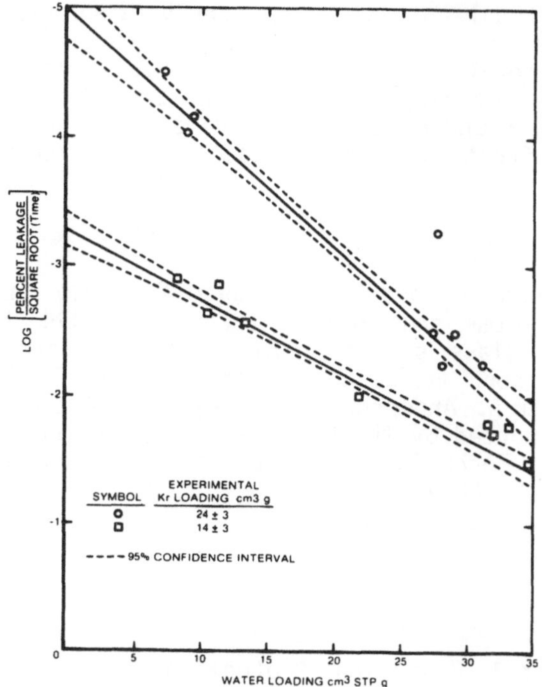

Figure 5: The Effect of Water and Krypton Loadings on Sodalite Leakage at 210°C.

krypton under similar conditions. However the leakage is higher for sodalite 5A or "Thirsty" Vycor which haven't completely formed an amorphous structure or sintered, respectively.

No effect due to particle size of glass ground and sieved to 1.5, 0.9, and 0.4 mm was observed in krypton leakage at 298 and 527°C for 84 days. The effect of water and krypton loadings on long-term krypton leakage has not yet been studied. The theoretical leakage behaviour such as shown by equation (1) has not yet been ascertained.

CONCLUSION

Based on experimental loading and leakage tests, zeolite 5A and "Thirsty" Vycor glass are preferred to sodalite as a medium for immobilizing krypton-85 for long-term storage, with respective loadings of 50 and 16 cm³ STP/g. If the long-term leakage results measured at 260°C and 62 days is extrapolated to 10 years storage using equation (1), about 0.05 and 0.2% krypton leakage would be observed with "Thirsty" glass and zeolite 5A, respectively. For a total estimated krypton-85 production from a 2000 metric ton commercial fuel reprocessing plant of 190 m^3 at STP of 6% 85Kr in Kr, total annual volumes of immobilized krypton would be 6.5 and 4.5 m^3 of "Thirsty" glass and zeolite 5A, respectively. Further tests are under way to evaluate krypton loading conditions and the effect of absorbed water and temperature on long-term leakage.

AKNOWLEDGEMENTS

Special thanks are due N. S. Graham for help in building the encapsulation system and to G. L. Tingey of PNL for providing "Thirsty" Vycor glass samples. Also, appreciation is expressed to L. L. Dickerson, E. S. Dickerson, and J. R. Delmastro for analyses, and to A. Anderson for carrying out the encapsulation runs. Work supported by the U.S. Department of Energy Assistant Secretary Energy Technology, Nuclear Waste Management under DOE Contract Number DE-AC07-79ID01675.

REFERENCES

1. "Environmental Radiation Protection Standards for Nuclear Power Operation", Federal Register, 42 No. 9, Part VII, 2858 (Jan. 13, 1977).

2. D. A. Knecht, An Evaluation of Methods for Immobilizing Krypton-85, ICP-1125 (July 1977).

3. R. W. Benedict et al., Technical and Economic Feasibility of Zeolite Encapsulation for Krypton-85 Storage, ENICO 1011 (September 1979).

4. R. D. Penzhorn et al., "Long-Term Storage of Krypton-85 in Zeolites," Proc. Int. Symp. Management of Gaseous Wastes from Nuclear Facilities, IAEA-SM-245, Vienna (February 1980).

5. G. L. Tingey et al., "Solid State Containment of Noble Gases in Sputter Deposited Metals and Low-Density Glasses," Proc. Int. Symp. Management of Gaseous Wastes from Nuclear Facilities, IAEA-SM-245/31, Vienna (February 1980).

6. D. J. Schipper et al., "Thermal Decomposition of Sodalites," Amer. Ceramic Soc. Journ., 56, No. 10, 523-525 (1973).

THE LASL EXPERIMENTAL ENGINEERED WASTE BURIAL FACILITY: DESIGN CONSIDERATIONS AND PRELIMINARY PLAN

Gerald L. DePoorter

Los Alamos Scientific Laboratory
Post Office Box 1663, MS-495
Los Alamos, New Mexico 87545

INTRODUCTION

The LASL Experimental Engineered Waste Burial Facility is a part of the National Low-Level Waste Management Program on Shallow-Land Burial Technology. It is a test facility where basic information can be obtained on the processes that occur in shallow-land burial operations and where new concepts for shallow-land burial can be tested on an accelerated basis on an appropriate scale. The purpose of this paper is to present some of the factors considered in the design of the facility and to present a preliminary description of the experiments that are initially planned. This will be done by discussing waste management philosophies, the purposes of the facility in the context of the waste management philosophy for the facility, and the design considerations, and by describing the experiments initially planned for inclusion in the facility, and the facility site.

WASTE MANAGEMENT PHILOSOPHIES

For properly operated (no surface spills, no broken packages, and proper administrative controls) low-level waste burial facilities, the primary modes of radioactive contamination have been through leaching and transport of radionuclides by water and by the release of radioactive gases to the atmosphere.[1,2] Because low-level radioactive waste burial sites may also contain hazardous materials, they must also be designed and engineered to prevent the escape of these components into the environment. Emphasis here will be on the problems associated with the generation and possible release of contaminated solutions from the facility intended to retain and control the waste form. The solutions can arise from interactions of infiltrating water with the waste, or they may be part of the wastes.

Waste management philosophies can be expressed in two ways: solve the problem once it happens, or prevent the problem in the first place. Since the purpose of the experimental waste burial facility is to demonstrate and substantiate new techniques, the waste management philosophy adopted is based on preventing problems rather than solving them. Three general waste management philosophies, expressed in terms of prevention rather than correction, are outlined in Table I.

The guaranteed 100% safe philosophy is unrealistic from a technical and cost viewpoint, but may have some political advantages. Planned dispersion, from a general political point of view, is probably not acceptable. However, this is a viable alternative if dispersion can result in a concentration level that is innocuous. State-of-the-art containment is probably the best approach from all

Table I. WASTE MANAGEMENT PHILOSOPHIES

GUARANTEED 100% SAFE

Engineer the waste disposal facility so that water cannot reach the emplaced waste, thereby eliminating the possibility of contaminants being mobilized. Water from the waste degradation or liquids contained in the waste that gets through the packaging and other preventative measures will be channeled, collected and passively treated.

STATE-OF-THE-ART CONTAINMENT COMBINED WITH ENGINEERED CHANNELING OF THE EFFLUENT

Recognize that some water is going to get into the burial pit or that there is going to be some leakage from liquid wastes. Design the pit so that infiltrating water is diverted and collected before it can reach the emplaced wastes. As a backup, and to manage liquids in or generated in the waste, engineer the system so that if any leachate or effluent is formed, it can be collected and properly treated with a passive system.

CONTROLLED DISPERSION

Provides specific access channels for the water so that contaminants are mobilized and dispersed in a controlled manner. If liquid waste forms are present, their release is controlled similarly.

viewpoints. It is the most honest approach because it is probably impossible to guarantee 100% containment or absolute isolation from water. With an understanding of the processes that occur in mobilization and migration of contaminants, based on quantitative experimental results, this approach can also be made politically acceptable. State-of-the-art containment combined with engineered channeling of the discharge is the waste management philosophy to be used in the design of the experimental engineered waste burial facility.

In discussing waste management philosophies we must also consider the possible intrusion of plants and animals into the buried waste after closure with the resulting mobilization or transport of contaminants out of the disposal area into the environment. In this situation we must consider two alternative waste management philosophies, guaranteed 100% safe from plant and animal intrusion or state-of-the-art closure with detailed monitoring to detect intrusion. Again, in this case, state-of-the-art closure is the philosophy adopted because the guaranteed 100% safe waste management philosophy is not realistic from either a technical or cost viewpoint.

PURPOSES OF THE EXPERIMENTAL ENGINEERED WASTE BURIAL FACILITY

State-of-the-art containment combined with engineered channeling of the discharge can only be successful if enough experimental results are available that the waste disposal facility can be constructed in a completely engineered environment. The experiments to be done in this facility will provide this collection of data.

To clarify the purposes of the facility the concept of a completely engineered environment must be expanded a little. In an engineered environment, data on migration and mobilization of contaminants, interaction of the contaminants with the fill material, water infiltration through the cap system, and through the disposal facility walls or liner systems, leakage rates from engineered containers, and the general chemistry and hydrology of the system are used to design the waste disposal facility so that liquid movement can be predicted, and so that the amount and composition of any leachate or liquid in the system is known. Any liquid formed will be passively treated to minimize release to the environment. For a waste management philosophy of controlled release, in addition to the above, the release rate and the composition of the released material will be controlled.

A summary of the general purposes of the experimental engineered waste burial facility is given in Table II. This list covers the full range of experiments that are necessary if future shallow-land burial facilities are to be constructed in a completely engineered environment. The emphasis is to obtain experimental results that can be used

Table II. GENERAL PURPOSES OF EXPERIMENTAL WASTE BURIAL FACILITY

TEST METHODS FOR CONSTRUCTING AND OPERATING WASTE DISPOSAL FACILITIES

Determine need for liners.
Evaluate burial pit liner systems, if needed.
Evaluate burial pit cap systems.
Evaluate backfill materials.
Evaluate burial pit drain systems.

DEVELOP AND TEST MONITORING SYSTEMS TO MEASURE BURIAL PIT PERFORMANCE

Measure infiltration rate of water into the burial pit.
Measure leach rate and leachate compositon in burial pit.
Measure water and leachate movement out of the burial pit.
Monitor heat flow into and out of the burial pit.
Measure evaporation and transpiration of water.

CONTROL CLIMATIC CONDITIONS AT THE BURIAL PIT

DETERMINE EFFECT OF BIOLOGICAL ACTIVITY ON MATERIAL CONTAINED IN THE BURIAL PIT

DETERMINE SCALING FACTORS TO BE USED IN BURIAL PIT DESIGN

EVALUATE BIO-BARRIERS

EVALUATE ARID SITE CLOSURE PROCEDURES

EVALUATE REMEDIAL ACTION PROCEDURES FOR ARID SITES

in the design of future low-level waste disposal facilities. In addition, the experimental results can be used to validate models for predicting long-term behavior of the facilities and will therefore be useful in convincing the public about the safety of the facility and design.

DESIGN CONSIDERATIONS

The experiments outlined in Table II involve hydrological, chemical, mechanical, and biological factors. In order to separate these various factors in the experiments and to extrapolate the experimental results to actual facilities, experiments should be performed on several different scales.

Three general scales have been chosen for experiments in this facility: isolated variable experiments, intermediate scale experiments, and integrated experiments. The isolated variable experiments will be performed in caissons or lysimeters, which will be more completely described in the section on the initial experiments. The intermediate scale experiments will be performed in experimental burial pits about 15 feet on a side and of variable depth. The integrated experiments will be performed in burial pits with dimensions typical of those encountered in commercial low-level radioactive waste disposal facilities.

Although the isolated variable experiments are quite chemical in nature, the intermediate scale experiments are designed to provide information on mechanical effects and a combination of mechanical and chemical effects. The leachate-liner-fill interactions are both mechanical and chemical, while the gas channeling and collection and the burial pit drainage systems are more mechanical. The integrated experiments will address such large scale mechanical problems as pit settlement, cap cracking, and the effects of the pit filling operation and backfilling and capping on the liner and drain systems.

Another important factor in the design considerations is the desired capability of accelerated testing of the experimental system performance. Since the majority of the problems results from interactions of the system with water, accelerated testing will be done by adding extra water to the system. This will also give information on the time dependence of weathering phenomena which will again be useful to substantiate models for shallow-land burial systems.

INITIAL EXPERIMENTS PLANNED FOR THE FACILITY

The range of experiments possible in the LASL Experimental Engineered Waste Burial Facility and an indication of an appropriate scale for these experiments have been presented in Table II and the text. Given the large number of materials and configurations possible for these experiments, the difficult task is selecting the experiments to be done and the order in which they should be done. In our case, specific requirements of the National Low-Level Waste Management Program on Shallow-Land Burial Technology have made the choice easier. Initial experiments will be in the areas of migration barriers, remedial action testing, arid site closure, and biological intrusion barriers.

Migration barriers for both water and radionuclide transport will be evaluated in isolated variable experiments. The isolated variable studies will be performed in experiment clusters, each of which consists of six experimental caissons clustered around a central access and instrument caisson. The instrument and access caissons will be about 9 to 10 ft (about 3 m) in diameter and deep enough for the experiment being performed. The experiment caissons can be any size up to the same size as the central caisson or can be larger if used in place of two smaller caissons. The access caisson will allow samples to be taken in a horizontal direction in any of the experimental caissons at any elevation without disturbing the surface of the caisson or allowing vertical access by water to the packing material.

The use of multiple experimental caissons around a central instrument and access caisson will allow a large number of separate experiments per unit. The caisson provides isolation of the experimental areas and also prevents the horizontal influx of water, allowing more precise control of the environment in each of the experimental areas. Different types of fill materials can be used in these caissons. The experiment cluster described here is a modification of a similar design by Phillips et al.[3]

Based on previous work on this program,[4,5,6] the most promising candidate natural barrier materials for water and radionuclide migration will be chosen for testing. A liner of this chosen material will be placed in a caisson and then backfilled. Appropriate tracer materials will be placed in the fill materials. Sufficient artificial rainfall will be applied to the caisson to mobilize the tracer and transport them to the fill-liner interface. The retarding effect of the liner material on water and radionuclide transport through the liner will be measured. As many different liner materials as the budget of the program allows will be tested.

Monitoring methods will include gamma probes to measure the movement of the tracer material, neutron probes to measure the movement of water independent of the tracer, and temperature and bulk density measurements. In addition, samples of leachate solution will be collected with porous cups.

The remedial action testing experiments are designed to provide solutions to possible problems that might occur in a closed shallow-land burial facility in an arid environment. These problems include surface water infiltration, surface erosion by wind or water, contaminant uptake by plants and animals, and upward migration of radionuclides due to moisture cycling.

Several configurations of integrated cap systems will be constructed and tested. These integrated cap systems will be multifunctional and will be designed to prevent water infiltration, plant infiltration, and wind and water erosion. Measurements will be made on the experiments so that the reasons for the success or failure of an integrated cap system will be known and documented. These experiments will be performed on the intermediate scale.

The upward migration experiments will be performed in a smaller individual caisson. Tracer materials will be placed in typical fill materials at a depth determined to be in the region of moisture cycling. The system will be monitored to determine if tracer material is brought to or near the surface of the caisson due to moisture cycling.

The arid site closure experiments will be designed to field test, on an appropriate scale, methods for closing shallow-land burial facilities in an arid environment. These tests will include both the physical methods used to close the facilities and the monitoring methods to evaluate and confirm the performance of the closure procedures used. The purpose of these experiments is to provide field tested, well documented procedures for arid site closure so that the problems described in the remedial action testing section are anticipated and prevented while the site is operational and being closed.

These experiments will be similar to but distinct from the remedial action testing experiments. In these experiments, all phases of burial pit operation can be considered so that problems like pit subsidence can be approached early in the closure procedure. They will consist of complete model burial pits constructed on the intermediate scale. They will have liners, drain systems, caps, and appropriate monitoring systems to evaluate their performance.

The biological intrusion barrier experiments will be designed to field test trench cover configurations that will prevent the growth of deep rooted plants and the intrusion of burrowing animals into the buried waste materials. A variety of cover configurations will be tested in small lysimeters (about 1 ft in diameter) and on the intermediate scale. The cover configurations to be tested consist of various combinations and depths of soil and biobarriers such as cobble, clay, and backfill.

SITE DESCRIPTION

The LASL Experimental Engineered Waste Burial Facility will be located on a mesa top on DOE land in Los Alamos County, New Mexico. The 20 acre site is about 2 miles west of the active low-level radioactive waste disposal facility for the laboratory. The soils and geology at both sites are similar.[7,8] The soils are of the Hackroy Series which consists of a surface layer of brown sandy loam, or loam, about 10 cm thick with a subsoil of reddish brown clay, gravelly clay, or clay loam about 20 cm thick. The depth to tuff bedrock varies from 20 to 50 cm. The native vegetation is mainly pinyon pine, one-seed juniper, scattered ponderosa pine, and blue grama.

ACKNOWLEDGEMENTS

The author gratefully acknowledges helpful and stimulating discussions concerning the waste management philosophies discussed in this paper and the experiments to be placed in this facility with Drs. J W. Nyhan, M. L. Wheeler, T. E. Hakonson and Mr. J. G. Steger. This work was performed under the auspices of the U.S.D.O.E.

REFERENCES

1. Panel on Land Burial, Committee on Radioactive Waste Management, Commission on Natural Resources, NATIONAL RESEARCH COUNCIL, "The Shallow-Land Burial of Low-Level Radioactively Contaminated Solid Wastes," National Academy of Sciences, Washington, DC (1976).
2. D. G. Jacobs, J. S. Epler, and R. R. Rose, "Identification of Technical Problems Encountered in the Shallow-Land Burial of Low-Level Radioactive Wastes," ORNL/SUB/13619/1 (1980).
3. S. J. Phillips, A. C. Campbell, M. D. Campbell, G. W. Gee, H. H. Hoober, and K. O. Schwarzmiller, "A Field Test Facility for Monitoring Water/Radionuclide Transport Through Partially Saturated Geologic Media: Design, Construction, and Preliminary Description," PNL-3226 (1979).
4. D. E. Daniel and R. E. Olson, "Geotechnical Aspects in Design of Disposal Sites for Low-Level Radioactive Wastes," Geotechnical Engineering Report GR80-6, The University of Texas, Austin, Texas (1980).
5. M. Pertusa, "Materials to Line or to Cap Disposal Pits for Low-Level Radioactive Wastes," Geotechnical Engineering Report GR80-7, The University of Texas, Austin, Texas (1980).
6. E. S. Takamura, "Disposal of Low-Level Radioactive Wastes: Alternate Methods and Improvements to Shallow-Land Burial," EHE-79-04, Environmental Health Engineering, University of Texas at Austin (1979).
7. J. W. Nyhan, L. W. Hacker, T. E. Calhoun, and D. L. Young, "Soil Survey of Los Alamos County, New Mexico," LA-6779-MS (1978).
8. M. A. Rogers, "History and Environmental Setting of LASL Near-Surface Land Disposal Facilities for Radioactive Wastes (Areas A, B, C, D, E, F, G, and T)," LA-6848-MS, Vol. I (1977).

MELTING PROCESS TO CONDITION DECLADDING HULLS GENERATED BY THE REPROCESSING OF LWR AND FBR SPENT FUELS

R. Bonniaud, N. Jacquet-Francillon,
A. Jouan, and C. Sombret

Commissariat à l'Energie Atomique
France

INTRODUCTION

The reprocessing of spent fuels generates several types of wastes. Among these wastes, the decladding materials viz points, spacers, hulls, grids, springs, etc., must be stored under stringent conditions. Particular attention has to be paid more especially to the hulls because these materials pertain to TRU wastes as well as $\beta\gamma$ waste owing to the activation of some elements, to the presence of undissolved fuel, and to the diffusion of fission products, tritium included, through the matrix. In addition, a fire hazard is associated with the hulls related to the light water reactors.

The most generally operated storage system lies in filling silos with the decladding materials under water. This procedure must be considered as an interim one, and a conditioning process providing waste materials having properties suitable enough to allow a safer and more compact disposal has to be elaborated.

A number of techniques are at present under investigation in several nuclear countries, in particular mechanical compaction associated with an embedding in concrete or lead, and melting. This latter has been investigated in the U.S. and in France. The French work carried out by the Commissariat à l'Energie Atomique at Marcoule, and sponsored under contract by the Commission of the European Communities is presented in this paper. Both stainless steel and Zircaloy hulls were taken into consideration.

LABORATORY RESEARCH STAGE

Zircaloy Melting

As the melting point of Zircaloy alloys used for the cladding of spent fuels is high (1800°C range) it was attempted to decrease it in order to make easier a further industrial operation.

Fe, Ni, Cu were chosen as basic additives because they make low melting point eutectics with zirconium. Inconel and stainless steels ought therefore to be used too insofar as their Cr content is not too high.

Experiments were carried out, submitting 200 g of mixed materials to various temperatures. They were kept at a given temperature for 4 hours under argon atmosphere. The examination consisted of visual inspection to determine whether the product had been melted and whether a homogeneous ingot was formed.

For Zircaloy 4, a 1200°C melting point was reached with the addition of (1) Cu (21.5%), (2) Ni (17.0%), (3) stainless steel 316 L (15-17.5 and 20%), and (4) Inconel 600 (15%).

Stainless Steel Melting

The melting points of stainless steels involved in cladding are lower than those of Zircaloy, but high enough to justify an attempt to decrease them. Two ways were investigated:

1. eutectic formation with C, Si, B or ferroboron; and
2. dissolution in a low melting point alloy (Sb, Cu).

Experiments were conducted in the same way as for Zircaloy.

A 1050°C melting point was reached with:

1.	stainless steel 305 70 to 80%	+	90% Sb and 10% Cu 20 to 30%
2.	stainless steel 316 L 50 to 70%	+	90% Sb and 10% Cu 30 to 50%
3.	Inconel 601 70 to 80%	+	90% Sb and 10% Cu 20 to 30%
4.	Inconel 601 80%	+	76.5% Sb and 23.5% Cu 20%

A 1250°C melting point was reached with:

1.	stainless steel 316 L 94 to 98%	+	C 2 to 6%

2. stainless steel 316 L + B (under ferroboron form)
   95 to 96% — 4 to 5%
3. stainless steel 316 L + silicon
   75 to 85% — 15 to 25%

Cinetics of melting are in most cases less than an hour.

Corrosion Resistance of Various Refractory Materials

Refractory materials which will be used in industrial equipment will have to withstand the effect of molten material during a period long enough not to hinder a normal operation.

Experiments were therefore undertaken to provide a prior selection of suitable refractory products.

Zircaloy. Eleven materials were submitted to specific tests. The basic components were:

1. MgO (trademarks: C 104 and Anker D1),
2. Al_2O_3 (trademarks: Magmalox and Alcor),
3. Al_2O_3 + ZrO_2 (trademarks: ZAC 1681 and ZAC 1711),
4. Cr_2O_3 (trademarks: C 1215 and Cr 100),
5. C (graphite and CSi), and
6. CaF_2.

The alloy involved in the experiment was Zr 80% - Cu 20%.

The tests consisted of putting samples in the molten metal contained in zirconia crucibles. These crucibles were put in a furnace under argon atmosphere and at a temperature of 1200°C for 100 hours.

It was found after visual inspection and microscopic examination that every kind of refractory material had behaved satisfactorily apart from CaF_2 and CSi. Graphite even appeared to be absolutely inert.

Stainless Steel. Two alloys were taken into account: (1) stainless steel 316 L + Si (15%), and (2) stainless steel 316 L + ferroboron (20%).

Stainless steel 316 L is the material used for the cladding of the French FBR Phenix fuels.

Eleven materials were submitted to a test similar to the above mentioned one. The basic components of these materials were:

1. MgO (trademarks Electrex F 55, Anker D 1 and Magnorite Mn 197),

2. Al_2O_3 (trademarks Alcor, Superstar, Isostal G and pure sintered alumina),
3. ZrO_2 (trademark ZFE),
4. $Al_2O_3 + ZrO_2$ (trademark: ZAC 1681),
5. Cr_2O_3 (trademark C 1215), and
6. C (graphite).

The difference from the test related to Zr Cu alloy was the temperature: 1270°C instead of 1200°C.

The results pointed out:

- Roughly structured ceramics were the materials most subjected to the corrosion effect.
- Chromium oxide was dissolved in the boron bearing alloy.
- Graphite was dissolved in both alloys.
- A material containing Cr_2O_3 and another containing graphite as secondary components were greatly corroded by the boron bearing alloy.
- Zirconia behaved well but was fragile and inclined to crack.
- The boron bearing alloy was more aggressive towards most refractory materials than the silicon bearing one.
- Among the materials involved in the test, the ceramic ISOSTAL G (Al_2O_3 55%, C 30%, SiO_2 15%) seemed to be the most corrosion resistant. Nevertheless the presence of C seemed to be a precaution in order to predict satisfactory behavior for a much longer period than that of the experiment.

Decontamination

It was thought interesting to carry out decontamination during the same stage of the process. Therefore preliminary investigations were carried out to estimate the degree of physical feasibility of such treatment.

Glass and molten salts were chosen as decontamination agents in order to be collected as slags at the end of the process. Tests were first carried out on nonradioactive stainless steel and Zircaloy-4 hulls.

Several glasses and particularly a borosilicated one appeared as suitable materials for stainless steel. On the other hand, these products are not to be recommended for the decontamination of Zircaloy because their components can be reduced by zirconium. In this case only the use of molten mineral salts (CaF_2 and MgF_2, for example) gives a clear two-phase separation.

Taking into account the mode of contamination when examining the results, the tests are extremely encouraging for stainless steel 316 L, but negative as for Zircaloy 4 where only the use of molten fluoride-type salts could be envisaged.

STUDY OF SEMI-INDUSTRIAL LOADS

This study was undertaken on a larger scale (30 to 50 kg loads) in order to study concrete technological problems.

The equipment used for the fusion of the loads comprised an induction furnace and crucible within a sealed metal glovebox type container which enabled the study to be carried out in a controlled atmosphere. The furnace had a capacity of 50 kg and was cleaned by tilting.

Zircaloy Fusion

Four types of alloy have been formulated from Zircaloy 4.

1. Zircaloy + 21.5% copper,
2. Zircaloy + 17% nickel,
3. Zircaloy + 21% stainless steel 304 L, and
4. Zircaloy + 15% inconel 600.

The copper bearing ingot was obtained by preliminary fusion of the total copper content followed by the addition of the Zircaloy in the form of tubes simulating the cut hulls. For the other alloys the tests were begun by loading mixture of constituents of a total weight of 3 kg. Then by continuing to add the mixture to the molten bath until full, the temperature was constantly below 1200°C and the ingots obtained were always homogeneous.

Stainless Steel Fusion

Ingots of stainless steel without additives and alloys were thus elaborated from 304 L ingots have been formed using:

1. stainless steel + 2.4 + 6% carbon,
2. stainless steel + 15% silicon,
3. stainless steel + 3.4% boron made either by adding boron or by adding ferroboron, and
4. stainless steel dissolved in a Sb-Cu 50% alloy.

The experiments were carried out by melting 3 kg of alloy constituents in the crucible, then by continuing to feed the molten bath until full. Only the Sb-Cu alloy was handled differently by preliminary fusion of the Sb-Cu and introduction of pieces of stainless steel into the molten bath.

The functioning of the tests was always satisfactory, the working temperatures were on the order of 1380°C for the carbon alloy, 1250°C for the silicon alloy, and 1300°C for the boron alloy. All the products thus obtained may be esteemed satisfactory, with the exception of the boron alloy obtained by the addition of boron because it marries badly with stainless steel.

Corrosion of Refractory Materials

Zircaloy. Tests have only been carried out on the Zircaloy copper alloy containing 22% copper.

It was found that:

- Graphite-base crucibles suffered practically no effects of corrosion from the bath; pure graphite, Superstar and Isostal G are the best choices, but Isostal G has, however, a slight tendency to crack.
- Al_2O_3 crucibles are not attacked but are not very resistant to thermal shock (e.g., Alcor, Isocreu).
- Magnesium base crucibles are deeply corroded (Magnorite Mn 197).

Stainless steel. Tests were only run on stainless steel 316 L allied with 15% silicon. Only one vitreous silica crucible was put into operation, then suddenly cracked after 12 alloy fabrication tests. Its fragilization was probably due to change of allotropic state.

Other more promising materials are at present being studied.

Decontamination Tests

Stainless steel. The feasibility of the operation was examined by first using noncontaminated hulls, then uranium contaminated ones.

A series of eight tests was run on voluntarily contaminated hulls. 316 L steel was used in the form of hulls with the addition of granulated silicon. A single borosilicated glass was used as the decontaminating agent. Pure UO_3 contamination was used for three tests and a uranium rich solution (nitrate) was used in the subsequent ones.

The operation was begun by fusing the steel and silicon mixture, then the glass was fed in followed by agitation before casting in the first five tests. For the subsequent tests the glass was placed under the initial load. The temperature of the bath varied from 1300 to 1500°C. The tests showed that it was preferable not to place the whole glass load in the bottom of the crucible since it tends to float on the surface and impede the loading. The ingots obtained have a satisfactory appearance and the vitreous slags separate quite easily.

The decontamination factor thus obtained is in the region of 200. The uranium content of the cast or recuperated slag varies from 2400 to 10,500 ppm, and there only remains from 1 to 10 μg/g of uranium in the metal.

Zircaloy. Given the problems encountered in the laboratory only the feasibility aspect has been considered. The alloy worked out was Zr-Cu with 22% copper, and the decontaminating agents used were phosphate glasses, a mixture of CaF_2 and MgF_2, and a mixture of $CaCl_2$ and CaF_2.

During the eight tests carried out the operation was always begun by first melting the copper alone, then introducing the Zircaloy into the molten copper, the decontaminating agent being poured into the molten bath at the end of the experiment.

The tests run confirmed that since it is practically impossible to use a glass as a decontaminating agent, the $CaF_2 + MgF_2$ mixture is relatively difficult to use due to volatilization, and the metallic ingot obtained by using a $CaCl_2 + CaF_2$ mixture presented a relatively rough surface state.

CONCLUSION

The fusion compaction of metallic waste from spent fuel hulls is shown to be easily feasible for both Zircaloy and for stainless steel, and volume reduction factors in the region of 5 to 7, corresponding to the theoretical density of the alloy obtained, are arrived at quite easily.

The Zircaloy copper alloy, put into use to lower the fusion point of the Zircaloy, appears extremely interesting both as to the ease with which it can be used and the possibility which it offers of working at temperatures always lower than 1250°C.

The decreasing of fusion temperature is less spectacular with stainless steel; only the use of silicon enabling the lowering of the temperature to around 1200°C appears really feasible.

The use of decontaminating agents either during or at the end of the fusion operation seems to be a promising technique, especially in the case of stainless steel where the use of a borosilicated glass is easy. The choice of decontaminating agent is more difficult for Zircaloy which reduces the principal oxide components of glasses and makes necessary the use of molten salts mixtures, the composition of which has not yet been defined.

The decontamination factors obtained during the tests run on steel are encouraging although they were obtained using artificially contaminated hulls; they should therefore be considered with precaution and be confirmed by further tests in hot cells using real hulls. It is at this stage that it will also be possible to seriously study the treatment of elements brought by the gas produced during fusion.

This study has made it possible to determine the principal parameters needed to set up an industrial furnace project. The realization of fusion compaction units for waste from fuel hulls generated by future reprocessing plants seems to be a real short-term possibility.

ALKALI AND ALKALINE EARTH ELEMENT STUDIES AT OKLO

Douglas G. Brookins

Department of Geology
University of New Mexico
Albuquerque, NM 87131

INTRODUCTION

The Oklo Natural Reactor was discovered by French Scientists in 1970. These studies are summarized in the first International Symposium on the Oklo Phenomena sponsored by the International Atomic Energy Agency (IAEA, 1975). Two years later a follow-up symposium (IAEA, 1977) was held with greater emphasis on geochemistry, fine structure of fluence calculations, material balance between fissiogenic and normal crustal amounts of many elements present in the Oklo ores, and related studies.

Of great significance was the early determination by workers in France and in the United States that many fission products remained in place during the $\underset{\sim}{\sim}$ 500,000 year duration of the reactor's lifetime or else migrated locally. While data are now available for the actinides (including some transuranics), lanthanides, most noble metals, many chalcophile elements and noble gases, data for the alkali and alkaline earth elements (Rb, Sr, Cs, Ba) have not received the study they warrant. Because the Oklo host rocks have acted as a fuel rod, migration terrain, and repository for many fissiogenic elements; and because 90-Sr and 137-Cs are two very important radionuclides produced in man made reactors, study of the suite Rb-Sr-Cs-Ba has been carried out in an attempt to see if limits on, and mechanism for, migration can be determined. The presence of fissiogenic 88-Sr in the Oklo reactor has been documented by the works of Havette et al. (1975), Lancelot et al. (1975) and Brookins et al. (1975). Havette showed, by ion imagery, the presence of 88-Sr in grains of pitchblende which were known to contain depleted 235-U and enrichments of other fissiogenic elements. No such findings were observed for non-reactor ore. Brookins et al. (1975) independently undertook to study alkali and alkaline earths from Oklo ore powders provided by the French CEA to the Los Alamos Scientific Laboratory (LASL) in 1974. Samples of high grade ore for which low (235/238)-U ratios and total U contents were known were available for study. All of the powders represented high grade ores and some were splits of samples from which Havette et al. (1975) had demonstrated the presence of 88-Sr. Brookins et al. (1975) showed the presence of up to 5-10 percent fissiogenic Rb and Sr in some samples, most concentrated in pitchblende grains. Vein fillings and secondary mineral coatings present on the pitchblende grains were removed by leaching and found to contain Rb and Sr of normal isotopic composition. It was not possible to test for local versus widespread migration of Rb or Sr.

Brookins et al. (1975) also examined the isotopic composition of Ba for the same samples. Ba is present in large quantities, such that calculated amounts of 138-Ba-f (fissiogenic) would probably not be resolved from 138-Ba-N in some pitchblende and its presence is suspect in at least two samples studied by Brookins et al. (1975). Of more interest, however, was the study of (135/137)-Ba. Both are the stable products from (135/137)-Cs-f decay respectively, which are produced in near equal quantities (Brookins et al. 1975), by 235-U fission. Since the ratio of (135/137)-Ba in crustal material unaffected by fission is about 0.583 (Brookins et al. 1975) then ratios well in excess of 0.583 may be taken to indicate the presence of fissiogenic Cs which has decayed to Ba. Four of five samples yielded 135/137 ratios far in excess of 0.585 and, interestingly, a crude correlation between increasing 135/137 ratio with decreasing (85/87)-Rb. These data sufficed to demonstrate the presence of fissiogenic Cs (135, 137). Further, since the half life of 135-Cs is 2.6 m.y. whereas the half life for 137-Cs is only 30 y., then both may have been retained in these Oklo samples for more than 25 m.y. The presence of 138-Ba-f is, by inference, indicated by these data.

The obvious need for better samples with which to work is apparent, yet by late 1977 mining had recommenced at Oklo and the mine floor lowered some 50 meters. A few samples from the edges of reactor zones and from areas between zones of depleted 235-U have been made available for study. Of these, several have been analyzed which indicate the presence of fissiogenic Rb, Sr, Ba, and Ba from Cs. Since the suite contains both high grade ore (reactor ore), low grade ore (no fission) and barren rock near ore zones, then presence of any of the fissiogenic elements of all three types of samples may allow an attempt for a material balance based on calculated fission yield and the total amount of fissiogenic alkali and alkaline earths present. It must be emphasized that the suite of samples studied are very well documented in terms of total uranium content, 235-U content, and location (See

Table 1). Analytical methods are described in Brookins et al. (1975).

DISCUSSION

The samples in Group I. have been discussed in detail by Brookins et al. (1975). The insoluble residues contain fissiogenic 88-Sr, (85/87)-Rb; a fact independently arrived at by Lancelot et al. (1975) for Sr and Rb and by Havette et al. (1975) for 88-Sr. The presence of 138-Ba is suggested by Naudet (1974) and Brookins et al. (1975) but large amounts of diluent prevent any attempt at a material balance. The total Rb and Sr contents of both the insoluble residues and leachates vary within powdered samples thus

Table 1. Uranium Systematics and Locations*

Sample	Total U(%)	^{235}U(%)	Location
Group I. Samples from High Grade Reactor Ore			
401, A,B	39	0.4	Reactor Zone 2
1178	52	0.52	2'P' Traverse
1179	50	0.507	2'P' Traverse
1185	46	0.413	2'P' Traverse
1186	48	0.42	2'P' Traverse
1187	52	0.423	2'P' Traverse
Group II. Samples from the Edges of Zones of Reactors			
SC53-1763	31.3	0.7190	Reactor Zone 3
SC18-1801	1.4	0.7195	Reactor Zone 1
SC55-1844	26.8	0.6904	Reactor Zone 3
SC55-1852	47.1	0.6855	Reactor Zone 3
SC55-1856	20.8	0.6864	
SC55-1860	12.3	0.6845	Reactor Zone 3
SC55-1864	2.9	-	Reactor Zone 3
SC54-1876	6.6	0.7177	Reactor Zone 3
SC56-1877	47.7	0.6930	Reactor Zone 3
SC47-1878	30.1	0.7170	Edge;Reactor Zone 2
Group III. Samples from the Edges and Between Reactor Zones			
SCO-2252	3.2	0.7201	Between Zones 1 and 2
KN224B	3.0	0.7190	Edge of Zone 4
KN236B	1.5	0.72	Edge of Zone 4
KN241B	8.3	0.7200	Edge of Zone 4
KN244B	2.7	0.7205	Between Zones 6/5 and 3
KN255B	5.3	0.7197	2 m. east of Zone 5
KN258B	2.0	0.7196	Edge of Zone 5
KN226B	8.5	0.7204	Between Zones 6/5 and 3
KN270B	13.3	0.7189	Between Zones 1 and 2

*Notes: (1) Data for Group I from Brookins et al. (1975).

(2) Data for Group II,III from Naudet (written commun., 1977).

attesting to their inhomogeneity. Use of atomic (85/87)-Rb and (86/88)-Sr ratios suggests the presence of fissiogenic Rb and Sr in amounts from less than one percent to as much as ten percent, but it must be remembered that (a) the total U and %235-U were determined on separate aliquots from those used for Rb and Sr determination and (b) Rb and Sr are not uniformly distributed in the samples analyzed. Of more interest of the Group I samples is the excess of (135/137)-Ba (Brookins et al., 1975) which formed from fissiogenic (135/137)-Cs. The fact that 133-Cs is slightly depleted (IAEA, 1975) in the reactor zones may be due to migration after a significant duration of earth history (i.e. at least 26 million years) based on the barium data (Brookins et al., 1975).

The group II. samples consist of those from the edges of reactor zones and from a traverse drilled across part of reactor zone 3 (SC55) with total U varying from 27% to 3%; all samples contain slightly 235-U depleted uranium (Table 1). Four of the other samples contain essentially normal 235/238 ratios although the total uranium contents vary from one to 31 percent. All of the insoluble residues except one contain normal Sr and eight of ten contain normal Rb. The leachates, however, show the opposite effect as four of nine contain fissiogenic Sr and eight of nine fissiogenic Rb. It is not possible to estimate how much fissiogenic Rb and Sr is present based on weight percent due to the dilution of fissiogenic material with normal material. If the amount of dilution is high due to dissemination of both fissiogenic Rb and Sr then the percent fissiogenic Rb and Sr calculated from atomic ratios (i.e. <1 to 10 percent) may be considered minimum values. The barium data for Group II. samples are difficult to interpret, in part as the data (Table 2) are incomplete. However, four of six leachates show 135/137 ratios above normal values as do four of five insoluble residues. Since the barium background values are higher (i.e. barite) and Ba does not leach as readily as Rb and Sr, then the presence of small (?) amounts of fissiogenic Ba in both the leachates and insoluble residues is not unexpected. Barite would keep fissiogenic 135 and 137 in the insoluble phases while these isotopes would be leached if incorporated into exchangeable sites in clay minerals.

The group III. samples are of extreme interest in that all contain normal (235/238)-U ratios with total uranium varying from 1.5 to 13 percent. Further, these samples were taken at the edges of and from distances of 2 to 4 meters. For this reason it is especially noteworthy that, despite normal 235/238 ratios, the leachates from this group contain fissiogenic Rb (three of nine) and Sr (six of seven). Only one (of nine) insoluble residue contains fissiogenic Rb whereas three of six of the insoluble residues contain fissiogenic 88-Sr. Total Rb and Sr data show the

Table 2. Rubidium, Strontium, and Barium Isotopic Data*+

Sample	Leachate 85/87 Rb	Leachate 86/88 Sr	Leachate 135/137 Ba	Insoluble Residue 85/87 Rb	Insoluble Residue 86/88 Sr	Insoluble Residue 135/137 Ba
Group I						
401A	2.5866	---	0.5892	---	---	---
401B	2.5973	---	---	---	---	---
1178a	2.497	---	---	2.578	0.1133	0.5898
1178b	---	---	---	---	0.1169	---
1179	2.595	0.1192	0.5885	2.574	0.1176	0.5902
111185	2.594	0.1194	0.5893	2.596	0.1168	0.5898
1186	2.595	0.1188	---	2.586	0.1162	0.5868
1187a	2.595	0.1190	0.5874	2.583	0.1145	0.5876
1187b	---	---	---	---	0.1171	---
Group II						
SC53-1763	2.575	0.1185	---	2.579	0.1201	---
SC18-1801	2.550	0.1160	0.5877	2.593	0.1186	---
SC55-1844	2.520	0.1196	0.5890	2.564	0.1205	---
SC55-1852	2.551	0.1195	0.5842	2.573	---	0.5843
SC55-1856	2.578	0.1167	---	2.600	0.1203	0.5868
SC55-1860	2.566	0.1185	0.5900	2.610	0.1200	0.5880
SC55-1864	2.525	---	0.5826	2.604	0.1188	0.5875
SC54-1876	2.555	0.1169	---	2.593	0.1187	---
SC56-1877	2.587	0.1171	0.5860	2.607	0.1176	---
SC47-1878	---	0.1200	---	2.598	0.1188	---
-2223	---	---	---	---	---	0.5890
Group III						
SCO-2252	2.577	---	---	2.590	0.1185	---
SCO-2252b	---	---	0.5865	---	---	---
KN224B-a	2.541	0.1163	0.5831	2.587	0.1172	---
KN224B-b	2.550	0.1164	---	2.588	0.1168	---
KN236B	2.589	---	0.5897	2.587	0.1180	---
KN241B	2.554	0.1165	0.5836	2.598	0.1143	---
KN244B	2.591	0.1173	---	2.560	---	---
KN255B	2.596	0.1154	---	2.592	0.1186	0.5878
KN258	---	---	0.5731	---	---	---
KN258B	2.591	0.1164	---	2.594	---	---
KN266B	2.586	0.1186	---	2.568	---	---
KN270B	2.592	0.1183	---	2.601	---	---

Notes: (1) Data for Group I. from Brookins et al. (1975). (2) Suffixes -a,-b indicate separate aliquots of samples and not replicate mass spectrometric determinations.

*Natural isotopic ratios of 85/87 Rb = 2.5926; of 86/88Sr = 0.1194 and of 135/137Ba = 0.5843.

+Error Bars for isotopic ratios are ± 0.2 for 85/87Rb, ± 0.25 for 86/88 Sr and 0.3 for 135/137 Ba (all 1 σ).

background values to range from 5-34 and 2-14 ppm respectively, and leachates to contain much more Rb than Sr (Table 3). In these samples the effect of dilution of fissiogenic Rb and Sr with the normal elements is important. Even for only several ppm Rb or Sr, to find fissiogenic Rb or Sr in zones of normal 235/238 uranium, indicates migration. Since the amount of reactor ore is small

Table 3. Rubidium, Strontium, and Barium Abundance Data: Group III*

Sample	Leachates(mg/ml) Rb	Sr	Ba	Insoluble Residues(ppm) Rb	Sr	Ba
SCO-2252B	0.73	0.33	15.2	5.08	2.29	106.8
KN224B	0.73	0.45	9.5	5.08	3.17	68.2
KN236B	0.73	0.32	15.8	5.08	2.52	110.2
KN241B	0.73	0.45	24.8	5.08	3.15	173.2
KN244B	0.73	0.27	5.2	10.2	3.74	73.5
KN255B	2.4	1.03	99.5	33.6	14.4	1395.0
KN258B	0.73	1.1	14.8	5.08	7.70	103.2
KN266B	0.73	0.24	9.5	11.6	3.84	156.0
KN270B	0.73	0.24	4.8	5.10	1.68	33.2

Notes: Data by atomic absorption spectrophotometry (J.W. Husler, analyst). Data precise to ± 10%(1 σ).

relative to the total uranium present, and the amount of normal Rb and Sr is very large relative to that produced by fission, then the atomic (85/87)-Rb and (86/88)-Sr ratios are indeed minimum values. It is not possible, for example, to state with certainty the uranium content and degree of fission (burnup) which might have produced fissiogenic Rb and Sr which later migrated to sites of no fission. If a sample with UO_2=5% underwent 5% burnup and the fissiogenic Rb migrated and mixed with approximately 45 ppm normal Rb, then the resultant 85/87 ratio of 2.54 could represent 60 percent fissiogenic retention instead of the much lower ratio based on just the atomic ratio. This is important to note as the implication of the small but significant amounts of fissiogenic Rb and Sr in the leachates argues for local migration rather than for widespread migration of these elements. Further, that fissiogenic Rb and Sr are more obvious in the leachates of Group III compared to the insoluble residues of Group I which suggests migration of the fissiogenic elements after reactor shutdown concomitant with the formation of new minerals. The Rb and Sr data are shown in Figure 1.

The Barium data for Group III. samples are very sparse. Only two of five leachates show high 135/137 ratios as does the one insoluble residue analyzed. Of more interest, perhaps, is sample KN258B which shows a low 135/137 ratio suggesting that some loss of 135-Cs occurred after 137-Cs had decayed to 137-Ba. At present all that can be said for barium in Group III. samples is that fissiogenic barium is noted and thus local migration has occurred. Since the barium background varies from 90 to about 1900 ppm the dilution of fissiogenic by normal elements masks detection of all but very small amounts of the former. Collectively, the data for the Group I, II and III samples argue for the following conclusions: (1) Group I insoluble residues contain small amounts of fissiogenic Rb, Sr and Ba. (2) Group I leachates contain normal Rb and Sr arguing for post-reactor dilution in the form of veins, fracture fillings and grain coatings. (3) Group II leachates are

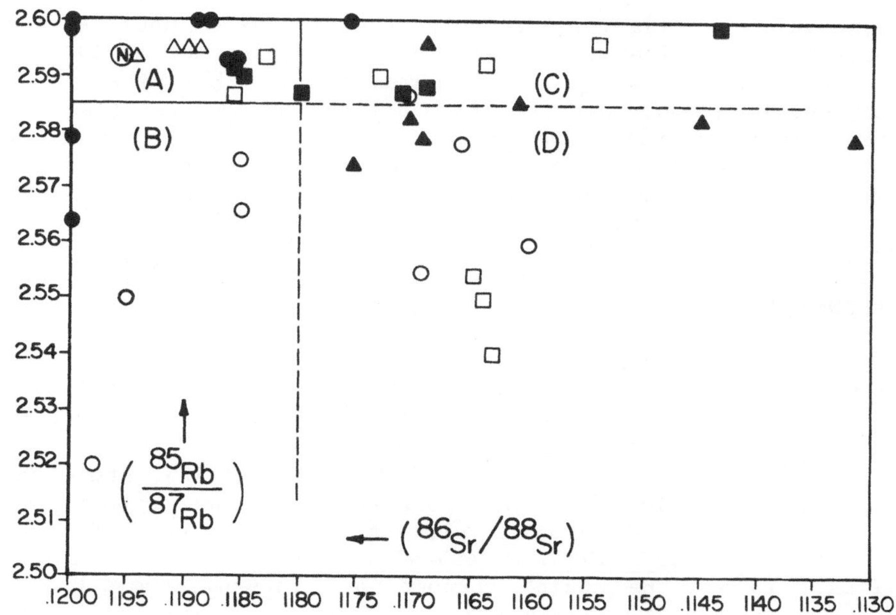

Fig. 1. (85/87)-Rb Versus (86/88)-Sr Ratios for Oklo Samples

Notes: 1. Area A contains samples with Rb and Sr of normal isotopic composition. 2. Area B contains samples with normal Sr but some fissiogenic rubidium. 3. Area C contains samples with normal Rb but some fissiogenic Sr. 4. Area contains samples with some fissiogenic Rb and Sr. 5. Legend: Insoluble residue samples from Groups I, II, and III are shown by closed triangles (I) solid circles (II), solid squares (III); leachates are shown by open triangles (I), open circles (II) open squares (III). See text for details. The dotted lines represent the range, including errors, which separate the various zones. Normal Rb,Sr would plot at $^{85}Rb/^{87}Rb$ = 2.5926 and $^{86}Sr/^{88}Sr$ = 0.1194. The errors for the data are ± 0.2%(1 σ) for Rb and ± 0.25% (1 σ) Sr isotopic data.

more enriched in fissiogenic Rb and Sr than the corresponding insoluble residues. (4) Group III leachates are more enriched in fissiogenic Sr, and some in fissiogenic Rb, versus the insoluble residues which contain normal Rb and some which contain fissiogenic Sr. (5) Loss of fissiogenic Rb and Sr from high grade reactor ore and fixation of an unknown, but possibly large, quantity of fissiogenic Rb and Sr in the periphery of the reactor zones is indicated. (6) The barium data (groups I, II, III) argue for retention of (135/137)-Cs in the overall system until at least 26 million years after reactor shutdown; the presence of 138-Ba is inferred because of the enrichments of (135/137)-Ba, but background barium is too high to allow an accurate assessment of barium retention versus migration. (7) A material balance for fissiogenic alkali and alkaline earth elements is not possible at this time due to lack of adequate samples from edges of, and in between high grade reactor zones.

REFERENCES

Brookins, D. G., Lee, M. J., Mukhopadhyay, B., and Bolivar, S. L., 1975, Search for fission-produced Rb, Sr, Cs and Ba at Oklo: in IAEA 1975, p. 401-414.

Havette, A., Naudet, R., and Slodzian, G., 1975, Etude par analyse ionique de la repartition et des proportions isotopiques de certains elements dans des echanntillons d'Oklo: in IAEA 1975, p. 463-478.

IAEA, 1975, The Oklo Phenomenon: Intern. Atomic Energy Agency Sym. 204, (Libreville, Gabon; June-July 1975), 647 p.

IAEA, 1977, Natural Fission Reactors: Intern. Atomic Energy Agency Proc. Sym. 119 (Paris, 1977), 754 p.

Lancelot, J. R., Vitrac, A., and Allegre, C. J., 1975, The Oklo natural reactor: age and evolution studies by U-Pb and Rb-Sr systematics: Earth Plan. Sci. Lttrs., v. 25, p. 189-196.

Naudet, R., 1974, Summary report on the Oklo phenomenon, French CEA report: Bull. Infor. Scien. Tech., v. 193, p. 7-85 (Eng. trans.).

DURABILITY OF RHYOLITIC OBSIDIAN GLASS INFERRED FROM HYDRATION DATING RESEARCH

J. E. Ericson

Peabody Museum of Archaeology and Ethnology
Harvard University
Cambridge, Massachusetts

INTRODUCTION

Archaeological materials provide samples for studying natural processes of chemical and physical degradation. They can provide important information about natural processes. Generally, due to time constraints, processes of material degradation are simulated and studied under controlled conditions in the laboratory. If the process of natural degradation is relatively simple, then it is simple to simulate. However, if the number of parameters are extensive and interactive, then simulation is difficult to impossible. Such appears to be the interaction between obsidian and its environment. Recent results of obsidian hydration dating research in the field of archaeology have a direct bearing on our understanding of the durability of natural glass. Since obsidian-like glasses are being considered as containment media for high activity nuclear fuel waste, it is important that both the natural and laboratory results are well understood.

In this paper the preliminary results of studying the natural and laboratory-induced hydration of obsidian sources are presented.

OBSIDIAN AS A NATURALLY OCCURRING GLASS

Obsidian is a naturally occurring volcanic glass which has been extruded or intruded at very shallow depth from the crustal interior under high pressure and temperature. Although a generally rare phenomenon, obsidian occurs within the circum-Pacific tectonic belt, along major faults and within other volcanic areas.

Prehistorically, obsidian, having the importance of modern steel, was much sought after and widely traded. The use of obsidian by prehistoric peoples and subsequent discarding of this material in many environments, for a period of at least 30,000 years, provides an important record of the natural degradation of obsidian. The actual degradation of similar man-made glass can be evaluated from this data.

There are a number of different chemical types of obsidian occurring worldwide which parallel the major chemical divisions of crystalline rocks. Rhyolitic obsidian has greater than 66 wt % silica. Dacitic obsidian has less than 66 and greater than 52 wt % silica. Finally, tachylite (basaltic glass) has less than 52 wt % silica.

Twenty-eight rhyolitic obsidian samples found in California, western Nevada and southern Oregon were chemically analyzed by x-ray fluorescence. The amounts of major oxides such as SiO_2, Al_2O_3, Na_2O, K_2O, CaO, MgO, Fe_2O_3 and TiO_2 were determined. The chemical compositions of rhyolitic obsidians are similar to normal granites and other rhyolitic rocks except they are more siliceous. The results of the analysis are presented in Table 1.

Table 1. Mean Chemical Composition of Rhyolitic Obsidian Compared to Granite (after Ericson et al. 1975)

	Granite (mol %)	Obsidian (mol %)
SiO_2	77.18	83.00 ± 5.23
Al_2O_3	10.13	8.25 ± 0.52
CaO	2.36	0.98 ± 0.98
Na_2O	4.37	3.86 ± 0.36
K_2O	3.17	2.85 ± 0.27
Fe_2O_3	1.11	0.62 ± 0.31
MgO	1.25	0.28 ± 0.28
TiO_2	0.38	0.11 ± 0.07

The process of degradation of obsidian forms perlite or hydrated obsidian. Obsidian, being inherently unstable in the natural environment, forms perlite and, in turn, clays and other minerals. It is this process of degradation of obsidian to perlite through hydration that is important to nuclear waste disposal.

OBSIDIAN HYDRATION DATING: PRIOR RESEARCH

Atmospheric water is chemically absorbed on the surface of the obsidian. This water diffuses into the interior of the obsidian as functions of time and temperature (Friedman and Smith 1960). It is also feasible on theoretical grounds (Ericson 1973; Ericson, MacKenzie and Berger 1976) that the water also reacts with the structure which causes the hydration rate to deviate from the diffusion model proposed by Friedman and Smith (1960) (Ericson 1981).

The obsidian hydration dating technique has been shown to be useful to archaeology (Katsui and Kondo 1965; Michels 1967, 1973; Meighan et al. 1968; Johnson 1969; Suzuki 1973; Bell 1977; Singer and Ericson 1977) and geology (Friedman 1968; Friedman et al. 1973). The hydration phenomenon involves the development of a measurable birefringence stress layer through a sequence of processes which are not totally understood. In the absence of a complete understanding of the hydration process and the variables controlling the rate of hydration, there has been a considerable debate over the mathematical form based on archaeological evidence (Clark 1961a, 1961b, 1964; Neighan et al. 1968, 1970; Johnson 1969; Friedman and Smith 1960), the physical mechanism of the hydration process (Marshall 1961; Haller 1963; Friedman, Long and Smith 1966; Ericson 1975; Ericson, MacKenzie and Berger 1976), and the variables which influence the hydration rates (Friedman and Smith 1960; Aiello 1969; Ericson 1973, 1975; Kimberlin 1971; Ericson and Berger 1976; Kimberlin 1976; Friedman and Long 1976; Ambrose 1976). As originally formulated by Friedman and Smith (1960) the obsidian hydration dating technique relied on a general diffusion equation having two variables, namely time and temperature. To facilitate the application of the technique, broad temperature zones were established after the work of Chang (1957), within which a zonal hydration rate was to be used. Later, based on archaeological evidence, Clark (1961a, 1961b, 1964) and Meighan, Foote and Aiello (1968) suggested that the proposed diffusion model did not fit the empirical hydration data. In support of their original thesis, Friedman et al. (1966) defended their diffusion model of hydration with the results of a four-year induced hydration experiment. The impact of these findings was to suggest to researchers that tighter geographical control was definitely necessary in order to resolve the hydration problem. A summary discussion of subsequent regional studies has been presented in Ericson, MacKenzie and Berger (1976).

Even with increased geographical control, yet another form of variability was observed in hydration rate formation. Although Friedman and Smith (1960) did demonstrate hydration rate differences between tachylitic and rhyolitic obsidian families, they did not suggest the degree of importance of chemical factors within each family of obsidians. As a result, a series of papers now suggests the importance of chemical composition in affecting hydration rates (Aiello 1969; Ericson 1973; Ericson and Berger 1976; Kimberlin 1971, 1976; Susuki 1973; Layton 1973; Ambrose 1976; Friedman and Long 1976).

In summary, prior research has continued to refine the obsidian hydration dating technique by determining and controlling variables of the hydration process which have been defined as time, temperature, and chemistry of the obsidian.

NATURAL HYDRATION RATES

Source-specific hydration rates were determined using archaeological data. Unit-level association of radiocarbon dates and obsidian artifacts from regionally diverse sites, sampled over a wide range of time, provided data to determine the hydration rates. The trace element chemistry of each obsidian artifact was determined using instrumental neutron activation analysis. These results were compared to chemically known obsidian source samples and, thus, identified as to their origin. Fourteen source-specific obsidian hydration curves were determined by using the chemical characterization data to stratify the hydration measurements and associated radiocarbon data. The results are presented in Table 2.

LABORATORY-INDUCED HYDRATION RATES

Induced hydration rate experiments were conducted to control the environmental parameters. Experiments were performed at elevated temperatures: 150°C, 163°C, 172°C, 192°C and 200°C for varying periods up to 3-28 days. Samples of each obsidian were freshly chipped and placed in a reaction vessel made of 3/4" ID stainless steel tubing with SWAGELOK tube fittings containing doubly distilled water filled to 75% of the volume. The activation energies and diffusion constants were calculated using the Arrhenius equation. The results are presented in Table 2.

Table 2. Comparisons of Diffusion Rate Constants ($\mu^2/10^3$ years) and Observed Mean Natural Effective Temperature (°C) with Expected Temperature (°C) Using Natural and Laboratory Data

Source	N	Natural Rate	Laboratory Rate(25°C)	Natural Temp.	Expected Temp.
Annadel	6	11.86	43.8	22	22
Bodie Hills	7	6.54	20.6	22	15
Borax Lake	10	5.54	---	23	--
Buck Mt.	10	3.16	0.789	15	34
Casa Diablo	10	25.3	---	21	--
Coso	14	28.6	0.005	25	74
Modoc Glass Mt.	11	3.15	0.616	15	36
Mono Craters	4	28.6	0.426	24	54
Mono Glass Mt.	6	12.3	---	23	--
Mt. Konocti	4	2.29	1.422	25	28
Napa Glass Mt.	23	4.73	0.021	22	56
Obsidian Butte	2	---	---	25	--
Mt. Hicks	3	27.7	0.005	21	74
Pine Grove Hills	2	---	0.367	22	--

VARIABLES INFLUENCING THE HYDRATION PROCESS

To examine the rank-order importance of each variable in influencing the observed variation of hydration rates among the sources, the chemical and physical properties of 28 obsidian sources in California, western Nevada and southern Oregon were determined by x-ray fluorescence analysis. The density, Vickers hardness and percent crystallinity were measured by standard procedures. The internal and saturated water concentrations of at least two samples from each source were measured by a nuclear reaction technique, performed at the California Institute of Technology. Seven additional properties were created by calculations based upon the original physical and chemical properties.

Multivariate regression analysis was used to determine the rank-order importance of the above variables presented in Table 3. The data appears in an extended work (Ericson 1981).

Table 3. Variables of the Hydration Process Using Laboratory Hydration Data

Variable	Mean	Standard Deviation	Partial Correlation	Step	Stepwise Multiple R
SiO_2	77.38	3.17	0.42	5	0.94
Al_2O_3	12.77	0.51	-0.34	-	-
CaO	0.80	0.35	-0.58	1	0.58
Na_2O	3.51	0.45	0.02	-	-
K_2O	4.50	0.34	0.22	-	-
Fe_2O_3	1.40	0.61	-0.30	-	-
MgO	0.12	0.07	-0.55	-	-
TiO_2	0.11	0.08	-0.28	2	0.81
Final H_2O	4.27	1.14	-0.38	-	-
Initial H_2O	0.67	0.27	0.36	4	0.92
S Factor	0.423	0.003	0.50	-	-
Density	2.38	0.03	-0.24	3	0.89
Hardness	739.	101.	0.27	6	0.98
A Factor	1.48	0.59	-0.56	-	-
Cal. Density	2.39	0.02	0.33	-	-
Crystallization	11.80	24.18	-0.33	-	-
Activat. Energy	26.80	7.49	-	-	-

CONCLUSIONS

Archaeological materials can provide important data for analyzing the processes of material degradation. Rhyolitic obsidian can serve as a model of man-made glass which is used for containment of high activity nuclear waste.

Natural obsidian hydration rates were determined using artifacts and archaeological data. This was followed up by inducing the hydration of the same source samples under controlled laboratory conditions. A comparison of the two data sets provides a model of the natural and induced processes. It appears that the diffusion coefficients of the laboratory results are very much less than those observed for natural hydration. This is interpreted to mean that laboratory conditions do not simulate well the processes present in the natural environment. The effective environment of obsidian is a complex of chemical and physical parameters, of which one is the inward diffusion of water. Obsidian is much more unstable than indicated by laboratory experiments.

The intrinsic variables influencing the hydration rate are presented. Among them, CaO, TiO_2, density and H_2O concentrations appear to be the most important variables. CARD Reprint No. 10.

REFERENCES

Aiello, P. V. (1969), The Chemical Composition of Rhyolitic Obsidian and Its Effect on Hydration Rate: Some Archaeological Evidence, M.A. Thesis, Department of Anthropology, UCLA.

Ambrose, W. (1976), "Intrinsic hydration rate dating of obsidian," Advances in Obsidian Glass Studies, edited by R. E. Taylor, Noyes Press, New Jersey.

Bell, R. E. (1977), "Obsidian hydration studies in highland Ecuador," American Antiquity, 42(1):6878.

Chang, J. (1957), "Global distribution of the annual range in soil temperature," Trans. Amer. Geophys. Union, 38(5):718-723.

Clark, D. L. (1961a), The Application of the Obsidian Dating Method to the Archaeology of Central California, Ph.D. Dissertation, Stanford University.

Clark, D. L. (1961b), "The obsidian dating method," Current Anthropology, 2:111-116.

Clark, D. L. (1964), "Archaeological chronology in California and the obsidian hydration method: Part I," UCLA Archaeological Survey, Annual Report, 1963-64.

Ericson, J. E. (1973), On the Archaeology, Chemistry and Physics of Obsidian, M.A. Thesis, Department of Anthropology, UCLA.

Ericson, J. E. (1975), "New results in obsidian hydration dating," World Archaeology, 7:151-159.

Ericson, J. E. (1981), California Obsidians: The Results of Tracing and Dating, British Archaeological Reports, International Series (in press).

Ericson, J. E. and R. Berger (1976), "Physics and chemistry of the hydration process in obsidians, II: Experiments and measurements," Advances in Obsidian Glass Studies, edited by R. E. Taylor, Noyes Press, New Jersey.

Ericson, J. E., J. D. MacKenzie and R. Berger (1976), "Physics and chemistry of the hydration process in obsidians, I: Theoretical implications," Advances in Obsidian Glass Studies, edited by R. E. Taylor, Noyes Press, New Jersey.

Ericson, J. E., A. Makishima, J. D. MacKenzie and R. Berger (1975), "Chemical and physical properties of obsidian: A naturally occurring glass," J. Noncrystalline Solids, 17:129-142.

Friedman, I. (1968), "Hydration rind dates of rhyolitic flows," Science, 159(3817):878.

Friedman, I. and W. D. Long (1976), "Hydration rate of obsidian," Science, 191:347-352.

Friedman, I., K. L. Pierce, J. D. Obradovich and W. D. Long (1973), "Obsidian hydration dates glacial loading?" Science, 180(4087): 733.

Friedman, I. and R. L. Smith (1960), "A new dating method using obsidian, I: The development of the method," American Antiquity, 25(4):476:522.

Friedman, I., R. L. Smith and W. D. Long (1966), "The hydration of natural glass and the formation of Perlite," Geol. Soc. Amer. Bull., 77:323-360.

Haller, W. (1963), "Concentration-dependent diffusion coefficient of water in glass," Physics and Chemistry of Glasses, 4(6):217-220.

Johnson, L., Jr. (1969), "Obsidian hydration rate for the Klamath Basin of California and Oregon," Science, 165(3900):1354-1355.

Katsui, Y. and Y. Kondo (1965), "Dating of stone implements by using the hydration layer in obsidian," Japanese Jour. Geol. Geog., 46(2-4):45-60.

Kimberlin, J. (1971), Obsidian Chemistry and the Hydration Dating Technique, M.A. Thesis, UCLA.

Kimberlin, J. (1976), "Obsidian hydration rate determinations of chemically characterized samples," Advances in Obsidian Glass Studies, edited by R. E. Taylor, Noyes Press, New Jersey.

Marshall, R. R. (1961), "Devitrification of natural glass," Geol. Soc. Amer. Bull., 72(10).

Meighan, C. W., L. J. Foote and P. V. Aiello (1968), "Obsidian dating in West Mexican archaeology," Science, 16 (3832).

Meighan, C. W. and C. V. Haynes (1970), "The Borax Lake site, revisited," Science, 167:1213-1221.

Michels, J. W. (1967), "Archaeology and dating by hydration of obsidian," Science, 158(3798).

Michels, J. W. (1973), Dating Methods in Archaeology, Seminar Press, Inc., N.Y.

Singer, C. A. and J. E. Ericson (1977), "Quarry analysis at Bodie Hills, Mono County, California: A case study," Exchange Systems in Prehistory, edited by T. K. Earle and J. E. Ericson, Academic Press, N.Y.

Suzuki, M. (1973), "Potential of obsidian hydration dating," Jour. Faculty Sci., Univ. Tokyo, Section V, 4:241.

BACKFILL BARRIERS: THE USE OF ENGINEERED BARRIERS BASED ON GEOLOGIC MATERIALS TO ASSURE ISOLATION OF RADIOACTIVE WASTES IN A REPOSITORY

John A. Apps[+] and Neville G. W. Cook[*]
[+]Earth Sciences Div., Lawrence Berkeley Lab., Berkeley, Ca. 94720; [*]Department Materials Science and Mineral Engineering, University of California, Berkeley, Ca. 94720

INTRODUCTION

The disposal of high level radioactive wastes by deep geologic burial involves questions that cannot be answered on the basis of human experience. Assurance that an underground repository will provide adequate isolation of wastes from the biosphere for unprecedented periods of time will have to be based upon predictive modeling (NRC, 1979). Verification of the performance of a repository by *in-situ* measurement is not practicable over the long term (NRC, 1979). Geologic media are inherently variable and their properties can rarely be determined *in-situ* with the degree of certainty and confidence perceived to be necessary for disposal of wastes. Therefore, uncertainties must exist in the analysis and prediction of the performance of a repository, the magnitude of which increases with time. The practical and the theoretical difficulties in resolving this problem are very great; large amounts of time and effort could be devoted to improving methods for determining properties and conditions and their variability *in-situ* with no guarantee of success. Any other means by which uncertainties in the prediction of the performance of a repository may be diminished are, therefore, of great potential importance.

A means by which these uncertainties may be diminished is through the use of engineered barriers. Engineered barriers may be made from materials known to exist geologically, the properties of which are, or can be, fully understood in the long term. The control exercised over engineered barriers should resolve the questions of variability and uncertainty, so that these will be less than those inherent in the geologic media within which a repository is developed. Thus, predictions concerning the performance of engineered barriers should be relatively precise.

Accordingly, it becomes important to identify geologic analogs of materials that may be used as barriers, and to assess the potential of such materials to provide adequate assurance of the performance of a repository in the long term, regardless of the geologic media.

Ringwood (1978) and others already are addressing the concept of synthesizing waste forms consisting of minerals similar to those known to occur stably in nature. This concept can be expanded to include the canister, overpack and backfill. Copper, iron, nickel-iron alloys, and nickel-iron-cobalt alloys occur in nature. Fyfe (1977) has investigated copper as a canister material, and this has been incorporated in the design of a repository in Swedish bedrocks (KBS, 1978). Another canister material proposed by Fyfe (1977) is a nickel-iron alloy known to occur naturally in some ultrabasic rocks.

In this paper, a preliminary assessment is made to show that canisters fabricated of nickel-iron alloys, and surrounded by a suitable backfill, may produce an engineered barrier where the canister material is thermodynamically stable with respect to its environment. As similar conditions exist in nature, the performance of such systems as barriers to isolate radionuclides can be predicted over very long periods, of the order of 10^6 years.

ORIGIN AND OCCURRENCE OF IRON AND NICKEL-IRON ALLOYS IN NATURE

Iron and nickel-iron alloys have been observed on many occasions as secondary alteration products resulting from the serpentinization of ultrabasic rocks, particularly dunites and peridotites. These rocks consist primarily of olivine, $(Mg,Fe)_2SiO_4$ and enstatite, $MgSiO_3$. They alter in the presence of water to antigorite or chrysotile, $Mg_3Si_2O_5(OH)_4$, together with smaller amounts of tremolite, $Ca_2(Mg,Fe)_5Si_8O_{22}(OH)_2$, and occasional diopside, $CaMgSi_2O_6$. The range of temperatures over which these processes take place has been estimated to vary between 25°C to 500°C (Wenner and Taylor, 1973). During alteration of ultrabasic rocks, the coexisting aqueous phase becomes extremely alkaline, particularly in the absence of dissolved carbonates. The pH of groundwaters emanating from partially serpentinized bodies ranges from 11.2 to 12.0 (Barnes and O'Neil, 1969). In the presence of carbon dioxide, the pH is lowered, but is still alkaline (pH = 7.8-9.0), (Barnes and O'Neil, 1969). The ferrous iron released from the decomposing olivine under these conditions is immediately precipitated as an oxide, usually magnetite, driving down the oxidation potential:

$$3FeO_{(ol)} + H_2O = Fe_3O_4 + 2H^+_{(aq)} + 2e_{(aq)} \qquad (1)$$

Because the FeO component of the olivine and enstatite is usually less than 10 weight percent, the proton production is offset easily by hydroxyl ion production resulting from hydrolysis by the magnesia component of these two minerals:

$$MgO_{(ol)} + H_2O = Mg(OH)^+_{(aq)} + (OH)^-_{(aq)} \qquad (2)$$

The uptake of electrons can be accomplished only through the reduction of the metal ions, sulfur, carbon dioxide, nitrogen or water. Ultrabasic rocks contain between 5 and 10 weight percent FeO, and small amounts of nickel, (0.2 percent) (Nickel, 1959; Azais _et al._, 1968).

Field observations indicate that reduction of iron and nickel occurs with the formation of αFe (kamacite), γ(Fe,Ni) (taenite) with a composition of approximately Ni_2Fe, and possibly Ni_3Fe, commonly referred to as awaruite. Associated with these metallic phases in serpentinite are native copper, pyrrhotite or troilite, FeS; heazlewoodite Ni_3S_2; millerite, NiS; and pentlandite, $(Ni,Fe)_9S_8$. Evidence for thermodynamic equilibrium among these phases is as usual, inconclusive.

This summary description and interpretation of the genesis of nickel-iron alloys during the alteration of ultrabasic rocks must be followed by an evaluation of these natural alloys as a part of an engineered barrier for the isolation of wastes. The next step involves a closer examination to identify any deficiencies in knowledge that may hamper the adequate determination of the use of such an engineered barrier. Such an examination includes: a thermodynamic evaluation of the chemistry of the processes taking place; field studies to obtain estimates of the reaction rates and chemical fluxes in the system; field and laboratory studies to determine the physical properties of serpentinite, and design and testing to establish whether or not practicable barrier systems can be engineered replicating the natural analog.

THERMODYNAMIC ANALYSIS OF THE STABILITY OF NICKEL-IRON ALLOYS

The preliminary analysis presented below suggests that mineralogical and field observations published in the literature are essentially consistent with theoretical predictions, and that chemically, an engineered barrier could be constructed consisting of a nickel-iron canister surrounded by a backfill of olivine and serpentine.

Eh-pH Conditions Arising from the Alteration of Peridotite

To establish whether or not Eh-pH conditions could be attained that are compatible with the formation of iron-nickel alloys and consistent with field observations of groundwaters issuing from dunites and peridotites, the course of chemical reaction of a hypothetical ultrabasic rock (consisting of 40 mole percent olivine, 40 mole percent enstatite and 20 mole percent diopside, reacting with almost pure water at 25°C and one atmosphere pressure) has been simulated, Figures 1 and 2. Details of this type of simulation and the thermodynamic data used have been given by Helgeson _et al._, (1970).

The phases formed during the alteration of the ultrabasic rock are listed in Figure 1. All except talc and fayalite persist until native iron is formed at the termination of the simulation.

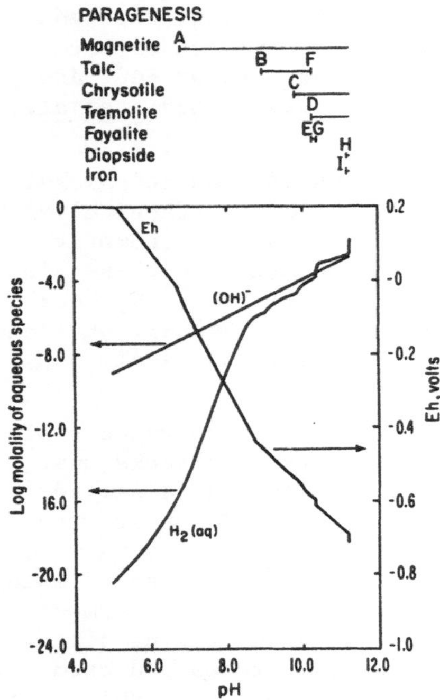

Figure 1. Chemical alteration of peridotite to serpentine: variation of $(OH)^-$, $H_{2(aq)}$ and Eh as a function of pH.

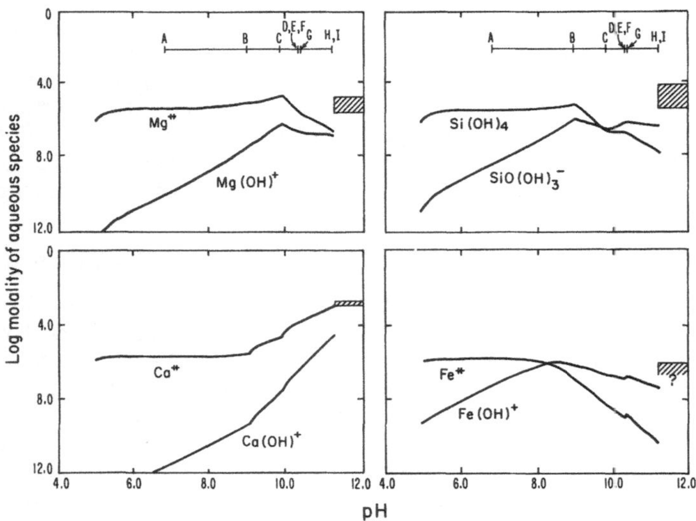

Figure 2. Variation of Mg, SiO_2, Ca, and Fe^{++} species in solution as a function of pH during the serpentinization of peridotite.

The phases produced are generally consistent with those observed in the field (Barnes and O'Neil, 1969). Figure 1 shows that the alteration leads to a pH of 11.25 and Eh of -700 millivolts S.H.E. upon formation of native iron. A dissolved hydrogen concentration of approximately 34 ppm is attained at that point.

Figure 2 illustrates the variation of the principal ions of magnesium, silica, calcium and iron in solution, as a function of pH. Comparisons are made with reported ranges of values of these elements in groundwaters of the Ca^{2+}-$(OH)^-$ type issuing from dunities and peridotites (Barnes and O'Neil, 1969). In general, the theoretical results compare favorably with chemical analyses of groundwaters.

The significance of this simulation is that the conditions of Eh-pH necessary for the formation of metallic iron can be achieved during the alteration of ultrabasic rocks.

Stability Relations Between Nickel-Iron Alloys and Coexisting Sulfides in the Presence of Water

The free energy data needed to calculate the stability relations are incomplete so both estimates and approximations must be made through a graphical evaluation of phase relations in the system Fe-Ni-S at 25°C and one atmosphere using known thermochemical data (Garrels and Christ, 1965; Robie *et al.*, 1978; Wagman *et al.*, 1969). Those phases for which free energy data are not known have been estimated from experimental and field information.

The relations between metallic and sulfide phases, and coexisting oxides in the presence of water have been examined through construction of an Eh-pH diagram for the system Fe-Ni-S-O-H, at 25°C and one atmosphere, covering the region of interest, Figure 3. The stability fields of various metal alloys, sulfides and oxides are located within the ranges Eh -1000 to -200 millivolts S.H.E., and pH 8 to 14. The total sulfur activity has been set at 10^{-6} molal in concordance with field conditions (Barnes and O'Neil, 1969).

The stability fields of coexisting phases are largely consistent with both laboratory investigations and field observations by numerous investigators. However, taenite γ(Fe,Ni) may be metastable below 350°C (Goldstein, 1973). They demonstrate further that regions can be attained where nickel-iron alloys are thermodynamically stable, particularly when: the pH is greater than 10; sufficient $FeO_{(ol),(en)}$ is present to produce the desired reducing conditions; total sulfur activity is below 10^{-6} molal, and relatively high concentrations of dissolved hydrogen, that is, up to 0.1 molal (200 ppm) exist.

FURTHER FIELD AND LABORATORY STUDIES

Field studies of nickel-iron alloys in altered ultrabasic rocks provide incidental observations that bear on their use as an engineered barrier, particularly as they relate to the thermodynamic stability of the metal canister, and the behavior of the backfill in the event that

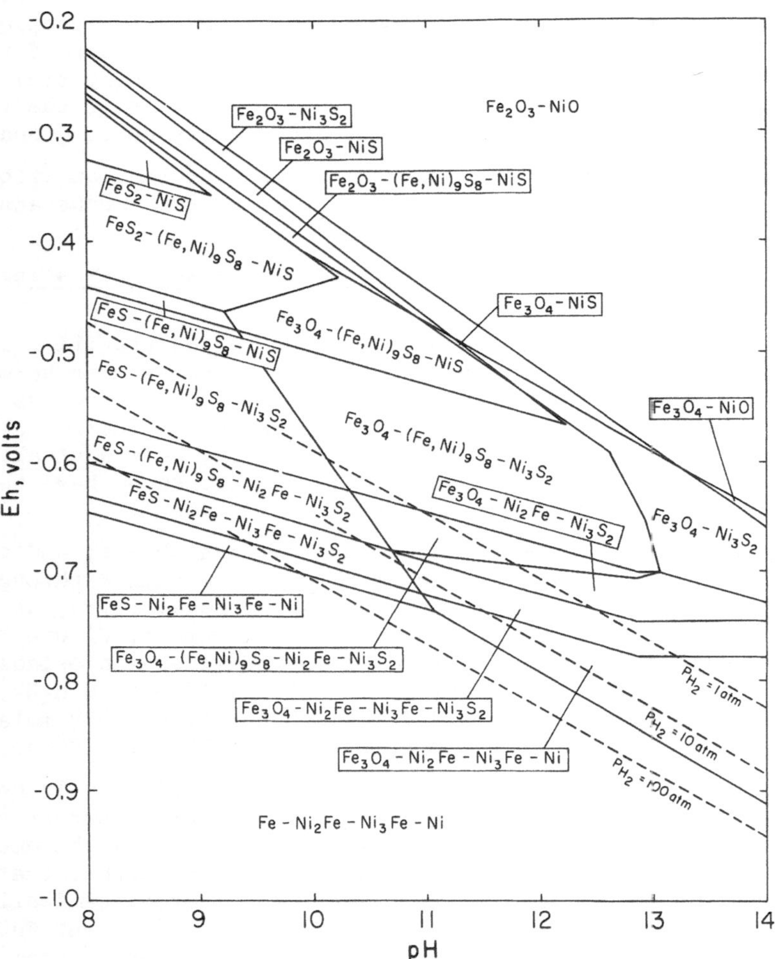

Figure 3. A preliminary Eh-pH diagram of part of the system Fe-Ni-S-O-H at 25°C and 1 atmosphere and $S = 10^{-6}$m.

a canister leaks. The geochemical environment within an altering ultrabasic rock is different usually from surrounding host rocks, causing chemical potential gradients to be set up. Petrological, geochemical and physical studies, conducted from the interior to exterior of an ultrabasic body to establish the extent of chemical migration and conditions of Eh-pH during alteration, yield semi-quantitative estimates of chemical migration rates over time spans up to millions of years (Chamberlain *et al.*, 1965; Azais *et al.*, 1968).

Several physical aspects to the use of altered ultrabasic rocks as an engineered barrier merit further study. The stoicheometry of reactions proceeding during serpentinization of ultrabasic rocks has not been resolved (Turner and Verhoogen, 1960), so that the volume change during alteration is not known. A backfill that swells during alteration would be advantageous. Little is known about the porosity, permeability or rheologic properties of partially or completely serpentinized rocks, but serpentine can flow plastically, a desirable property.

CHEMICAL BEHAVIOR OF ALTERING SERPENTINITE TO RADIONUCLIDE MIGRATION

Although canisters should not leak during the first 1000 years in consideration, the migration of radionuclides through the serpentinite matrix of the backfill must be examined. Cesium would be mobile, and unless either chrysotile or antigorite adsorb $Cs^{+}_{(aq)}$ a secondary barrier would be necessary to contain it, such as some natural zeolite that is stable under alkaline conditions.

It is clearly beyond the scope of this paper either to explore the many other fascinating concepts regarding a nickel-iron alloy canister - serpentinite backfill barrier system or to advocate it as the best system. However, further evaluation is merited if only because a clear connection exists with an engineered barrier system, and because this or some other system may provide predictable long term isolation of radionuclides.

RFEFERENCES

Azais, H., J. Bouladon, P. Picot, and P. Sainfeld (1968). Le probleme du nickel dans les serpentines de Corse, Bull. B.R.G.M., (Ser 2), Sect. 2, No. 1, pp. 55-93.

Barnes, I. and J. R. O'Neil (1969). The relationship between fluids in some fresh alpine-type ultramafics and possible modern serpentinization, Western United States, Geological Society of America Bulletin, V. 80, pp. 1947-60.

Chamberlain, J. A., C. R. McLeod, R. J. Traill, and C. R. Lachance (1965). Native metals in the Muskox Intrusion, Canadian Journal of Earth Sciences, V. 2, pp. 1188-215.

Fyfe, W. S. (1977). Container stability and permeability control, Abstract with Programs 1977 Annual Meetings, The Geological Society of America, V. 9, No. 7, p. 983.

Garrels, R. M. and C. L. Christ (1965). Solutions, Minerals, and Equilibria, Harper & Row, New York, 450 pp.

Goldstein J. I. (1973) Fe-Ni(Iron-Nickel) in Metals Handbook, 8th Edn., V. 8, Metallography, Structures and Phase Diagrams, American Society for Metals, Metals Park, Ohio.

Helgeson, H. C., T. H. Brown, A. Nigrini and T. A. Jones (1970) Calculation of mass transfer in geochemical processes involving aqueous solutions, Geochim. et Cosmochim. Acta, V. 34, pp. 569-592.

Kärn-Bränsle-Säkerhet, (1978). Handling and Final Storage of Unreprocessed Spent Nuclear Fuel, V. 1 - General, 111 pp.

National Research Council (1979). Implementation of Long-Term Environmental Radiation Standards: The Issue of Verification, National Academy of Sciences, Washington, D. C.

Nickel, E. H. (1959). The Occurrence of Native Nickel-Iron in the Serpentine Rock of the Eastern Townships of Quebec Province, The Canadian Mineralogist, V. 6., pp. 307-319.

Ringwood, A. E. (1978). Safe Disposal of High Level Nuclear Waste, a New Strategy, Australian National University Press, Canberra, Australia, 64 pp.

Robie, R. A., B. S. Hemingway, and J. E. Fisher (1978). Thermodynamic Properties of Minerals and Related Substances at 298.15 K and 1 Bar (10^5 Pascals) Pressure and at Higher temperatures, Geol. Survey Bulletin 1452, 456 pp.

Turner, F. J., and J. Verhoogen (1969). Igneous and Metamorphic Petrology, McGraw Hill Book Company, Inc., 2nd edition, New York 694 pp.

Wagman, D. D., W. H. Evans, V. B. Parker, I. Halow, W. M. Bailey, and R. H. Schumm (1969). Selected Values of Chemical Thermodynamic Properties. Tables for Elements 35 through 53 in the Standard Order of Arrangement, NBS Technical Note 270-4, 141 pp.

Wenner, D. B., and H. P. Taylor, Jr., (1973). Oxygen and Hydrogen isotope studies of the serpentinization of ultramafic rocks in oceanic environments and continental ophiolite complexes, Am. Journal of Sci., V. 273, pp. 207-239.

NATURAL ANALOGUES FOR CRYSTALLINE RADIOACTIVE WASTE FORMS, PART II: NON-ACTINIDE PHASES

Richard F. Haaker and Rodney C. Ewing

Department of Geology
University of New Mexico
Albuquerque, New Mexico 87131

INTRODUCTION

Synroc[1] and supercalcine[2] are proposed crystalline radioactive waste forms that may be regarded as highly radioactive "artificial rocks". In these proposed waste forms, oxides of transition metal elements, alkali elements, alkaline earth elements and additives (e.g. SiO_2, Al_2O_3, TiO_2) react when sintered or hot pressed to form phases which have mineral analogues. In synroc, these phases include perovskite ($CaTiO_3$), zirconolite ($CaZrTi_2O_7$) and "hollandite" - priderite ($BaAl_2$-Ti_6O_{16}). In supercalcine important phases include apatite ($(AE,RE)_5(SiO_4)_3O$), nepheline ($NaAlSiO_4$), pollucite ($CsAlSi_2O_6$), scheelite ($AEMoO_4$) and sodalite ($AE_2(NaAlSiO_4)_6(MoO_4)_2$). The behavior of analogues to perovskite, zirconolite and apatite have been previously summarized.[3] The purpose of this paper is to summarize relevant geologic literature on mineral analogues to important non-actinide phases in synroc and supercalcine. This paper emphasizes: (1) solid solution chemistry, (2) occurrence and (3) alteration effects. A special effort is made to identify stable and unstable mineral associations, as this may be important if hydrothermal conditions occur in the waste repository. It is not possible to select a "compatible" waste form-wall rock combination so that thermodynamic equilibrium prevails. Nonetheless, some waste form-wall rock assemblages may be more reactive than others.

This paper is a summary of PNL Report #3035 prepared for Battelle Pacific Northwest Laboratories with support from the Department of Energy (EY-76-C-06-1830). The report also summarizes the behavior of natural analogues to the actinide phases in synroc and supercalcine. A review of the behavior of mineral analogues to actinide phases has been abstracted from that report[3].

NEPHELINE

Nepheline ($Na_3KAl_4Si_4O_{16}$) has a distorted tridymite structure. Fe_2O_3 will substitute for Al_2O_3 and is normally present in amounts of 0.2 - 1.5 weight percent in natural nephelines. Excess silica is common and framework stoichiometries in the range $Al_{3.8}Si_{4.2}O_{16}$ to $Al_{4.0}Si_{4.0}O_{16}$ are typical. The Na:K ratio is variable but is usually close to 3:1. Synthetic nephelines can contain considerable amounts of Ca, but natural nepheline has less than 1.5 weight percent CaO. Natural nepheline does not contain significant amounts of Cs, Rb, Sr or Ba. The nepheline phase in supercalcine is not expected to contain large amounts of alkalis other than Na (or K if present), but several weight percent CaO and minor amounts of SrO and BaO may be present.

Occurrence and Alteration

Nepheline is the most common feldspathoid and is a characteristic mineral in high alkali-low silica rocks. It does not occur in silica saturated rocks such as granites, as it is unstable with the following saturated silicate minerals: quartz (and other silica polymorphs), orthopyroxene ($(Mg,Fe)SiO_3$) and pigeonite ($(Mg,Fe,Ca)(Mg,.Fe)Si_2O_6$). In nepheline syenites, it occurs with alakli feldspars ($(Na,K)AlSi_3O_8$), alkali ferromagnesian minerals such as sodic pyroxenes ($(Na,Ca)(Fe,Mg)Si_2O_6$) and amphiboles, feldspathoids (sodalite group, cancrinite), apatite ($Ca_5(PO_4)_3(OH,F)$), zircon ($ZrSiO_4$) and titanite ($CaSiTiO_5$). Nepheline occurs in high calcium-low silica rocks with melilite ($(Ca,Na)_2(Mg,Fe,Al,Si)_3O_7$), monticellite ($CaMgSiO_4$) and wollastonite ($CaSiO_3$). In high potassium-low silica volcanic and low pressure igneous rocks, nepheline occurs with leucite ($KAlSi_2O_6$).

Nepheline is often altered. Frequently mentioned alteration products include analcite (a zeolite, $NaAlSi_2O_6$), natrolite (a zeolite, $Na_2Al_2Si_3O_{10}\cdot 2H_2O$) phillipsite (a zeolite, $(Ca,Na_2,K_2)_5$-$Al_{10}Si_{22}O_{64}\cdot 20H_2O$), thomsonite (a zeolite, Na_4Ca_8-$(AlSiO_4)_{20}\cdot 24H_2O$), cancrinite (a feldspathoid, $Na_6Ca(AlSiO_4)_6CO_3$), hydrous aluminum oxides, muscovite ($KAl_3Si_3O_{10}(OH)_2$), paragonite (a mica, $NaAl_3Si_3O_{10}(OH)_2$), sodalite, alkali feldspars, plagioclase feldspar ($(Na,Ca)Al_{2-x}Si_{2+x}O_8$) and kaolinite ($Al_2Si_2O_5(OH)_4$).

SCHEELITE

Scheelite ($CaWO_4$) and powellite ($CaMoO_4$) have structures that are derived from the tetragonal-I zircon structure. Heavy atoms of zircon and scheelite-powellite have the same fractional coordinates. There is complete solid solution in the following scheelite systems: $CaWO_4$-$CaMoO_4$, $CaWO_4$-$SrWO_4$ (825°C) and $CaWO_4$-$PbWO_4$ (815°C). No more than 2.5 mole percent $BaWO_4$ is soluble in $CaWO_4$ at 1100°C. Above 900°C there is extensive solid solution in the system $CaWO_4$-$Sm_2(WO_4)_3$. Above 1025°C there is extensive solid solution in the systems $CaWO_4$-$La_2(WO_4)_3$ and $CaWO_4$-$NaLa(WO_4)_2$.[5,6]

Occurrence and Alteration

Scheelite occurs in hydrothermal deposits, pegmatites, ore veins associated with granites, contact metamorphic rocks and placers. Quartz and calcite are associated with vein deposits; beryl ($Be_3Al_2Si_6O_{18}$) and fluorite (CaF_2) are common associations in pegmatites. Depositional temperatures range from 100 to 500°C. Powellite ($CaMoO_4$) is much less common than scheelite. It most often occurs as an oxidation product of molybdenite (MoS_2) which is associated with scheelite. Mo-bearing scheelite forms in an environment with a higher oxygen fugacity than does scheelite + molybdenite. Scheelite has a larger field of stability than powellite. Both appear to be stable at surface conditions, but scheelite is stable to lower oxygen fugacities and to higher sulfur fugacities than powellite[7].

$CaMoO_4$ is more soluble than $CaWO_4$. At constant temperature the solubilities of both increase with chloride concentration. Both have positive temperature coefficients of solubility. At 350°C the limit of solubility of $CaMoO_4$ in 39 weight percent NaCl is 160 ppm.[8]

Powellite is not an important ore mineral; therefore, alteration effects have not been studied in detail. Since supercalcine "scheelite" is in large part $(Sr,Ba)MoO_4$, scheelite ($CaWO_4$) and powellite ($CaMoO_4$) are not ideal natural analogues. $SrMoO_4$, $SrWO_4$, $BaMoO_4$ and $BaWO_4$ have scheelite structures but are not known to occur as minerals.

POLLUCITE

Pollucite has a typical zeolite type structure. Natural pollucite does not occur as the end member $CsAlSi_2O_6$, instead it occurs as a solid solution with analcime (a zeolite, $NaAlSi_2O_6 \cdot H_2O$). Natural pollucite contains at least 25 mole percent of the analcime molecule. Rb is present in trace amounts. Pollucite is the only Cs mineral and the major source of Rb. Pollucite has more silica than indicated by the ideal formula. The framework stoichiometry $Al_{0.9}Si_{2.1}O_6$ is typical. Alkaline earth elements do not occur in significant amounts. The synthetic pollucite of supercalcine is anhydrous while natural pollucite is hydrated.

Occurrence and Alteration

Because of the extremely low crustal abundance of Cs, pollucite occurrences are restricted to highly differentiated Li-rich granitic pegmatites. As pegmatite forming fluids crystallize, the Cs becomes concentrated in the fluid phase until its activity is high enough to allow crystallization of pollucite.

Pollucite weathers by removal of alkalis and alters to clays. Cerny[9] has identified illite ($K_y(Al,Mg,Fe)_2(Al_ySi_{4-y})O_{10}(OH)_2$, y>1), kaolinite ($Al_2Si_2O_5(OH)_4$) and smectites ($(1/2\ Ca,Na)_{0.7}$-$(Al,Mg,Fe)_4(Si,Al)_8O_{20}(OH)_4 \cdot nH_2O$) in clay pods derived from large pollucite crystals. Cerny determined 0.24 weight percent residual Cs_2O in clays derived from pollucite. Vlasov[10] indicates that removal of Cs from pollucite is rapid and the Cs content will even diminish in samples in tailings piles. Cerny reports pollucite altering to smectites after 6 months exposure to surface conditions.[9]

SODALITE

Sodalite group minerals ($(Na_8(AlSiO_4)_6Cl_2)$ and $(Na,Ca)_{4-8}$-$(AlSiO_4)_6(SO_4,S)_{1-2}$)) are feldspathoids which are closely related to the zeolite group in structure and physical properties. Sodalite is distinguished from other members of the sodalite group by having Cl as an essential constituent. In natural sodalite, CaO and K_2O are rarely present in amounts above 1 weight percent. Fe_2O_3 substitutes for Al_2O_3 but normally less than 1 weight percent is present. S and SO_3 are present in amounts of 1 weight percent or less. Nosean ($Na_8(AlSiO_4)_6(S,SO_4)$) and haüyne ($(Na,Ca)_{8-4}(AlSiO_4)_6(SO_4,S)_{1-2}$) form a series with variable Na:Ca ratios and amounts of sulfate. Haüyne with 5.4 weight percent K_2O is known, but nosean-haüyne usually has less than 1 weight percent. Nosean-haüyne and sodalite have similar Fe_2O_3 contents.

Occurrence and Alteration

Sodalite occurs in nepheline syenite, sodalite syenite and other silica undersaturated intrusive and volcanic rocks. Igneous and volcanic mineral associations of sodalite include alkali feldspars, nepheline, leucite, cancrinite and melanite (a garnet, $Ca_3(Fe,Ti)_2Si_3O_{12}$). Sodalite also occurs in metamorphosed calcium carbonate bearing rocks that are in contact with alkaline igneous assemblages. The nosean-hauyne series occurs chiefly in volcanic rocks. Nosean also occurs in silica undersaturated syenites. Sodalite group minerals never occur in granites due to the thermodynamic instability of quartz-sodalite assemblages.

Mechanisms for the alteration of sodalite group minerals have not been examined in detail, but many hydrothermal alteration products have been identified. Sodalite ($Na_8(AlSiO_4)_6Cl_2$) altering to other zeolites has been mentioned most frequently. Thomsonite, natrolite, gismondine ($Ca(AlSiO_4)_2 \cdot 4H_2O$), kaolinite (a clay mineral), disapore ($AlO(OH)$), gibbsite ($Al(OH)_3$) and muscovite are reported as alteration products. Sodalite altering to cancrinite ($(Na,Ca,K)_{6-8}(AlSiO_4)_6(CO_3,SO_4,Cl)_{1-2}$) has been reported by Deer et al.[4] Nosean and hauyne transform to cancrinite at 480°C in the pressure range 1000 to 2000 bars PH_2O. This transformation may explain the almost exclusive occurrence of nosean and hauyne in volcanic rocks, while cancrinite is the common sulfate bearing tectosilicate in plutonic rocks. Cancrinite has the same framework stoichiometry as sodalite.

PRIDERITE - "HOLLANDITE"

Hollandite type phases have the general formula $A_xM_8(O,-OH)_{16}$ ($x \leq 2$). The a-site cations can be alkali and alkaline earth elements, Ag, Tl and Pb. M-site cations can be Mg, Zn, Ga, In, Si, Ge, Sn, Sb and many of the first row transition metal elements.[11] The crystal chemistry of hollandite type phases is complex and these phases are frequently nonstoichiometric. The hollandite structure is a framework of MO_6 octahedra and a series of large nonintersecting tunnels parallel to the c-axis (tetragonal). A number of manganese oxide minerals have closely related structures. The structures of pyrolusite (beta-MnO_2, rutile structure), ramsdellite (MnO_2), hollandite ($(Ba,K)_xMn_8O_{16}$) and psilomelane ($(Ba,K,Mn^{+2},Co)_2Mn_5O_{10} \cdot xH_2O$) differ in the relative proportion of corner and edge sharing of MO_6 octahedrons. This in turn affects the tunnel dimensions.[12] In the above MnO_2 minerals, complex intergrowths having dimensions of only a few unit cells are common. Thus, x-ray diffraction is not an adequate tool for the characterization of hollandite type structures.[13]

Occurrence and Alteration

Priderite, a rare titanate with the hollandite structure is probably the best mineral analogue for synroc "hollandite". Priderite from the West Kimberly area, Western Australia has the composition $(K_{0.87}Ba_{0.32}Na_{0.14})(Fe^{+3}_{1.14}Al_{0.33}Ti_{6.48})O_{16}$. Optical properties of priderite closely resemble those of rutile, and on the basis of optical microscopy it can be mistaken for rutile. In Western Australia it occurs in low-silica volcanic rocks. Associated with priderite are leucite ($KAlSi_2O_6$), rutile (TiO_2), amphiboles, serpentine ($Mg_3Si_2O_5(OH)_4$), calcite ($CaCO_3$), barite ($BaSO_4$), zeolites, phlogopite ($K_2(Mg,Fe^{+2})_6Al_2Si_6O_{20}(OH,F)_4$, accessory wadeite ($K_4Zr_2Si_6O_{18}$), perovskite and apatite (Ca_5-$(PO_4)_3(OH,F)$). In Leucite Hills, Wyoming priderite occurs as an accessory mineral in leucite-sanidine lavas. Other minerals associated with this occurrence include diopside ($Ca(Mg,Fe)Si_2O_6$), apatite, glass and phlogopite. Priderite is one of the minerals which can occur in low-silica/low-sodium rocks with mole percent $K_2O>Al_2O_3$. Alteration of priderite is not described in the geologic literature, and the phase may be durable. The mineral is too rare to conclude that it is resistant in most weathering environments.

CONCLUSIONS

It is inappropriate to conclude that one radioactive waste form "mineral assemblage" is better than another unless it is with reference to a specific set of conditions. The "compatibility" of the waste form and the wall rock are of key importance in determining the long term stability of any crystalline waste form at high temperatures. Unfortunately much of the thermodynamic and kinetic data needed to make long term predicitions are not yet available. Nonetheless, it is sometimes possible to predict whether a phase will be "stable" in a particular set of conditions. Without kinetic data it is not possible to predict whether or not thermodynamic instability will be of any significance. One can, from the geologic literature, only suggest which alteration trends may be expected. The most easily predicted alteration trends result from variations in the silica activity.

For equilibrium to exist among two or more phases the activities of all chemical species in all phases must be equal. This is impossible for waste form-wall rock assemblages, as none of the isotopes of concern (Cs^{137}, Sr^{90}, Tc, other radioactive fission products, Pu and other TRU) exist in significant amounts in rocks. If the waste form had an extremely low waste loading and was compositionally and mineralogically very similar to the surrounding rock, a "near equilibrium" situation might be considered to exist.

Even though it is obvious that supercalcine or synroc would not be thermodynamically stable in any repository, it is useful to tabulate phases for which the mineral analogues are "stable" or "unstable". Since high-Na supercalcine formulations contain feldspathoids (nepheline and sodalite), they would be unstable in rocks of granitic compositions (i.e. silica saturated). Alkali feldspars are stable with nepheline + sodalite, thus nepheline syenite or rocks with a similar silica content would be "stable" with high-Na supercalcine formulations. Low-Na supercalcine formulations have little or no nepheline or sodalite but do contain "scheelite". Scheelite and nepheline react to form sodalite at high temperatures,[14] thus low-Na supercalcines may be unstable in rocks that contain a feldspathoid (nepheline). Scheelite is normally associated with granitic rock types. Thus, silica-saturated rock types would be "stable" with low-Na supercalcine formulations. Of the supercalcine phases--scheelite, sodalite, nepheline and pollucite--only scheelite can be considered durable at surface or near surface conditions.

Synroc phases are stable in rocks having a lower SiO_2 content than that required for supercalcine phases. The presence of perovskite is synroc precludes thermodynamic stability with rocks having silica polymorphs or alkali feldspars (i.e. most types of rocks). Priderite has been reported only in high K_2O-TiO_2 and low Na_2O-Al_2O_3-SiO_2 rocks. Priderite may not be stable in other rock types, this would explain its rare occurrence. Its occurrence with K-feldspar indicates that it may be stable with higher silica rocks than perovskite. Synroc "Ba-hollandite" contains less K and more Ba than priderite and may be "stable" in rock types that do not contain large amounts of K. On the other hand, there are no known Ba-titanate hollandite type minerals. Thus, this composition may not be stable in common crustal rocks.

In conclusion, supercalcine (high Na) is recommended for burial in rocks of nepheline syenite compostion and supercalcine (low Na) would be recommended for burial in rocks of granitic composition. Synroc is recommended for burial in low silica, feldspar free rocks.

REFERENCES

1. A. E. Ringwood and S. Kesson, Immobilization of High Level Nuclear Wastes in Ceramic Materials - A New Approach, in: "Proceedings of International Symposium on Ceramics in Nuclear Waste Management, American Ceramic Society, Cincinnati, April 30 - May 2, 1979," (1979).
2. G. J. McCarthy, High-level Waste Ceramics: Materials Considerations and Product Characterization, Nuclear Technology 32:92 (1977).
3. R. F. Haaker and R. C. Ewing, Uranium and Thorium Minerals: Natural Analogues for Radioactive Waste Forms, in: "Proceedings of the Second International Symposium on the Scientific Basis for Nuclear Waste Management, Materials Research Society, Boston, November, 1979" 2:281 (1980).

4. W. A. Deer, R. A. Howie and J. Zussman, "Rock Forming Minerals, volume 4: Framework Silicates," Longman Group Limited, London (1963).
5. L. L. Y. Chang, Rare Earth Substitution in Scheelite, Journal Inorganic Nuclear Chemistry. 31:2003 (1969).
6. L. L. Y. Chang, Solid Solutions of Scheelite with other $R^{II}WO_4$-Type Tungstates, American Mineralogist 52:427 (1967).
7. L. C. Hsu, Effects of Oxygen and Sulfur Fugacities on the Scheelite-Tungstenite and Powellite-Molybdenite Stability Relations, Geological Society of America, Abstr. Programs, 7:1123 (1977).
8. L. F. Yastrebova, A. F. Borina and M. I. Ravich, Solubility of Calcium Molybdate and Tungstate in Aqueous Potassium and Sodium Chlorides at High Temperatures, Russian Journal Inorganic Chemistry 8:105 (1963).
9. P. Cerny, Alteration of Pollucite in some Pegmatites of Southeastern Manitoba, Canadian Mineralogist. 16:89 (1978).
10. K. A. Vlasov, editor, "Mineralogy of Rare Elements, Volume II," Israel Program for Scientific Translations, Jerusalem, (1966).
11. H. Pentinghaus, Crystal Chemistry of Hollandites, $A_xM_8(O,OH)_{16}$, $x \leq 2$, Physics and Chemistry of Minerals 3:85 (1978).
12. L. A. Bursill, Structural Relationships between beta-gallia, Rutile, Hollandite, Psilomelane, Ramsdellite and Gallium Titanate-Type Structures. Acta Crystallographica B35:530 (1979).
13. S. Turner and P. Buseck, Manganese Oxide Tunnel Structures and Their Intergrowths, Science 203:456 (1979).
14. G. J. McCarthy, "Crystal Chemistry and Phase Formation in Developmental Supercalcines, MRL Report COO-2510-14," The Pennsylvania State University (1978).

GEOCHEMICAL STUDY OF A LAMPROPHYRE DIKE NEAR THE WIPP SITE

D. G. Brookins

Department of Geology
University of New Mexico
Albuquerque, NM 87131

INTRODUCTION

Tertiary igneous dikes which intruded evaporite rocks in the area near the WIPP site have been known for many years (See discussion in SAND, 1978). In brief, in the northern Delaware Basin near the WIPP site, dikes (and sills?) have been reported in outcrop from 70 km southwest of the WIPP site which extend in the subsurface in a northeasterly trend passing some 16 km to the northwest of the WIPP site based on exposure in the Kerr-McGee Potash Mine, a possible exposure in the Mississippi Chemical Company Mine, and from drilling records elsewhere in the area. The total length of the dike system is about 125 km and its greatest width of 15 km is recorded from the Kerr-McGee Potash Mine (SAND, 1978).

This dike system is of extreme interest as it is lamprophyric in composition and intrudes the evaporite sequence at depths in the Salado Formation at about the levels proposed for the pilot storage of radioactive waste. Since the emplacement temperature of the dike system is estimated to be between 600°C to 800°C (Loehr, 1979), then a study of contact effects in the evaporite sequence due to dike emplacement can be, with caution, used as an analog for a canister with a temperature significantly above ambient emplaced in the evaporite sequence. More specifically, since a main concern of salt repositories is canister-brine-evaporite reactions then the contact zone of the dike-evaporite system assumes a more important role as the effects of brine (± vapor) in a zone of much higher temperatures than planned for canister emplacement can be treated as a worse case scenario for possible elemental migration redistribution from a breached canister.

Further, although the dike composition is very different from HLW or TRU radwastes (± canister material) use of lanthanides and U, Th and other dike trace elements as tracers can be applied to the dike-evaporite system as lamprophyres are typically enriched in lanthanides, Th, U and other key elements relative to evaporites by several orders of magnitude (See further discussion of this and related topics in Register et al., 1980).

The mineralogy of the lamprophyre dike has been covered in detail (See SAND, 1978) and only a brief summary will be given here. The rock is usually porphyritic and consists of altered plagioclase and pyroxene phenocrysts set in a ground mass of biotite, K-feldspar (predominantly orthoclase), and minor amounts of apatite, anatase, ilmenite, pyrite, ± baddelyite (?) and ± Ti-magnetite. The plagioclase (An-30-40) is commcnly altered to clay minerals of variable composition and the pyroxene is commonly replaced by antigorite and/or siderite. The rock contains numerous amygdules which are commonly filled with evaporite minerals (halite, sylvite?), carbonates (calcite siderite) and, occasionally, natrolite. Apophyses of the dike extend into the evaporite and the entire dike system exhibits pinch-and swell behavior and is not thought to be continuous over its 125 km length. For additional details on the background information of this dike system the reader is referred to Calzia and Hiss (1978) and Loehr (1979) and references cited therein.

The mineralogy of the Salado Formation evaporites has been discussed in great detail by numerous investigators (See summary in SAND 1978). A general description of the Salado Formation (SAND, 1978) follows: "The Salado Formation is composed of rock salt, anhydrite, and potassium rocks with varying amounts of other evaporites and fine grained rocks. Rock salt constitutes about 85-90 percent of the formation except in the western part of the area where percolating ground water has dissolved and removed some of it. The next most abundant rock in the formation is anhydrite. The remainder of the formation is chiefly polyhalite and other potassium and magnesium bearing minerals with minor amounts of sandstone, and claystone." Halite is present as two varieties, one with small but significant amounts of clay minerals and as a relatively pure variety. The clay minerals have been described by Bodine (1978) and Register (1979). Polyhalite and anhydrite occur as interlayers in both halite types. Glauberite, sylvite, carnallite, kieserite and other minor minerals are locally common.

In the vicinity of the lamprophyre dike in the Kerr-McGee Potash Mine, the Salado Formation away from the dike is much like that described above. Near the dike pronounced melting, partial melting, dissolution, plastic flow (?) and reprecipitation have taken place (See discussions in Calzia and Hiss, 1978; and Loehr, 1979).

K-Ar AGE DETERMINATIONS

K-Ar age determinations for the lamprophyre dike have been reported by Calzia and Hiss (1978) and Brookins (1980a), and one polyhalite K-Ar age determination of 21.5 $\pm$ 0.8 MYBP from the contact zone has been reported by Brookins et al. (1980). The lamprophyre dike ages range from 32.2 ± 1.0 MYBP (million years before the present) to 34.7 ± 1.4 MYBP including samples taken from the interior of the dike to the edge of the dike. No difference in age amoung these samples is obvious and, further, Calzia and Hiss (1978) report identical ages and radiogenic 40-Ar content from untreated lamprophyre and a basaltic dike treated with boiling distilled water to remove inclusions of soluble K-salts. Kaneoka (1972) studied hydration effect on volcanic rocks and concluded that significant radiogenic 40-Ar loss may occur for samples with greater than one percent H_2O. The lamprophyre dikes contain approximately 2.85 percent H_2O and no radiogenic argon loss is significant for either the treated sample (See Calzia and Hiss, 1978) or from the contact-zone lamprophyre (Brookins, 1980a).

CONTACT EFFECTS

The most pronounced contact effects are those involving melting, partial melting, and recrystallization effects of the evaporite minerals. In the zone of dike-evaporite direct contact, many of the evaporite minerals have been melted and reprecipitated. The dike-evaporite contact zone is extremely complex with some knife edge contacts, some inclusions of evaporite minerals in lamprophyre and lamprophyre embaying and in places embayed by reprecipitated evaporite minerals. Calzia and Hiss (1978) and Loehr (1979) have discussed many of the obvious contact effects and the reader is referred to their studies for more detail. Recrystallization effects reported by Loehr (1979) in the Kerr-McGee Potash Mine are most pronounced from the contact outward to about 4 meters, although some perturbations of the evaporite mineralogy some 10 meters removed from the contact are presumed due to lamprophyre emplacement. Loehr (1979) notes that within 1.5 meters of the contact the clay mineral assemblage consists of 1M-d phlogopitic clay ± saponite ± irregularly distributed saponite rich mixed layer saponite-chlorite ± a talc-silicate mixture. Between 1.5 and 7 meters from the contact the typical assemblage reported by Loehr (1979) consists of mixed layer chlorite-saponite ± illite + chlorite + talc, and at distances greater than 7 meters from the contact she reports a normal assemblage of partially ordered corrensite + illite + talc + chlorite. This last assemblage is typical of the clay mineral-rich parts of the Salado Formation throughout the general area (See Register, 1979; Bodine, 1978).

Fluid inclusion studies of the contact zone are incomplete. Roedder (commun. to Loehr, 1978; cited in Loehr, 1979; p. A-46) reported normal two-phase fluid inclusions with maximum homogenization temperatures ranging from 116°C (1-2 cm. from contact) to 110°C (0.2 m from contact) to 88°C (2.5 m from contact). She (p. A-46) further reports a maximum temperature of homogenization of 150°C in the immediate contact zone. Since the nature of the brines formed near the dike and the vapor phases generated during the dike emplacement are unknown, then these preliminary data must be used with caution. The temperatures are apparently too low as partial melting effects are noted or suggested up to several meters from the contact. What is probably more important about these observations, especially in view of the heater experiments on fluid migration conducted by Sandia Laboratories (Lambert, 1980), is that fluid movement was most pronounced only very close to the dike edge. No apparent disturbance of fluid inclusions past ∿10 meters of the dike-evaporite contact have been noted. Loehr's (1979) data for clay mineral changes as a function of distance from the contact can be indirectly approached by use of activity diagrams at different temperatures for the systems $HCl-H_2O-Al_2O_3-CO_2-MgO-K_2O-SiO_2$ and $HCl-H_2O-Al_2O_3-CO_2-MgO-Na_2O-SiO_2$ in temperature range 60°C to 300°C (Helgeson et al. 1969). The diagrams do support Loehr's (1979) and other evidence (cited elsewhere in this report) that the temperatures with in meters of the dike contact were below 150°C and possibly 100°C.

The chemical data reported by Calzia and Hiss (1978) for the lamprophyre dike, as well as the summary of other data by Loehr (1979), are sufficient to demonstrate a relatively homogeneous chemistry even in the contact zones. Unfortunately, due to the complex mineral reactions in the contact zones and the presence of small amounts of clay minerals in these zones, it is not easy to distinguish between possible elemental migration between dike and contact rocks as opposed to elemental redistribution within the contact zones caused by the thermal effects of the dike but not necessarily involving any of the dike material proper. To test the problem of possible elemental migration from dike into contact rock we have analyzed composite samples of lamprophyre from close to the center of the dike and from the dike edge as well as polyhalite and anhydrite from the contact zones. The samples were analyzed by neutron activation analysis although halite-rich samples could not be analyzed with our facilities due to irradiation-induced interference from sodium.

NEUTRON ACTIVATION ANALYSES

The neutron activation analyses for polyhalites and two lamprophyres are shown in Table 1. The data are limited but do

Table 1. Chemical data (ppm except where noted)

Element	MB 76-34 lamp.	MB 76-23 lamp.	MC2B MB121 polyh.	MB 76-22 polyh.	Average polyh. (m = 12)	Average anhy. (m = 4)
Na_2O (%)	3.19	3.98	0.13	1.63	0.66	0.25
K_2O (%)	11.1	10.6	23.4	22.0	16.23	0.02
Fe_2O_3	11.4	11.3	ND	0.006	0.19	0.15
Cs	ND	ND	ND	ND	ND	NA
Ba	ND	ND	ND	ND	NA	ND
Sc	7.97	8.30	ND	ND	0.23	ND
Zr	ND	ND	ND	ND	26.2	60.0
Hf	ND	ND	ND	ND	0.34	ND
Ta	3.40	3.23	ND	ND	0.04	ND
Cr	ND	ND	ND	ND	ND	ND
Rb	81.6	60.0	ND	ND	7.54	11.8
Co	18.3	23.2	ND	ND	0.42	0.28
Ni	ND	ND	ND	ND	9.2	9.5
Ag	NA	NA	NA	NA	---	NA
Sb	ND	ND	ND	ND	0.05	0.09
U	1.45	1.05	ND	ND	ND	NA
Th	8.22	5.50	ND	ND	0.11	0.05
Gd	ND	ND	ND	ND	ND	NA
La	75.2	79.3	ND	ND	0.07	ND
Ce	228.6	136.0	ND	ND	0.06	ND
Nd	87.8	ND	ND	ND	0.63	1.85
Sm	16.1	16.6	ND	ND	0.25	ND
Eu	3.60	4.14	ND	ND	0.01	0.06
Tb	1.27	1.30	ND	ND	0.07	0.09
Tm	ND	ND	ND	ND	0.01	0.06
Yb	1.93	2.38	ND	ND	ND	ND
Lu	ND	ND	ND	ND	ND	ND
Total REE	409.4	239.7	ND	ND	---	---

point out that the polyhalites contain small amounts of impurities. Since hematitic staining is commonly noted in many evaporite samples the presence of Fe, Sc, Ni, Co, is not surprising. Small amounts of the rare earth elements are noted in several samples; this is probably due to small amounts of original clay minerals and/or hematite as the rare earth content for marine evaporites is several orders of magnitude lower than for other sedimentary rocks such as shales (See discussion in Register, 1979 for further details). The NAA data for the two lamprophyres are of extreme interest. Fig. 1 shows the REE/chondrite abundance plotted on log scale, versus the relative position of the REE following conventional display. The two curves are very typical of those for lamprophyres or alkali basalts with pronounced enrichment of the light REE relative to the heavy REE. Further, both Ce and Eu fall on the curves and do not show any evidence for loss (i.e. a pronounced negative Ce/chondrite anomaly). More importantly these two samples were selected from lamprophyre in direct contact and close to evaporite minerals which would be most susceptible to REE loss were there any exchange with evaporite minerals. It is also significant that the polyhalite from the contact with the lamprophyre, MB76-22, contains the lowest iron content, 60 ppm, as well. Collectively these data argue for little if any chemical exchange between the lamprophyre and intruded evaporite sequence. This, in turn, implies that the brine (± vapor) generated by melting induced by dike emplacement was more effective in causing alteration of the evaporite + clay mineral sequence near the dike as opposed to attack on the dike. Further, canister material (± overpack) should remain relatively stable in the evaporites (Brookins, 1980b). The studies by Calzia and Hiss (1978), Loehr (1979) and Brookins et al. (in progress and this report) show evaporite melting and other metamorphic-metasomatic events to be restricted to the evaporite contact zone.

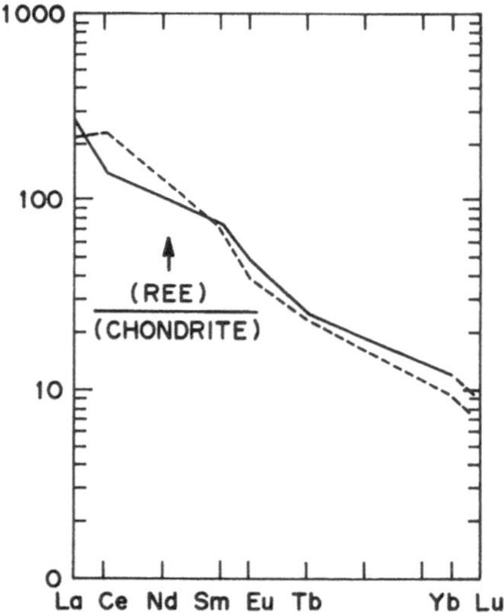

Fig. 1. REE/chondrite distribution plots for lamprophyres. Solid curve, sample MB76-23; dashed curve, sample MB76-34. The curves for the composite samples are identical within the limits of analytical uncertainty.

REFERENCES

Bodine, M., 1978 Clay-mineral assemblages from drill core of Ochoan evaporites Eddy County: NM Bur. Mns. Min. Res. Circ. 159, 21-32.

Brookins, D. G., 1980a (in press), K-Ar age of lamprophyre dike from the Kerr-McGee Potash Mine, southeastern New Mexico: Isochron/West.

Brookins, D. G., 1980b, Near-field reactions in evaporites considered for waste repositories: Jnl. Ariz. -Nev. Acad. Sci., v. 15, p. 45.

Brookins, D. G., Register, J. K., and Kruegar, H. W., 1980, Potassium-argon dating of polyhalite in southeastern New Mexico: Geochim. Cosmochim. Acta, v. 44, p. 635-638.

Calzia, J. P., and Hiss, W. L., 1978, Igneous rocks in northern Delaware Basin, New Mexico and Texas: NM Bur. Min. Resources Circ. 159, p. 39-45.

Helgeson, H. C., Brown, T. H., and Leeper, R. H., 1969, Handbook of theoretical activity diagrams depicting chemical equilibria in geologic systems involving an aqueous phase at one atmosphere and 0° to 300°C: Freeman, Cooper Σ Co., San Francisco, 253 p.

Kaneoka T., 1972, The effect of hydration on the K/Ar ages of volcanic rocks: Earth Plan. Sci. Lttrs., v. 14, p. 216-220.

Lambert, S. J., 1980, Mineralogical aspects of fluid migration in the Salt Block II experiment: SAND79-2423 (Sandia Ntl. Labs.), 24 p.

Loehr C. A., 1979, Mineralogical and geochemical effects of basaltic dike intrusion into evaporite sequences near Carlsbad, New Mexico: M.S. Thesis, NM Inst. Min. Tech. 70 p.

Register, J. K., Jr., 1979 Rb-Sr and Related Studies of the Salado Formation, Southeastern New Mexico, Master Thesis University of New Mexico, Albuquerque, New Mexico. 119 p.

Register, J.K., Brookins, D.G., Register, M.E., and Lambert, S.J., 1980: Clay mineral brine interations during evaporite formation: lanthanide distribution in WIPP samples: Sci. Basis Nuc. Waste Mngmt., II. C.J. Northrop, Ed.; Plenum Press, p. 445 to 452.

SAND (1978) Geological Characterization report, waste isolation pilot plant (WIPP) site, southeastern New Mexico. Sandia Labs. Rpt. SAND-78-1596 350 p.

NUCLEAR WASTE GLASSES AND VOLCANIC GLASSES: A COMPARISON OF THEIR STABILITIES

Günter Malow and Rodney C. Ewing

Hahn-Meitner-Institut für Kernforschung Berlin Gmbh
Glienicker Strasse 100, D-1000 Berlin 39, Germany;
Department of Geology, University of New Mexico,
Albuquerque, New Mexico 87131, USA

INTRODUCTION

The long term stability of radioactive waste forms is an important criterion in their selection for the final disposal and isolation of radioactive waste; however, this long-term stability is difficult to assess in necessarily time-limited laboratory experiments. Although the compositions and thermal histories of glass waste forms and volcanic glasses are quite different, the natural analogues provide standards for comparison to the synthetic glasses. The volcanic glasses represent nature's long-term experiment that may serve as a check on the data of time-limited, laboratory experiments that are extrapolated to estimate long-term stability.

The thermal and chemical stabilities of two borosilicate glasses (F-SON 58.30.20, France; UK-209, United Kingdom) and a glass ceramic (D-B 1-3, Hahn-Meitner-Institute, Germany) are compared to those of three rhyolite glasses from northern California which are 500, 5,000 and 670,000 years old. Thermal stability was evaluated by differential thermal analysis and heat treatments up to 100 days at 500° to 800°C. Samples were analyzed before and after heat treatment by means of optical and scanning electron microscopy, electron probe microanalysis and x-ray diffraction. Chemical stability was evaluated by (1) the grain titration method DIN 12111[1], (2) soxhlet leaching at 100°C for 3 days in deionized water[1], (3) autoclave leaching at 200°C and 10 bars for 3 days in saturated carnallite brine solution[2], and (4) phase selective leaching of polished surfaces with a solution of pH 4 at 50°C for up to 6 hours.

SAMPLE DESCRIPTION

The two waste glasses and the glass ceramic were prepared at AERE Harwell following procedures outlined by Malow et al.[1]. Three volcanic glasses of rhyolitic composition were selected for the purpose of comparison. Samples (VG 0.5) of a rhyolite-dacite flow at Glass Mountain, Siskiyou County, California are less than 500 years old. Glass Mountain is a single flow of 1 km^3 consisting of rhyolite and dacite which is nearly free from vegetation or alteration[3]. C^{14} dates and hydration rim dates range between 100 and 400 years[4]. Obsidian Mound, Mono County, California, (VG 5.0) is a recent rhyolitic flow. Although a radiometric age is not available, the age of 1,000 to 5,000 years is based on the stratigraphic relation of Obsidian Mound to nearby, dated flows. The rhyolite is extensively fractured and minor alteration is apparent. Samples (VG 670) from Long Valley Caldera, Mono County, California, are 670,000 ± 14,000 years old (K/Ar dating)[5]. Alteration effects are minimal. Electron microprobe analyses of the synthetic glasses and the glass matrix of the volcanic glasses are summarized in Table 1.

It is important to note that, in contrast to the waste glasses, volcanic glasses have an important crystalline component. All of the volcanic glasses contain phenocrysts of plagioclase, clinopyroxene, orthopyroxene and olivine. X-ray powder diffraction patterns of the as-cast glasses showed only a few weak traces for the F-SON glass; however, SEM and EPMA investigations revealed microscopic inhomogeneties for all glasses except D-B 1-3. Spots of strong enrichment of Ru, Pd, Rh and Te were detected. The solubility of the noble metals in the glasses is obviously extremely low and appears to be independent glass composition.

RESULTS

Thermal stability

Fig. 1 shows SEM-micrographs of the heat treated waste products. The glass F-SON 58.30.20 is strongly crystallized with various phases as shown in Fig. 1a, whereas in the glass UK-209 only small quantities of crystalline phases formed (Fig. 1b). Fig. 1c shows the glass ceramic and Fig. 1d the same product after heat treatment 100 days at 800°C. Figs. 2a-2, b-2 and c-2 show volcanic glasses heated for 100 days at 800°C. Whereas the waste glasses and the glass ceramic crystallized and transformed respectively; volcanic glasses already contained crystals, but did not change structurally during heat treatment. Although the thermodynamically unstable parent glass of the glass ceramic was transformed into a more stable glass ceramic by nucleation at 610°C for 3 hours and a subsequent crystallization at 800°C for 10 hours, its structure

Table 1. Compositions of waste glasses, the glass ceramic and volcanic glasses (weight percent).

	F-SON	UK 209	D-B 1-3	VG 0.5	VG 5.0	VG 670
SiO_2	43.6	50.9	28.0	74.72	74.67	75.15
B_2O_3	19.0	11.1	6.4			
Al_2O_3	0.1	5.1	12.8	14.08	14.03	14.41
Li_2O		4.0	2.4			
K_2O				4.30	5.33	2.66
Na_2O	9.4	8.3	3.8	4.29	4.50	3.65
MgO		6.3	1.2	0.19	0.01	0.02
MnO				0.04	0.05	0.06
CaO			4.0	1.26	0.68	0.10
BaO	1.0	0.38	15.5			
ZnO		0.44	3.6			
Rb_2O	0.23	0.10	0.16			
Cs_2O	1.79	0.77	1.75			
SrO	0.66	0.32	0.62			
Y_2O_3	0.37	0.17	0.35			
La_2O_3	0.94	0.44	0.68			
CeO_2	2.16	0.99	1.34			
Pr_6O_{11}	0.92	0.43	Pr_2O_3 0.61			
Nd_2O_3	3.82	1.83	2.48			
TeO_2	0.45		0.26			
ZrO_2	3.12	1.43	3.38			
MoO_3	4.06	1.77	2.42			
RuO_2	1.85	0.68				
PdO	1.24	0.44	0.07			
UO_2	3.60	0.06	U_3O_8 0.47			
Fe_2O_3	0.60	2.73	1.51			
FeO				1.57	1.33	1.35
NiO	0.10	0.36	0.24			
TiO_2				0.28	0.10	0.10
Cr_2O_3	0.20	0.56	0.43			
total				100.73	100.70	97.50

The waste glasses and glass ceramic contain minor amounts of the following elements:
(1) F-SON: P_2O_3, CdO, Ag, Sb_2O_3, SnO_2
(2) UK 209: PO_4^{-3}, $SO_4^{=}$
(3) D-B 1-3: TiO_2, MnO_2, Co_2O_3, As_2O_3, CdO, Ag_2O, Sb_2O_3, SnO_2

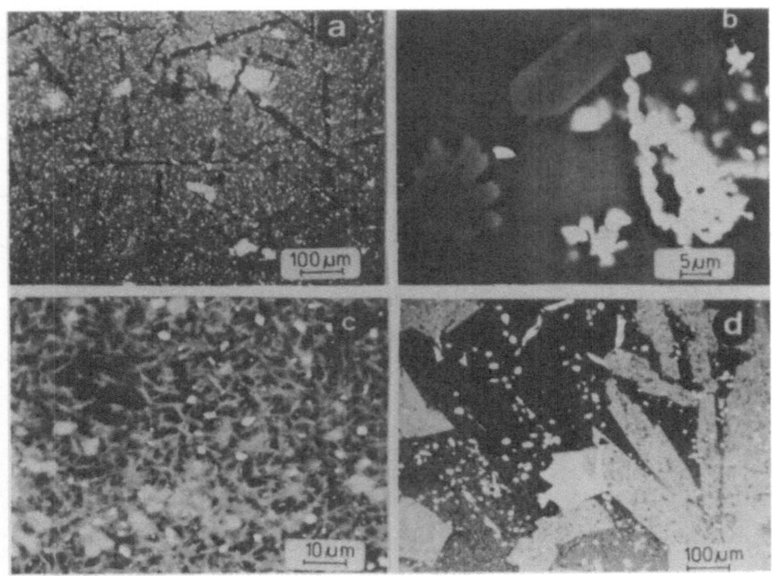

Fig. 1. SEM micrographs of heat treated waste glasses: (a) F-SON, 100 days at 800°C; (b) UK-209, 100 days at 800°C, (c) D-B 1-3, unheated, (d) D-B 1-3 heated 100 days at 800°C.

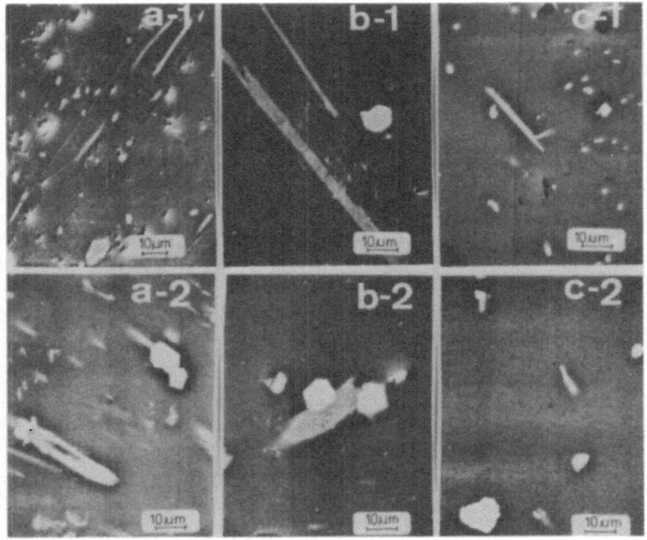

Fig. 2. SEM micrographs of unheated (1) and heated (2) volcanic glasses for 100 days at 800°C: (a) VG-0.5; (b) VG-5; (c) VG-670.

further coarsened during heat treatment. Thus, volcanic glasses are more stable against heat treatment than the waste glasses.

Chemical Stability

The results of the DIN 12111[1] leach test after heat treatment at 800°C are summarized in Fig. 3. Only the glass F-SON changed its hydrolytic resistance after heat treatment. All volcanic glasses belong to the chemically resistant glasses, whereas the waste glasses fall into different hydrolytic classes. The grain titration method covers only the alkali release of the glasses. Therefore, the more comprehensive soxhlet and autoclave leaching methods for measuring weight losses have been applied (Table 2). Leach rates were calculated from the results of the grain titration method with the following assumptions: (1) the particles are spherical, (2) the roughness factor of the glass powder for calculating the surface area is $\sim$ 2.5, and (3) the whole glass is leached at the same rate. All leach rates for the volcanic glasses are lower than those of the waste glasses. For autoclave leaching the differences between waste and volcanic glasses are smallest, due to the destruction of the glass network. This can be seen in Fig. 4 which shows SEM-micrographs of the hydrothermal leached surfaces of the volcanic glasses. The crystalline phases seem to be more stable than the glass phase. Similar stable crystalline phases were caused by heating the waste glasses. Fig. 5 shows SEM-micrographs of the leached surfaces of the waste glasses. Whereas the residual glass phase was leached, the crystal phases were not attacked. Fig. 5a shows that the glass F-SON formed a gel layer, but glass UK-209 (Fig. 5b) did not show any effects due to leaching. The waste glasses showed the same sequence in their leach resistance as listed in Table 2.

CONCLUSIONS

1. Naturally occurring volcanic glasses are more "stable" (chemically and thermally) than the three waste form glasses.

2. The greater "stability" of the volcanic glasses is due to their higher silica content (74 weight percent vs. 28-50 weight percent) which results in a higher temperature of formation ($\sim$1500°C vs. $\sim$1100°C).

3. Despite the range in ages ($\sim$500 years to 670,000 years), there was no apparent, time-dependent change in the alteration or devitrification among the three volcanic glasses; however, altered glasses are a well-recorded part of the geologic record[6], and these particular specimens were selected because they were fresh and unaltered.

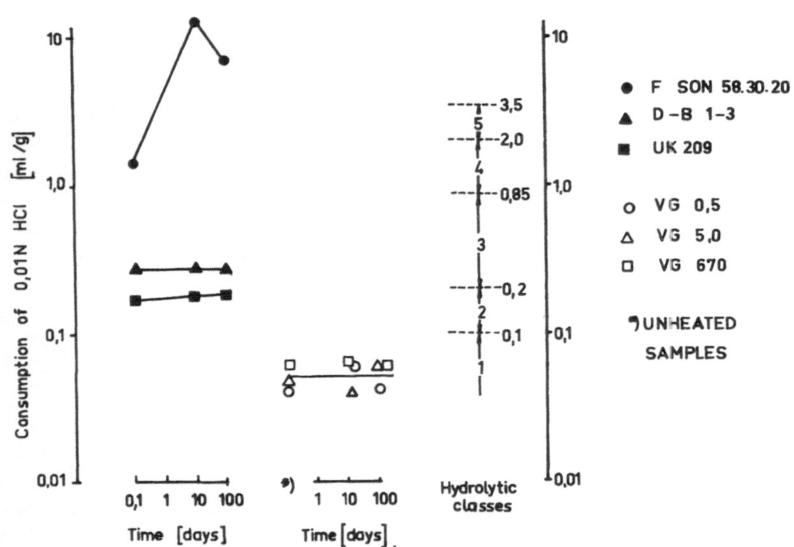

Fig. 3. Hydrolytic resistance of waste glasses and volcanic glasses, as measured by DIN 12111 leach test.

Table 2: Measured leachabilities of waste glasses, the glass ceramic and volcanic glasses

	Leach Rate [mg/cm² day]					
	F-SON	UK 209	D-B 1-3	VG 0.5***	VG 5.0***	VG 670***
DIN 12 111*	2.5	0.1	0.1	0.03	0.04	0.07
Soxhlet	3.1	0.3	0.6	0.01	0.05	0.01
Autoclave**	3.7	1.7	1.8	0.7	1.1	0.4

* Calculated from base release with special assumptions described in text.
** 200°C and 10 bars for 3 days in saturated carnallite brine solution.
*** Leach rates for volcanic glasses are based on total sample (glass + crystalline component).

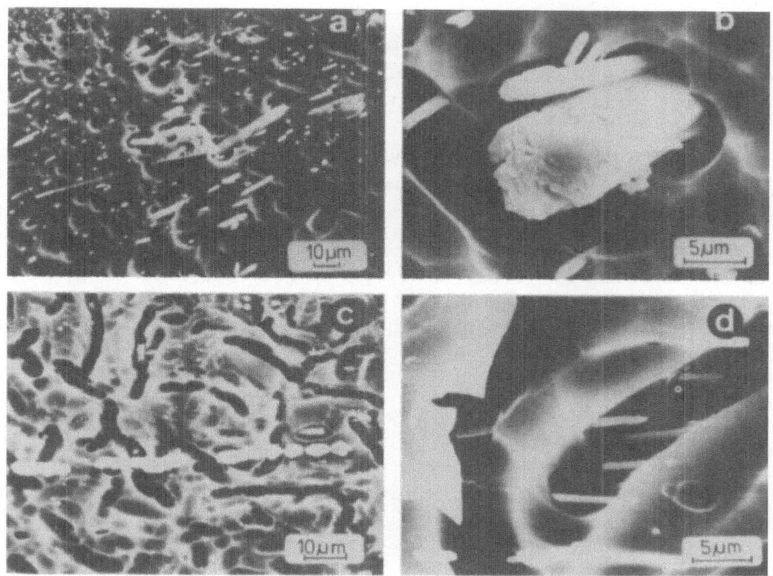

Fig. 4. Hydrothermal leaching of volcanic glasses at 200°C and 12 bars for 3 hours. (a) and (b) = VG 0.5; (c) and (d) = VG 670.

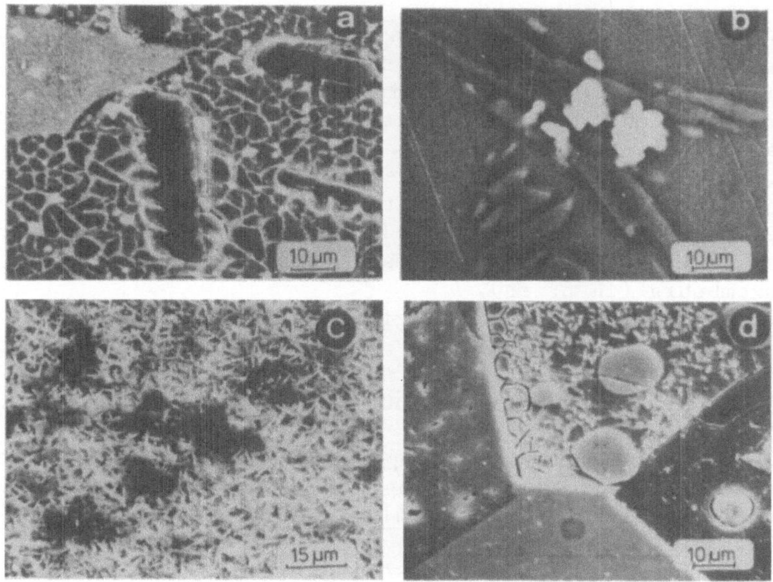

Fig. 5. Phase selective leaching of waste glasses and the glass ceramic. (a) F-SON and (b) UK 209, heated 100 days at 800°C, etched 45 min; (c) D-B 1-3, etched 3 hours; (d) D-B 1-3, heated 100 days at 800°C, etched 15 min.

4. To the extent that elevated temperatures (800°C) can be used to accelerate time-dependent reactions, the waste glasses show dramatic changes in degree of crystallinity and the formation of new phases, while the volcanic glasses show essentially no change.

5. As most volcanic glasses are less than 2 million years old[6], this suggests a <u>maximum</u> long-term "stability" of the waste glasses of not more than 10^6 years. This comparison assumes ambient, surface conditions and does not take into consideration effects peculiar to the waste form and the waste repository (e.g. unique chemistry, radiation effects or the presence of brine solutions), which may reduce the long-term "stability" of the glass waste forms.

ACKNOWLEDGEMENTS

The characterization of the waste form glasses was in part supported by the European Atomic Energy Community, under contract 029-77-1 WASD. R. Ewing collected the samples of volcanic glass while holding a NORCUS research appointment at Battelle PNL. J. Taylor of the Institute of Meteoritics at the University of New Mexico provided the microprobe analyses of the volcanic glasses.

REFERENCES

1. G. Malow, V. Beran, W. Lutze, J. A. C. Marples, J. T. Dalton, A. R. Hall, A. Hough, K. A. Boult, "Testing and Evaluation of the Properties of Various Potential Materials for Immobilizing High Activity Waste," <u>Nuclear Science and Technology</u>, EUR 6213 (1978).
2. J. A. C. Marples, W. Lutze and C. Sombret, The Leaching of Solidified High Level Waste Under Various Conditions <u>in</u>: "Proceedings of the First European Community Conference on Radioactive Waste Management and Disposal," May, 1980, Luxembourg (in press).
3. J. C. Eichelberger, "Origin of Andesite and Dacite: The Petrographic and Chemical Evidence for Volcanic Contamination," Ph.D. Dissertation, Stanford University (1974).
4. I. Friedman, Hydration Rind Dates Rhyolite Flows, <u>Science</u>, 159:878 (1968).
5. R. A. Bailey, G. B. Dalrymple and M. A. Lamphere, Volcanism, Structure, and Geochronology of Long Valley Caldera, Mono County, California, <u>Journal of Geophysical Research</u>, 81:725 (1976).
6. R. C. Ewing and R. F. Haaker, "Naturally Occurring Glasses: Analogues for Radioactive Waste Forms,"Battelle PNL Report 2776/UC-70 (1979).

SURFACE ANALYSIS--ITS USES AND ABUSES IN WASTE FORM EVALUATION*

G. L. McVay and L. R. Pederson

Pacific Northwest Laboratory

Richland, Washington 99352

INTRODUCTION

Investigations of the chemical durability of waste forms have been conducted for many years and large quantities of data have been gathered,[1,2] particularly for glass. However, more recently the emphasis has shifted from data gathering to mechanistic understanding. This shift in emphasis has occurred primarily because of the need to develop reliable, long-time predictive capabilities for waste-form behavior in aqueous environments. These predictive models will be incorporated into larger models for the entire repository and should be acceptable to the scientific community and to NRC before repository licensing will be feasible. A key element in the acceptance of predictive models is a defendable mechanistic interpretation of waste form-water interactions.

Surface and near-surface analytical techniques offer a means to investigate this interaction by analyzing the reaction (gel) layer which forms on the solid waste form during exposure to an aqueous environment. Surface analytical tools such as Auger, ESCA, SIMS, Rutherford Backscattering (RBS), and Nuclear Microprobe Analyses (NMA) are being increasingly utilized to study this reaction layer, particularly for glass. Although surface analysis offers the possibility of greater understanding, it also can result in misinterpretations if not properly executed.

*Work supported by the U.S. Department of Energy under Contract DE-AC06-76RLO 1830.

RESULTS AND DISCUSSION

The reaction layer, which usually develops on waste forms exposed to water, is a complex substance to investigate. It is formed not only by the removal of elements, but also by the addition of elements to the waste form via the aqueous solution. The chemistry of elements on the surface can be analyzed by ESCA techniques. However, it is usually desirable to know the elemental concentration profiles in the gel layer. This can be done with varying degrees of success by all of the surface analytical techniques.

Both the RBS and NMA techniques involve striking the sample with energetic particles and analyzing either the scattered incident particles or the scattered radiation from the resultant nuclear reactions, respectively. Both techniques usually lack good depth resolution and the ability to simultaneously analyze many elements. However, they are isotope (mass) selective and quantitative. The other techniques all involve removal of the gel layer by sputtering to determine elemental profiles. This can present problems for insulators, such as most waste forms, because of charging effects resulting from sputtering with ions. These charging effects can cause mobile ions in the gel layer to move if the sample is not cooled during analysis.[3] Auger analyses are particularly sensitive to the effects of charging.

One of the most successful techniques for elemental depth profiling of gel layers is ESCA analysis used in conjunction with ion milling. An example of this type of analysis is given in Fig. 1 for 76-68 glass, a complex borosilicate waste glass, which has been exposed to 90°C distilled water for 14 days with a sample surface area to solution volume ratio of 10 m^{-1}. As can be seen from the figure, the profiles for a large number of elements are routinely attainable. With this type of analysis it is possible to determine which elements are accumulating in or on the gel layer and which elements are being depleted. Additionally, when used in conjunction with solution analysis, this technique readily determines the degree to which congruent dissolution has occurred. There are three common ways to display the elemental profile data. The first, used in Fig. 1, is to normalize the signal intensity of each element, after sputtering to bulk glass, to the known concentration of that element in the bulk glass. This technique assumes that the signal intensity varies directly with concentration. Information about variations in density through the reaction layer can thus be obtained, which is particularly useful when evaluating reaction layers on waste containment glasses. Wide variations in the density of gel layers from that of bulk glass have been reported for these materials.[4] Concentrations within the gel layer can be slightly overestimated by this method, since electron escape depths can be increased with decreasing density, resulting in greater analysis volumes in the gel layer than in the bulk glass. The second

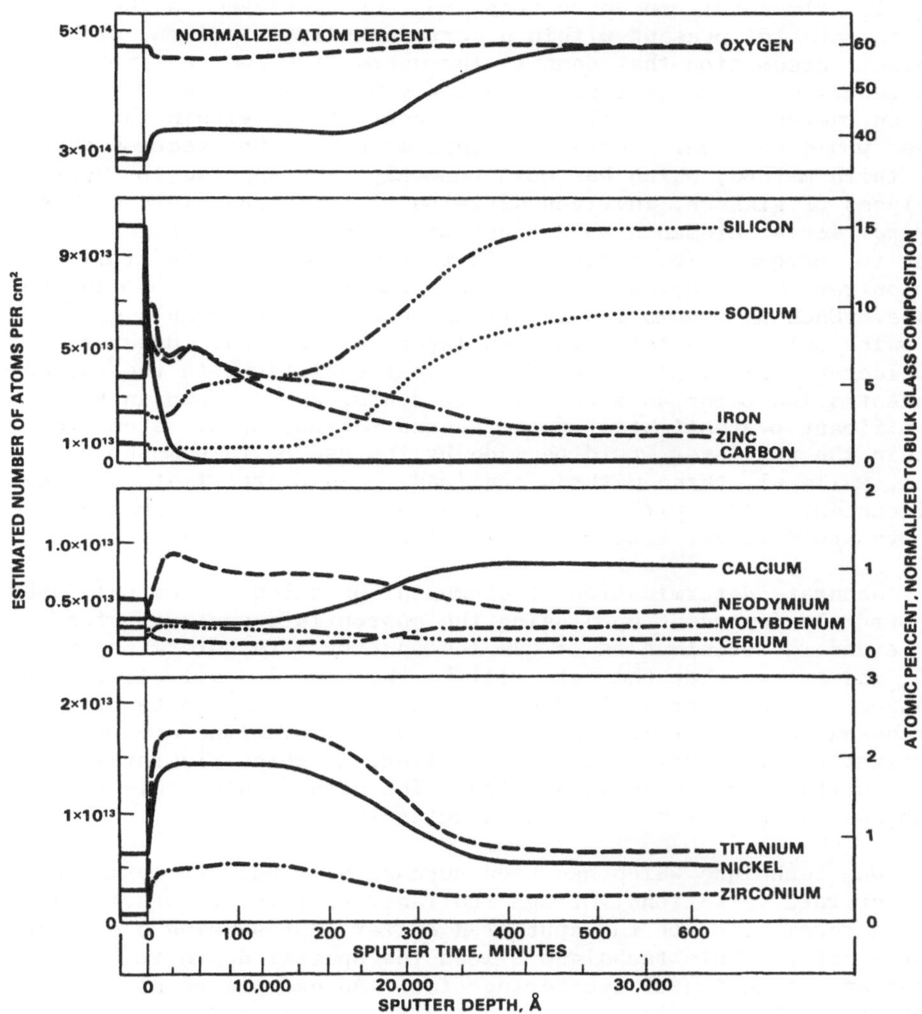

Fig. 1. Elemental concentration profiles through the reaction (gel) layer of a polished 76-68 glass sample exposed to 90°C distilled water for 14 days with a sample surface area to solution volume ratio of 10 m^{-1}.

common display method involves division of the signal intensities as a function of sputter depth by a set of sensitivity factors, which were predetermined from standard spectra.[5] An example of this calculation is illustrated by the dashed line for oxygen in Fig. 1. Elemental concentrations are thus displayed as the fraction of the material present within a certain analysis volume, with an implicit assumption that density throughout the sputter profile is constant. On waste glasses where low density gel layers are formed, strong overestimates of elemental concentrations within the gel layer would be made. Therefore, this method is not recommended. The third method, which has most commonly been applied to Auger analyses of glasses, involves division of the signal intensities for particular elements as a function of depth by the signal intensity for oxygen. It is assumed in this method that the atomic concentration of oxygen is invariant from the gel layer to the bulk glass. Such an assumption is not unreasonable for simple glasses on which relatively thin reaction layers are formed. However, considerable error is introduced if real variations in the oxygen concentration occur as a function of depth, as is shown in Fig. 1. Significant overestimates of the concentrations of all elements within the gel layer would be made by the use of this method. Thus, while all three methods can lead to an overestimate of elemental concentrations within a low density gel layer, the smallest errors would result from the application of method one.

Accurate determination of elemental profiles in the gel layer is strongly dependent upon using the appropriate sputter rates. The standard technique is to use the sputter rate determined for SiO_2 and assume that the rate will be the same for all silica-based glasses. This is normally done because SiO_2 films can be made or purchased in known thicknesses and thus, calibrations are relatively straightforward. However, the gel layer, in which elemental profiles are important, can vary in density. The effect which this variation has on sputter rates must be determined.

One technique which has been successfully used to evaluate sputter rate variations through the reaction layer on waste glasses is a determination of the sputtered crater depths using a surface profilometer. This technique allows the sputter depth to be determined as a function of sputtering time. An example of sputter rates through a gel layer is given in Fig. 2 for a 76-68 waste containment glass. This glass had identical treatment to the sample used for Fig. 1. As can be seen from Fig. 2, the sputter rate within the gel layer is significantly larger than that in the bulk (unleached) glass. The sputter time versus depth is shown by the solid lines, and the sputter rate is illustrated by the dashed line. In the initial (lowest density) portion of the layer, the sputter rate was 2.2 times larger than the bulk rate. In the remaining portion of the gel layer, the sputter rate progressively decreased until the bulk sputter rate was achieved. The bulk rate was indistinguishable from that of SiO_2. Significant errors can be introduced in elemental profiling by assuming SiO_2 sputter rates through the reaction layer. However, the assumption of SiO_2 sputter rates for unleached silicate glasses appears to be valid.

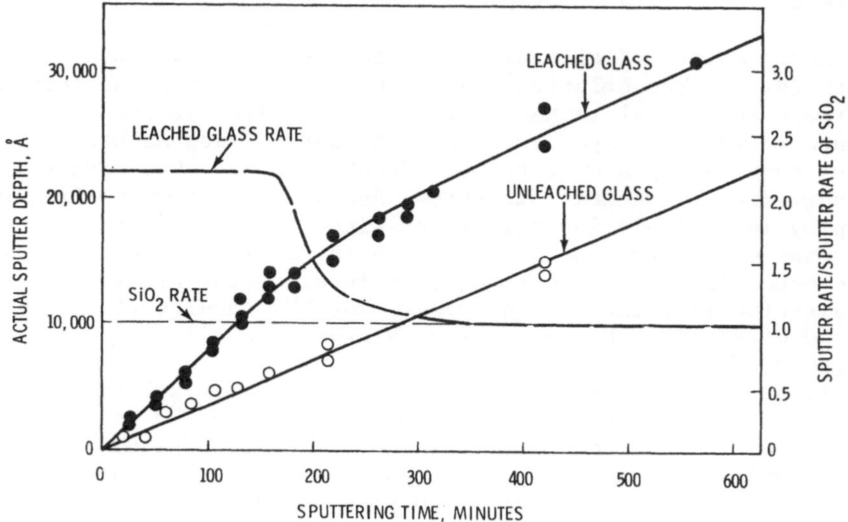

Fig. 2. Comparison of the sputtering characteristics of 76-68 glass leached 14 days at 90°C in deionized water and unleached glass to that of SiO_2. The dashed line displays sputter rates and the solid lines display sputter depths.

Surface analysis is useful in evaluating the effects of procedures used in sample preparation or data gathering. For example, the use of HF acid to prepare sample surfaces for leaching[6] and to progressively remove gel layers for elemental profiling[7,8] have been reported in the literature. The effects of HF acid on a sodium-silicate glass are shown in Fig. 3 for both a dilute (1%) and a concentrated (48%) acid solution and are compared to an as-prepared sample. As illustrated in Fig. 3, the HF acid selectively removes the sodium ions from the glass. The dilute HF depletes sodium to a greater depth than concentrated HF. Presumably, this occurs because the concentrated acid attacks the silicate structure more rapidly than the dilute HF acid. Even the as-prepared sample has a small sodium depletion layer. The samples exposed to HF also have fluorine incorporated into their structure. If leach samples were prepared by etching the surfaces in HF, then a significant error could result due to initial depletion of elements and to incorporation of fluorine into the glass structure. Additionally, the use of HF etching to progressively remove the gel layer for elemental concentration determinations is not a good procedure. Since the alkali ions are selectively removed, a distorted elemental profile could be obtained.

Incorporation of elements into the waste form which are present in both the aqueous solution and the waste form can be differentiated by surface analytical techniques. If one desires to know which elements in the gel layer came from the water (such as hydrogen or oxygen) then isotopes must be used (such as D or ^{18}O). Rutherford Backscattering or Nuclear Microprobe Analysis are good techniques to employ in isotopic investigations since they are sensitive to small mass variations and quantitative in nature. Investigations of the mechanisms of waste form-water interactions are examples of the gainful utilization of isotopes. The NMA results of time-dependent ^{18}O uptake from $H_2{}^{18}O$ water by a soda-lime-silica glass are illustrated in Fig. 4. The samples were prepared by exposing the glass to 50°C $H_2{}^{18}O$ with a sample surface area to solution volume ratio of 20 m^{-1}. As can be seen from Fig. 4, the ^{18}O is taken into the glass up to five days, after which an apparent saturation occurs. This data may be interpreted as diffusion of water into the glass matrix, balanced at times greater than five days by surface dissolution to remove the reaction layer. Information such as this, when used in conjunction with D uptake is a valuable key to unlocking the secrets of glass-water interaction mechanisms.

CONCLUSIONS

Surface and near-surface analytical techniques are significant aids in understanding waste form-aqueous solution interactions. They can be beneficially employed to evaluate reaction layers on waste forms, to assess surface treatments prior to and after leaching, and to identify interactions with waste forms. Surface analyses are best used in conjunction with other types of analyses, such as solution analyses, in order to obtain a better overall understanding of reaction processes. In spite of all the benefits to be gained by using surface analyses, misinterpretations can result if care is not taken to properly obtain and analyze the data. In particular, the density variations through a reaction layer must be accounted for in both sputtering and data analysis techniques.

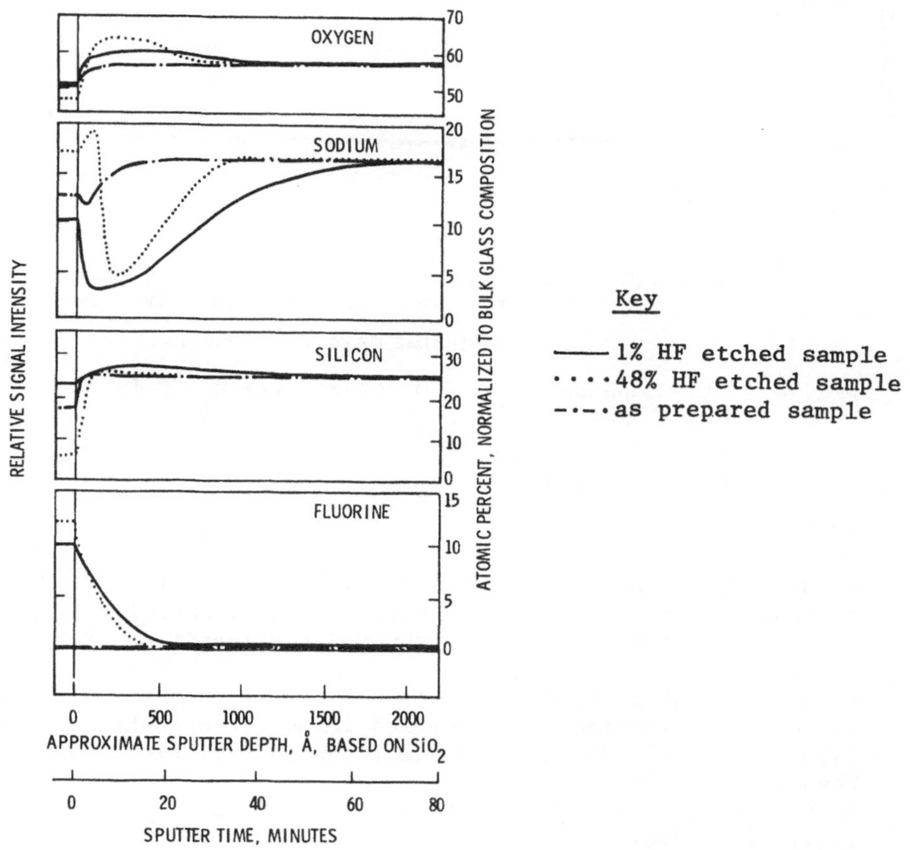

Fig. 3. Elemental depth pofiles through the reaction (gel) layer formed on a $Na_2O \cdot 3SiO_2$ glass by exposure to HF acid.

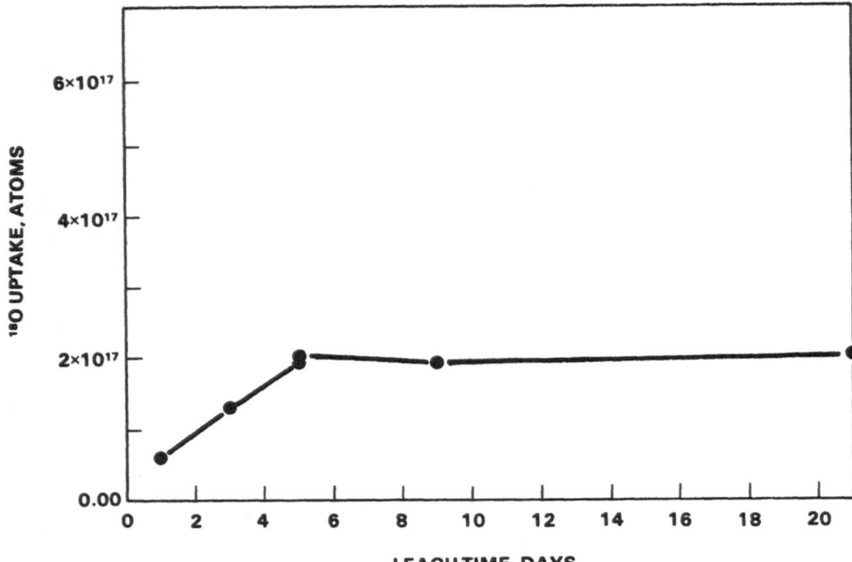

Fig. 4. Time dependence of ^{18}O uptake from $H_2{}^{18}O$ by a soda-lime-silica glass at 50°C with a sample surface area to solution volume ratio of 20 m^{-1}.

REFERENCES

1. W. Heimerl, H. Heine, L. Kahl, H. W. Levi, W. Lutza, G. Malow, E. Schiewer and P. Schubert, "Research on Glasses for Fission Product Fixation," HMI-B109, Hahn-Meitner Institute, Berlin, Germany (1971).
2. W. A. Ross and J. E. Mendel, "Annual Report on the Development and Characterization of Solidified Forms for High-Level Wastes: 1978," PNL-3060, Pacific Northwest Laboratory, Richland, Washington (1979).
3. C. G. Pantano, Jr., D. B. Dove and G. Y. Onoda, Jr., JVST 13(1):414 (1976).
4. K. A. Boult, J. T. Dalton, A. R. Hall, A. Hough, and J. A. C. Marples, The Leaching of Radioactive Waste Storage Glasses, in: "Ceramics in Nuclear Waste Management," T. D. Chikalla and J. E. Mendel, eds., National Technical Information Service, Springfield, Virginia.
5. C. D. Wagner, et al., "Handbook of X-ray Photoelectron Spectroscopy," Perkin-Elmer Corporation, Physical Electronics Division, Eden Prarie, Minnesota (1979).
6. G. R. Holdren, Jr., and R. A. Berner, Geochim. Cosmochim. Acta 43:1161 (1979).
7. A. F. White, Geol. Soc. Am. Annu. Meet. 11(7):539 (1979).
8. Z. Boksay, G. Bouquet, and S. Dobos, Phys. Chem. Glasses 8:140 (1967).

ARE SOLUBILITY LIMITS OF IMPORTANCE TO LEACHING?

Allen Ogard, Glenn Bentley, Ernest Bryant, Clarence Duffy, Jane Grisham, Edward Norris, Carl Orth, and Kimberly Thomas

Los Alamos Scientific Laboratory
P. O. Box 1663, MS-514
Los Alamos, New Mexico 87545

This project developed from the Oklo natural fission reactor studies. It had been determined in the Oklo studies that many fission products and actinides remained in the reactor site during the periods of their radioactive decay following formation in the reactor zone two billion years ago. An explanation for this retention of fission products and actinides uses the extreme insolubility of uraninite (UO_2) in very reducing water environments. One can estimate from available thermodynamic data that the concentration of uranium in equilibrium with uraninite in pH 7 water that is free of dissolved oxygen is $\sim 10^{-11}$M. This low value suggested that the reducing conditions that can occur in deep geologic burial would result in a very slow leaching of spent fuel elements in contact with water since spent fuel elements are largely sintered UO_2.

During our studies on the leaching of spent fuel elements we found it difficult to readily duplicate the reducing conditions of deep geologic burial. This result we inferred from the relatively high uranium concentrations that were found in the leachants rather than the low value listed above. However, it was observed that under our reducing condition as well as under an oxidizing atmosphere, that some of the rare earth fission products and actinides behaved differently than the uranium; a behavior we attribute to a precipitation of the rare earths and actinides.

Results of our uranium leaching experiments are shown in Figure 1. The dissolution was carried out at 25°C under the oxidizing and reducing atmospheres for 65 days. The temperature was then raised to 70°C and the dissolution was continued for an additional 125 days. The results in oxidizing atmosphere are shown as dark lines with the letter O, and those in reducing atmosphere as open

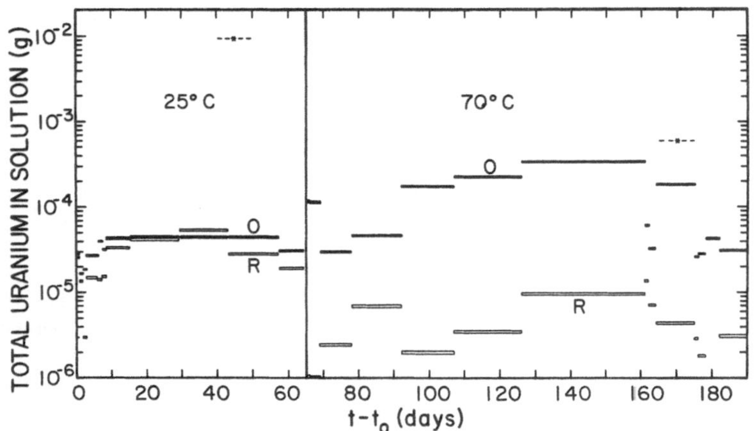

Fig. 1. Uranium per 5 mℓ leachant as a function of time.

boxed lines with the letter R. Each line on the graph represents a new volume of leachant that was in contact with a slice of the spent fuel element for the period of time shown. The leachant was then removed through a glass fritted disk and analyzed for radioactive elements. The atmospheres were set by bubbling CO_2-free air or 94% argon-6% hydrogen through the leachants. Platinum gauze was also present in the reducing atmosphere experiments. Fission product and actinide data are shown in Figure 2.

Under oxidizing conditions the dissolution of the UO_2 in the spent fuel element appears to be straightforward. At 25°C the concentration of uranium is essentially constant after about 7 day leach periods. Whereas, at 70°C it continues to increase with increasing length of contact time to 35 days. On Fig. 1 are drawn two values (---X---) of the solubility limit of shoepite, $UO_3(H_2O)_x$ as reported by Holland[1] for 25°C and 90°C. Shoepite is considered to be the uranium oxide in equilibrium with water under oxidizing conditions. The uranium concentrations that we measure at 25°C are far below this solubility limit. This is most likely due to slow kinetics of dissolution at 25°C. However, at 70°C the concentrations are much closer to being at the solubility limit. Note that a negative temperature coefficient of solubility was reported by Holland.

With reducing conditions the concentration of uranium was also essentially constant for contact periods over 1 week in length at 25°C. After the temperature was increased to 70°C there was a drop in uranium concentration which increased again as the length of contact time increased. However, after 92 days, there was another

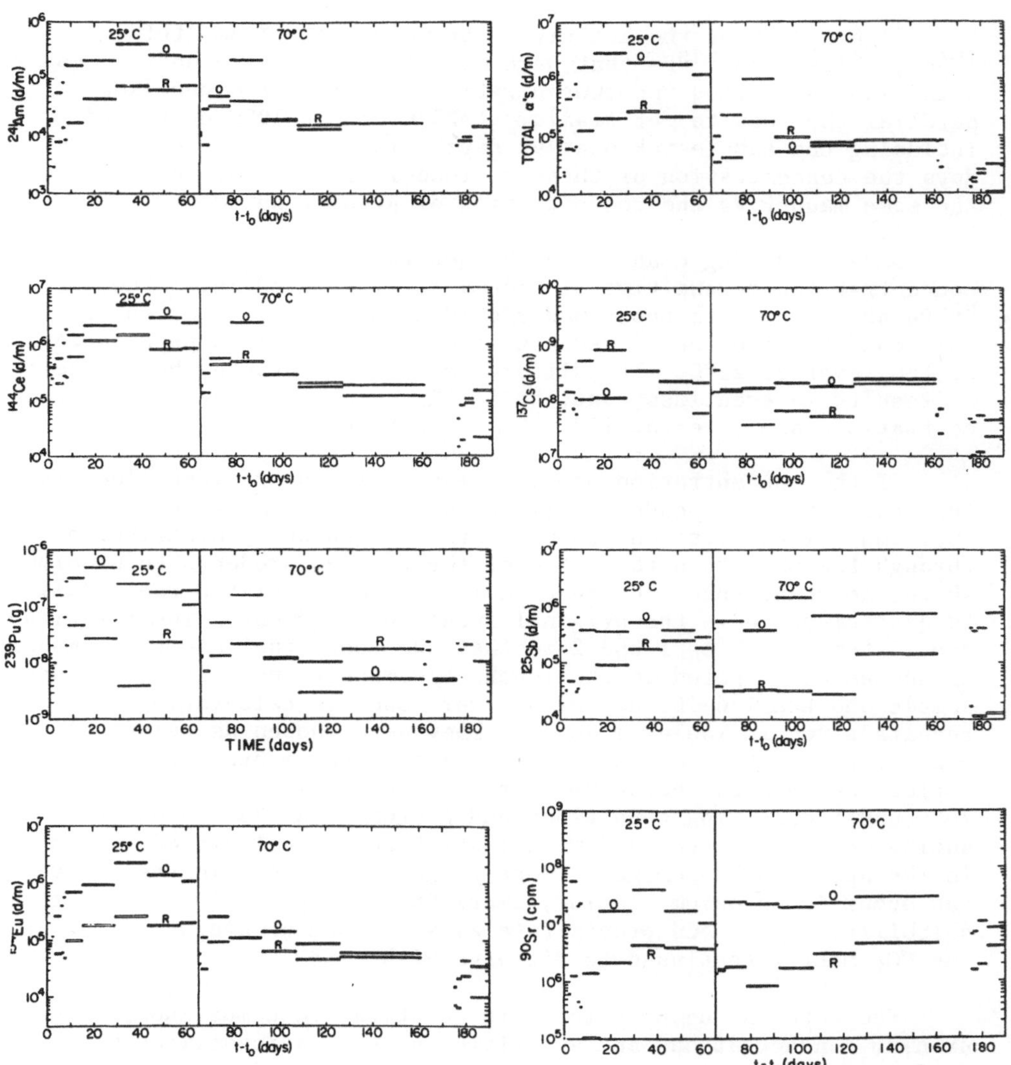

Fig. 2. Fission product and actinide quantities as a function of time.

drop in concentration with time. It is tempting to attribute this erratic behavior to analysis scatter but the same general shape of the graph is shown in the fission product concentrations (Fig. 2) whose analyses were done on entirely different aliquots.

Figure 2 shows the activities or weight of ^{241}Am, total α's, ^{154}Eu, ^{144}Ce, and ^{239}Pu that were measured in the leachants. Under oxidizing conditions the concentrations of these radionuclides parallel the results for uranium at 25°C and at 70°C up to and including the sample taken at 92 days. In samples taken after 92 days the concentration of these radionuclides did not increase in the same manner as the concentration of uranium but decreased.

Under reducing conditions the general shape of the plots of concentrations versus time of ^{241}Am, total α's, ^{154}Eu, ^{144}Ce, and ^{239}Pu are similar to those under oxidizing conditions except that the concentrations under reducing conditions are perhaps a factor of ten lower at 25°C. The difference at 70°C between the two sets of results is even less, but, as in the oxidizing results, the concentrations are lower at 70°C than at 25°C.

If the concentration of uranium and all the radionuclides in the leachant are dependent only on the dissolution of the matrix UO_2, and all the fission products are homogeneously distributed through the UO_2, then the ratio of the fission products to uranium in the leachant should be comparable to the calculated radionculide to uranium ratio in the original spent fuel element. These ratios are listed on Tables 1 and 2 so that the spent fuel element compositions can be compared to the leachant compositions. The 29-43 day sample and the 126-161 day sample were used in calculating the results given in Tables 1 and 2. They were chosen as being the most representative of the 25°C and 70°C experiments. Of these ratios the ones for europium, cerium, americium, and plutonium at 70°C under oxidizing conditions and plutonium at 25°C under reducing conditions are very much lower than the calculated ratios in the spent fuel element. These large differences in the ratios can occur if europium, cerium, americium, and plutonium are at their solubility limits and precipitate as some form of hydrous oxide as the UO_2 matrix continued to dissolve with time.

The data can be used for more quantitative comparisons. The uranium concentrations and activities of the radionuclides were used to calculate isotope concentrations. These isotope concentrations were then converted to elemental concentrations by using the calculated isotopic composition of the H. B. Robinson spent fuel element[2]. The elemental concentrations are listed in Tables 1 and 2 for the same samples whose ratios were listed. Based on the discussion and the data in the Tables, the solubility limits for some of the fission products are listed in Table 3. These numbers assume kinetics did not prevent precipitation at either 25° or 70°C. Also,

TABLE 1

ELEMENT CONCENTRATIONS
REDUCING CONDITIONS

Element	Moles/Liter 25°C	Moles/Liter 70°C	Ratio to U 25°C	Ratio to U 70°C	Ratio to U in Fuel
U	4.6 E-5	8.0 E-6			
Eu	7.2 E-9	1.3 E-9	1.6 E-4	1.6 E-4	1.9 E-4
Ce	1.1 E-7	1.4 E-8	2.3 E-3	1.7 E-3	3.6 E-3
Am	1.0 E-8	2.2 E-9	2.2 E-4	2.7 E-4	5.0 E-4
Pu	5.0 E-9	2.8 E-8	1.1 E-4	3.5 E-3	1.3 E-2
Cs	6.6 E-6	4.0 E-6	1.4 E-1	4.9 E-1	4.4 E-3
Sr	1.9 E-7	5.6 E-7	4.2 E-3	7.0 E-2	2.0 E-3
Sb	4.9 E-10	5.2 E-10	1.1 E-5	6.5 E-5	2.1 E-5

TABLE 2

ELEMENT CONCENTRATIONS
OXIDIZING CONDITIONS

Element	Moles/Liter 25°C	Moles/Liter 70°C	Ratio to U 25°C	Ratio to U 70°C	Ratio to U in Fuel
U	3.8 E-5	2.8 E-4			
Eu	6.3 E-8	1.6 E-9	1.7 E-3	5.8 E-6	1.9 E-4
Ce	3.8 E-7	8.9 E-9	9.9 E-3	3.2 E-5	3.6 E-3
Am	5.4 E-8	<1 E-9	1.4 E-3	<3 E-6	5.0 E-4
Pu	3.0 E-7	4.8 E-9	7.9 E-3	1.7 E-5	1.3 E-2
Cs	6.6 E-6	3.5 E-6	1.7 E-1	1.0 E-2	4.4 E-3
Sr	1.7 E-6	6.5 E-7	4.5 E-2	2.3 E-3	2.0 E-3
Sb	1.5 E-9	2.2 E-9	4.0 E-5	7.8 E-6	2.1 E-5

Note: Computer notation is used in the Tables. 2.0 E-5 = 2.0×10^{-5}.

TABLE 3

SOLUBILITIES AT pH 4, DEIONIZED WATER

Element	25°C	70°C	Condition
Eu	≧3 E-8M	2 E-9M	O & R
Ce	≧3 E-7	1 E-8	O & R
Am	≧5 E-8	2 E-9	O & R
Pu	3 E-7	1 E-8	O & R
U	>5 E-5	>3 E-4	O
		~2 E-6	R

at 25°C the calculated numbers are largely lower limits since there was not the evidence of precipitation from differences in the spent fuel-leachant ratios. For comparison, we can use the compilations of Baes and Mesmer[3] and the calculated solubilities of Newton et al.[4] Under oxidizing conditions and pH 4 uranium oxide would have a solubility of 10^{-1} to 10^{-2}M and under reducing conditions 10^{-7} to 10^{-11}M depending on which solid, the amorphous or the crystalline, is in equilibrium with the liquid. Similarly the solubility of plutonium oxide would be 10^{-3} to 10^{-7}M for oxidizing and 10^{-7} to 10^{-14}M in reducing conditions. The solubility is not only dependent on the crystal form but also on the actual Eh of the reducing condition. These comparisons are probably as good as one could expect from such a heterogeneous and multicomponent system as a spent fuel element dissolving in water.

The fission products ^{137}Cs, ^{90}Sr, and ^{125}Sb behave differently than the other fission products (Fig. 2). Under reducing conditions and at both 25° and 70°C ^{137}Cs, ^{90}Sr, and ^{125}Sb act similarly to uranium except that, as noted in Table 1, the concentrations of these isotopes in solution are far higher than expected from the isotope to uranium ratio in the spent fuel element. The ratios to uranium at 25° and 70°C for cesium and strontium are very high, especially the cesium to uranium ratios. Katayama et al.[5] in leaching experiments in air of H. B. Robinson spent fuel elements also found a higher fraction of some radionuclides, especially cesium, in the leachant than in the spent fuel element. They used this higher concentration as indicating an absence of congruent dissolution. Higher concentrations of radionuclides is not by itself sufficient reason to dismiss congruent dissolution. Post-irradiation examinations of spent fuel elements have shown that the radionuclides can be inhomogeneously distributed through the fuel, especially the more volatile elements such as cesium and iodine. If the cesium were concentrated at the grain boundaries of the fuel and at the colder exterior of the fuel pellet but were still largely incorporated into the UO_2 matrix, its concentration in the leachant would be controlled by the dissolution of the UO_2 matrix but at a higher cesium to uranium ratio. Figs. 1 and 2 indicate the behavior of cesium, strontium. and uranium to be very similar but at different concentration levels.

Conclusions

The solubilities of some radionuclides, especially rare earths and actinides, may be an important and controlling factor in leaching of waste forms. These solubilities should be measured accurately as a function of pH and not as a part of a multicomponent system.

Although the amount of data is small it is interesting to postulate that a negative temperature coefficient of solubility is being exhibited by the actinides and rare earths in Figs. 1 and 2. Individual solubilities should be measured as a function of temperature to determine if a kinetic effect is being observed in the data. A negative temperature coefficient of solubility for actinides and rare earths in water would have important consequences for nuclear reactor safety and for the management of nuclear wastes.

References

1. H. D. Holland and L. H. Brush, "Uranium Oxides in Ores and Spent Fuels," NUREG-CP-0005, 597-615, 1978.

2. J. Grisham, "T-2 Calculated Source Term for H. B. Robinson-2 Fuel Element B05-E-14," Los Alamos Scientific Laboratory office memorandum, Sept. 15, 1978.

3. C. F. Baes, Jr. and R. E. Mesmer, "The Hydrolysis of Cations," John Wiley & Sons, N.Y., 1976.

4. T. W. Newton, R. D. Aguilar and B. R. Erdal, "Estimation of U, Np, and Pu Solubilities vs. Eh and pH," in "Laboratory Studies of Radionuclide Distribution Between Selected Groundwater and Geologic Media," B. R. Erdal, Ed., LA-8339-PR, 1980.

5. Y. B. Katayama, D. J. Bradley and C. O. Harvey, "Status Report on LWR Spent Fuel IAEA Leach Tests," PNL-3173, 1980.

LEACH AND CORROSION TESTS UNDER NORMAL AND ACCIDENT CONDITIONS ON CEMENT PRODUCTS FROM SIMULATED INTERMEDIATE LEVEL EVAPORATOR CONCENTRATES

G. Rudolph, P. Vejmelka, and R. Köster

Kernforschungszentrum Karlsruhe GmbH

7500 Karlsruhe 1, Federal Republic of Germany

INTRODUCTION

For safety considerations in the disposal of radioactive waste products in salt formations, it is necessary to investigate, in addition to improving the leach rates of cemented solidification products under normal conditions (ambient temperature, atmospheric pressure), their leach and corrosion behavior at elevated temperatures and pressures (accident conditions).

These investigations are aimed at establishing mathematical expressions which describe the time-dependent release of radionuclides from waste products by determining the leaching and corrosion mechanism. These expressions will allow extrapolation of real waste product behavior for extended periods of time.

LEACH TESTS AT AMBIENT TEMPERATURE AND ATMOSPHERIC PRESSURE

Leach tests at ambient temperature and pressure have been conducted for three years. The specimens have been prepared from various types of cement (portland cement, slag cement, and pozzolanic cement); many of them contain additives such as bentonite to improve the retention of cesium and strontium. All cement forms were made using a simulated intermediate level liquid waste with the waste salts making up 10% of the total weight. The leachants used were distilled water, tap water, saturated sodium chloride solution, and saturated salt brine.*

---

*Composition (wt %): 24.7 $MgCl_2$, 2.3 $MgSO_4$, 1.9 NaCl, 3.3 KCl, and 67.9 H_2O.

The leach tests are performed in accordance with the IAEA recommendations.[1] Some data on the influence of cement type, additives, and leachant on the leach rates have been reported in an earlier paper in this series.[2] The change in leach rates over extended periods of time supplies valuable information for predicting the long-time release of radionuclides due to leaching and corrosion effects.

Cesium Leaching in Tap Water

There is no visible change on any sample stored in tap water which would indicate any corrosion has occurred. This is in agreement with the generally held opinion that natural water is not aggressive towards cement. A white deposit has formed, however, on some samples, especially those made with portland cement, perhaps consisting of carbonates of calcium and magnesium.

A plot of the leached fraction of cesium vs the square root of time generally yields an S-shaped curve which Fig. 1 shows a typical example. It indicates that leaching is not controlled by a diffusion mechanism with a constant diffusivity. It should be kept in mind that the cement specimens under investigation differ from

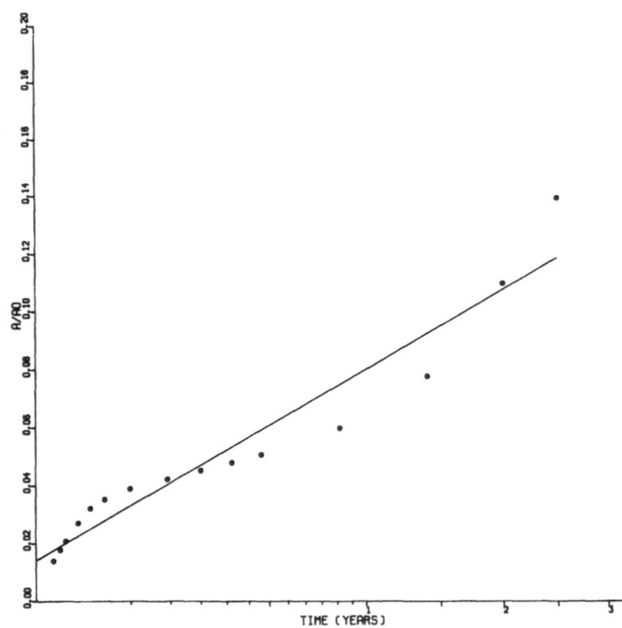

Fig. 1. The fraction of cesium released to tap water from Portland cement without additives as a function of the square root of time.

conventional cement products in their high salt content. The leaching of contained salt may cause chemical changes of an as yet unknown nature which express themselves in the irregular shape of the leach curves.

Cesium Leaching in Salt Brine

Salt brine, containing both magnesium and sulfate ions, causes heavy corrosion of cementitious products. The rate of corrosion is markedly dependent on the composition of the product. Specimens containing leach-retarding additives in excess of 10% by weight were completely disintegrated into a soft voluminous sludge after about 1 year of leaching. On the other hand, all of the samples containing 5 or 10% bentonite have retained their integrity, with only superficial changes being noticeable. This is also true for specimens made of portland cement or pozzolanic cement without additives.

The leach rates in salt brine are in the same order of magnitude as in tap water. The shape of the leach curves generally conforms with that predicted by diffusion theory, i.e., there is a straight line if plotted as in Fig. 2. If disintegration occurs, this is generally reflected in a sharp upward bend of the leach curve.

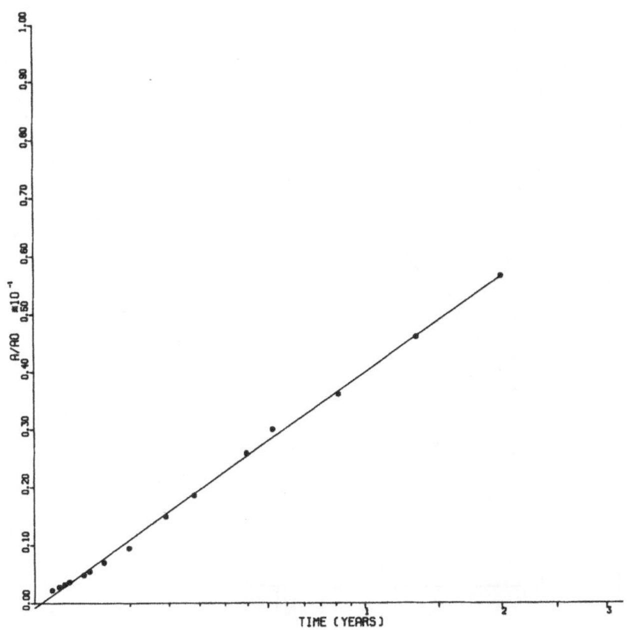

Fig. 2. The fraction of cesium released to salt brine from Portland cement without additives as a function of the square root of time.

Cesium Leaching in Distilled Water and Sodium Chloride Solution

Neither of these leachants has caused corrosion so far on any of the specimens. In a few cases there is a superficial bright discoloration. The leach rates are higher than in tap water or salt brine as reported earlier.[2] The leach data are on straight lines when plotted vs $\sqrt{t}$; in some cases there are deviations in either direction without apparent regularity.

Strontium Leaching

The leach results for strontium are limited to an observation period of about one year, since ^{85}Sr is used which decays with a halflife of 64.9 days. Within the one-year period, strontium leach curves are generally shaped as shown in Fig. 3, regardless of specimen composition or leachant. The leach rates in tap water and salt brine are lower than in distilled water and in sodium chloride solution as mentioned above for cesium.

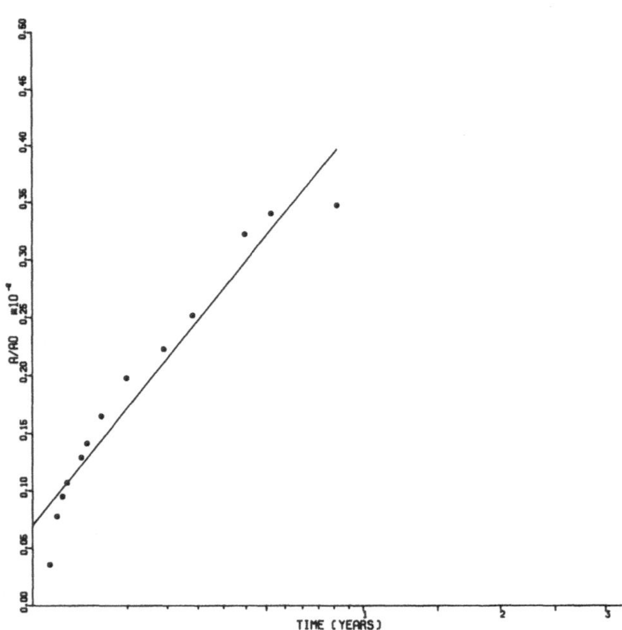

Fig. 3. The fraction of strontium released to salt brine as a function of the square root of time.

LEACH AND CORROSION TESTS AT ELEVATED TEMPERATURES AND PRESSURES

Cesium Leaching in Salt Brine

The investigation of cemented waste products under accidental conditions was started in 1979. In addition to cemented simulated intermediate level evaporator concentrates containing inactive $CsNO_3$, salt free cement specimens and grouts to be used for solid waste fixation are being investigated. The cement types used are slag cement and portland cement. The influence of porosity on leaching and corrosion shall be investigated by using well defined conditions of preparation.

For experimental reasons the leachant changing frequency recommended by IAEA[1] is not useful for autoclave experiments. To determine the time dependent release of cesium a varied changing frequency was applied.*

Systematic investigations were started using specimens prepared from portland cement (high sulfate resistance) and slag cement. The specimens were prepared to investigate the influence of porosity on cesium leaching and cement corrosion. The leach time for these experiments will be at least 6 months at 40°C and 0.1 and 10 MPa using only the highly corrosive salt brine.

Figure 4 shows the cesium leaching from cement specimens with a water/cement ratio (W/C) = 0.32 and containing 8 wt % salt. The specimens were compacted by vibration during preparation. Similar results were obtained by leaching specimens prepared without and with compaction and with degassing.

The results of these experiments indicate that under the described conditions and at short leaching times, cesium leaching is independent of pressure but dependent on the cement type. Specimens prepared from slag cement show lower cesium leachability than those prepared from portland cement. The continuation of these experiments will show whether there is an influence of corrosion on cesium leaching and whether corrosion will be influenced by different preparation conditions.

The independence of cesium leaching on preparation conditions was also shown by the leaching of specimens core-drilled from inactive 200-l monoliths (portland cement, W/C = 0.4, sand-to-cement = 0.5, 6 wt % salt). The results are shown in Fig. 5.

---

*Changed after 5, 15, 30, 50 and 90 days, then every 90 days.

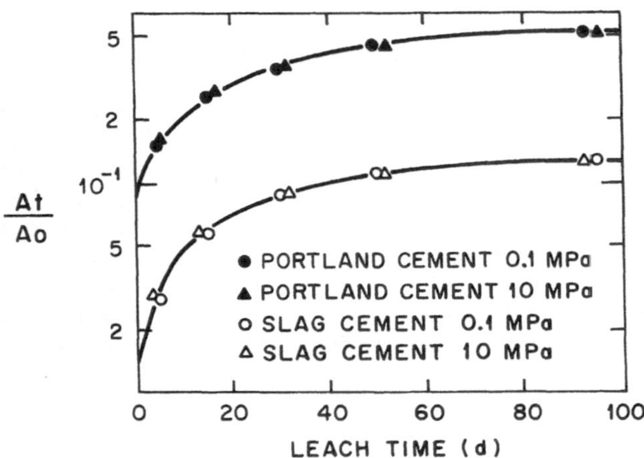

Fig. 4. Cesium leachability from cement specimens into salt brine at 40°C and 0.1 and 10 MPa (W/C = 0.32, 8 wt % salt, surface-to-volume ratio = 2.25 cm^{-1}).

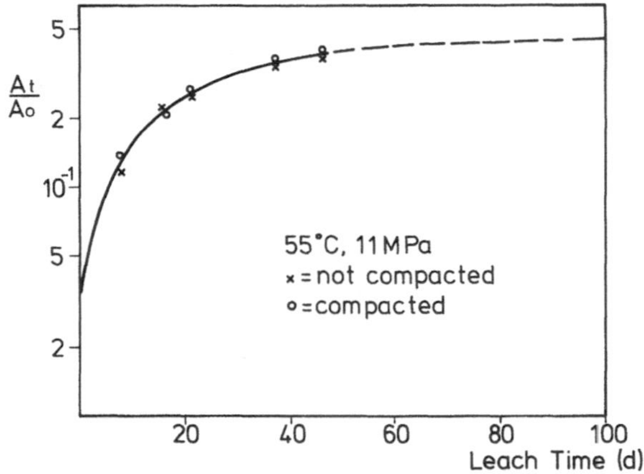

Fig. 5. Cesium leachability from Portland cement specimens into salt brine (core drilled from 200-L monoliths, W/C = 0.4, sand-to-cement ratio = 0.5, 6 wt % salt, surface-to-volume ratio = 1.2 cm^{-1}).

Corrosion Test at Elevated Temperatures and Pressure

Beside the leach experiments described, preliminary investigations were conducted to test the corrosion behavior of cemented evaporator concentrates and salt free grouts in the highly corrosive salt brine at 100°C and 0.1 and 10 MPa. Because of the preliminary character of these experiments, the obtained results will be discussed only briefly.

In these tests the products made from slag cement turned out to be more stable than those from portland cement. The stability of portland cement products improves with increasing pressure; in the case of slag cement products this is true to a lesser degree. For either type of cement, the products containing solidified evaporator concentrates are more stable than salt-free specimens.

Under elevated pressure the corrosion products form, unlike the tests at normal pressure, a dense layer firmly adhering to the concrete surface and obviously retarding leaching and corrosion. This phenomenon was investigated further.

$CaSO_4$ has been found in corrosion products by means of x-ray diffraction; Ca, Mg, S (from SO_4), and Cl (from chloride) have been detected using x-ray fluorescence. For complete identification of the reaction products further investigations are required and are under way.

Systematic investigations at higher temperatures and pressure will be started using a nondestructive method for measuring the dynamic modulus of elasticity of the cement specimens.

CONCLUSIONS

Cesium leach tests at ambient temperature and pressure show that tap water, distilled water, and sodium chloride solutions cause no or only very slight corrosion of cemented evaporator concentrates prepared from various types of cement with or without various additives. Salt brine is very corrosive to cement products; the corrosion rate depends on the composition and cement type. Cesium leaching follows a diffusion controlled mechanism in distilled water, sodium chloride solution, and salt brine but only to the point where disintegration occurs. In tap water cesium leaching does not follow a single diffusion mechanism. Leach tests at elevated temperatures and pressures show that cesium leaching in salt brine for relatively short leach times is independent of preparation conditions and pressure but depends on the cement type. The corrosion resistance of cemented evaporator concentrates is better than that of salt-free cements.

ACKNOWLEDGMENTS

We wish to thank Mrs. I. Boch, Mr. P. Jakobs, Mr. R. Gebauer, and Mr. R. Götz for their experimental contribution to this work.

REFERENCES

1. E. D. Hespe, "Leach Testing of Immobilized Radioactive Waste Solids, A Proposal for a Standard Method," Atomic Energy Review 9:195 (1971).

2. G. Rudolph, R. Köster, "Immobilization of Strontium and Cesium in Intermediate-Level Liquid Wastes by Solidification in Cement" in Scientific Basis for Nuclear Waste Management, Vol. 1, pp. 467-470, Plenum Press, New York (1979).

STANDARD LEACH TESTS FOR NUCLEAR WASTE MATERIALS*

D. M. Strachan, B. O. Barnes, and R. P. Turcotte

Pacific Northwest Laboratory
Battelle Blvd.
Richland, WA 99352

INTRODUCTION

In October 1979, the U.S. Department of Energy established the Nuclear Waste Materials Characterization Center at the Pacific Northwest Laboratory. The center has the charter to develop standard tests for the characterization of the components of the waste package, which include spent fuel, waste forms, overpacks, canisters, and barriers, and to publish a Nuclear Materials Handbook. Development of these tests will result in a more fundamental understanding of these materials as well as unifying test procedures. The tests will evaluate chemical and physical properties, including thermal stability, leaching, mechanical strength, radiation stability, corrosion and others. These tests are submitted by the MCC to the Materials Review Board for acceptance. Data generated by the MCC and others using the approved tests are also submitted to the Materials Review Board for acceptance. These data are then published by the MCC in the Materials Handbook.

Five related leach tests have been proposed by the Center to study time dependent leaching of waste forms. The first four tests include temperature as a variable and the use of three standard leachants. Three of the tests are static and two study leaching under dynamic conditions. All tests require multielement analyses and measurement of the pH of the leachate. Bulk chemical analysis before leaching is required and other solid-state analyses are strongly suggested. This suite of tests satisfies the needs for waste form comparison and for comprehensive information about the leaching behavior of waste forms under various conditions.

*Work supported by U.S. Department of Energy under Contract DE-AC06-76RLO 1830.

In all tests, normalized elemental mass losses must be reported. This unit is given by the equation

$$NL_i = \frac{M_i}{F_i \cdot SA} ,$$

where

NL_i = normalized elemental mass loss ($g \cdot m^{-2}$),

M_i = mass of element "i" in the leachate (g),

F_i = fraction of element "i" in the unleached waste form (unitless),

SA = specimen surface area (m^2).

In this paper, we discuss the waste-form leach tests in some detail and report representative data that have been collected. Systems tests which test the waste package assembly are in the early stages of design and are not discussed in this paper.

STATIC TESTS

MCC-1 Static Low Temperature Test

Test MCC-1 has, in essence, been used by a number of investigators for some time (Bradley, 1980; Chick, 1980; Rusin, 1979) and as reported by Turcotte and Strachan (1980). In this test a monolithic sample with an approximate surface area of 400 mm^2 is totally immersed in leachant of 4×10^4 mm^3 ($SA/V = 1 \times 10^{-2}$ mm^{-1}). The leach container is made from PFA Teflon. Distilled or deionized water, 80% saturated WIPP Brine A and an abbreviated Tuff groundwater are the standard leachants but the use of other groundwaters is acceptable.

A minimum test matrix is required for all studies to assist in data comparison, for quick comparison of waste forms and to assist the Center and the Materials Review Board in data verification. For MCC-1, this matrix requires 28 days to complete, the use of the three standard leachants, and a test temperature of 90°C. Leachate analyses are required. The results for 28 d in water are shown in Fig. 1 for PNL 76-68 glass. Leach results at 28 d are about the same in the silicate leachant as the leach results in water. In brine, Sr and Ce leach at substantially higher rates (∿100 times) than in the other two leachants.

A round robin is currently under way to verify the practicality and reproducibility of this procedure.

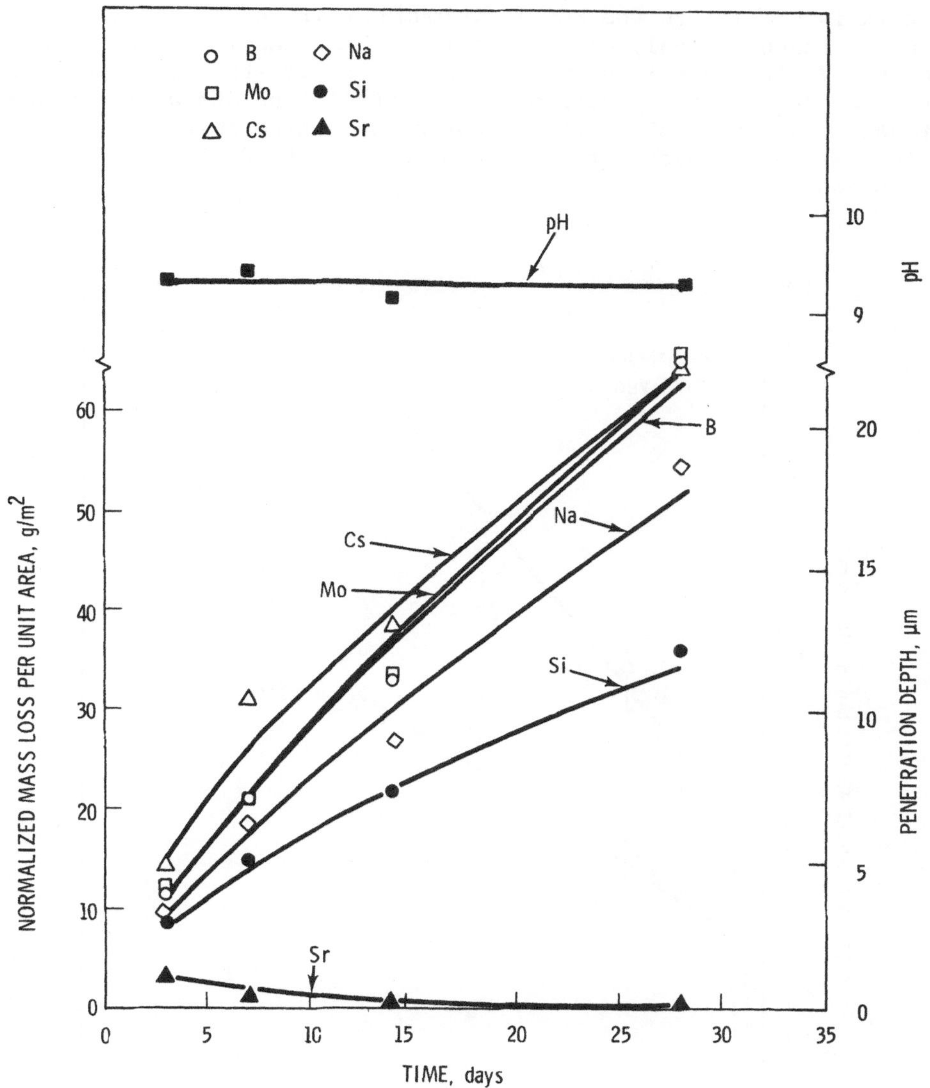

Fig. 1. Normalized elemental mass loss vs time for PNL-76-68 glass leached in deionized water at 90°C using the MCC-1 test.

MCC-2 High Temperature Static Test

This test extends the MCC-1 test temperature to above 100°C. Testing can be done in autoclaves or in inexpensive Teflon-lined acid digestion vessels. A monolithic sample is used with the surface area such that the ratio of SA/V is 1×10^{-2} mm^{-1}. A minimum test matrix is required for the same reasons outlined under MCC-1. This matrix is similar to the MCC-1 minimum matrix and is performed at 150°C. Leachate analyses are required. An example of the results for PNL-76-68 glass is shown in Fig. 2. Solid state analyses such as x-ray diffraction, depth profiling, ESCA, and SEM analyses are strongly recommended. Although not shown, these initial data indicate a marked change in the leaching of PNL-76-68 glass between 150°C and 250°C.

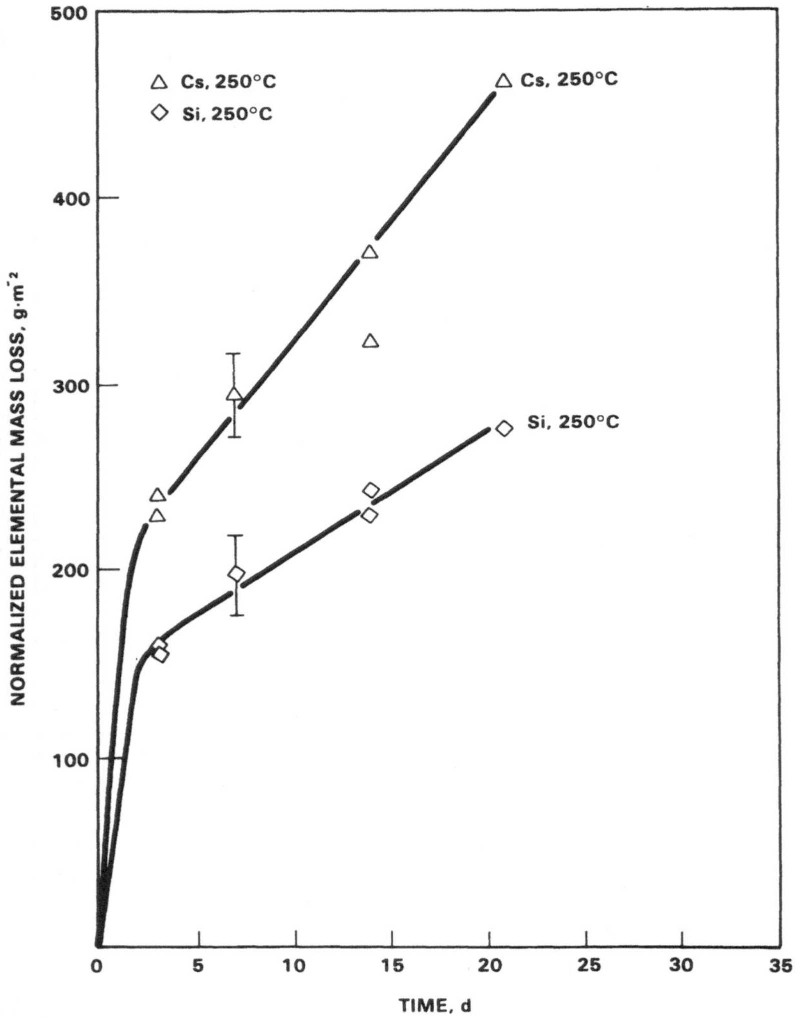

Fig. 2. Normalized elemental mass loss vs time for PNL-76-68 glass leached in deionized water at 150°C using the proposed MCC-2 test.

MCC-3 Solubility Test

This test is performed on powdered samples to determine the maximum solubility of the waste form in the leachant of interest. In this test the waste-form sample is ground to pass through a 325 mesh screen. The sample is then allowed to react with a volume of leachant equal to ten times the sample mass until no change in the leachate composition is observed. As indicated in Fig. 3 even at 250°C and 14 days with powdered PNL-76-68 glass the leachate composition has

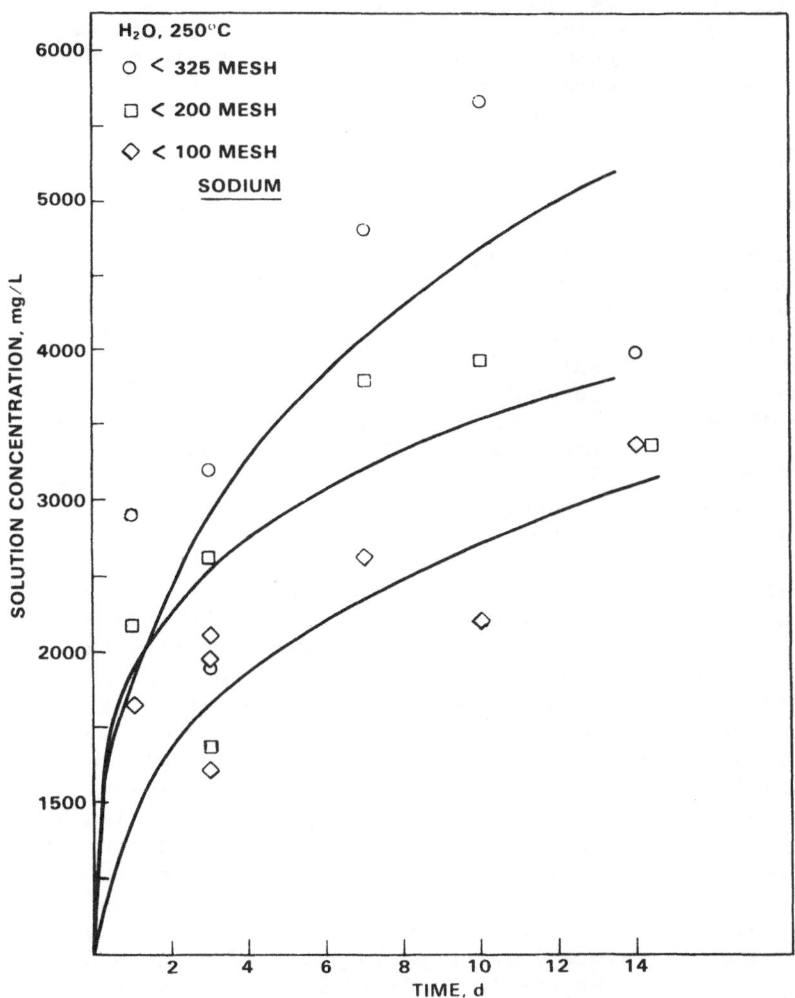

Fig. 3. Leachate solution concentration vs time for powdered PNL-76-68 glass at 250°C using the proposed MCC-3 test.

not reached steady state. Stirring of the sample may be very important. The data summarized in Fig. 3 were obtained under basically static conditions but the sample was stirred four times daily. The ovens are being modified so that the vessels can be continuously stirred. Three mesh sizes were tested to see at what mesh size no change was noted.

Some materials will require lengthy times to equilibrate with the leachate and laboratory studies are continuing. However, the minimum test matrix at 150°C may require testing of samples at 3, 7, and 14 days in water and triplicate data obtained in the standard leachants at 14 days. Constant stirring of the sample may also be required. Data obtained using the Teflon-lined, acid digestion vessels will be corroborated using an autoclave from which leachate samples at temperature and pressure can be extracted.

FLOW TESTS

MCC-4 Low Flow Test

In principle, this test will most closely simulate repository conditions. It is similar to that developed by Coles et al., 1978, but uses the same leach container as MCC-1. In this test the leachant is pumped past a monolithic sample. Data are collected from the leachate after a single pass through the leach container. The minimum test matrix at 90°C extends to 28 days with flow rates of 0.1, 0.01 and 0.001 $mL \cdot min^{-1}$. Using the minimum test matrix, analyzable quantities of material were found in the unconcentrated leachate from 76-68 glass. However, at flow rates >0.1 $mL \cdot min^{-1}$ or with more durable waste forms, techniques for concentrating the leachate or samples with radioactive tracers may be needed.

Data are plotted as normalized cumulative amount of element "i" leached per unit area versus time. Since leachant is single pass and continuous and the leachate is sampled periodically, the cumulative amounts of each element must be calculated from each periodic sample. This plot is then differentiated either mathematically or graphically to obtain the slope of the curve at any time. The slope is the normalized instantaneous leach rate which can be plotted versus time as shown in Fig. 4. The conditions for the experiment shown in Fig. 4 were 90°C and 0.10 $mL \cdot min^{-1}$ using water as a leachant. Under these conditions a combination of ion exchange and corrosion mechanisms exist at the beginning of the leaching experiment. At about 10 days the dominant mechanism appears to be corrosion since such dissimilar elements as Cs, Si and Sr leach at nearly equivalent rates. The 28 d average leach rates obtained using MCC-1 are also shown in Fig. 4. The static conditions of MCC-1 yield leach rates which are lower by a factor of 2 for Si and 200 for Sr.

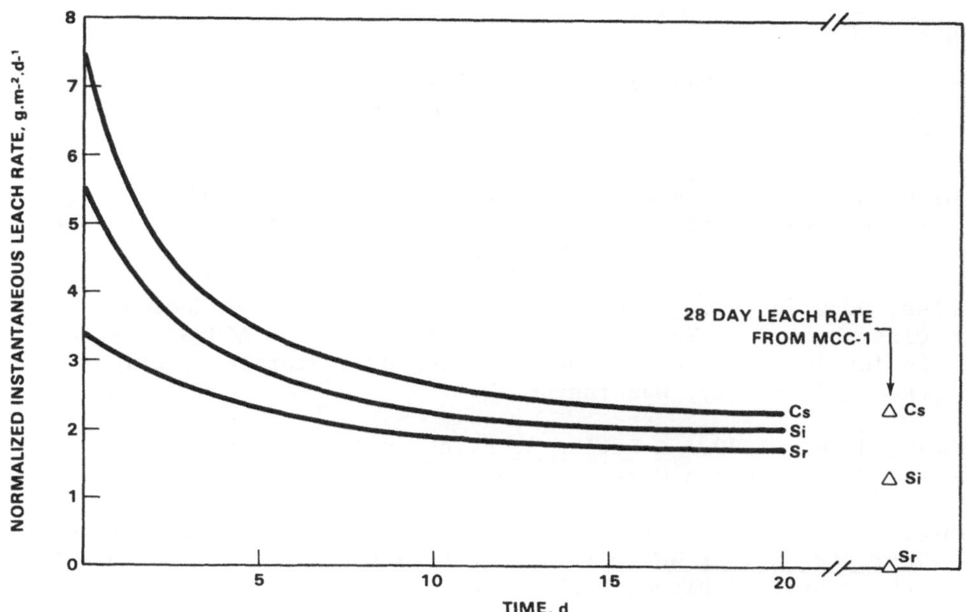

Fig. 4. Normalized instantaneous leach rate vs time for PNL-76-68 glass at 90°C and flow rate of 0.10 ± 0.01 mL min^{-1} using the proposed MCC-4 test

MCC-5 Soxhlet Test

This test was adopted because it provides a convenient method of assessing waste-form leach rates under conditions of infinite dilution. It has been in use for a long time by the waste form developers and is currently in popular use in Europe. Three features of the test as used in the past have been changed. A monolithic sample replaces the powdered sample. Solution analyses and an all-Teflon apparatus are required.

In the past the soxhlet apparatus was constructed of Pyrex. However, solution analyses are now required and Pyrex contributes to and interferes with the leachate concentrations. For instance, when a synthetic leachate is allowed to reflux for 14 days in the Pyrex soxhlet apparatus, elements such as Cs, Ca, Sr, and Mo decreased in concentration by up to a factor of 100. Elements such as B, Si, and Na were found to increase in concentration by up to a factor of 10. These problems should be alleviated by using an all-Teflon apparatus and by adding acid to the leachate, which then remains in the apparatus for 12 hours before sampling. Results from experiments using MCC-5 are not yet available.

REFERENCES

Bradley, T. D., McVay, G. L. and Coles, D. G. 1980. Leach Test Methodology for the Waste/Rock Interactions Technology Program, PNL-3326, Pacific Northwest Laboratory, Richland, Washington.

Chick, L. A. and Buckwalter, C. Q. 1980. Low Leach Rate Glasses Immobilization of Nuclear Wastes, PNL-3522, Pacific Northwest Laboratory, Richland, Washington.

Coles, D. G. et al. 1978. "Single Pass Leaching of Nuclear Melt Glass by Ground Water," paper presented at the ACS Symposium Series 100, Radioactive Waste in Geologic Storage, American Chemical Society, Washington, D.C.

Rusin, J. M. 1979. A Review of High-Level Waste-Form Properties, PNL-3035, Pacific Northwest Laboratory, Richland, Washington.

Turcotte, R. P. and Strachan, D. M. 1980. MCC-1 Standard Leach Test for Nuclear Waste Forms. PNL-SA-8737, Pacific Northwest Laboratory, Richland, Washington.

TIME AND TEMPERATURE DEPENDENCE OF THE LEACHING OF A SIMULATED HIGH-LEVEL WASTE GLASS

J. H. Westsik, Jr., and R. D. Peters

Pacific Northwest Laboratory (PNL)*

Richland, Washington 99352

INTRODUCTION

Exposure of a radioactive waste form to water and subsequent leaching is seen as the primary mode for potential release of radionuclides within a geologic repository. Should leaching occur during the first 1000 years of repository storage, the heat generated by radioactive decay could accelerate water/waste-form reactions. It is therefore important that we understand how temperature affects the leaching process.

High-temperature studies reported in the literature have been conducted in order to a) accelerate water/waste-form reactions to characterize geologic-time behavior at lower temperatures, and b) help provide an understanding of actual waste-form behavior in a geologic repository. Included in these studies are some time- and temperature-dependent data on the durability of simulated high-level waste glasses(1-7). The objective of this work is to expand the data base on the leaching of PNL glass 76-68 to include information obtained over longer time periods and at temperatures from 25° to 350°C.

EXPERIMENTAL APPROACH

The data discussed here were obtained through several procedures, but all procedures share common features. In each test, simulated high-level waste glass 76-68 was leached in deionized

*Operated by Battelle Memorial Institute for the Department of Energy under Contract DE-AC06-76RLO-1830.

water under static conditions. The nominal composition of the glass is described by McElroy(8). The ratio of glass surface area to leachate volume was maintained at nominally 0.1 cm^{-1}. At the end of each leaching period, the elemental composition of the leachates were measured using induction-coupled plasma spectroscopy. For tests conducted at 75°C and above, leachate samples and pH measurements were taken after the leachate had cooled to 60° to 70°C. Leachates from tests done at below 75°C were sampled at the test temperature. None of the solutions were filtered before the analyses, so the elemental compositions of the solutions may include suspended colloidal and precipitated materials.

Tests done to determine the temperature-dependent nature of glass-leaching were performed using Inconel- and Hastelloy-lined autoclave vessels. Glass cylinders 1.8 cm in diameter were leached for 3 days at 25°, 50°, 100°, 150°, 200°, 250°, 300°, and 350°C. Additional tests were conducted for 1, 12, 27, 48, and 64 days to obtain time-dependent data at 100° and 250°C. An inert gas was introduced into the autoclave system to maintain the operating pressures at or above 13.8 MPa.

Additional time-dependent data were obtained at 25°, 50°, 75°, and 90°C. Glass beads 0.8 cm in diameter were leached for times up to 341 days(9,10). The leachant was contained in sealed polypropylene bottles at atmospheric pressure.

RESULTS & DISCUSSION

Figure 1 shows the normalized releases (g glass/m^2) based on boron, molybdenum, sodium and silicon as a function of time for temperatures from 25° to 250°C. Normalized releases based on these four elements are approximately congruent, except in the early stages at 25° and 50°C. For tests lasting less than 10 days at 50°C and less than 32 days at 25°C, releases based on sodium are higher than for the other three elements. For these same times and temperatures the leachate pH<8, whereas at the higher temperatures and for longer times the pH values lie in the 9.3 to 9.8 range.

Releases based on other components of the waste glass are less than those based on B, Mo, Na, or Si. The differences in releases can span several orders of magnitude, depending upon the time, temperature and element in question. Westsik and Harvey show the time dependencies of normalized releases based on a number of elements leached from glass at 150° and 250°C(4). Their data for B, Mo, Na, and Si are included in Figure 1.

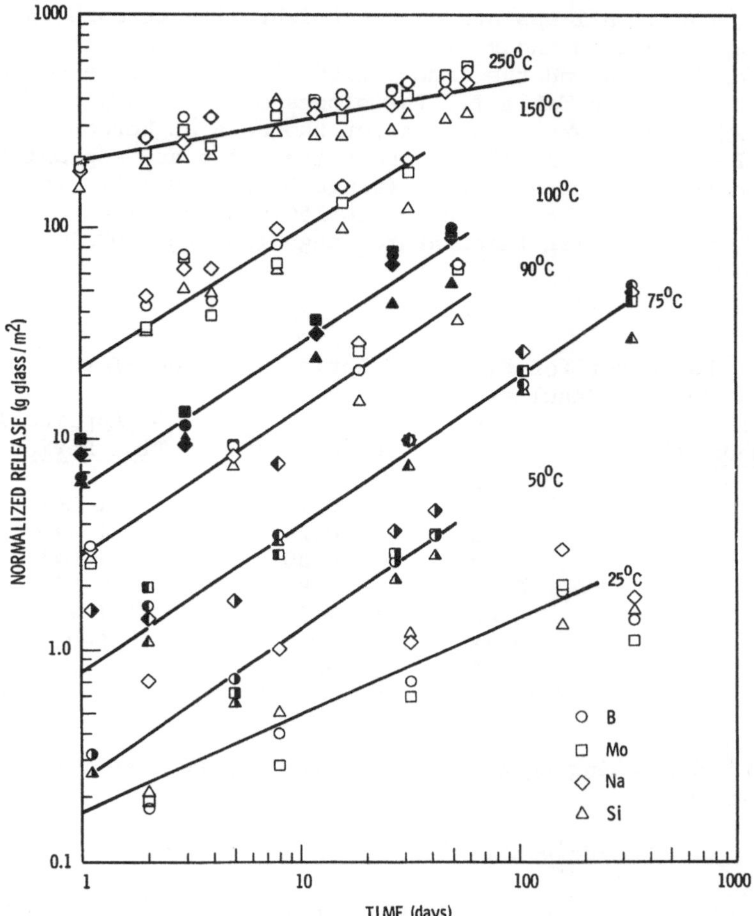

Fig. 1. Time-Temperature Dependence of Leaching of 76-68 Glass

An empirical equation can be fitted to the release data in Figure 1:

$$R(\text{g glass/m}) = kt^n \quad (1)$$

where R is the normalized release and k and n are temperature-dependent parameters(11). When shown on a log-log plot, as in

Figure 1, Equation 1 appears as a straight line through the data and has a slope of n and an intercept of k. Graphically determined values for k and n and the time periods for which they are applicable are listed in Table 1. The slopes of the lines through the 25° and 250°C data are less than for the data at between 50° and 150°C. Between 50° and 150°C, the values of n are approximately equal and their average is 0.67 (range 0.58 to 0.75). The lines in Figure 1 drawn through the 50° to 150°C data all have a slope of 0.67. Over the entire temperature range (25° to 250°C), k increases with temperature.

TABLE 1. Parameter for Fitting Equation 1 to Long-Term Normalized Release Results.

Temperature (°C)	n	k (g/m^2-d^n)	Applicable Time Period (day)
25	0.43	0.17	1-341
50	0.68	0.24	1-42
75	0.68	0.80	1-341
90	0.75	2.8	1-54
100	0.68	7.0	1-48
150	0.58	30	2-32
250	0.20	200	1-64

Differentiating Equation 1 with respect to time yields the rate equation

$$dR/dt = nkt^{n-1}$$

Since n is approximately constant from 50° to 150°C, k becomes a temperature-dependent rate constant. Figure 2 shows the relationship between k and temperature. The dependence of k upon the temperature can be expressed by the Arrhenius relationship

$$\ln k = E/RT + \ln A$$

where E is an apparent activation energy, R is the gas constant, T is the absolute temperature and A is a constant. For the 50° to 150°C temperature range, the apparent activation energy is 5.3×10^4 J/mol-°K (13 kcal/mol-°K).

Figure 3 shows normalized releases after 3 days based on B, Mo, Na, and Si. The depths of glass alteration caused by hydrothermal reactions are also shown. This reaction layer was visible on cross sections of the glass cylinders leached at temperatures above 150°. A reaction layer may also have formed on samples

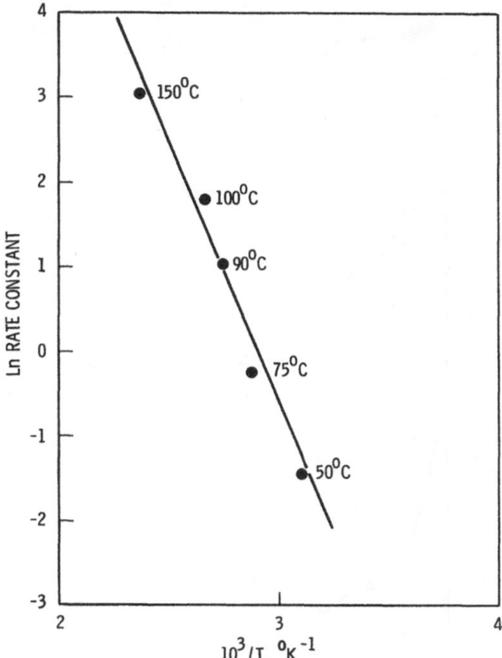

Fig. 2. Temperature Dependence of Apparent Rate Constant k.

leached at below 150°C, but it was not visible with the optical microscope.

It is interesting to see the correspondence between the normalized release data and the reacted-layer thicknesses. The slope of the normalized release curve from 150° to 250°C matches that of the layer-thickness curve from 150° to 300°C. Above 300°C the slope of the normalized release line is steeper than in the 150°-250°C region, although it is not as steep as the slope of the line for the reacted-layer thicknesses at above 300°C. Similar temperature-dependent behavior was observed with hydrothermal reaction layers of three other PNL waste glasses. Note in Figure 3 that the silicon releases at 300° and 350°C do not increase as the Na, Mo, and B releases do. Below 150°C the curve through the normalized data again becomes steeper as the temperature decreases.

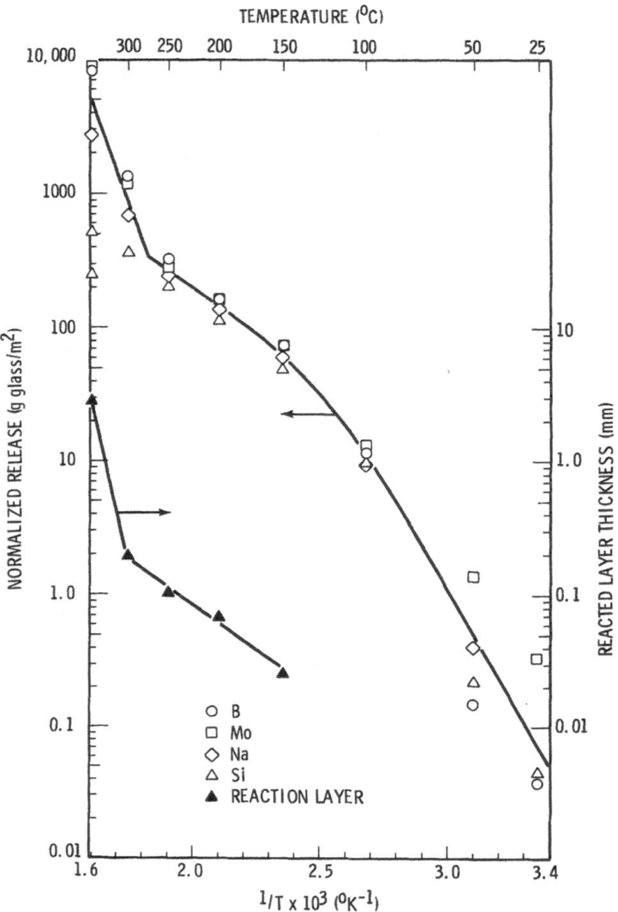

Fig. 3. Temperature Dependence of Normalized Releases Based on B, Mo, Na, and Si and of Reacted Layer Thickness After 3 Days of Leaching.

These results imply that the time and temperature dependencies of leaching cannot be characterized by any one process. Below 250°C the leaching process is described by three phases that proceed one after the other. The first phase is exemplified by our 25° and 50°C data at short times. The pH of the leachates in tests at these conditions change with time, and congruent leaching is not observed. In the second phase; B, Mo, Na, and Si appear to be released congruently and the normalized releases are described by Equation 1 with the time exponent n=0.67 (range 0.58 to 0.75). If all elements in the waste glass leached congruently, the time exponent would be expected to be 1.0. In this second phase, the pH of the leachate solution is in the range of 9.3 to 9.8 and is constant with time. Results obtained at 250°C illustrate the third phase in the leaching process. Releases are less dependent on time at this temperature--i.e., the exponent in Equation 1 is small (n=0.2 for 250°C).

Yet to be determined is whether, at longer times than were used in our experiments, lower-temperature (50° to 150°C) release curves will flatten and reach slopes similar to the results for the 250°C tests. Extended leaching times at temperatures in the 50° to 150°C range are needed to resolve this question.

Above 250°C the results imply that there is a change in the glass/water reactions. This is manifest in the larger reaction layers and in larger normalized releases than would be expected by extrapolating lower-temperature results.

CONCLUSIONS

In this study we have investigated how time and temperature affect the leaching of 76-68 waste glass. Two temperature-dependent regions were identified in the results. Above about 250°C, hydrothermal alteration and releases of B, Na, and Mo occur faster than would be expected from the results of tests at lower temperatures. At 250°C and below, leaching appears to proceed through three steps. Characteristically, in the first step, the pH of the leachate changes with time; in the second, the pH is constant and normalized releases based on B, Mo, Na, and Si are described by the empirical equation

$$\text{Release (g glass/m}^2) = kt^{0.67(\text{range } 0.58-0.75)}$$

where k follows an Arrhenius temperature dependence (apparent activation energy = 5.3×10^4 J/mol-°K); and in the third step, so far observed only at 250°C, normalized releases show a weaker time dependence than they do in the second step.

REFERENCES

1. K. F. Flynn, L. J. Jardine, and M. J. Steindler, Resistance of High-Level Waste and Canister Materials to Dissolution in Aqueous Media, in: "Scientific Basis for Nuclear Waste Management," vol. 2, Plenum Press, New York, New York, (1980).
2. S. Komarneni, W. P. Freeborn, B. E. Scheetz, W. B. White, and G. J. McCarthy, "Reaction and Devitrification of a Prototype Nuclear Waste Storage Glass with Hot Magnesium-Rich Brine," ONWI-72, The Pennsylvania State University, University Park, PA 16082, (1979).

3. G. J. McCarthy, B. E. Scheetz, S. Komarneni, D. K. Smith, and W. B. White, Hydrothermal Stability of Simulated Radioactive Waste Glass, in: "Solid State Chemistry: A Contemporary Overview," American Chemical Society, Washington, D.C.
4. J. H. Westsik, Jr. and C. O. Harvey, "High-Temperature Leaching of an Actinide-Bearing, Simulated High-Level Waste Glass," PNL-3172, Pacific Northwest Laboratory, Richland, WA 99352, (1980).
5. J. H. Westsik, Jr., J. W. Shade, and G. L. McVay, Temperature Dependence of Hydrothermal Reactions with Waste Glasses and Ceramics, in: "Scientific Basis for Nuclear Waste Management," vol. 2, Plenum Press, New York, New York, (1980).
6. J. H. Westsik, Jr. and R. P. Turcotte, "Hydrothermal Reactions of Nuclear Waste Solids-A Preliminary Study," PNL-2759, Pacific Northwest Laboratory, Richland, WA 99352, (1978).
7. E. Lue Yen-Bower, D. E. Clark, and L. L. Hench, An Approach Towards Long-Term Prediction of Stability of Nuclear Waste Forms, in: "Ceramics in Nuclear Waste Management," CONF-790420, Technical Information Center, Springfield, VA. 22161, (1979).
8. J. L. McElroy, "Quarterly Progress Report, Research and Development Activities, Waste Fixation Program, October through December 1976," PNL-2264, Pacific Northwest Laboratory, Richland, WA 99352, (1977).
9. D. J. Bradley, C. O. Harvey, and R. P. Turcotte, "Leaching of Actinides and Technetium from Simulated High-Level Waste Glass," PNL-3152, Pacific Northwest Laboratory, Richland, WA 99352, (1979).
10. D. J. Bradley, G. L. McVay, and D. G. Coles, "Leach Test Methodology for the Waste/Rock Interactions Technology Program," PNL-3326, Pacific Northwest Laboratory, Richland, WA 99352, (1980).
11. E. Ewest, Calculations of Radioactivity Release Due to Leaching of Vitrified High-Level Waste, in: "Scientific Basis for Nuclear Waste Management, vol. 1, Plenum Press, New York, New York, (1979).

THE MECHANISMS FOR HYDROTHERMAL LEACHING OF GLASS AND GLASS-CERAMIC NUCLEAR WASTE FORMS

F. K. Altenhein, W. Lutze, and G. Malow

Hahn-Meitner-Institut für Kernforschung Berlin
Glienicker Strasse 100
D 1000 Berlin 39

INTRODUCTION

The leaching mechanisms for nuclear waste forms in contact with aqueous solutions depends on parameters such as glass composition, temperature, pressure, chemical composition, flow rate, concentration, and volume of the liquid phase. Most parameters have only recently been assessed within the frame of safety studies for selected geologic HLW repositories. Some parameters typical of worst case accidental conditions in a rock salt repository have been assumed for the purpose of this study. These are a maximum temperature of 200°C, a maximum hydrostatic pressure of 130 bars, and varying volumes of water having access to the waste form.

EXPERIMENTAL

The investigated waste forms were a borosilicate glass with 20 wt % LWR-type (30,000 MWd/t_{HM}) simulated waste oxides, and a glass-ceramic of the same chemical composition.[1] The samples were in the form of beads (0.5 cm diam, weight about 0.2 to 0.3 g) having the size as produced by the PAMELA process.[2] The solutions were deionized water, rock salt, and carnallite brine.[3] The solutions were saturated at 28°C. The carnallite brine was received from the ASSE II salt mine in West Germany. The leaching experiments were performed in autoclaves at 200°C, some at lower temperatures. Most of the experiments were run for 3 days; some were extended up to 1 year. The pressure was varied from 15 bars (equilibrium pressure) to 130 bars. The volume of the liquid was varied between 3 and 300 cm^3/cm^2 of exposed sample surface. All experiments were performed in stagnate solutions. The samples were washed and dried after the experiment and their weight loss was determined. The leached surfaces were analyzed by light microscopy, scanning electron microscopy (SEM), and by electron microprobe analysis (MPA) without pretreatment of the samples. The solutions were quantitatively analyzed for all constituents having concentrations $>0.05 \times 10^{-6}$ g/cm^3 by means of multielement emission spectroscopy (ICP).

RESULTS AND DISCUSSION

The net specific weight loss (g/cm^2) was used to verify leaching effects as a function of the following parameters:

Temperature. Temperature has a major influence on leaching. The increase of the specific weight loss with temperature for four different waste glasses and a glass-ceramic in deionized water was found to be in the range of one to two orders of magnitude when increasing the temperature from 50 to 100°C.[1] At 100°C the weight loss data for 3 days were in the order of 10^{-3} g/cm^2. An extrapolation to 200°C would yield, however, leach data far beyond the actually measured data which are still in the order of 10^{-3} g/cm^2 as shown in Fig. 1a. According to the extrapolated specific weight loss, the samples should have dissolved completely. They showed, however, only a thin reacted area and about 1% decrease in diameter. This finding could be explained by formation of a protective layer on the sample surface and will be discussed later in this paper.

Pressure. An influence of pressure on the leaching results was not detected in the range of 15 to 130 bars at 200°C. Therefore, the pressure dependence was not considered here and the experiments were performed at equilibrium pressure.

Time. The reaction rate decreased during the first 10 days at 200°C to almost zero. Thereafter, only a very small increase in the fractional release was observed. Three glass and ceramic samples were each tested for 1 year in carnallite brine. Their specific weight loss was found to be the same as after 30 and 90 days and was 5×10^{-3} to 6×10^{-3} g/cm^2.

Composition of solution. The different solutions had a strong influence on the leaching process. The formation of a 20- to 50-μm thick gel layer was observed on the glass sample in water and in salt brine (Fig. 2a,b), whereas there was no gel layer in the case of glass in carnallite. The surface of the glass-ceramic is covered by a reaction layer which contains unattacked crystals. The reaction layers of the glass-ceramic (Fig. 2c) look alike in all liquids. The behavior of glass and glass-ceramic in deionized water was studied in more detail in order to proceed from the "simpler" to the more complex system.

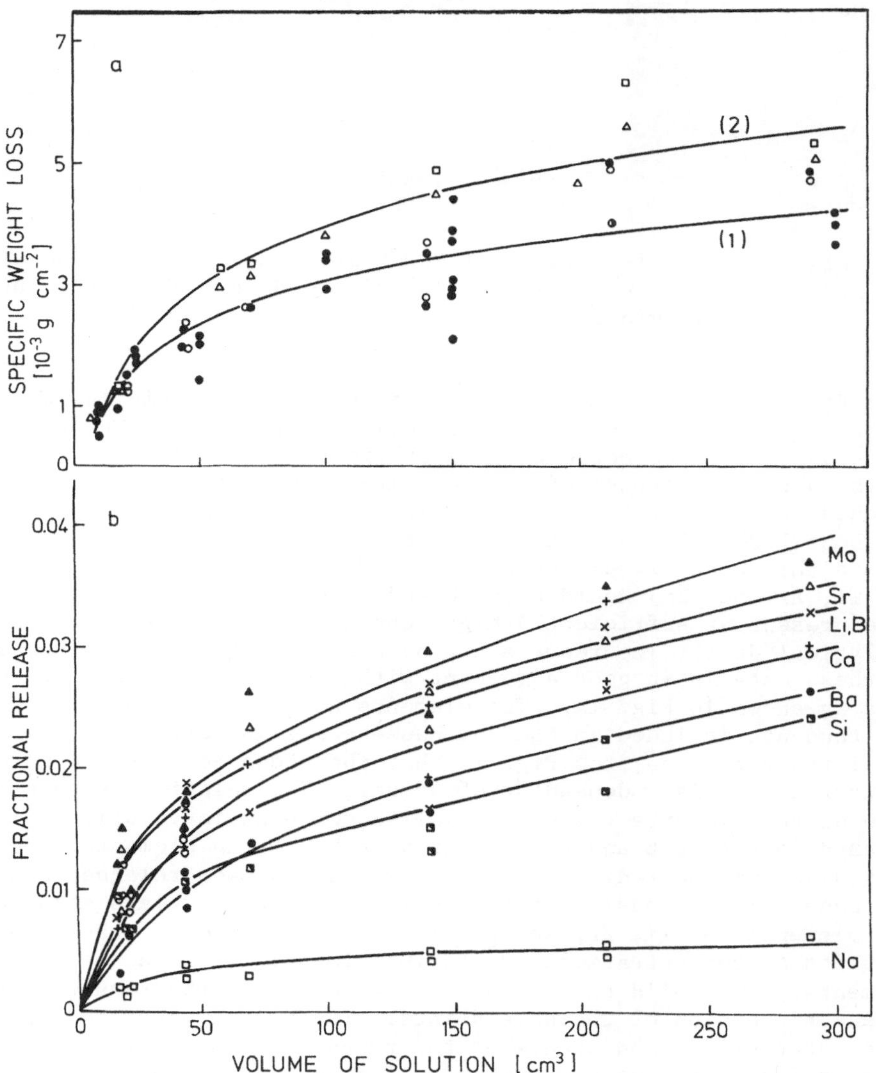

Fig. 1. (a) Specific weight loss curves for glass (1) and glass-ceramic (2) vs volume of liquid after 3 days at 200°C in deionized water. (1) ●, (2) Δ = direct weight measurements; (1) o, (2) ◘ = integrated from ICP measurement. (b) Fractional release curves for detectable elements in solution (ICP)-glass/water only.

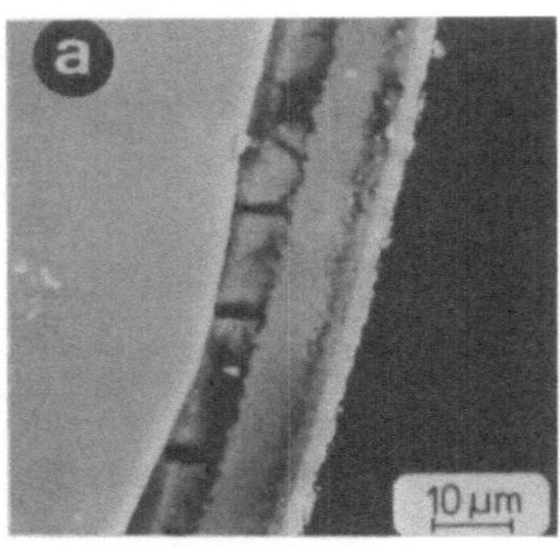

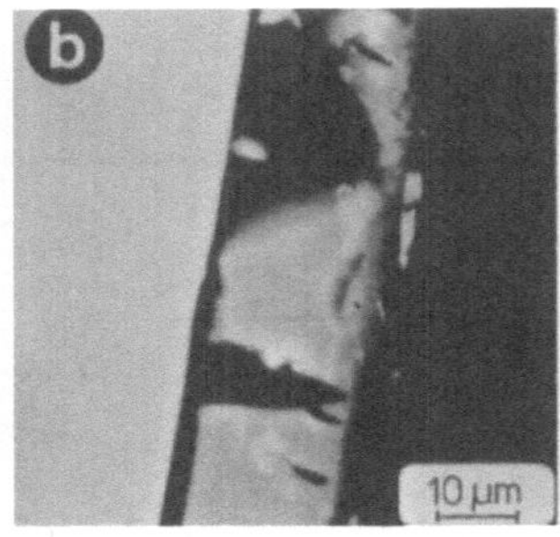

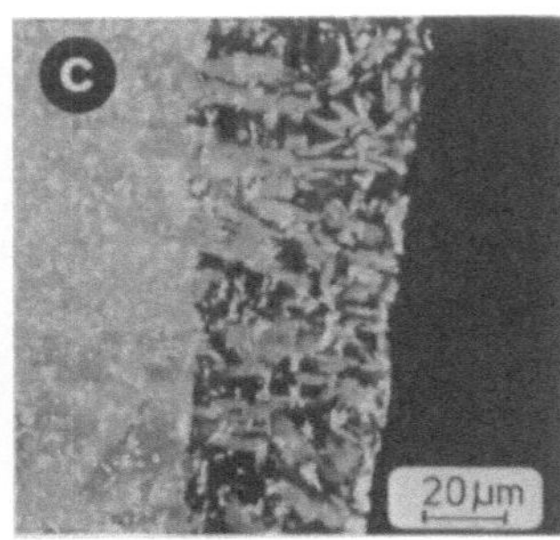

Fig. 2. SEM-micrographs of 200°C leached surfaces of beads. (a) glass 3d in deionized water; (b) glass 3d in rock salt solution; and (c) glass-ceramic la in carnallite brine.

Volume. The volume of the liquid has considerable influence on the fraction released from the solid after 3 days. The results for glass and glass-ceramic in water are shown in Fig. 1. Figure la shows that the increase of liquid volume causes an increase of the leached fraction of the solid. This effect disappears at S/V ratios of about 1/300 cm. An analysis of the solution for 20 elements showed that about 14 elements (Si, B, Na, Li, Ca, Sr, Ba, Nd, Al, Fe, Ti, Zr, Mo, Zn) could be detected. However, only 8 elements were present at sufficiently high concentration to yield reliable analytical data. The results can be seen in Fig. lb. The summation of these data is in good agreement with the measured specific weight loss as shown in Fig. la. The elements not detected in solution and thus not included in the sum cause a maximum error of 10%. It can further be seen from Fig. lb that the sequence of the element concentrations is independent of volume. The relative release of some of the elements reflects a well-known behavior. Silicon is leached less than B and Li and the alkaline earth elements. Only Na behaves unexpectedly showing a remarkably lower release than Si. Molybdenum is obviously least tightly bound in the glass structure. The elements Fe, Ti, Zr, Nd, Al, and Zn were found as traces in solution at concentrations 10^2 to 10^4 times lower than the eight elements. The solid curves in Fig. la represent the result of a linear regression of the data points. Tests were performed to check statistical confidence of the trend functions and a strong correlation of the data was found. The results of these tests are positive. Thus, a dependency of the specific weight loss on volume does exist and is best fitted by a semilogarithmic function. It is obvious from the above findings that simple dissolution of the glass can be excluded as a mechanism. Instead, a highly selective leaching process is indicated.

An investigation of the leached layer by means of SEM and MPA was performed and some results are shown in Figs. 3 and 4.

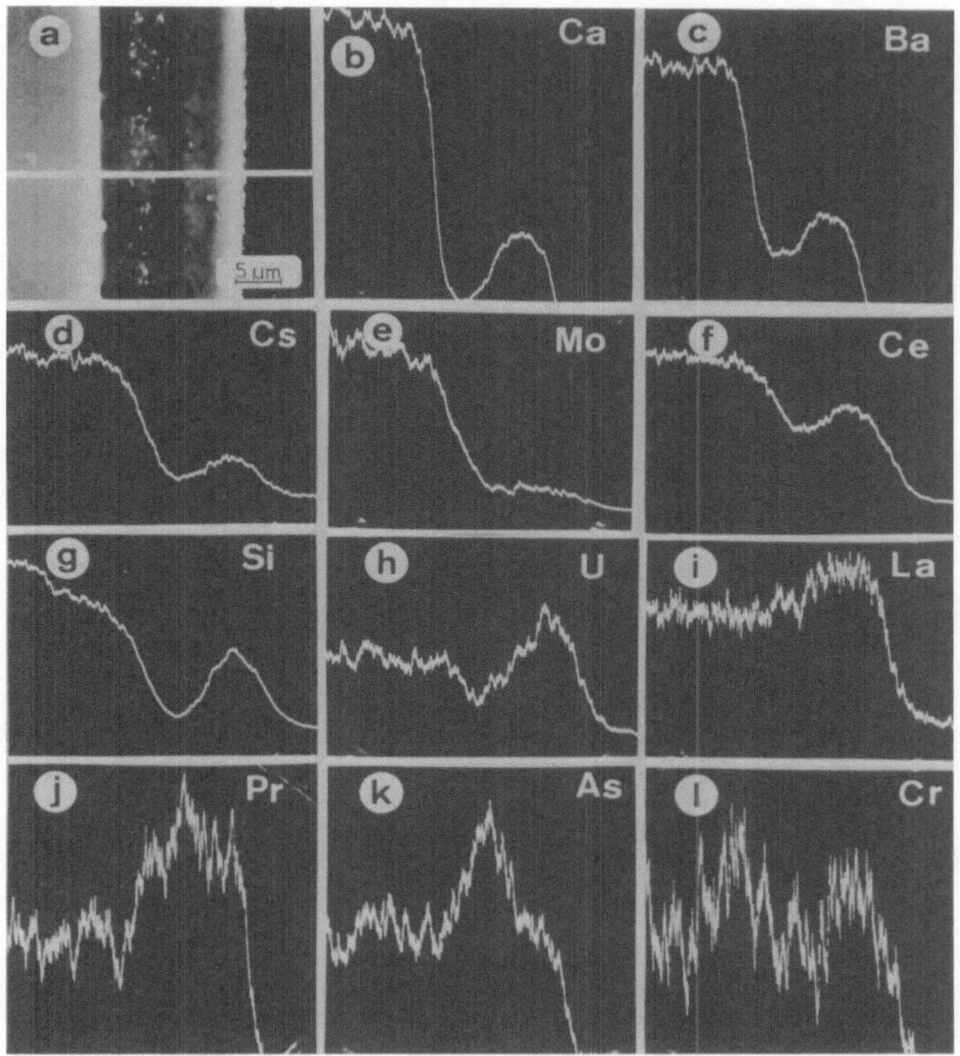

Fig. 3. SEM-micrograph and x-ray line scans of a glass bead leached 3 days at 200°C in deionized water. (a) SEM-micrograph with line scan position, (b) Ca-, (c) Ba-, (d) Cs-, (e) Mo-, (f) Ce, (g) Si-, (h) U-, (i) La-, (j) Pr-, (k) As-, and (l) Cr-line scans.

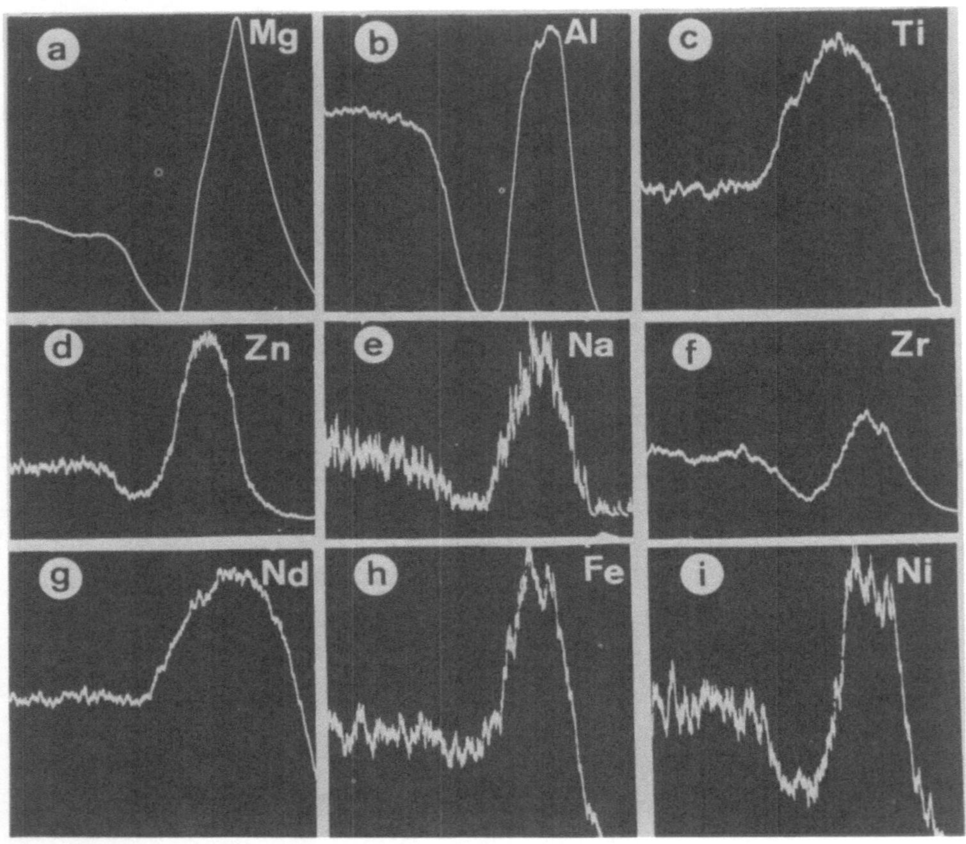

Fig. 4. X-ray line scans as in Fig. 3a for (a) Mg, (b) Al, (c) Ti, (d) Zn, (e) Na, (f) Zr, (g) Nd, (h) Fe, and (i) Ni.

Figure 3a shows a SEM-micrograph of a glass bead leached 3 days at 200°C in H_2O. The leached sample is covered by a gel layer. The relative concentrations in the glass and in this layer are shown in Fig. 3b to 4i in the form of x-ray line scans, each obtained at the white line in Fig. 3a. As seen, the elements missing in solution are retained in the surface layer (U, La, Mg, Al, Ti, Zn, Zr, Nd, Fe, Ni) including Na, whereas the elements found in solution are depleted in the layer (Ca, Ba, Cs, Mo, Si). Boron and Li were not detectable by MPA and Cs not by ICP. The leached zone consists of at least two layers. The innermost layer (Fig. 2a) looks like a typical gel layer with drying cracks, whereas the outer one looks dense and not cracked. The chemical composition of both layers is different. Molybdenum is not contained in the layers (Fig. 3e). The glass network former Si and the typical network modifiers Na,

Cs are strongly depleted in the inner layer relative to the glass. The rare earth elements and U, Ti, Cr, and As prevail in this layer. The outer layer is characterized by strong enerichment of several elements relative to the glass. The enriched elements are U, Mg, Al, Ti, Zn, Zr, Fe, Ni, the rare earths, and Na. Hence, this layer consists mostly of elements known as oxide-hydrate formers. Their respective compounds are known to be insoluble in water. The enrichment of Na is not typical of any leaching mechanism as yet.

It was found that the total thickness and the relative thicknesses of the layers depend on time. After 5 hours only one layer could be detected, the thickness being ~ 2 µm. Its chemical composition corresponds at this early state already to that of the outer layer, except for Na, Mg, and Al. These elements are still depleted rather than enriched. After 30 days the overall layer thickness is still the same as after 3 days; however, the outer layer has grown at the expense of the inner. The thickness and the element distribution within the layers do not depend on the volume of water.

The strong decrease of the leach rate with time and the relatively low specific weight loss at 200°C, compared to that extrapolated from data obtained at $\leq$100°C, suggest that the outer layer in Fig. 2a is a protective coating which impedes the further leaching of the glass. As this layer increases with time, it must be concluded that the underlying glass is increasingly protected against leaching as long as it does not become mechanically unstable. The strong depletion of Na, Mg, and Al at the very beginning of the gel layer formation, compared to their later enrichment in the outermost layer, could be an indication for the formation of compounds within the layer. This could, in particular, explain the enrichment of sodium. However, this needs further investigation and experimental evidence. The formation of compounds at the phase boundary solid-liquid is favored by locally high concentrations. These are rather likely to occur in the isothermal autoclave experiment. If one assumes that convection is low in smaller volumes of liquid, a diffusion process will control the release of dissolved constituents into the solution. In larger volumes, however, thermal convection is more likely to occur due to local temperature differences. This would increase the concentration gradient at the phase boundary and thereby the release of material as shown in Fig. 1. These assumptions can be supported by performing experiments with more than one bead in one autoclave. It was clearly demonstrated that the released fraction is independent of the number of beads (two to six were used) at any volume. Thus, each bead's release depends only on its .ocal concentration gradient and not on the average concentration in solution.

CONCLUSIONS

The very different chemical constituents contained in the waste glass lead to a specific leaching behavior under hydrothermal conditions. The results are still difficult to interpret in detail but the mechanism in deionized water resembles the one where a protective layer is formed by some of the glass constituents which determine the further release of material from the sample surface. Magnesium, Al, Zn, and Ti were found to contribute considerably to the layer formation in addition to waste constituents such as rare earths and U, Zr. The former elements have already been used for glass frit formulations in order to better digest the waste oxides. They might also be helpful, if used at appropriate concentration levels, to increase the leach resistance of the waste form.

ACKNOWLEDGMENTS

The authors are indebted to Messrs. P. Schubert, H. Matiske, V. Beran, and Mrss. L. A. Mertens and H. Fuchs for their excellent technical work.

REFERENCES

1. G. Malow et al., "Testing and Evaluation of the Properties of Various Potential Materials for Immobilizing High Activity Waste," European Applied Research Reports, Vol. 2, No. 2, pp. 453-515 (1980).
2. J. van Geel et al., "Solidification of High-Level Liquid Wastes to Phosphate Glass-Metal Matrix Blocks," Symp. on Management of Rad. Wastes from the Nuclear Fuel Cycle, Vienna, March 22-26, 1976, IAEA-SM-207/83, pp. 341-359.
3. W. Lutze, J. Borchardt, A. K. Dé, "Characterization of Glass and Glass-Ceramic Nuclear Waste Forms," in Scientific Basis for Nucl. Waste Management, Vol. 1, G. M. McCarthy, Ed., Plenum Press, New York (1979), pp. 69-81.

THE INFLUENCE OF SURFACE PROCESSES IN WASTE FORM LEACHING

A. J. Machiels and C. Pescatore

Nuclear Engineering Program
University of Illinois
Urbana, Illinois 61801

ABSTRACT

A leaching model incorporating bulk diffusion and surface processes is presented. The short-term leaching behavior of waste forms under dynamic leach testing conditions is discussed. The proposed model is qualitatively and semi-quantitatively consistent with available experimental leach data obtained at moderate temperatures and based on the behavior of alkali and alkaline earth radionuclides.

INTRODUCTION

In order to assess the potential radiological consequences of geologic disposal of solidified nuclear wastes, mathematical models of the leaching of radionuclides from high level waste forms (source term) and their migration through the geosphere (transport term) must be developed.

Several models relying on mass transport theory have been used to describe leaching by diffusion of ions through the solid waste and by dissolution of the matrix.[1,2] They are especially successful in the midrange of time when diffusion is the rate-controlling process and for test samples which are large enough such that a one-dimensional approach can be used. However:

(i) during the early period of contact between the waste form and the water, leaching usually decreases too rapidly to be fit by a diffusion model; depending upon the temperature and the species involved, the transient behavior may last from a few hours to a few years[3];

(ii) extrapolation to long periods of time is quite uncertain even when matrix dissolution is taken into account because the chemical processes occurring at the leachant - waste form interface are not accounted for.

LEACH MODEL

Recognizing that a waste form surface is usually markedly different in composition from the bulk,[4] a leaching model is proposed based on the applicability of:

a) Fick's law to describe the time and space-dependent concentrations of radionuclides in the solid waste form as they diffuse to the outer surfaces contacted by the water; and

b) chemical kinetics rate equations to describe the exchange processes occuring at the leachant - waste form interface between species absorbed on the waste form and in solution.

The model assumes that the release of any given species from the solid waste form to the leachant solution proceeds by a series of steps, the last one being a surface process. Adopting a one-dimensional geometry and neglecting network dissolution for simplicity, the model is illustrated in Fig. 1.

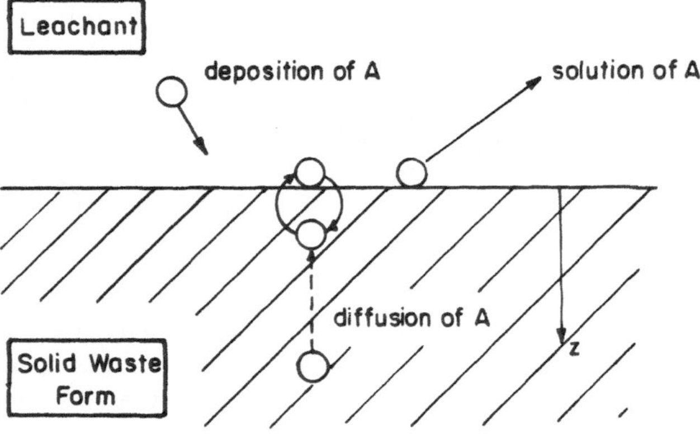

Fig. 1 Physical representation of the leach model.

Leaching of any given species A is assumed to proceed by the following steps:

$$A_{(bulk)} \overset{D}{\rightleftharpoons} A_{(surface\ layer)} \tag{1}$$

$$A_{(surface\ layer)} \overset{H}{\rightleftharpoons} A_{(surface)} \tag{2}$$

$$A_{(surface)} \underset{k_d}{\overset{k_\ell}{\rightleftharpoons}} A_{(solution)} \tag{3}$$

Step (1) represents the diffusion of A in the solid waste form in terms of a diffusion coefficient, D. The physical interpretation of D depends on the nature of the waste form. For waste forms such as glasses, in which ions of A in the solid are replaced by hydrogen (or hydronium) ions, $D = \tilde{D}$, where $\tilde{D}$ is an interdiffusion coefficient describing the ion exchange process between the ions of A and hydrogen. For an ideal solution:[5]

$$\tilde{D} = \frac{D_A\ D_{H+}}{x_A\ D_A + x_{H+} D_{H+}} \tag{4}$$

where D_A, D_{H+}, x_A and x_{H+} represent the diffusion coefficients and mole fractions of A and H^+ ions in the waste form.

The equilibrium of step (2) can be expressed in terms of a solubility coefficient H relating the surface concentration of A atoms, n(t), with the bulk concentration in the outer surface of the waste form, C(0,t):

$$C(0,t) = Hn(t). \tag{5}$$

Step (3) represents the release of species A into the leachant in terms of a rate constant k_ℓ and the redeposition of A on the surface in terms of a mass transfer coefficient k_d.

Based on the behavior of the radionuclide A, the leach rate per unit surface area, L(t), is:

$$L(t) = \frac{W}{\alpha}\left[k_\ell n(t) - k_d C'_{sol}(t)\right] \tag{6}$$

where W represents the weight of the test specimen, α is the amount of radionuclide A present in the test specimen, and $C'_{sol}(t)$ is a specific concentration of species A in the leachant.

SHORT-TERM LEACHING BEHAVIOR

The influence of the surface step can be illustrated simply by considering the short-term leaching behavior of waste forms under dynamic leach testing conditions (i.e. $C'_{sol} \sim 0$). For that case, L(t) is known if n(t) is known. The latter can be obtained by using eq. (5) and solving the diffusion equation:

$$\frac{\partial C(z,t)}{\partial t} = \frac{\partial}{\partial z} D \frac{\partial C(z,t)}{\partial z} - \lambda C(z,t) \tag{7}$$

where C(z,t) represents the concentration of species A in the waste form and λ is the decay constant. One boundary condition results from geometric considerations while the other boundary condition is given by the balance of species A on the waste form surface:

$$\frac{dn}{dt} = - k_{\ell} n + D\left(\frac{\partial C}{\partial z}\right)_{z=0} - \lambda n \tag{8}$$

or

$$\frac{dC(0)}{dt} = - k_{\ell} C(0) + HD\left(\frac{\partial C}{\partial z}\right)_{z=0} - \lambda C(0) \tag{9}$$

by using eq. (5).

Assuming a uniform initial distribution of A in the waste form, the initial condition is written:

$$\left.\begin{aligned} &C = C_o \\ \text{and}\quad &n = n_o = \frac{C_o}{H} \end{aligned}\right\} \quad \text{for } t = 0 \tag{10}$$

Analytical solutions for the system of eqs. (7), (9), and (10) are compiled in ref. (6) for one-dimensional geometries such as slabs, infinite cylinders and spheres when D, k_{ℓ} and H are constant and the following substitution introduced:

$$C(z,t) = C'(z,t)\, e^{-\lambda t} \tag{11}$$

The initial leaching rate is given by:

$$L(0) = \frac{W}{\alpha} k_{\ell} n_o \;, \tag{12}$$

and the short-term leaching behavior is dictated by the relative magnitude of the first two terms of the right-hand side of eq. (8). Radioactive decay of A plays a negligible role during the initial duration of the leach test: $\lambda n \sim 0$, and the assumption of uniform concentration results in:

$$\left(\frac{\partial C}{\partial z}\right)_{z=0} = 0 \text{ for } t = 0, \tag{13}$$

which leads to:

$$\frac{dn}{dt} \simeq - k_\ell n \text{ for } t = 0. \tag{14}$$

Therefore, depletion of the surface species occurs with a rate which depends upon the value of k_ℓ. As a result, a concentration gradient between the surface and the region of the solid in contact with the surface is created leading to a diffusive flux, the rate of which depends upon the value of D.

If k_ℓ is such that:

$$k_\ell n >> D\left(\frac{\partial C}{\partial z}\right)_{z=0} \tag{15}$$

for a measurable amount of time, τ_1, then eq. (8) becomes:

$$\left.\begin{aligned} \frac{dn}{dt} &= - k_\ell n \\ \text{or} \quad n &= n_o e^{-k_\ell t} \end{aligned}\right\} \quad \text{for } t \leq \tau_1, \tag{16}$$

i.e.: initial leach rates follow an exponential behavior given by:

$$L(t<\tau_1) = \frac{W}{\alpha} k_\ell n_o e^{-k_\ell t} \tag{17}$$

After a time, τ_2, a quasi steady state is eventually attained corresponding to:

$$\frac{dn}{dt} = - k_\ell n + D\left(\frac{\partial C}{\partial z}\right)_{z=0} \simeq 0 \tag{18}$$

leading to a leach rate which is controlled by the rate of diffusion of A from the bulk of the waste form to the surface:

$$L(t>\tau_2) = \frac{W}{\alpha} D\left(\frac{\partial C}{\partial z}\right)_{z=0} \tag{19}$$

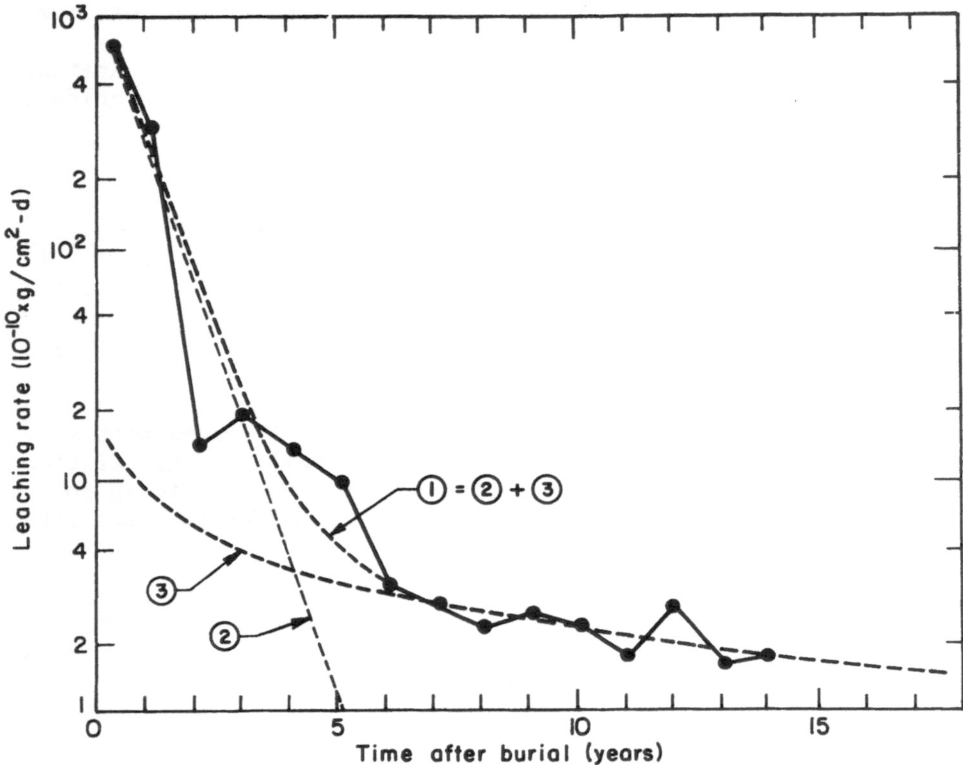

Fig. 2 Experimental time-dependence of leach rates based on the behavior of ^{90}Sr [After ref. (7)].

When k_ℓ is very large, τ_1 and τ_2 become too small to be conveniently measured and leach rates are diffusion-controlled from the start.

EXPERIMENTAL DATA

Available experimental evidence qualitatively supports the model. Leach tests performed at ambient temperatures[2,7,8] can be considered to be approximately exponential, while leach tests performed at temperatures of 50°C or higher yield diffusion-controlled behaviors after short periods of time.

Fig. 2 reproduces the revised[9] data reported in ref. (7). A good overall fit of the leach rate points based on the behavior of ^{90}Sr can be obtained by adding the surface and bulk diffusion contributions represented by curves (2) and (3). The time dependence of curve (2) yields $k_\ell \sim 3\ yr^{-1}$ by using eq. (17).

Assuming that the release of ^{90}Sr into the leachant can be modeled by an elementary process such as:

$$k_{\ell} = \nu e^{-E/RT} \tag{20}$$

where:

ν is the frequency factor, $\nu \simeq 10^{13}$ sec^{-1};

E is the activation energy required to transfer an atom from the solid surface into the liquid; the upper limit of E is represented by the strength of the bond between the surface substrate and the atom; for monovalent and divalent surface species, the single bond strength is retained for steric considerations;

R is Boltzmann's constant;

T is the surface absolute temperature.

For the leach rate data shown in Fig. 2: $\overline{T}_{av} \sim 6°C$, which yields E ~ 110 kJ/mole. This value is to be compared to the single bond strength of Sr in strontium oxide equal to ~ 134 kJ/mole.[10]

Similarly leach rates of glass[2] and SYNROC waste forms[8] based on the behavior of ^{137}Cs and ^{90}Sr at 25 and 75°C are qualitatively and semi-quantitatively consistent with the proposed model and the formulation of k_{ℓ} given by eq. (20).

CONCLUSION

The proposed model provides a basis for interpreting dynamic leaching rate data. It predicts that, at room temperatures, initial leach rates follow an exponential behavior characterized by a rate constant which depends on the strength of the bond between the surface species and the solid substrate. As depletion of the surface species occurs, diffusion gradually becomes rate-controlling. Available experimental information based on the short-term leaching behavior of alkali and alkaline earth nuclides at moderate temperatures can be correlated semi-quantitatively by the model. For simplicity, network dissolution has not been considered; however it may have to be included to properly interpret leach rate data for some waste forms and/or under high temperature conditions.

REFERENCES

1. H. W. Godbee, E. L. Compere, D. S. Joy, A. H. Kibbey, J. G. Moore, C. W. Nestor, O. U. Anders, R. M. Neilson, Application of Mass Transport Theory to the Leaching of Radionuclides from Waste Solids, Nuclear and Chemical Waste Management, 1:29 (1980).

2. J. R. Wiley, Leach Rates of High Activity Waste From Borosilicate Glass, Nucl. Technol., 43:268 (1979).
3. A. J. Machiels, Short-Term Leaching Behavior of Waste Forms, Workshop on Alternate Nuclear Waste Forms and Interactions in Geologic Media, Gatlinburg, Tennessee, May 13-15, 1980.
4. L. L. Hench, D. E. Clark, and E. L. Yen-Bower, Surface Leaching of Glasses and Glass-Ceramics, Proceedings of the Conference on High-Level Radioactive Solid Waste Forms, Denver, Colorado, December 19-21, 1978.
5. G. H. Frischat, Ionic Diffusion in Oxide Glasses, Trans. Tech. Publications (1975).
6. H. C. Carslaw and J. C. Jaeger, "Conduction of Heat in Solids," Oxford University Press (1959).
7. W. F. Merritt, High-Level Waste Glass: Field Leach Test, Nucl. Technol., 32:88 (1977).
8. D. G. Coles, F. Bazan, Continuous-Flow Leaching Studies of Crushed and Cored SYNROC, Workshop on Comparative Leaching Behavior of Radioactive Waste Forms", Argonne National Laboratory, Sept. 3-4, 1980.
9. W. F. Merritt, personal communication, June 4, 1980.
10. K. H. Sun, Fundamental Condition of Glass Formation, Journal of American Ceramic Society, 30:277 (1947).

PROBABLE LEACHING MECHANISMS FOR UO_2 AND SPENT FUEL

Rong Wang and Y. B. Katayama

Pacific Northwest Laboratory
Richland, Washington 99352

INTRODUCTION

This paper presents the current progress in identifying the probable leaching mechanisms for UO_2 and spent fuel based on work done at the Pacific Northwest Laboratory. We have been generating spent fuel leach data since 1975[1]. Since leach data alone does not allow a clear understanding of the spent fuel leaching process, a series of experiments were initiated in 1979 with UO_2, the spent fuel matrix, to study the oxidation and dissolution mechanisms without encountering the very high radiation levels associated with spent fuel[2].

Leaching mechanisms for UO_2 were investigated based on single crystal surfaces in deionized water, $NaHCO_3$ (0.03M) and WIPP "B" brine solutions by autoclave dissolution and electrochemical dissolution experiments. The effects of radiation, in terms of H_2O_2 production, on the UO_2 surface were studied. Electrochemical methods were applied to study a light-water reactor spent fuel for comparisons of the dissolution rates and surface oxidation with the UO_2 surfaces.

EXPERIMENTAL PROCEDURES

Single crystals of UO_2 were selected from fused UO_2 materials available at PNL in kilogram quantity. The spent fuel samples, in the form of fuel fragments, were removed (about February 1977) from light-water reactor spent-fuel bundles discharged from the HB Robinson II reactor on June 6, 1974, after an average burnup of 28,000 MWd/MTU.

Electrochemical dissolution experiments for single crystal UO_2 and spent fuel were based on measurements of the dissolution current density (i_o) from Tafel plot, and open circuit potential (E_o) for determinations of the dissolution rate and surface oxidation respectively. The oxidation and dissolution behaviors for both the UO_2 and spent fuel were compared based on polarization curves obtained by potentio-dynamically scanning at a rate of 1 mV/s.

Autoclave dissolution experiments were based on leaching of single crystal UO_2 samples in Ti capsules containing test solutions of 10X the surface area and 200 ppm dissolved oxygen level from pressurized pure oxygen at both 75 and 150°C. After a designated time period, test solution was transferred for analysis of the uranium content by isotopic dilution/mass spectrometer technique. The Ti capsules were refilled with fresh solution after each sampling period. After 11 , 30 and 60 d, a small sample of about 50 mg, placed previously with the large sample in the capsules, was removed for surface characterization by SEM, x-ray diffraction and x-ray elemental analysis.

RESULTS AND DISCUSSIONS

Initial Dissolution Mechanisms

Both the single crystal UO_2 and spent fuel were studied for their initial leaching mechanisms by electrochemical methods. We found that the E_o value of the UO_2 surface was very sensitive to the oxidation state of that surface. A clean UO_2 surface had a low E_o value between -0.5 to 0 V(SCE), and the E_o value increased as the surface was oxidized either by anodic oxidation or by addition of H_2O_2. We also found that the i_o values for the clean UO_2 surfaces were rather high, in the order of 1 $\mu A/cm^2$ for three test solutions. Based on Faraday's law, this dissolution current density is equivalent to a dissolution rate of UO_2 in the order of 1 g/m^2-d. Increasing solution temperature from 25 to 75°C enhanced the dissolution rate for $NaHCO_3$ solution and brine solution but had little effect for the deionized water, Figure 1.

Polarization curves of UO_2 surface in deionized water and $NaHCO_3$ solution, as shown in Figure 1, indicated that the surface film, if formed, had little passive nature to reduce the dissolution rate. This may be due to the thinness and the porosity of the surface film. However, there was indication that the surface film grown on the UO_2 in WIPP "B" brine solution had passive nature.

Formation of thick surface films up to 1 μm thick on UO_2 surface was observed in two conditions: with accelerated dissolution of UO_2 surface for several hundred hours, and with 50-500 ppm of H_2O_2 addition to the deionized water. Prolonged accelerated dissolution to several hundred hours resulted in a surface containing area of

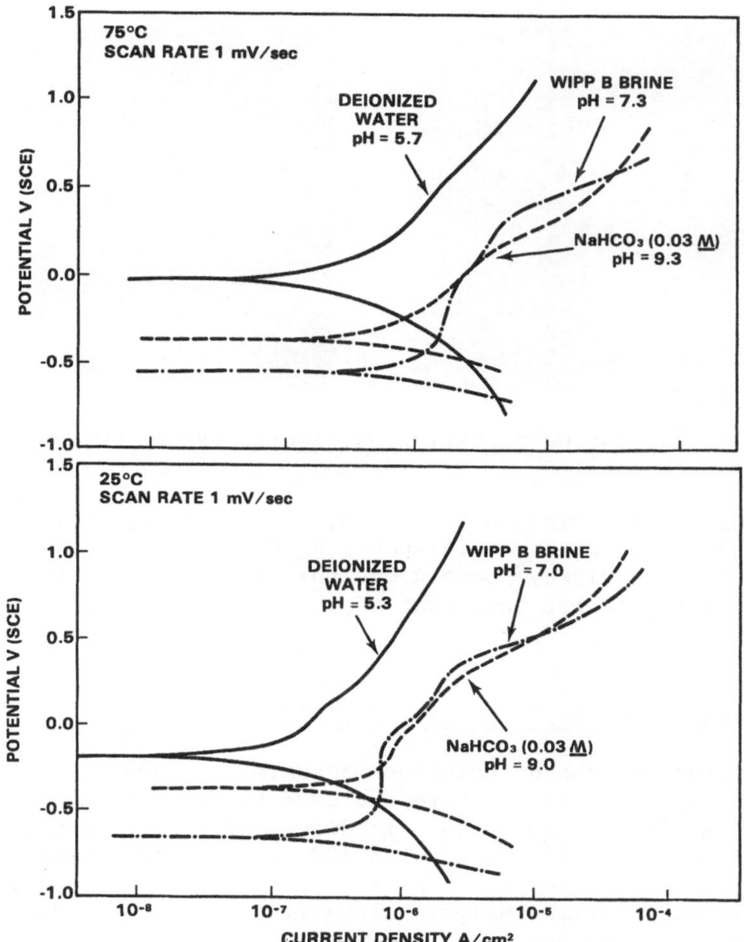

Fig. 1. Cathodic and anodic polarization behaviors or single crystal UO_2 surface.

UO_3-hydrate coating and corroded surface. Often, the thick film contained many cracks and was loosely attached to the surface as seen by SEM. The effects of H_2O_2 were investigated due to the expected presence of O_2 and H_2O_2 from radiolysis of water. A surface film of UO_3-hydrate was clearly observed after dipping of the UO_2 surface in 50 ppm H_2O_2 solution for 24 hours. Electrochemical accelerated reaction at 0.5 V(SCE) for 26 hours in deionized water containing 500 ppm of H_2O_2 resulted in a nearly 1 μm thick $UO_2(OH)_2$ coating as shown in Figure 2. The film can easily be removed from the surface and thus exposed the smooth surface of UO_2.

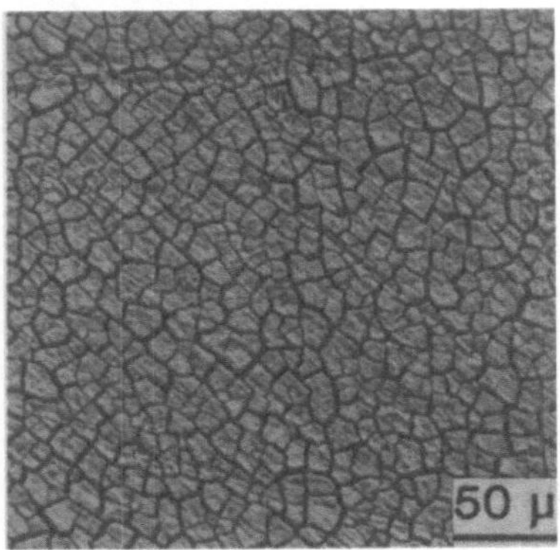

Fig. 2. $UO_2(OH)_2$ coating on single crystal UO_2 surface.

Oxidation and dissolution of UO_2 surface in deionized water containing H_2O_2 were rapid as indicated by the formation of the surface hydrated films. Hiskey[3] points out that the effect of H_2O_2 was about 100 times faster than that of the dissolved oxygen in carbonate solution.

Dissolution of UO_2 in Oxygenated Solutions

For studying the dissolution mechanisms of UO_2 at higher temperatures than the room temperature, we used the autoclave experiments at both 75 and 150°C up to 80 days in deionized water, $NaHCO_3$ and WIPP "B" brine solutions saturated with 200 ppm dissolved oxygen.

Based on solution analysis alone (including suspended solid phase) there was no apparent temperature effect on the dissolution rate of UO_2 at 75 and 150°C. The solution analysis based UO_2 dissolution rates were highest in $NaHCO_3$ solution (5×10^{-1} g of U/m^2-d), about 2 order of magnitude lower in deionized water (2×10^{-3} g of U/m^2-d) and an additional one order of magnitude lower in WIPP"B" brine (2×10^{-4} g of U/m^2-d). However, analysis of the uranium content in the solution cannot represent the amount of uranium dissolved from the single crystal surfaces because thick UO_3-hydrate coatings were observed on the UO_2 and container surfaces. In the case of $NaHCO_3$ solution nearly 100 times more uranium was found plated-out on the container wall at 150°C than at 75°C. Characterization of the UO_2 surface by SEM clearly indicated the difference between the dissolution rate for these two temperatures. For all the tests done at 75°C, no noticeable surface film was grown at the UO_2 surface, although some plate out was found at the container wall. For tests at 150°C, the surface of UO_2 crystal in the deionized water and $NaHCO_3$ solution resulted in nearly 1 μm size of uranyl hydrates while the brine solution resulted in thin film and some loose deposit, Figure 3.

Fig. 3. UO_3-hydrate coatings on single crystal UO_2 surfaces after 30 days at 150°C in A. deionized water, B. $NaHCO_3$ (0.03M) and C. WIPP "B" brine solution.

The low dissolution rate and low surface hydrate deposition agrees with the electrochemical polarization curves for UO_2 surface in brine solution.

Oxidation and Dissolution Mechanisms of UO_2

The probable oxidation and dissolution mechanisms for single crystal UO_2 surface based on electrochemical and autoclave dissolutions are summarized in Figure 4 as a surface oxidation-dissolution-hydrates deposition mechanism. The detailed mechanisms may vary as a function of leaching solution, pH, temperature and oxidizing species. However, a general leaching mechanism may be described as follows:

A clean UO_2 surface exposed in solution containing oxygen or other oxidizing chemical species will be rapidly oxidized. The initial surface oxidation will produce a thin oxide layer of several A. The oxidized surface in contact with the dissolved oxygen will rapidly dissolve to form uranyl ions or complex ions depending on the solution chemistry. For deionized water, one of the reactions will be:

$$2\ U_4O_9 + 16\ H^+ + 3\ O_2 \rightleftharpoons 8\ UO_2^{+2} + 8\ H_2O \qquad (1)$$

The dissolution of the oxidized UO_2 surface thus favors low pH values. As the surface region becomes saturated with the uranyl ions, the reactions for hydrolysis of uranyl ions will take place[4] as:

$$\text{at } 25\text{-}75°C, \quad UO_2^{+2} + 2H_2O \rightleftharpoons UO_2(OH)_2 + 2H^+ \tag{2}$$

$$\text{or at } 150°C \quad UO_2^{+2} + 3H_2O \rightleftharpoons UO_3 \cdot 2H_2O + 2H^+ \tag{3}$$

The formation of UO_3-hydrates from hydrolysis of uranyl ions will result in coating of the UO_2 surface with hydrate films, the container wall plated with hydrate particles and the solution containing suspended particles. The overall oxidation and dissolution mechanisms now can be considered as a phase transformation of bulk UO_2 into particulate UO_3-hydrates via liquid media as shown in Figure 4. In the presence of dissolved oxygen, the dissolution of UO_2 surface will continue and the formation of the hydrates will increase while the dissolved concentration of uranyl ions will remain nearly constant[5].

The UO_3-hydrate coating may have some passivation nature to reduce the dissolution of UO_2, such as was shown in the brine solution. However, as the coating grew thick, it had a tendency to crack and spall from the surface. The instability of the surface coating will create a major problem for investigating the long term leaching behaviors for UO_2 and spent fuel.

Leaching Mechanisms for Spent Fuel

The data from the electrochemical leaching of spent fuel indicated that the initial leaching and dissolution of spent fuel matrix material was similar to that of the UO_2 surface. The spent fuel had a higher E_o value than that of UO_2. Polishing of the spent fuel

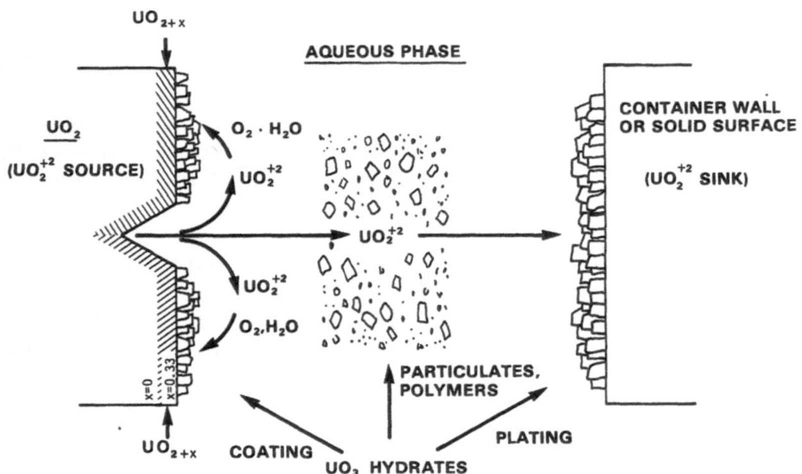

Fig. 4. Oxidation and dissolution mechanisms for UO_2.

surface reduced the E_o values which indicates that the spent fuel surface was oxidized during the period of storage. The spent fuel showed initial dissolution currents in the order of 10^{-7} A/cm^2 or equivalent to a dissolution rate of UO_2 at 0.1 g/m^2-d based on Faraday's law.

The potentiodynamic polarization of the spent fuel in deionized water, $NaHCO_3$ and WIPP "B" brine shown in Figure 5 was similar to those observed for UO_2. It also indicated that the surface film formed in brine solution for spent fuel had some passive nature as observed for UO_2. The spent fuel, after accelerated dissolution for 1 h at 0.5 V(SCE) above E_o value, showed little change in E_o and i_o which indicated that the surface of the spent fuel was covered with thick oxide or hydrate.

The electrochemical methods have demonstrated that initial dissolution behavior of spent fuel is similar to that of the UO_2. Thus study of UO_2 may provide the understanding of the spent fuel. The electrochemical methods for investigation of long term leaching of spent fuel needs further improvements and detailed investigations in terms of different types of spent fuel under various conditions to allow comparison to the UO_2. Spent fuel leach data and dissolution mechanism data may be determined from this method.

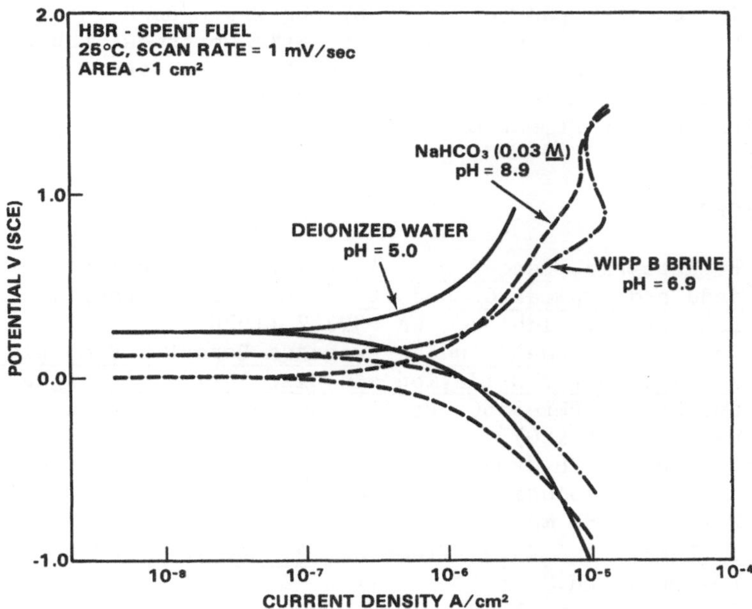

Fig. 5. Polarization behaviors of a HB Robinson spent fuel.

CONCLUSIONS

The oxidation and dissolution mechanisms for UO_2 and spent fuel will be quite similar based on this preliminary work with electrochemical leaching of UO_2 and spent fuel. In solutions containing oxygen or other oxidizing species, the UO_2 surface will be rapidly oxidized and dissolved following the transformation of uranium from U(IV) to U(VI). The hydrolysis of dissolved uranyl ions forms solid UO_3 hydrates or related complex compounds deposited onto the UO_2 surface, or other surfaces, as thin or thick coatings. Depending on the pH, temperature and time, the various kinds of porosity and the mechanical properties of the hydrate coatings will control the dissolution rate. The effects of radiation, in terms of generation of H_2O_2 will enhance the dissolution kinetics.

Electrochemical methods may be useful for determining the surface conditions, dissolution rate and accelerated dissolution behavior for UO_2 and spent fuel. Electrochemical methods can rapidly generate much information in terms of dissolution rate and surface film properties such as thickness, porosity and oxidation state, in-situ during the progress of the leaching process.

ACKNOWLEDGEMENTS

This research was supported by the Waste/Rock Interaction Technology Program being conducted by Pacific Northwest Laboratory. This program is sponsored by the Office of Nuclear Waste Isolation which is managed by Battelle Memorial Institute under contract DE-AC06-76RLO 1830 with the Department of Energy.

REFERENCES

1. Y. B. Katayama, Leaching of Irradiated LWR Fuel Pellets in Deionized and Typical Groundwater. BNWL-2057, Pacific Northwest Laboratory, Richland, WA 99352 (1976).
2. R. Wang, Spent Fuel Studies Progress Report: Probable Leaching Mechanisms for Oxidation and Dissolution of Single Crystal UO_2 Surfaces. PNL-3566, Pacific Northwest Laboratory, Richland, WA 99352 (1980).
3. J. B. Hiskey, "Hydrogen peroxide leaching of uranium oxide in carbonate solutions," presented at Symposium of Hydrogen Peroxide ASM Meetings, Las Vegas, NV (1980).
4. R. J. Lemire and P. R. Tremaine, J. Chemical and Eng. Data, to be published.
5. H. D. Holland and L. H. Brush, "Uranium oxides in ores and spent fuels," Proceedings of the Conference on High-Level Radioactive Solid Waste Forms, December 19-21, 1978, Denver, CO. NUREG/CP-0005 (1979).

STABILITY OF POLLUCITE IN HYDROTHERMAL FLUIDS

Sridhar Komarneni and William B. White

Materials Research Laboratory
The Pennsylvania State University
University Park, PA 16802

ABSTRACT

Pollucite, $CsAlSi_2O_6$, a ceramic host structure for cesium was reacted with deionized water, with several brine solutions, and with several salt solutions under hydrothermal conditions at 100, 200, and 300°C. Pollucite is stable in the presence of deionized water under all conditions. Cesium can be extracted from pollucite by ion exchange at high activities of Na^+, K^+ and Ca^{++}. Solutions with high Mg^{++} activity breakdown the pollucite structure with the concurrent precipitation of layer silicate.

INTRODUCTION

Pollucite, $CsAlSi_2O_6$, has a three dimensional framework structure of corner-sharing AlO_4 and SiO_4 tetrahedra characteristic of the zeolite family. The large Cs^+ ion occupies a cage site within the framework. Most naturally occurring pollucites contain some sodium and some water so that the composition is better expressed as $Cs_{1-x}Na_xAlSi_2O_6 \cdot xH_2O$ where x is on the order of 0.3. There is thus a close chemical and structural relationship between pollucite and analcite, $NaAlSi_2O_6 \cdot H_2O$.

As the only naturally occurring compound for cesium, pollucite was early recognized(1) as a potential host compound for the immobilization of ^{137}Cs from high level waste. It appears in the mixed oxide (supercalcine) ceramics(2,3) and in other ceramic waste forms. Experiments on the ceramic processing showed that pollucite is stable at processing temperatures with little cesium loss(4-6).

Investigations of the interactions between a variety of cesium compounds and clay minerals and shale rocks showed quite clearly that pollucite is formed by direct reaction between Cs^+ in hydrothermal solutions and the aluminosilicate minerals[7]. Likewise pollucite is a common reaction product when various prototype waste forms are reacted with shale rocks in the presence of hydrothermal water[8].

Pollucite is found in nature in alumina-rich pegmatite deposits where it can be shown to have crystallized from hydrothermal solutions in the temperature range of 150 to 600°C at pressures of 2-4 kilobars[9].

The fact that pollucite can be prepared under ceramic firing conditions and the observation that aluminosilicate minerals act as scavengers for Cs^+ with the formation of pollucite under low temperature conditions suggest that pollucite has much to recommend it as a crystalline host in nuclear waste ceramics. However, some questions remain. There have been claims that pollucite weathers rather easily under ambient conditions[9]. The stability of the pollucite structure in the presence of concentrated brine solutions such as might be expected either in bedded salt or in some sedimentary repository rocks is unknown. The possibility that cesium could be leached by solutions containing high concentrations of other alkali ions by direct ion exchange must be investigated. Finally, the dissolution kinetics ("leach rate") of pollucite under ambient to mild hydrothermal conditions is not known in any quantitative way.

The present paper addresses the question of the hydrothermal stability of pollucite in the presence of various brine solutions. The question of dissolution kinetics will be taken up elsewhere.

EXPERIMENTAL

Materials

A synthetic pollucite prepared by the gel method[4] was used in these stability studies. Chemical analysis of this pollucite by a lithium borate fusion technique[10] (Table 1) shows that the total impurities amounted to <0.1 wt%. The composition normalized to 100% $Cs_2O + Al_2O_3 + SiO_2$ gives the formula of the pollucite to be $Cs_{0.90}Al_{0.98}Si_{2.04}O_6$ on the basis of six oxygen atoms per formula unit. This pollucite specimen was found to be phase-pure by x-ray diffraction. Sieved fractions (-100 + 200) of this pollucite were used in all of the experiments.

Brine solutions were prepared by dissolving reagent grade chemicals in deionized water. Table 2 lists the fluid compositions.

TABLE 1. Chemical Analysis of Pollucite

	Synthetic Pollucite	Theoretical ($CsAlSi_2O_6$)
Cs_2O**	42.11*	45.2
Al_2O_3	16.72	16.3
SiO_2	41.12	38.5
Fe_2O_3	<0.02	--
MgO	<0.02	--
CaO	<0.02	--
Na_2O	<0.02	--

*All analyses in weight percent.
**K_2O and Rb_2O which were each less than 0.1 wt% are included in the Cs_2O value.

TABLE 2. Fluid Compositions (%)

	Bicarbonate Brine	NaCl Brine	NBT-6a Bittern	3N Salt Solutions			
				NaCl	KCl	$CaCl_2$	$MgCl_2$
NaCl	0.25	Saturated + excess NaCl	5.0	∿17	--	--	--
Na_2SO_4	0.22	--	--	--	--	--	--
$NaHCO_3$	0.41	--	--	--	--	--	--
KCl	--	--	5.0	--	∿21	--	--
$CaCl_2$	--	--	10.0	--	--	∿16	--
$MgCl_2$	--	--	10.0	--	--	--	∿14
H_2O	99.12	Remainder	70.0	∿83	∿79	∿84	∿86

Methods

A 20 mg sample of pollucite was treated with 200 µl of each of the fluids (except the NaCl brine) in sealed gold capsules at 100, 200, and 300°C for 4 and/or 12 weeks under a confining pressure of 30 MPa either in autoclaves or cold-seal vessels. In the case of the NaCl brine, a 20 mg pollucite sample was mixed with 200 mg of NaCl and 200 µl of deionized water was added so that the solution would be saturated and excess solid NaCl would remain when treated under the above conditions. All the hydrothermal experiments were conducted in duplicate.

At the completion of an experiment the solid and solution phases from the cooled run capsules were separated as follows: The gold capsule was cut open in a long glass vial and a 20 ml aliquot of 0.1 N KCl was added. The glass vial was then capped and the contents were mixed in a shaker for about 15 minutes. The gold capsule free of sample was then removed from the vial with tweezers. The vials were centrifuged and a portion of the supernatant liquid was collected in polyethylene bottles without disturbing the sediment. This procedure extracts the cesium present in solution and probably any cesium that was weakly adsorbed on the solids. The use of 0.1 N KCl serves three purposes: (i) it brings any easily exchangeable cesium into solution; (ii) it helps in the flocculation of the sample during centrifugation, and (iii) it removes the ionization interference during cesium analysis by atomic absorption spectrophotometry. The solid samples were then washed once with water and acetone mixtures and thrice with 95% acetone to remove the KCl and were then dried. The solid samples were analyzed by x-ray diffraction. The solutions in the polyethylene bottles were analyzed for cesium by Atomic Absorption Spectrophotometry using a Perkin Elmer PE403 instrument and for silicon and aluminum by Atomic Emission Spectroscopy using a computer interfaced Spectra-Metrics SpectraSpan III instrument.

RESULTS

The results of solution analyses for the reaction of pollucite at 100, 200, and 300°C with various fluids are given in Table 3. That pollucite is extremely stable up to 300°C in deionized water under hydrothermal conditions is evident from the small concentrations or percentages of cesium released into solution. Pollucite seems to be quite stable in all fluids at 100°C. At 100°C the stability of pollucite in deionized water and brines decreases as follows:

Deionized water >> NBT-6a brine = Bicarbonate brine > NaCl brine

Among the equinormal salt solutions, monovalent salt solutions seem to be more effective in replacing Cs than the divalent

TABLE 3. Average Concentration and Percent of Cesium in Capsule Fluid.

Fluid*	100°C		200°C		300°C	
	μg/ml**	%†	μg/ml	%	μg/ml	%
Deionized water	21	0.05	17	0.04	32	0.08
Bicarbonate brine	88	0.22	380	0.95	920	2.31
NaCl brine	141	0.36	760	1.92	8100	20.38
NBT-6a brine	67	0.17	950	2.40	16300	40.97
3N NaCl	124	0.31	1060	2.66	7000	17.61
3N KCl	136	0.34	1290	3.25	9400	23.66
3N $CaCl_2$	60	0.15	330	0.83	9400	23.69
3N $MgCl_2$	81	0.20	1290	3.25	22400	56.43

*When determining the stability sequences of pollucite in various fluids, the ± deviation that results from Cs determination was taken into consideration.

**Concentration in original capsule fluid.

†Expressed as percentage of the cesium originally present in pollucite.

salt solutions at 100°C. In the presence of equinormal salt solutions at 100°C, the stability of pollucite decreases as follows:

$$3\underline{N}\ CaCl_2 = 3\underline{N}\ MgCl_2 > 3\underline{N}\ NaCl = 3\underline{N}\ KCl$$

No new solid reaction products were detected by x-ray diffraction.

Interaction of various fluids with pollucite at 200°C indicates that up to 3.25% of cesium from pollucite was released into solution (Table 3). At 200°C, the stability of pollucite in deionized water and various brines decreases as follows:

Deionized water >> Bicarbonate brine > NaCl brine > NBT-6a

Among the equinormal salt solutions, $CaCl_2$ is the least effective in removing cesium from pollucite at this temperature. The stability of pollucite in equinormal salt solutions decreases as follows:

$$3\underline{N}\ CaCl_2 > 3\underline{N}\ NaCl > 3\underline{N}\ KCl = 3\underline{N}\ MgCl_2$$

No new crystalline solid phases were detected.

Cesium was released significantly from pollucite at 300°C by interaction with various fluids except with deionized water, but the stability sequence was the same as at 200°C. Among the 3N salt solutions, NaCl is the least effective while $MgCl_2$ is the most effective at taking cesium into solution and the stability of pollucite in these salt solutions decreases as follows:

$$3\underline{N}\ NaCl > 3\underline{N}\ KCl = 3\underline{N}\ CaCl_2 > 3\underline{N}\ MgCl_2$$

At 300°C, substantial fractions of cesium were released into brine and salt solutions while little or no silicon or aluminum were detected in the solutions (Table 4). This result suggests that the silicon and aluminum are retained in solid phases. X-ray diffraction of solid interaction products produced at 300°C revealed pollucite-analcite solid solution with NaCl, pollucite-leucite solid solution with KCl (Figure 1) and pollucite and probably layer silicates (Figure 2) were detected by the interaction of pollucite with $MgCl_2$ and NBT-6a brine. The interaction of $CaCl_2$ with pollucite did not reveal any new mineral phases (Figure 2) which may suggest a pollucite-wairakite solid solution as a result of ion exchange. A solid solution of pollucite with leucite ($KAlSi_2O_6$) is indicated by shifted peaks (Figure 1) but a solid solution of pollucite with analcite or wairakite ($CaAl_2Si_4O_{12}, 2H_2O$) is indistinguishable from pollucite by x-ray diffraction.

TABLE 4. Average Aluminum and Silicon Concentrations in Capsule Fluids.

Fluid	Al Concentration			Si Concentration		
	100°C	200°C	300°C	100°C	200°C	300°C
Deionized water	2	<0.1	<0.1	60	200	360
Bicarbonate brine	<0.1	<0.1	<0.1	80	80	80
NaCl brine	<0.1	<0.1	<0.1	30	20	20
NBT-6a brine	25	38	16	14	9	3
3N NaCl	<0.1	<0.1	<0.1	50	30	40
3N KCl	<0.1	<0.1	<0.1	70	13	11
3N $CaCl_2$	27	21	19	20	<1	13
3N $MgCl_2$	1	<0.1	<0.1	50	20	20

All analyses are given in μg/ml.

To determine the exact nature of the solid interaction products at 300°C when divalent salt solutions and NBT-6a brine interact with pollucite, pollucite was treated with 3N $CaCl_2$, 3N $MgCl_2$ and NBT-6a brine for 12 week duration runs. X-ray diffraction analysis of the solid reaction products did not reveal any new mineral phases when pollucite was treated with 3N $CaCl_2$ even after 12 weeks (Figure 3).

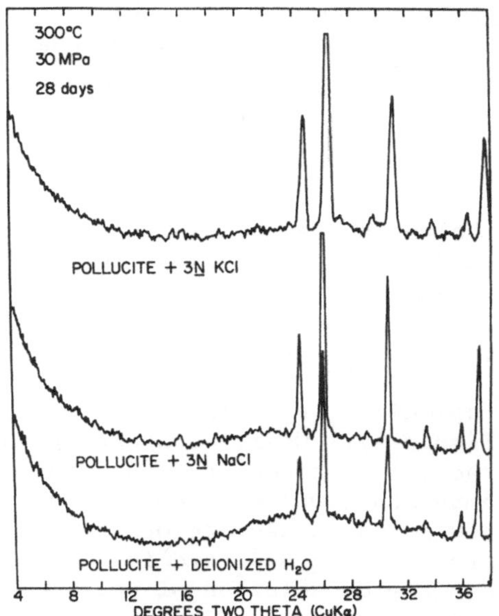

Figure 1. X-ray diffraction patterns of interaction products.

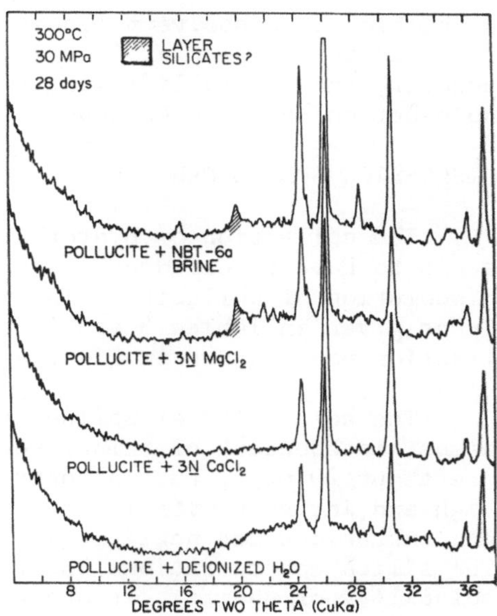

Figure 2. X-ray diffraction patterns of interaction products.

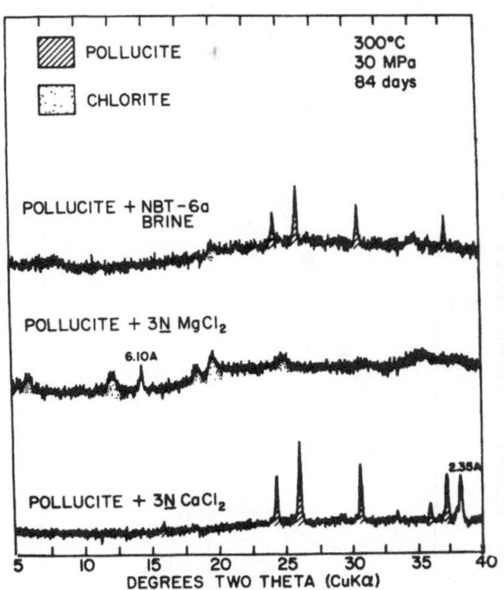

Figure 3. Generation of chlorite in Mg-rich brines.

The absence of new solid phases along with the presence of strong pollucite x-ray peaks even after 12 weeks of treatment suggests a solid solution of pollucite-wairakite by ion exchange process. The absence of substantial amounts of silicon or aluminum in solution when pollucite was treated with $CaCl_2$ solution (Table 4) also lends credence to this hypothesis. The peak of 2.35 A in the interaction products of pollucite-$CaCl_2$ solution belongs to gold which was dissolved by $CaCl_2$ solution at 300°C but recrystallized upon cooling. When pollucite was treated with 3N $MgCl_2$ for 12 weeks chlorite, a layer silicate mineral, crystallized while pollucite disappeared (Figure 3). There is an unidentified peak at 6.10 A in this interaction (Figure 3). Treatment of pollucite with NBT-6a brine for 12 weeks did not dissolve pollucite completely. This interaction resulted in one peak at 4.50 A which is probably chlorite (Figure 3). However, the crystallization of chlorite cannot be verified since no other chlorite peaks appear in the x-ray diffractograms.

DISCUSSION AND CONCLUSIONS

The collection of hydrothermal solvents examined in this study seems to have revealed at least three different mechanisms for the dissolution of pollucite. These are deduced from the solution analyses given in Tables 3 and 4 and the appearance of new x-ray diffraction peaks shown in Figures 1-3.

The solubility of pollucite in deionized water is low and almost independent of temperature (Figure 4). Cesium concentrations are about 20 mg/l, but in contrast the silica concentrations are high and increase with increasing temperature whereas the aluminum concentrations are negligible. No new solid phases are observed. The silica concentrations observed are about as expected for the solubility of quartz. It appears that the dissolution of pollucite in water is incongruent and that aluminum must be plating out.

Fluids containing high Na^+, K^+ and Ca^{++} activities produce a cation exchange interaction in which the high activity of alkali ion in the fluid drives out the Cs^+ ion from the pollucite. The cesium release in these fluids becomes substantial at 300°C. To a rough fit (Figure 4) the concentrations of cesium extracted by the various fluids increases exponentially with increasing temperature. The solid product following exchange is a pollucite-analcite solid solution (for Na^+), a pollucite-leucite solid solution (for K^+) and a pollucite-wairakite solid solution (for Ca^{++}). The silica concentrations are much lower than those in pure water which suggests that some species other than quartz is buffering the reaction. Again, however, the aluminum concentration is very low. It is not clear from these experimental results alone why the ion exchange should suppress the breakdown of the aluminosilicate framework but this is what the analyses indicate.

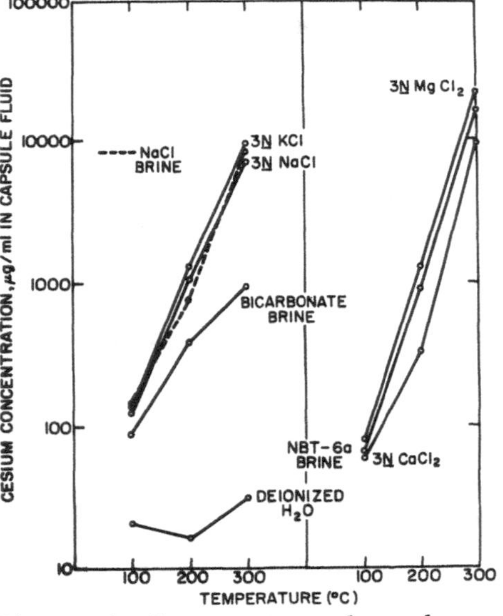

Figure 4. Temperature dependence of cesium loss.

The exchange of Cs^+ from pollucite upon treatment with 3 $\underline{N}$ NaCl, KCl and $CaCl_2$ solutions at 300°C/300 bars may be given as follows:

$$CsAlSi_2O_6 + Na^+ \rightleftarrows Cs_{0.82}Na_{0.18}AlSi_2O_6,\ 0.18H_2O + 0.18Cs^+$$

$$CsAlSi_2O_6 + K^+ \rightleftarrows Cs_{0.76}K_{0.24}AlSi_2O_6 + 0.24Cs^+$$

$$CsAlSi_2O_6 + Ca^{2+} \rightleftarrows Cs_{0.76}Ca_{0.12}AlSi_2O_6,\ 0.24H_2O + 0.24Cs^+$$

The fluids with high Mg^{2+} activity produce a unique chemistry. These fluids break down the aluminosilicate three-dimensional framework structure and replace it with a two-dimensional layer structure. This is clearly seen with both the NBT-6a brine and the 3$\underline{N}$ $MgCl_2$ solution leaving no doubt about the ion responsible. The magnesium-rich fluids create the largest cesium concentrations in the fluid, but again the silica concentrations are comparable, if not lower, than the silica concentrations produced by ion exchange phenomena. Again the aluminum concentrations are very low. The x-ray diffraction results indicate that the apparent reaction is between the pollucite and the magnesium-rich solution and the presence of chlorite can be clearly seen in the pattern. The crystallization of chlorite by the dissolution of pollucite in $MgCl_2$ may be given by the following formula:

$$2CsAlSi_2O_6 + 5MgCl_2 + 8H_2O \rightleftarrows Al_2Si_3Mg_5O_{10}(OH)_8 + SiO_2 + 2CsCl + 8HCl$$

The SiO_2 formed at 300°C/300b after 4 weeks was too little to be detected by x-ray diffraction.

Overall, the stability of pollucite in a repository environment depends very sensitively on the chemical composition of the circulating ground water. High silica activity and the presence of solid aluminum phases such as are characteristic of shale and other silicate rock ground waters drives the reactions to the left, causing any Cs^+ present in the water to be scavenged with the production of the observed pollucite phase. High alkali activities promote ion exchange. Only the magnesium ion appears to attack the silicate framework and it likely does this by creating a substantial lowering of the pH due to the precipitation of brucite, $Mg(OH)_2$ phase.

ACKNOWLEDGMENT

This work was supported by the U.S. Department of Energy through subcontract with the Office of Nuclear Waste Isolation, Battelle Memorial Institute, Contract E512-03400.

REFERENCES

1. R. E. Issacson and L.E. Brownell, "Ultimate Storage of Radioactive Wastes in Terrestrial Environments," in: "Management of Radioactive Wastes from Fuel Reprocessing," OECD Proceedings, Paris, 953 (1972).
2. G. J. McCarthy and M.T. Davidson, Ceramic Nuclear Waste Forms: I. Crystal Chemistry and Phase Formation, Ceram. Bull 54: 782 (1975).
3. G. J. McCarthy, High Level Waste Ceramics: Materials Considerations, Process Simulation and Product Characterization, Nucl. Tech. 32:92 (1977).
4. S. A. Gallagher and G.J. McCarthy, "Preparation and Characterization of Pollucite ($CsAlSi_2O_6$)," Dept. of Energy Topical Report COO-2510-138 (1977).
5. S. A. Gallagher, "Cesium Aluminosilicates for Nuclear Materials Application," Ph.D. Dissertation, The Pennsylvania State University (1979).
6. D. M. Strachan and W.W. Schulz, "Characterization of Pollucite as a Material for the Long Term Storage of ^{137}Cs, "ARH-SA-294 (1977).
7. S. Komarneni and W.B. White, Hydrothermal Interactions of Clay Minerals and Shales with Cesium Phases of Spent Fuel Elements, Clays & Clay Minerals (in press).
8. W. P. Freeborn, M. Zolensky, B.E. Scheetz, S. Komarneni, G.J. McCarthy and W.B. White, "Shale Rocks as Nuclear Waste Repositories: Hydrothermal Reactions with Glass Ceramic and Spent Fuel Waste Forms," in: "Scientific Basis of Nuclear Waste Management," C.J. Northrup, ed., Plenum Press, New York, pp. 499-506 (1980).
9. P. Cerny, "Pollucite and its Alteration in Geological Occurrences and in Deep-Burial Radioactive Waste Disposal," in: "Scientific Basis for Nuclear Waste Management," G.J. McCarthy, ed., Plenum Press, New York (1979).
10. J. H. Medlin, N.H. Suhr and J.B. Bodkin, "Atomic Absorption Analysis of Silicates Employing $LiBO_2$ Fusion," Atomic Absorp. Newsletter, 8:25 (1969).

XPS AND ION BEAM SCATTERING STUDIES OF LEACHING IN SIMULATED WASTE GLASS CONTAINING URANIUM*

D. P. Karim, P. P. Pronko, T. L. M. Marcuso, D. J. Lam, and A. P. Paulikas

Materials Science Division
Argonne National Laboratory
Argonne, IL 60439

ABSTRACT

Glass samples (consisting of 2 mole % UO_3 dissolved in a number of complex borosilicate simulated waste glasses including Battelle 76-68) were leached for varying times in distilled water at 75°C. The glass surfaces were examined before and after leaching using x-ray photoemission spectroscopy and back-scattered ion beam profiling. Leached samples showed enhanced surface layer concentrations of several elements including uranium, titanium, zinc, ion, and rare earths. An experiment involving the leaching of two glasses in the same vessel showed that the uranium surface enhancement is probably not due to redeposition from solution.

INTRODUCTION

Critical evaluation of the various proposals for the emplacement of radioactive waste in geological repositories must, to a large extent, be based on an understanding of radionuclide release from high-level waste (HLW) forms. Complex borosilicate glasses appear to be viable media for long-term immobilization of radioactive waste products, but the detailed mechanisms of HLW leaching from these glasses and the factors influencing the chemical species being leached are, as yet, incompletely understood. The mechanisms are especially important if one wants to use leaching parameters derived from relatively short-term testing to model and predict long-term leaching behavior.

*Work supported by the U.S. Department of Energy.

This study examines chemical and physical changes occurring at or near the borosilicate glass surface upon leaching, using x-ray photoemission spectroscopy (XPS), and ion beam scattering (IBS). The objective is to develop sufficient understanding of the leaching process to be able to provide a scientific basis for the formulation of actinide and fission product release models.

We have chosen to examine a number of complex silicate glasses containing dissolved UO_3 including Battelle simulated waste glasses 76-68, 3008, and 76-101.

EXPERIMENTAL

X-ray Photoemission Spectroscopy (XPS)

This technique involves irradiating the sample to be analyzed with monoenergetic soft x-rays. (In this study, the $A\ell K_\alpha$ line, $h\upsilon = 1486.6$ eV, was used.) The kinetic energy (K.E.) of photoemitted electrons is measured and the binding energy (B.E.) of the electrons in the sample is inferred from energy conservation via the relation, B.E. = $h\upsilon$ - K.E. The region of the sample analyzed is limited by the escape depth of the electrons (typically ~30 Å). Elemental compositional analysis is performed by measuring the intensity of electrons emitted from characteristic atomic core states. More detailed chemical and physical information comes from measurements of line widths (and detailed line shapes), absolute and relative line shifts, doublet splittings, satellite line structure and splitting, etc.

The system used in this study was a PHI model 548 spectrometer which has been installed in a glovebox. The system included a sample fracture mechanism for exposing fresh surfaces under high vacuum ($\sim 10^{-10}$ torr) conditions allowing spectra to be taken which are characteristic of bulk unleached glasses.

Ion Beam Spectroscopy (IBS)

In order to develop models for the mechanisms involved in leaching, it is often important to be able to interpret XPS results in light of compositional variations over a broader depth range. One might then be able to judge whether the detailed bonding information inherent in the XPS technique was representative of the entire gel layer or just the immediate surface (perhaps a surface layer laid down by redeposition from solution).

Analysis by this technique is accomplished by bombarding the sample with alpha particles accelerated to a fixed energy (typically 2.9 MeV) and energy analyzing the back-scattered particles. Most of the observed particles have undergone a single, large-angle elastic scattering event. For a given detection angle, the particle retains a fixed fraction of its kinetic energy in this scattering depending only on the ratio, R, of the mass of the scattering nucleus to the mass of the alpha particle. This fraction, known as the kinematic factor is given by

$$K = \left[\frac{(R^2 - \sin^2\theta)^{1/2} + \cos\theta}{R + 1}\right]^2$$

Most particles, however, are scattered from atoms at various distances beneath the surface. These particles have lost an additional amount of energy roughly proportional to the distance traveled through the sample. The spectra are readily interpretable as concentration profiles of the individual elements that have been shifted in energy and superposed. Typical distances analyzed are thousands of angstroms with resolutions of ∿500 Å.

Some of the features of Rutherford scattering spectrometry make its use in looking at actinides in complex glasses particularly attractive. In a complex glass as 76-68, the large number of components means the elemental edges are rather close, making determination of individual profiles difficult at best. However, actinide atoms, being substantially heavier than the other elements present, have step edges above and separated from those of the other components making profiling rather straightforward. In addition, the cross section for scattering from a nucleus goes as Z^2 so that actinides are observable in lower concentrations than the lighter atoms.

SAMPLE PREPARATION

The compositions of the simulated waste glasses used in this study are given in Table 1. Samples of 76-68, 3008, and 76-101 were powdered, mixed with 2 mole % UO_3 powder, and remelted. Samples cut from the glass ingots were cut to appropriate shape, polished, and cleaned with ethyl alcohol. Leaching was performed in individual polypropylene containers at 75°C. After drying, those samples which were to be examined by ion back-scattering were coated with ∿200 Å of carbon to minimize surface charging during ion bombardment. Unleached control samples were polished, cleaned, and carbon-coated in a similar manner. Rod-shaped XPS samples were also prepared for fracture under ultrahigh vacuum conditions.

RESULTS

Ion backscattering spectra of three different glass compositions before and after leaching for one day in distilled water at 75°C are shown in Figures 1-3. The figures represent raw data in the form of counts as a function of backscattered alpha particle energy. Because of the slow and smooth behavior of the energy loss rate dE/dx as a function of energy, the horizontal energy axis

Table 1. Glass Composition (wt %)

	76-101	3008	76-68
SiO_2	59.7	44.3	39.8
B_2O_3	14.2	11.4	9.5
Na_2O	11.2	15.7	12.8
CaO	3.0	3.0	2.0
ZnO	7.45	5.7	5.0
TiO_2	4.45	4.8	3.0
Fe_2O_3	-	12.3	10.3
NiO	-	2.4	0.21
Cr_2O_3	-	0.5	0.44
Nd_2O_3	-	-	4.7
CeO_2, ZrO_2	-	-	1-2
Cs_2O, RuO_2	-	-	1.1
MoO_3	-	-	2.4
P_2O_5, Pr_6O_{11}, La$_2$O$_3$, BaO, PdO			1.0>x>0.5
Gd_2O_3, Eu_2O_3, Sm_2O_3, TeO_2, CdO, Ag_2O, Rh_2O_3, Y_2O_3, SrO, Rb_2O			x<0.5

associated with a given elemental profile may be thought of as an epproximate depth scale.

Figure 1 shows data on the 76-101 + 2 mole % UO_3 glass. The profile edges of all of the elemental components of the glass except boron (and any hydrogen left after leaching) are apparent. Two features of the spectra are noteworthy. First, there is very little change in the sodium profile after leaching. This is in sharp contrast to a series of soda-lime-silicate and soda-alumina-silicate glasses examined in an earlier study (to be published). Second is a substantial enhancement of the titanium and uranium concentrations near the glass surface. The uranium peak is about 1.7 times as high as the level in the bulk glass and appears to be no broader than the energy resolution of the technique (equivalent to ∿500 Å). The relative enhancement of the titanium at the surface may be as great but the smaller titanium signal is obscured by the statistical noize.

Figure 2 shows profiles of 3008 + 2 mole % UO_3 glass. Titanium and uranium show concentration-enhanced surface layers comparable to those in the 76-101 based glass. Zinc also shows surface enhancement not readily apparent in the 76-101 based glass. Iron and nickel present in the 3008 based glass do not yield individual resolvable profiles but the composite Fe + Ni profile does show surface enhancement.

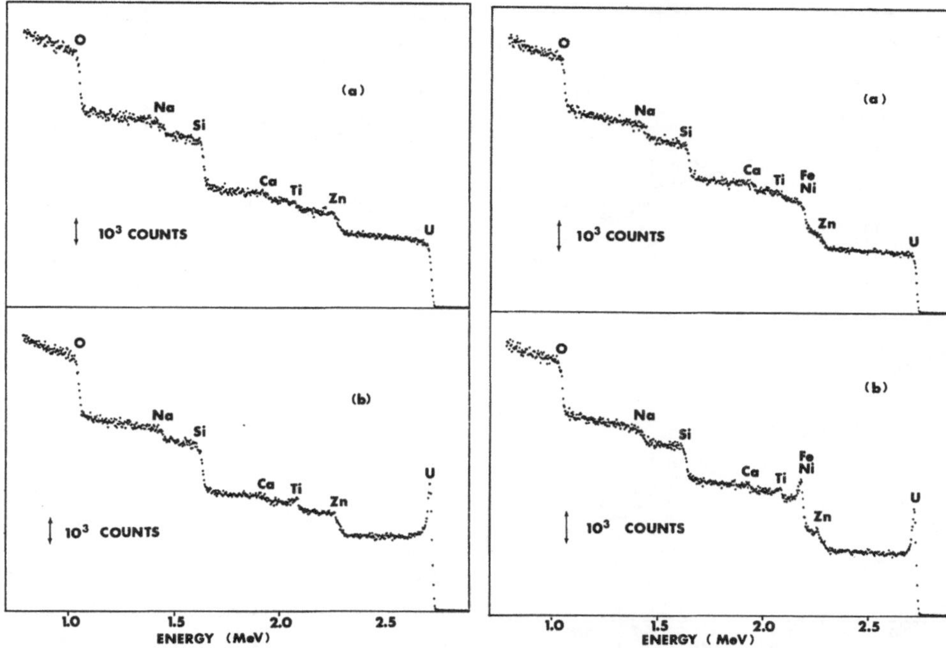

Fig 1. IBS profiles of 76-101 + 2% UO_3 (a) before (b) after leaching for 1 day at 75°C in distilled water.

Fig 2. IBS profiles of 3008 + 2% UO_3 (a) before (b) after leaching for 1 day at 75°C in distilled water.

Figure 3 shows profiles of 76-68 + 2 mole % UO_3 glass. The composition of this glass is quite complex and only the most prominent profile edges are noted. Many edges are weak and obscured but Fe + Ni enhancement is still clear. The rare earths are not individually distinguishable, but their composite profile, designated Ce, Nd on the figures, show surface enhancement. The uranium enhancement in Figure 3 is similar to that in the other glass compositions.

Figure 4 shows 500-electron-volt-wide XPS electra of 76-101 + 2% UO_3 glass for a freshly fractured surface (top) and after leaching in distilled water at 75°C for progressively longer periods. The intensity of a particular core line is proportional to the cross-section for photoemission from that core state times the concentration of that atomic species. The relative cross-sections can be determined by examination of the unleached sample which should be representative of the known bulk glass composition. For a given spectrum, concentrations can then be normalized to sum to 100%, thus determining the surface composition at different leach times. With this procedure, we find progressive enhancement in

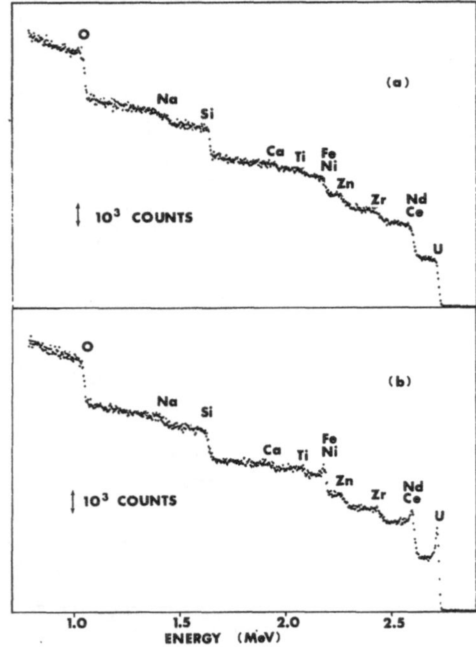

Fig 3. IBS profiles of 76-68 + 2% UO_3 (a) before (b) after leaching for 1 day at 75°C in distilled water.

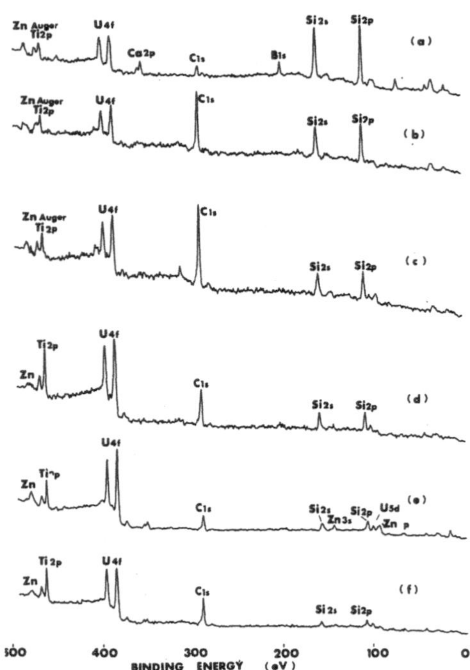

Fig 4. XPS spectra of 76-01 + 2% UO_3 (a) unleached, after leaching for (b) 1 hr (c) 3 hrs (d) 6 hrs (e) 1 day (f) 3 days at 75°C in distilled water.

TiO_2 and UO_3 and, to a lesser extent, in ZnO (to 33%, 15%, and 18%, respectively) as leach time increases (see Fig. 5). Necessarily, the SiO_2 concentration is reduced. In the earliest stages of leaching, there is an initial increase in SiO_2 presumably due to the even more rapid loss of Na and B.

The enhancement of U and Ti, as seen by XPS, is substantially greater than the peak enhancement as seen on the IBS profiles. The reason for this is that the IBS profile can only give the enhancement factor averaged over its depth resolution. The surface peaks in the true U and Ti profiles must be considerably narrower than the IBS profiles show and at least as high as the enhancement observed in the XPS spectra. Information from both techniques is valuable in studying the evolution of the leaching process; IBS to give the integrated excess U or Ti in the entire surface region, and XPS to give detailed compositional and chemical information about the 20-30 Å layer right at the glass-solution interface.

Surface enhancement could result from selective leaching or by redeposition from solution. An experiment was performed to determine which mechanism was responsible for the uranium enhancement in the 76-101 based glass. Two glass samples (one plain 76-101 and one containing UO_3) were placed in the same polypropylene container with ∿50 ml of distilled water and leached at 75°C for 24 hours as before. Gentle agitation was provided every few hours. If the uranium was being dissolved and redeposited, it should be detectable on both glass surfaces. XPS examination of the plain 76-101 glass did not reveal any uranium present indicating that the observed enhancement was due to selective leaching of other components and not redeposition from solution.

Figures 6a and 6b show XPS spectra of 76-68 + 2% UO_3 glass fractured in vacuum and leached for one day in distilled water at 75°C respectively. As the number of components increases, identification of many features in the spectra becomes more difficult. The relative surface concentrations of U, Si, Ti, and Zn seem to be similar to the 76-101 based glass. Zr shows substantial surface enhancement (rather ambiguous in the IBS profile). The Fe + Ni profile peak in the IBS spectrum would appear to be predominantly Fe from the XPS data.

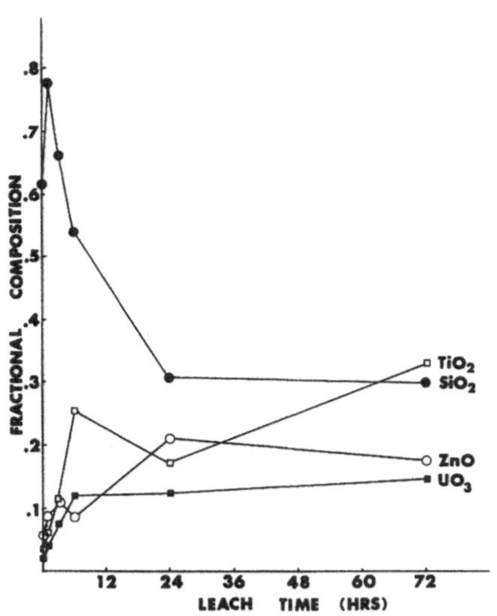

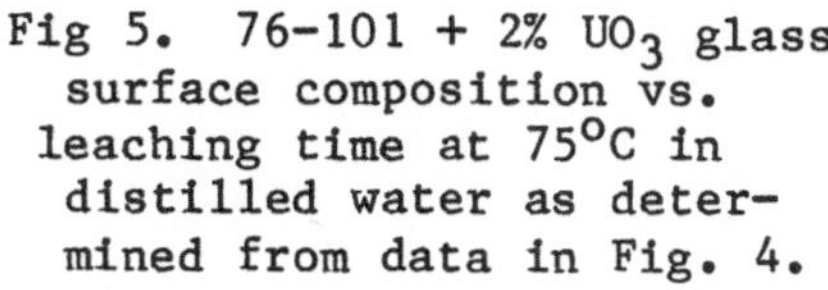

Fig 5. 76-101 + 2% UO_3 glass surface composition vs. leaching time at 75°C in distilled water as determined from data in Fig. 4.

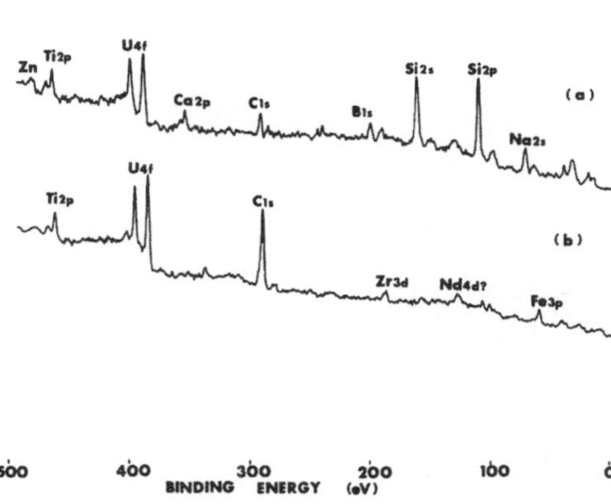

Fig 6. XPS spectra of 76-68 + 2% UO_3 (a) unleached; leached for 1 day at 75°C in (b) distilled water.

CONCLUSIONS

The surface composition of 76-101, 3008, and 76-68 simulated waste glasses containing UO_3 have been examined before and after leaching. Distilled water leaching in all of the glasses was characterized by the removal of Si, B, and Na from the surface region leaving a surface layer enhanced in Ti, Zn, Fe, the rare earths, and U.

The overall similarity in the behavior of the elements common to 76-101 and 76-68, especially uranium, is encouraging. XPS valence state identification in other actinides may rely on the analysis of weak satellite lines which might easily be obscured in the crowded spectra of 76-68. It would be much easier to study actinide behavior in the similar, but simpler, 76-101 glass.

ACKNOWLEDGEMENTS

This research was supported in part by the WRIT Program being conducted by the Office of Nuclear Waste Isolation, which is managed by Battelle Memorial Institute. The authors wish to thank A. W. Mitchell for his able technical assistance.

SPONTANEOUS FRAGMENTATION OF AN ALPHA-ACTIVE CERAMIC -- A MECHANISM FOR DISPERSION OF SOLID WASTE?*

F. W. Clinard, Jr., and D. L. Rohr

University of California
Los Alamos Scientific Laboratory
Los Alamos, New Mexico 87545

INTRODUCTION

$^{238}PuO_2$ is a short half-life (89 y) alpha-active ceramic often added to simulated solid nuclear waste to induce accelerated irradiation damage. Radiation effects in PuO_2 itself are also of interest, since this ceramic structurally resembles certain forms of nuclear waste. Some fabricated bodies of $^{238}PuO_2$ have been found to undergo spontaneous fragmentation,[1,2] resulting in a loss of material to the surrounding area. On the other hand $^{239}PuO_2$, with a half-life of 2.4 x 10^4 y, has not been observed to exhibit such behavior. If the mechanism responsible for fragmentation is operative in nuclear waste, the results could be dispersion of finely-divided waste. Even if the phenomenon is restricted to materials with greater rates of alpha decay than are characteristic of nuclear waste, fragmentation may cause misleading results from leach tests of waste forms doped with short half-life actinides. This paper reports on studies under way to characterize spontaneous fragmentation in $^{238}PuO_2$ and to determine the mechanism(s) responsible.

PROCEDURE AND RESULTS

Two forms of $^{238}PuO_2$ were available for this study. Fabrication methods for each were similar, involving hot-pressing in a graphite die at 1530°C and subsequent firing in an oxidizing atmosphere at 1440°C to restore full stoichiometry. The average density of each was ∿84% of theoretical. Nevertheless, the end product differed considerably: Material I was characterized by a

*Work performed under the auspices of the U. S. Department of Energy.

network of large cracks throughout the as-fired body, while Material II was essentially free of such cracks. All samples tested were fragments of larger bodies broken up earlier by impact testing as part of another study. Material I is also characterized by density gradients in the original body; thus random small pieces may differ in density from piece to piece.

All tests were carried out at room temperature; however, sample temperatures were slightly higher due to the considerable self-heat of $^{238}PuO_2$ (4 W/cm^3, for the 80% isotopic purity material studied here). Displacement damage and helium content build up in this ceramic with low-temperature storage after fabrication. In all but one case the history of samples studied here is such that the materials are considered "old", i.e., contain a high concentration of damage and helium. The exception is the material used for the weight change experiment, which was annealed just prior to initiation of the test.

Studies of Material I were carried out in ambient (glove box) air, while Material II was tested in a variety of controlled environments. Results are described in the following sections.

Material I -- Ambient Air

This form of $^{238}PuO_2$ undergoes spontaneous fragmentation readily during storage, as shown in Fig. 1. Particle sizes shown range from a few mm to the limit of detection with the unaided eye. An aqueous suspension technique and extraction replica electron microscopy were used to determine particle size distribution below 10 μm; the distribution was heavily weighted toward the finer fragments, with more than 90% being between 1 and 0.1 μm (the limit of detection at the magnifications used). In one instance a 10 μm particle was found to have broken up further during storage (Fig. 2), showing that fragmentation continues even in microscopic particles.

An experiment was carried out to determine roughly the magnitude of the rate of spallation. A collection of 70 small, irregular samples* of typical size 3 mm was annealed in flowing air for 6 h at 1600°C to remove accumulated displacement damage and helium.[3,4] A day later, periodic weighing of fines released by spontaneous fragmentation was begun. Fines were collected by dry-sieving through a 420 mesh sieve, using slight agitation. Figure 3 shows the cumulative weight of fines collected this way over a period of 266 days. A roughly linear increase of cumulative weight with time was observed, although a slight decrease in slope may have occurred at ∿120 days.

*Major non-actinide impurities in this material (in wt ppm) were: Ta 500; Mg 300; Si 230; Fe 140; Ca 100; and Ni 55.

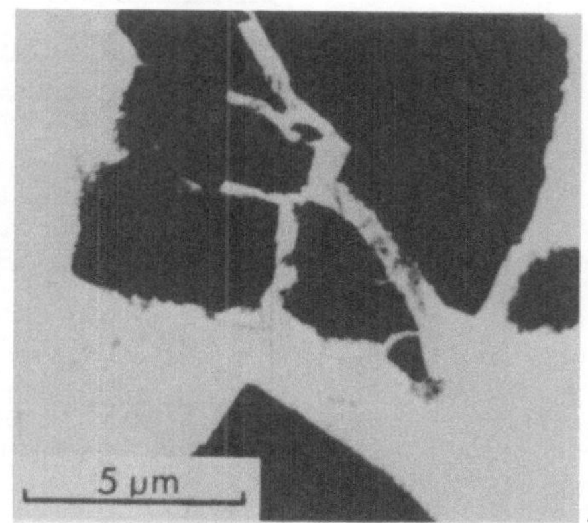

Fig. 1. (Left) Fragmentation product resulting from 4 months' storage of 23 chunks of $^{238}PuO_2$ (2 shown).

Fig. 2. (Right) Spontaneous breakup of a 10-micron fragment observed after several weeks' storage.

Surfaces of other samples from this batch (not annealed) were periodically replicated during the 266-day test to remove any loosely-adherent particles. Fines were extracted continually during this period, confirming that regular release with time does occur.

The possibility was considered that the significant self-heating power of $^{238}PuO_2$ might induce thermal stresses sufficient to cause spontaneous fragmentation. Finite element computer codes were used to evaluate thermal and stress fields in a cylinder 3 mm dia by 3 mm in height (roughly the size of the samples used in the weight change experiment).[5] Calculated stresses were low, with worst-case conditions* resulting in a maximum tensile stress of 6.4 MPa and a maximum compressive stress of 3.5 MPa. These stresses alone are not sufficient to cause fragmentation.

Material II -- Various Environments

In this study, five similar-sized pieces of Material II** (160 mg each) were stored in test tubes containing toluene, dry air, wet

*Vacuum environment, steel-on-steel thermal conductance at one end, and thermal conductivity of the ceramic reduced to 10% of its original value as a result of lattice damage.

**Major non-actinide impurities in this material (in wt. ppm) were: Ca 250; Si 230; Fe 200; Mg 150; Na 80; and Ni 65.

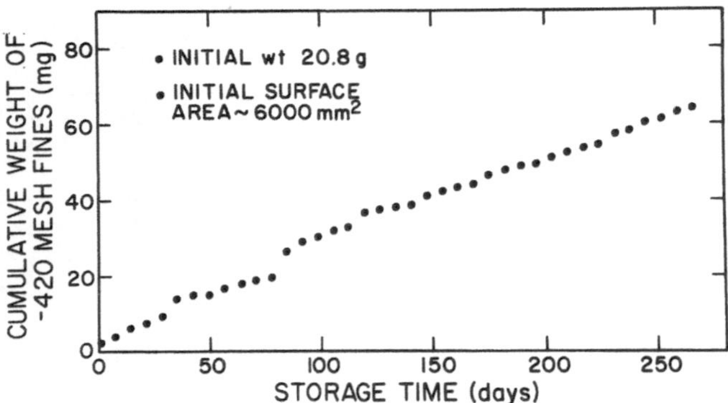

Fig. 3. Weight of fine particles (<34 μm) released by spontaneous fragmentation as a function of time.

air, distilled water, or brine for 71 days. Toulene is commonly used to achieve a moisture-free environment for static fatigue studies of ceramics and glasses. Dry air was obtained by the use of molecular sieve activated in vacuo at 250°C, while 100% relative humidity air was attained through use of a closed water bath. Salt content of the brine was 40 g per liter of distilled water. The amount of residue from each sample was determined by periodically removing the piece from its test tube to temporary storage area with the same environment, and then measuring the quantity of $^{238}PuO_2$ remaining in the tube by gamma counting the entire tube.

Results of this study to date are shown in Fig. 4. A strong environment-dependence is observed, with most samples exhibiting an approach to saturation. Relatively high rates of residue generation were observed for $^{238}PuO_2$ stored in an aqueous environment. This is apparently not due to thermal shock accompanying immersion of the samples, since exposure to toluene resulted in low release rates. As a comparison, if data from Fig. 3 (normalized either for weight or surface area of starting sample) were plotted on Fig. 4, the weight change curve would roughly superimpose on the distilled water curve, reaching a maximum value of $\sim 10^{-4}$ g. Some of the residue collected in the test tubes is visible to the naked eye, with ∿70% of the count in the distilled water being attributable to the presence of fragments. Particle size analysis of this residue will be carried out at the conclusion of the test.

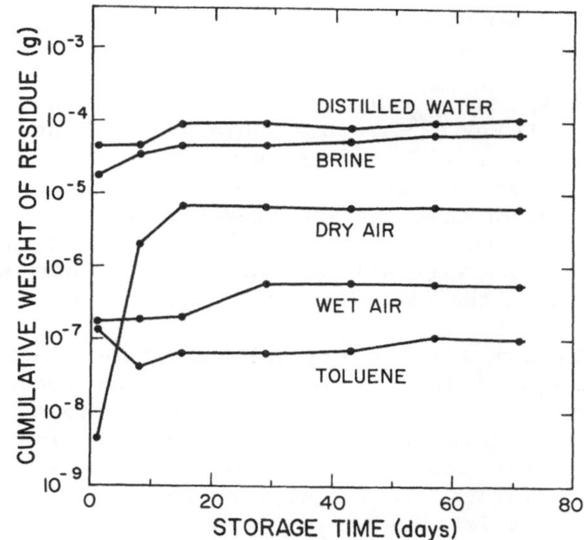

Fig. 4. Weight of residue accumulated during storage in various environments.

DISCUSSION

It seems likely, based on the fact that spontaneous fragmentation has been reported only in short half-life $^{238}PuO_2$, that a high rate of alpha decay is responsible. Given that assumption, * the question remains whether the phenomenon results from radiation-induced alterations to the ceramic itself or to its environment. With respect to the former possibility, structural degradation might result from thermal stresses, damage-induced lattice strain, or the effects of helium deposition. Calculations indicate that thermal stresses could not alone be responsible for fragmentation of mm-sized particles. Further, since such stresses decrease with sample size, they could not account for breakup of the

*Fragmentation studies of $^{239}PuO_2$ have been initiated to test this hypothesis.

10 μm particle shown in Fig. 2. Thermal stresses may, however, play a major role in fracture of large samples.

Lattice damage in $^{238}PuO_2$ is produced both by the 5.49 MeV alpha particle and the 93 keV ^{234}U recoil nucleus. The range of the latter can be estimated from calculations by Fleischer[7] to be 200 Å, while alpha particle range is expected to be ∿10 microns. Turcotte and Chikalla[3] have shown that lattice dilation builds up rapidly with time at room temperature, reaching 50% of saturation in ∿3 days and 90% at ∿10 days. Thus virtually all data reported here were obtained with the ceramic in the near-saturation condition.

Since PuO_2 has a cubic crystal structure, lattice dilation should be isotropic. Thus significant fracture-inducing stresses from this source are expected only near the surface, where damage gradients and therefore strain gradients exist. The thickness of this strained layer would be ∿10 microns at most. Since many fragmentation particles are larger than this, it may be concluded that damage-induced lattice strain is not directly responsible for the phenomenon in question. However, microcracks induced in this layer could lead to fragmentation of larger particles, if sufficient stresses were present from other sources to propagate the cracks.

Helium deposition does not saturate, but continues at a rate of 0.2 at.%/y. At elevated temperatures gas atoms can migrate to grain boundaries and form bubbles, thus compromising the structural integrity of the ceramic.[4] However, this is not expected at the low storage temperatures utilized here. In principle, accumulation of atomic helium will induce lattice strains as does displacement damage. In reality, the contribution of helium (and of ^{234}U atoms) is insignificant.[6]

We must also consider the possibility that alpha particle-induced radiolysis of the surrounding medium causes spontaneous fragmentation. By this argument, radiolysis would result in the formation of corrosive (that is, chemically active) products which attack the surface of the ceramic in such a way as to free small particles. The dependence of residue generation rate on environment (Fig. 4) supports this view.* Possible radiolysis products from the environments studied include: H, OH, and H_2O_2 from water, NO_2 from air, and HNO_3 from air-water systems. Several of these products

*In another study of environment-dependence, Fleischer and Raabe[1,7,8] investigated the release of clusters of atoms from PuO_2 powder during storage in simulated lung fluid, water, and vacuum. A strong dependence on the surrounding medium was found, with no release observed in vacuum. It was postulated that recoil nuclei-damaged microvolumes were released by exposure to water,but radiolysis was not considered in this work.

could prove corrosive to PuO_2. In studies of UO_2, Wang and Katayama[9] reported that the addition of H_2O_2 to water significantly increases the rate of attack of this ceramic, and could result in disintegration into fragments. It is noteworthy that neither water nor air is present in significant amounts in toluene, and that exposure to this medium in the present work resulted in the lowest observed accumulation of residue.

Fragmentation during storage may be enhanced by the presence of pre-existing microcracks (e.g., from earlier impact testing), which could propagate under the combined influence of a corrosive environment and internal stresses as in static fatigue. Possible sources of such stresses include thermal or compositional gradients, or strain gradients resulting from radiolysis-enhanced formation of a surface compound such as a hydrous oxide. The observation that Material I generated roughly two orders of magnitude more residue than did Material II under similar conditions (averaging wet air and dry air results for the latter) may correlate with the tendency of Material I to undergo cracking in fabrication. The trends toward saturation in amount of PuO_2 lost from Material II could then be attributed to the presence of microcracks on the surface of an otherwise relatively crack-free material.

SUMMARY

Results reported here show that:

- spontaneous fragmentation of $^{238}PuO_2$ generates a wide range of particle sizes, from a few mm to 1000 Å or less
- the phenomenon may continue with time or may saturate, depending on starting material
- the magnitude of the effect is dependent on storage environment.

Neither thermal stresses nor lattice damage appear to be solely responsible for fragmentation, but radiolysis of the environment could play an important role. Work is continuing in an effort to identify the controlling factors in this phenomenon.

ACKNOWLEDGMENT

The authors wish to thank G. M. Matlack at LASL for carrying out the radioanalyses reported here.

REFERENCES

1. R. L. Fleischer, On the "Dissolution" of Respirable PuO_2 Particles," Health Physics 29:69 (1975).
2. J. H. Patterson, G. M. Matlack, and F. J. Steinkruger (LASL), private communication.

3. R. P. Turcotte and T. D. Chikalla, Concentrated Defects in PuO_2, in: "Defects and Transport in Oxides," M. S. Seltzer and R. I. Jaffee, eds., Plenum Press, New York (1974).
4. B. A. Mueller, D. L. Rohr, and R. N. R. Mulford, Helium Release and Microstructural Changes in $^{238}PuO_2$, LASL Report LA-5524 (1974).
5. These calculations were carried out by G. A. Bennett, LASL.
6. T. D. Chikalla and R. P. Turcotte, Self-Radiation Damage Ingrowth in $^{238}PuO_2$. Rad. Effects 19:93 (1973).
7. R. L. Fleischer and O. G. Raabe, Fragmentation of Respirable PuO_2 Particles in Water by Alpha Decay--a Mode of "Dissolution," Health Physics 32:253 (1977).
8. R. L. Fleischer and O. G. Raabe, On the Mechanism of "Dissolution" in Liquids of PuO_2 by Alpha Decay, Health Physics 35:545 (1978).
9. R. Wang and Y. B. Katayama, talk presented at the Workshop on Alternate Nuclear Waste Forms and Interactions in Geologic Media, Gatlinburg, TN, May 1980.

EXPERIMENTAL STUDY OF STRUCTURAL DAMAGE IN CRYSTALLINE NUCLEAR WASTE PHASES FROM FISSION FRAGMENT IRRADIATION

E.R. Vance and K.K.S. Pillay

Materials Research Laboratory
The Pennsylvania State University
University Park, PA 16802, USA

ABSTRACT

Various uranium-doped candidate phases for crystalline nuclear waste forms were made up by ceramic techniques and were reactor-irradiated to a dose of $\gtrsim 10^{15}$ fission events$\cdot cm^{-3}$. Structural effects were studied by powder x-ray diffraction. Monazite, xenotime, perovskite, zirconolite, huttonite, and zircon, all possible TRU-fixing phases, were all rendered amorphous, whereas MgO, α-Al_2O_3, Fe_2O_3, ThO_2, Y_2O_3, and stabilized ZrO_2 were all apparently unaffected.

It was concluded that fission fragment irradiation is probably not a good simulator of α-recoil damage. However, a particular value of the results is that several candidate TRU-fixing phases can be readily made amorphous so the question of leachability and density differences of crystalline vs. amorphous phases should be able to be settled.

INTRODUCTION

After 10^6 yr., a waste form containing 20 wt.% of reprocessing waste would incur a radiation dose of $\sim 10^{20} \alpha \cdot cm^{-3}$ + 10^{13}-10^{14} fission events$\cdot cm^{-3}$, as well as β- and γ-radiation[1]. Such radiation effects would for instance render the refractory material zircon amorphous and cause a $\sim$16% increase in specific volume[2]. It has been generally held that by far the most destructive component of the radiation dose would be α-recoil nuclei, but it has been recently argued that in some cases purely ionizing radiation might have comparable effects[3].

To make laboratory studies which relate to the radiation response of a waste form for many years after solidification, it is clearly necessary to use accelerated testing. For α-recoil damage, this can be done most directly by incorporating short-lived α-emitters such as ^{236}Pu, ^{238}Pu or ^{244}Cm in the waste form.

However in the present work, we have elected to study the structural effects of fission fragment irradiation on candidate crystalline phases for ceramic waste forms. It has been demonstrated, for zircon at least, that the effects of fission fragment irradiation is qualitatively similar to natural α-recoil damage[4] and, as mentioned above, there will be actual fission fragment irradiation taking place in a waste form. From an experimental point of view, considerable lattice disruption can be induced in susceptible materials by small neutron fluences if enriched U is employed; in many cases, the radioactivity of the sample is negligible a few weeks after irradiation, and the cost of the experiments is quite modest.

The α-emitting nuclide host phases in The Pennsylvania State University tailored ceramics[5-7] are apatites, fluorite-structured dioxides, and monazites. Related phases are zircon and xenotime ($LnPO_4$), the monazite-structured huttonites ($ThSiO_4$), and the titanate hosts for Ln and An in the ceramic formulations of Ringwood[8,9], zirconolite ($CaZrTi_2O_7$) and perovskite ($CaTiO_3$). Of these, it is well known (10, for example) that the fluorite-structured dioxides are extremely radiation-resistant, but information on the radiation stabilities of the other minerals is fairly limited[11-15]. However, zircon has been studied in some depth and attempts made to interpret its radiation response at the atomic level[16,17].

EXPERIMENTAL

The various uranium-doped phases were made by mixing appropriate quantities of metal nitrate solutions, Ludox (a colloidal NH_3-stabilized suspension of SiO_2) and/or $NH_4H_2PO_4$ solutions. For the titanates, the TiO_2 was derived from anatase powder. The mixtures were dried at $\sim 100°C$, calcined at $\sim 600°C$ for several hours, finely ground, pelletized at a pressure of $\sim 3x10^4$ psi, and fired in air. Powders made from these materials were then irradiated in a nuclear reactor at a flux of $\sim 1x10^{13}$ $n/cm^2 \cdot sec$ (thermal) + $\sim 5x10^{12}$ $n/cm^2 \cdot sec$ (fast). The fission dose was calculated on the basis of thermal neutron fission of ^{235}U, fast fission of ^{238}U, together with thermal fission of ^{239}Pu derived from neutron capture on ^{238}U and subsequent β-emission. After irradiation, the specimens were stored for three months to allow the radioactivity to decay to levels which would not produce an appreciable background in the subsequent x-ray studies which were performed using a Siemens-Picker diffractometer and graphite-monochromated CuKα radiation.

RESULTS

The results are presented in Tables I-III. Most of the preparations contained impurity phases but space does not permit detailed discussions. No explicit lattice parameter measurements were made since, if broadening of the Bragg peaks occurs, the "lattice parameter" is not very meaningful[18,19]. In all cases, annealing at 1100°C was sufficient to remove the irradiation effects, as far as x-ray diffraction was concerned.

DISCUSSION AND CONCLUSIONS

It required fission fragment doses of $\sim$100 times those likely to be experienced in 10^6 yr. of storage for HLW derived from reprocessing to render amorphous several candidate crystalline hosts for α-emitting nuclei and other materials were essentially unaffected at these dose levels. Thus it seems that no gross structural effects on waste forms will derive from fission fragments alone.

For fission fragment irradiation to be a useful simulator of α-recoil damage, a necessary, but not sufficient condition would be that, in refractory materials, a fission fragment would be n times more damaging than an α-recoil nucleus, with n varying only slowly from material to material. However, whereas for zircon, $n \sim 10^4$ (4), for uraninite, $n < 1$ (20), so fission fragment irradiation is, in this sense, a poor simulator of α-recoil damage. The reason is probably as follows. For natural zircon, assuming 1 displacment per atom is necessary to produce amorphism, the number of displaced atoms per α-recoil can be found as $\sim 10^3$ (4), and this number agrees quite well with that calculated[21] from the Kinchin-Pease relationships[22]. However, performing the same analysis for fission fragment irradiation, we arrive at $\sim 10^7$ displaced atoms per fission event[4,10], compared with a figure of $\sim 10^4$ calculated from the Kinchin-Pease formula[1,21,23]. This discrepancy has been pointed out previously[10] for zircon, U_3O_8, and α-Al_2O_3 and must arise from ionization-related displacements not considered in the Kinchin-Pease discussion. A similar discrepancy occurs for the phase transformation in zirconia[24,25].

A more positive aspect of the work is that it appears that many materials can easily be rendered amorphous by fission fragment irradiation so that (limiting) density changes and leachability changes can be studied. For zircon at least, the density of a natural fully-metamict sample was unaffected by heavy fission fragment irradiation[26]. Utilization of ion bonbardment techniques[27,28] offers in principle a means of studying differential leaching effects, but only very thin layers can be irradiated; density changes are not amenable to study. It should be mentioned that the Ar ion irradiation effects[28] were observed at dose levels of

Table I. Fission Fragment Irradiation of Silicate and Phosphate Phases Which Could Accommodate Transuranic Elements.

Material	*Depleted (D) or Enriched (E) U	Irradiation time (hrs.)	Calculated Fissions cm^{-3} ($x10^{-14}$)	Firing Temp (°C)	Effect on diffraction pattern
ZIRCON					
$(Zr_{0.99}U_{0.01})SiO_4$	D	100	4	1400	Slight broadening
"	E	1	20	"	Broadening
"	E	3	60	"	Near-amorphism
$(Zr_{0.95}U_{0.05})SiO_4$	D	100	20	"	Broadening
XENOTIME					
$(Y_{0.99}U_{0.005}Ca_{0.005})PO_4$	D	100	2	1200	V.slight broadening
$(Y_{0.98}U_{0.01}Ca_{0.01})PO_4$	E	1	20	"	Broadening
"	E	3	60	"	Near-amorphism
$(Y_{0.9}U_{0.05}Ca_{0.05})PO_4$	D	100	20	"	Broadening
MONAZITE					
$(Ce_{0.98}U_{0.01}Ca_{0.01})PO_4$	D	100	4	1200	Broadening
"	E	1	20	"	Amorphism
"	E	3	60	"	Amorphism
$(Ce_{0.9}U_{0.05}Ca_{0.05})PO_4$	D	100	20	"	Amorphism
$(La_{0.98}U_{0.01}Ca_{0.01})PO_4$	D	100	4	"	Broadening
"	E	1	20	"	Amorphism
"	E	3	60	"	Amorphism
$(La_{0.9}U_{0.05}Ca_{0.05})PO_4$	D	100	20	"	Amorphism
HUTTONITE					
$(Th_{0.99}U_{0.01})SiO_4$	D	100	4	1400	Broadening
"	E	1	20	"	Near-amorphism
"	E	3	60	"	Amorphism
$(Th_{0.95}U_{0.05})SiO_4$	D	100	20	"	Amorphism
APATITE					
$Ca_2La_8(SiO_4)_6O_2$ + 1 mole % UO_2	D	100	0.5	1400	V.slight broadening?

*"Depleted" uranium was fully depleted. "Enriched" uranium contained 93% of ^{235}U.

Table II. Fission Fragment Irradiation of Titanate Ceramic Phases[8,9].

Material	*Depleted (D) or Enriched (E) U	Irradiation Time (hrs.)	Calculated Fissions cm^{-3} $(x10^{-14})$	Firing Temp (°C)	Effect on diffraction pattern
ZIRCONOLITE					
$Ca(Zr_{0.99}U_{0.01})Ti_2O_7$	E	1	15	1400	Broadening
"	E	3	45	"	Near-amorphism
"	D	100	3	"	Mild broadening
$Ca(Zr_{0.95}U_{0.05})Ti_2O_7$	D	100	15	"	Broadening
$Ca(Zr_{0.9}U_{0.01})Ti_2O_7$	D	100	30	"	Near-amorphism
$Ca(Zr_{0.8}U_{0.2})Ti_2O_7$	D	100	60	"	Amorphism
PEROVSKITE					
$CaTiO_3$ + 1 mole % UO_{2+x}	D	100	7	1400	Heavy broadening
$(Ca_{0.99}U_{0.01})(Ti_{0.98}Fe_{0.02})O_3$	E	1	30	"	Near-amorphism
"	E	3	100	"	Amorphous
$SrTiO_3$ + 1 mole % UO_{2+X}	D	100	7	"	Mild broadening
'HOLLANDITE'					
$BaAl_2Ti_5O_{14}$ + TiO_2 + 1 mole % UO_{2+x}	D	100	1.5	1400	Broadening

*"Depleted" uranium was fully depleted. "Enriched" uranium contained 93% of ^{235}U.

Table III. Fission Fragment Irradiation of Oxide Phases Which Could Possibly Constitute Parts of Tailored Ceramics.

Material	Fissions cm^{-3} ($x10^{-14}$)	Firing Temp (°C)	Effect on Diffraction Pattern
α-alumina*	20	1200	-----
Magnesia*	20	1200	-----
Monoclinic* Zirconia $(Zr_{0.99}U_{0.01})O_2$	10	1200	Slight broadening, possible slight formation of diffuse peaks from cubic ZrO_2
Stabilized* Zirconia $(Zr_{0.85}Y_{0.14}U_{0.01})O_{2-x}$	10	1400	-----
Yttria*	10	1200	-----
Thoria (impurity in huttonite preparation)	20-60	1400	-----
Rutile (impurity in hollandite and perovskite preparations)	1.5	1400	Broadening
	7	1400	Broadening
	100	1400	Amorphism

*For main phases investigated, doping was with 1 mole % of depleted UO_{2+x} and irradiation time was 100 hours.

several orders of magnitude above that expected in 10^6 yr. of HLW storage.

A more detailed discussion of the atomic-level changes in the materials studied is in preparation for publication elsewhere.

ACKNOWLEDGMENTS

This research was performed under subcontract with Rockwell International under contract DE-AC09-79ET41900 with the Department of Energy. The contract was administered by the Savannah River Operations Office. We thank R. Roy for helpful discussions.

REFERENCES

1. G. Malow and H. Andersen (1979), Helium Formation from α-decay and its Significance for Radioactive Waste Glasses, in Scientific Basis for Nuclear Waste Management, Vol. I, Ed. G.J. McCarthy, Plenum, NY, 109.

2. H.D. Holland and D. Gottfried (1955), The Effect of Nuclear Radiation on the Structure of Zircon, Acta Cryst. 8, 291.
3. L.W. Hobbs (1979), Application of Transmission Electron Microscopy to Radiation Damage in Ceramics, J. Amer. Ceram. Soc. 62, 267.
4. E.R. Vance and J.N. Boland (1975), Fission Fragment Damage in Zircon, Radiation Effects 26, 135.
5. G.J. McCarthy and M.T. Davidson (1975), Ceramic Nuclear Waste Forms: I, Crystal Chemistry and Phase Formation, Bull. Amer. Ceram. Soc. 54, 782.
6. G.J. McCarthy (1976), High-Level Waste Ceramics, Trans. Amer. Nuclear Soc. 23, 168.
7. G.J. McCarthy (1977), High-Level Waste Ceramics: Materials Considerations, Process Simulation and Product Characterization, Nucl. Tech. 32, 92.
8. A.E. Ringwood, S.E. Kesson, N.G. Ware, W. Hibberson, and A. Major (1979), Immobilization of High Level Nuclear Reactor Wastes in SYNROC, Nature 278, 219.
9. A.E. Ringwood, S.E. Kesson, N.G. Ware, W.O. Hibberson, and A. Major (1979), The Synroc Process: A Geochemical Approach to Nuclear Waste Immobilization. Geochem. Journal 13, 144.
10. R.M. Berman, M.L. Bleiberg, and W. Yeniscavish (1960), Fission Fragment Damage to Crystal Structures, J. Nucl. Mat. 2, 129.
11. A. Pabst (1952), The Metamict State, Amer. Mineral. 37, 137.
12. R.S. Mitchell (1973), Metamict Minerals: A Review, Min. Record 4, 177, 214.

13. R.F. Haaker and R.C. Ewing (1979), The Metamict State Radiation Damage in Crystalline Materials, Ceramics in Nuclear Waste Management, Eds. T.D. Chikalla and J.E. Mendel, DOE Publication, CONF-790420, 305.
14. W.A. Ross, R.P. Turcotte, J.E. Mendel, and J.E. Rusin (1979), A Comparison of Glass and Crystalline Waste Materials, Ceramics in Nuclear Waste Management, Eds. T.D. Chikalla, and J.E. Mendel, DOE Publication, CONF-790420, 52.
15. A.E. Ringwood and V. Oversby (1980), Effects of Radiation Damage in Long-Term Stability of Synroc Minerals, Scientific Basis for Nuclear Waste Management Vol. 2, Ed. C.J. Northrup, Plenum, NY, in press.
16. L.A. Bursill and A.C. McLaren (1966), Transmission Electron Microscope Study of Natural Radiation Damage in Zircon, Phys. Stat. Sol. 13, 331.
17. E.R. Vance, L. Efstathiou, and F.H. Hsu (1980), X-ray and Positron Annihilation Studies of Radiation Damage in Natural Zircons, Radiation Effects. In press.
18. B.S. Hickman and D.G. Walker (1965), Growth of Magnesium Oxide During Neutron Irradiation, Phil. Mag. 11, 1101.
19. M.A. Krivoglaz (1969), Theory of X-ray and Thermal-Neutron Scattering by Real Crystals, Plenum, NY, 249.
20. W.J. Weber (1980), Ingrowth of Lattice Defects in Alpha-Irradiated UO_2 Single Crystals, PNL-SA-8574.
21. F.P. Roberts, G.H. Jenks, and C.D. Bopp (1976), Radiation Effects in Solidified High-Level-Wastes -- Part I, Stored Energy, USERDA Report BNWL-1944, Battelle Pacific Northwest Laboratories, Richland, WA.
22. G.H. Kinchin and R.S. Pease (1955), The Displacement of Atoms in Solids by Radiation, Repts. Prog. Phys. 18, 1.
23. M. Antonini, F. Lanza, and A. Manara (1979), Simulations of Radiation Damage in Glasses, Ceramics in Nuclear Waste Management, Eds. T.D. Chikalla and J.E. Mendel, DOE Publication, CONF-790420, 289.
24. M.C. Wittels, J.P. Steigler, and F.A. Sherrill (1962), Radiation Effects in Uranium-doped Zirconia, Reactor Science and Technology (J. Nucl. Energy, Pts. A/B) 16, 237.
25. E.R. Vance and J.N. Boland (1978), Fission Fragment Irradiation of Single Crystal Monoclinic ZrO_2, Radiation Effects 37, 237.
26. E.R. Vance and B.W. Anderson (1972), Differences Among Low Ceylon Zircons, Min. Mag. 38, 721.
27. J.C. Dran, M. Maurette and J.C. Petit (1980), Radioactive Waste Storage Materials: Their α-Recoil Aging. Science 209, 1518.
28. E.H. Hirsch (1980), A New Irradiation Effect and its Implications for the Disposal of High-Level Radioactive Waste. Science, 209, 1520.

METAMICTIZATION BY HEAVY ION BOMBARDMENT OF α-QUARTZ, ZIRCON, MONAZITE AND NITRIDE STRUCTURES

L. Cartz, F.G. Karioris, R.A. Fournelle,
K. Appaji Gowda, K. Ramasami, G. Sarkar
Marquette University
Milwaukee, WI 53233 USA

M. Billy
Université de Limoges
87060 Limoges, France

The preparation of metamict forms of crystal structures has been carried out by heavy ion bombardments under well controlled conditions (1,2). The penetration of Kr^+ or Ar^+ ions, of energy 4 Mev, is about 1-2 μm, so a fine powder of such particle sizes needs to be used. Heavy ion bombardment has some advantages over the methods of fast neutron irradiation (3) or actinide doping in the preparation of metamict specimens. Only a few minutes of heavy ion bombardment is sufficient to produce metamict specimens in many cases in contrast to the extremely long times required by the other methods. In addition, the specimens do not become radioactive.

The heavy ion bombardments of the crystalline powders are carried out using the Dynamitron, Argonne National Laboratory. A horizontal beam of ions of energies up to 4 Mev, is spread out over an area, to irradiate a monolayer of particles of sizes <3 μm held onto a vertical metal plate. The irradiated specimen is of an area and size suitable for direct insertion into an x-ray diffractometer, so that x-ray diffraction patterns before and after irradiation can be compared. An extremely dilute rubber cement in benzene has been used to attach the crystalline particles to the metal plate, though later experiments have used sedimented powders with no binder of any sort being present. This latter method has permitted the preparation and collection, for example, of about 100 mg of metamict powder irradiated at 5 ions/nm^2 in about an hour.

Several categories of materials have been irradiated and

studied. These include the α-quartz structures (GeO_2, $FePO_4$, $AlPO_4$, SiO_2, $GaPO_4$), the zircon structures ($ZrSiO_4$, YPO_4, $CaCrO_4$), monazite and huttonite, several nitrides α and β Si_3N_4, AlN, ZrN and Si_2N_2O. In addition, several layer structures have been examined (Graphite, BN, MoS_2, talc), and some minerals such as perovskite, zirconolite and ThO_2.

The x-ray diffraction patterns obtained of the monolayer of particles consist of the most intense diffraction lines, sometimes only 3-5 in number. In a few cases where more adequate quantities of irradiated materials have become available, better x-ray diffraction patterns have been obtained.

For the nitrides and ThO_2, no changes are observed in the x-ray diffraction intensities at ~5000 ions/nm^2. For the other compounds listed in Table I, decreases are observed in the x-ray diffraction line intensities at fluences of ~5 ions/nm^2 though without any significant peak shifts or line broadening. In the case of layer structures where the specimen preparation is different, extensive peak shifts and broadening are observed and this is to be reported elsewhere. The specimens that become damaged seem to demonstrate a two-phase condition. The damaged powder consists of crystalline and non-crystalline fractions. This model for metamictization is essentially that a non-crystalline zone is created by the atomic collision cascade and at least a part remains fully non-crystalline after the quenching of the energy spike (4). Decrease of the diffraction intensities with fluence follows a first order law dependence where

$$\frac{I_F}{I_o} = \exp -(D_M.F)$$

$I_{F,o}$ being the x-ray diffraction intensities at fluence F and zero and D_M is the metamict damage cross-section. The ratio I_F/I_o represents the residual undamaged crystalline fraction at fluence F.

Metamict damage cross-sections are presented in Table I. In deriving these values, it was noted that the diffraction intensities decreased with fluence, attaining some constant low value. This constant value, indicating a small fraction of residual crystalline material, is believed to be due to some particles too large to be completely damaged by the heavy ion beam. This constant value has been subtracted from the observed intensities before preparing the semi-log plots which are linear and of slope $-D_M$. Some of the values of D_M are preliminary, particularly those of YPO_4 and $CaCrO_4$, due to difficulties in obtaining fine particle sizes. Semi-log plots of intensity against fluence are given in Fig. 1 for Huttonite and Monazite.

Table I

Metamict Damage Cross-Sections D_M of Various Crystal Structures Bombarded With Ar^+ (3 Mev) Ions

α-Quartz	D_M nm^2	Zircon	D_M nm^2
GeO_2	3.0 ± 0.3	$ZrSiO_4$	0.14 ± 0.02
$FePO_4$	2.7 ± 0.2	YPO_4	1.8 ± 0.1
$AlPO_4$	1.9 ± 0.4	$CaCrO_4$	0.5 ± 0.5
SiO_2	0.9 ± 0.3		
$GaPO_4$	0.7 ± 0.3		
Monazite Huttonite		Other Structures	
$(Ce,La,Y)PO_4$	0.46 ± 0.06	Si_2N_2O	0.03 ± 0.01
$ThSiO_4$	0.27 ± 0.03	$CaTiO_3$	0.25 ± 0.01

ThO_2, α and β Si_3N_4, AlN, ZrN show no perceptable change in x-ray diffraction intensities after 5000 ions/nm^2.

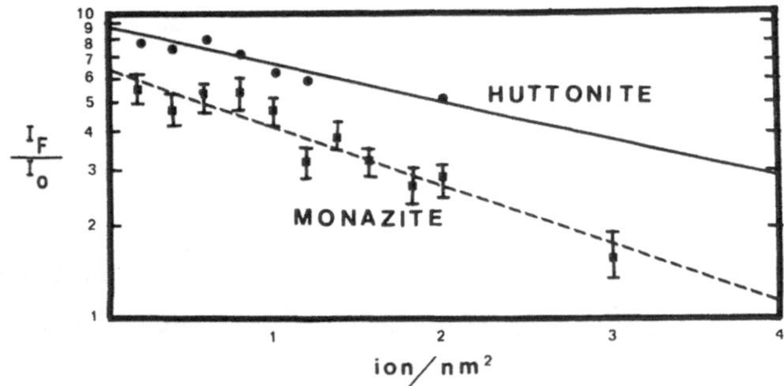

Figure 1 Semi-log plots of relative x-ray line intensity versus fluence of Ar^+ (3 Mev) ions for monazite and huttonite. The slope is $-D_M$.

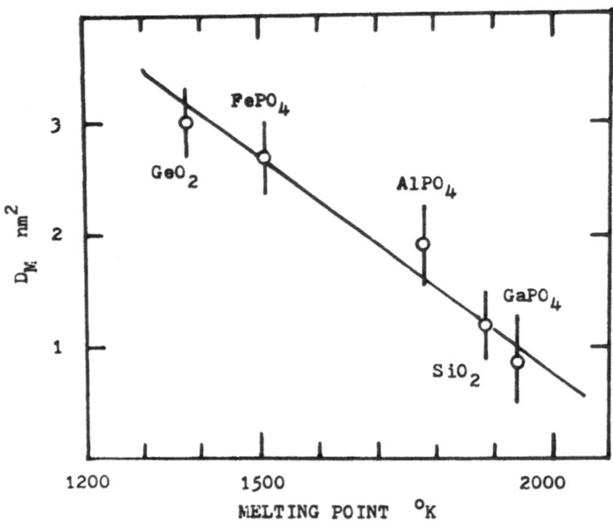

Figure 2 The observed metamict damage cross-section D_M plotted versus melting point for the α-quartz structure irradiated with Ar^+ (3 Mev) ions.

The α-quartz structures all become non-crystalline under heavy ion bombardment and D_M varies from 3.0 nm^2 for the lowest melting point material GeO_2, to 0.7 nm^2 for the highest m.p. material $GaPO_4$. Indeed, there is a linear relationship between D_M and the melting point such that $D_M = 8.6 - 4 \times 10^{-3}\ T_{m.p.}$ where D_M is in nm^2 and $T_{m.p.}$ is the melting point (Kelvin) for the α-quartz compounds examined. This relationship is shown in Fig. 2. This suggests that an α-quartz structure of m.p. above 2150K, would not become metamict since $D_M = 0$. Recrystallization studies are in progress and the first results indicate that GeO_2 anneals at ~650K, $GaPO_4$ at ~750K and the other compounds at intermediate temperatures. These annealing temperatures are somewhat less than 1/2 $T_{m.p.}$ in each case.

Zircon $ZrSiO_4$ becomes non-crystalline but with a smaller value of $D_M = 0.14$ nm^2, while YPO_4, with $D_M = 1.8$ nm^2, is comparable to the α-quartz structures. The recrystallization temperature (preliminary) of YPO_4 is ~600K which is much lower than that of $ZrSiO_4$ (~1270K).

Huttonite and Monazite both damage rather readily with values of D_M greater than that of $ZrSiO_4$ zircon (5). This is in contrast to the suggested stability of the monazite structure (6,7) though it should be noted that monazite recrystallizes after 20 hours at 570K and it may be this low temperature of annealing which makes natural monazite appear to be relatively stable to radiation damage. Similarly, the fact that the mineral xenotime YPO_4 is not generally found in a metamict condition (8) may be due to its relatively low annealing temperature.

Some observation on the crystal structures of zirconolite, and perovskite show that these materials all become metamict at relatively small fluence. On the other hand, nitride ceramics and also ThO_2 are stable to the effects of high fluence of heavy ion irradiations (2). Si_2N_2O is the only nitride here examined to become metamict though with a very small value of D_M (0.03 nm^2).

The values of D_M correspond to metamict damage radii R_M, defined by $D_M = \pi R_M^2$, of values ranging from 0 for the nitrides to ~10Å for GeO_2. These values may be related to the size of the non-crystalline region remaining after the passage of the bombarding ion, as well as to the type of crystal structure under irradiation and to the mp of the crystal. Studies are in progress using different bombarding heavy ions of energies up to 4 Mev on these same crystal types.

The different response of silicates, phosphates, oxides and nitrides to irradiation may be related to the different types of framework arrangements of the coordination polyhedra present in their crystal structures (2). Where bond angles can vary a flexible structure is obtained. Where bond angles are fixed, a rigid crystalline structure results. The α-quartz structure has a flexible framework where the Si-O-Si bond angle can vary so that neighboring silica tetrahedra can undergo cooperative tilting (9). The α and β Si_3N_4 have rigid framework structures since the Si-N-Si bond angle is fixed (10), and Si_2N_2O is intermediate with the Si-O-Si bond angles variable and the Si-N-Si bond angles fixed (11,12). Physical properties such as thermal expansion and compressibility are known to depend on the flexible or rigid nature of the framework arrangements (13,14,15,16). Radiation damage to a crystal structure with a flexible framework can result in regions with polyhedra linked together by a variety of bond angles and this will give rise to non-crystalline, glassy or metamict conditions. The rigid framework structures such as α and β Si_3N_4 do not permit any such non-crystalline condition to arise. The case of Si_2N_2O is intermediate so that this compound can become non-crystalline though with a metamict damage cross-section much smaller than that of α-quartz SiO_2. ThO_2 has the fluorite-type structure which is highly symmetrical with essentially fixed bond angles so that a non-crystalline condition cannot be achieved.

Metamict silicates are all observed to undergo a creep process under heavy ion bombardment at fluences ~200/nm^2 and this is also observed immediately in silicate glasses. The heavier the irradiation fluence, the more distended the specimens become (1,2). This glassy flow raises the question as to whether this is a process of melting under heavy ion bombardment or is due to the impact energy of the heavy ion on a very viscous glassy medium. However, the fact that low temperature annealing of metamict material occurs well below their melting points vitiates the possibility of a simple heating effect. This creep process under irradiation can explain several effects of sintering of powders that have been reported (17).

A host material for nuclear waste management needs to be stable to radiation damage effects. These studies have shown that the nitride compounds damage much less rapidly or not at all compared to silicate and other compounds with oxygen bonding. As a consequence, the nitride ceramics can be suggested as possible host materials.

REFERENCES

1. L. Cartz and R.A. Fournelle, Rad. Effects, 41, 211 (1979).
2. L. Cartz, F.G. Karioris and R.A. Fournelle, "Heavy Ion Bombardment of Silicates and Nitrides," Rad. Effects, (To be published.)
3. A.J. Leadbetter, P.A. Phillips and A.F. Wright, Phil. Mag., 34, 453-464 (1976).
4. G. Carter and R. Webb, Rad. Effects Letters, 43, 19-24 (1979).

5. F.G. Karioris, K. Appaji Gowda and L. Cartz, "Heavy Ion Bombardment of Monoclinic $ThSiO_4$, ThO_2 and Monazite," Rad. Effects Letters (submitted).
6. Unsigned Article, "Physics Today," April 1980, pp. 21-22.
7. R.A. Kerr, Science, 204, 289 (1979).
8. R.S. Mitchell, Mineralogical Record, 4, 177-182, 214-233 (1973).
9. J.D. Jorgensen, J. Appl. Phys., 49, 5473-5478 (1978).
10. S. Wild, P. Grieveson and K.H. Jack, "Special Ceramics," 5, 385, Editor, P. Popper, Brit. Ceram. Res. Assoc., (1972).
11. I. Idrestet and C. Brosset, Acta Chem. Scand., 18, 1879-1886, (1964).
12. S.R. Srinivasa, L. Cartz, J.D. Jorgensen, T.G. Worlton, R.A. Beyerlein and M. Billy, J. Appl. Cryst., 10, 167-171 (1977).
13. L. Cartz and J.D. Jorgensen, J. Appl. Phys. (to be published).
14. L. Cartz and J.D. Jorgensen, "Pressure and Temperature Behavior of Framework Crystal Structures," Proc. 7th Int. Thermal Expansion Conference 1979 (Plenum, to be published).
15. H.D. Megaw, Mat. Sci. Bull., 6, 1007-1018 (1971).
H.D. Megaw, "Crystal Structures; A Working Approach," (W.B. Saunders, 1973).
16. D. Taylor, Min. Mag., 38, 593-604 (1972).
17. A.C. Damask, Rad. Effects, 1, 95-100 (1969).

EXPERIMENTAL EVALUATION OF CHANGES IN PROPERTIES OF NATURAL MINERALS UNDER IRRADIATION

V. I. Spitsyn, V. D. Balukova, I. M. Kosareva, and S. A. Kabakchi

Institute of Physical Chemistry, USSR Academy of Sciences
Moscow, USSR

INTRODUCTION

The present paper is concerned with a further study of the influence of irradiation on natural minerals, which constitute the storage medium for radioactive wastes.[1] The influence of irradiation on the typical alumosilicate feldspar minerals microcline and oligoclase has been studied. The threshold registration doses of radiation effects were evaluated. The increased rate of supplied energy and optimum temperatures for storage in halite layers were determined for salt structures.

EXPERIMENTAL

The minerals were irradiated with γ-quanta ^{60}Co at a dose rate of 2.1 Mrad/hr up to the doses of 10^8, 10^9, and 10^{10} rad. Comparison of the normal crystal lattice energy of oligoclase and microcline with that obtained during irradiation at 10^9 rad shows that the latter accounted for 5-6% of the crystal lattice energy. Changes in the surface layers can be observed starting from a dose of 10^8 rad.

In the IR spectra of microcline, a change in the correlation of the intensity of absorption bands is observed within a range of wave numbers typical of the fluctuations of the Si-O bond, provided silicon enters into the tetrahedron SiO_4. This fact suggests the possibility of the formation of a compound with a different kind of bond from that found in amorphous silica. Analogous, but more pronounced changes were observed in the IR spectra of oligoclase.

Thermograms were used to determine the changes relating to the radiolysis of the water which is nonstructurally bound. Thermograms for minerals irradiated at 20 °C and nonirradiated, but previously annealed at 500 °C were identical. Radiolysis of water did not introduce any changes to the structure of minerals but did influence the coherence between separate grains (i.e., the rock dispersity). Microscopic investigation of irradiated samples showed a change in the color of the grains, from transparent to dim and dark. Similar changes in the color of the above-mentioned minerals took place during irradiation with an integral beam of neutrons at a dose of 10^{10} rad. An increase in the dose rate leads to further changes in the optical properties; amorphization and a decrease in density by several percent is observed. The samples gradually become x-ray amorphous. Under γ-irradiation the above-mentioned changes are observed starting with a dose of 10^8 rad. Further increases in the dose rate do not modify the nature of the changes. These changes in the physical properties and structure lead us to suppose that there exists a possibility of linear and volume deformations, with the latter most likely to occur in the subsurface layers.

The change in the color of the minerals as a result of irradiation is caused by the appearance of new radiation electron-hole centers. In feldspar these are represented by the Al-O-Al centers with two aluminum atoms, of which one is structural and the other is admixed. The possibility of the accumulation of energy in the lattice cannot be excluded. The data provided by the x-ray structural analysis of irradiated and nonirradiated samples show the appearance of new responses and a change in the intensity of the initial response. The latter process is small; therefore, it does not suggest the formation of a new phase. However, the above data on the physical properties plus indirect experiments on the evaluation of sorption properties on the surface of minerals after irradiation indicate changes in the surface and subsurface layers. As in the case of quartz, irradiation of minerals results in an increase of their sorption capacity with respect to microquantities of radioisotopes (cesium and strontium) as well as a decrease in the time of equilibrium distribution of isotopes during their sorption.

Thus, it is quite obvious that thermally stable (i.e., up to the temperature 1200°C) minerals belonging to the feldspar group have threshold doses of radiation stability within a range of 10^8-10^9 rad. This is also typical of quartz, as described in an earlier paper.[1] The similarity of the structure (frame-silicates) of quartz and feldspar permits the assumption to be made on the similarity of their thermal and radiation characteristics. The latter assumption is confirmed by IR spectra and indirect determinations of changes in sorption parameters.

The study of minerals with lamellar structure was done using muscovite as an example. The changes in the IR spectra and properties of this mineral can be registered at irradiation doses higher than those for feldspar. Secondary manifestation of the changes in the sorption parameters for stratified minerals is much less. However, the threshold dose of radiation stability also lies within the range 10^8-10^9 rad.

Taking into account the results of irradiation of the most widespread natural alumosilicate minerals quartz, feldspars, caolinite, chlorite, and others, we conclude that the coherence of minerals and particles in the structure formation depends, besides the radiolysis of water, on the changes which take place in the subsurface layers of the mineral grains at appropriate irradiation doses. These changes are capable of creating tension in the structure, modifying the density and mechanical stability, thus giving rise to dispersal. According to our data, a dose of 10^7-10^8 rad is the "threshold dose" marking the beginning of these effects; hence, it seems reasonable not to raise the radiation load on the alumosilicate structure higher than this value.

SALT STRUCTURES

In crystals of alkaline-halide compounds such as halite, the color centers are stable at room temperatures. A temperature rise leads to their decay, and the energy accumulated by the crystal in the form of color centers of various nature is liberated in the form of heat and light quanta. This accumulated energy represents one of the important characteristics of the irradiated crystal.[2] In particular, the resistance of the irradiated crystal to destruction is inversely proportional to the quantity of accumulated energy.[3] Liberation of energy as a result of heating irradiated crystals may be very dangerous. Hence, it seemed interesting to determine the quantity of accumulated energy during irradiation of minerals constituting the radioactive storage medium, as well as to find the temperatures at which this accumulated energy is liberated.

Accumulated energy in alkaline-halide crystals can be determined by two methods — either by calorimetry or by the dissolution of the irradiated crystal. We determined the accumulated energy by the second method. We measured the quantity of molecular hydrogen, which is formed during the dissolution of a mass unit of the crystal, irradiated at a given dose. There exists a correlation between the hydrogen quantity and accumulated energy $E/2\ NM = 4.6$, where E is the accumulated energy, eV/g, N is the Avogadro number, and M is the quantity of hydrogen formed during the dissolution of 1 g of halite irradiated at a given dose. It was found that accumulated energy increases linearly with the absorbed dose (up to 5 x 10^3 Mrad). The coefficient of the increase rate of accumulated energy is K = 3 x 10^{15} eV/g Mrad (or 1.2 x 10^{-4} cal/g Mrad).

The temperature at which the liberation of accumulated energy begins is determined by the temperature interval of annealing radiation defects (color centers). This interval (or intervals) can be determined by studying the thermoluminescence of irradiated halite samples. Figure 1 shows the curve for halite irradiated by a dose of 3.8 Mrad. There are two intervals where accumulated energy is liberated. At the lower release temperature (20-110°C) only a small part of the total accumulated energy is liberated. The liberation of the major part of energy takes place at the higher temperature (170-320°C). Neither peak depends on the dose absorbed by the sample (up to 10^4 Mrad). However, the efficiency of liberated energy, which depends on the activation energy of glowing luminescence, increases with an increase in the absorbed dose.

From the above data it follows that in order to prevent the liberation of accumulated energy, the temperature in halite layers should not exceed the lower limit of the high-temperature release interval of the liberation of accumulated energy (i.e., 170°C). However, during the disposal of radioactive wastes there is always a chance of a local temperature rise. A rise of temperature higher than 170°C is bound to cause an instantaneous liberation of accumulated energy, and consequently an additional temperature rise. If the temperature conditions in the storage are likely to be violated, the consequences of this process may be hazardous.

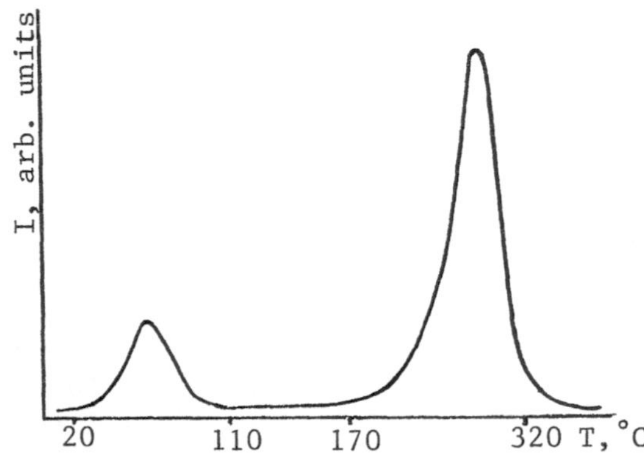

Fig. 1. Glowing curve of halite irradiated at 3.8 Mrad.

To prevent the possibility of unplanned temperature increases as a result of the liberation of accumulated energy, it is advisable to keep the temperature in disposal areas within the limits of the high-temperature release interval. It was experimentally established that irradiation of halite samples at an elevated temperature substantially influences the form of the thermoluminescence curve. If the samples are irradiated at 200-220°C, the glowing curve shows the absence of the low-temperature release interval. In addition, the intensity in the high-temperature interval is 10-15 times less. This is caused by the fact that during irradiation at an elevated temperature, radiation defects are annealed in the process of the irradiation itself. Thus, keeping the temperature in the storage at 200-220°C hinders the process of the accumulation of energy in halite. Such temperatures are safe as far as the mechanical properties of halite are concerned and contribute to the removal of water from microstructures and from mineral surfaces.

REFERENCES

1. V. I. Spitsyn, V. D. Balukova, I. M. Kosareva, and E. I. Evko, "Influence of Ionizing Irradiation on the Properties of Alumosilicate Mineral Rock. Repositories of Radioactive Wastes," p. 593 in Scientific Basis for Nuclear Waste Management, vol. 2, C. J. M. Northrup, Jr., Ed., Plenum Press, New York, 1980.

2. A. S. Marfutin, "Spectroscopiya, luministsentsiya i radiatsionnye tsentry v mineralakh," Izd. Nedra, Moskva, 1975.

3. E. L. Andronikashvili, N. Politov et al., Sbornik Radiation Damage in solids JAEA, Vienna, v. 3, p. 147, 1963.

4. G. M. Jenks, E. Sounder, C. D. Bopp, and J. R. Walton, J. Phys. Chem. 79:871 (1975).

RADIATION HARDENING OF ROCKSALT*

G. W. Arnold

Sandia National Laboratories†
Albuquerque, New Mexico 87185

INTRODUCTION

The hardening of natural (rocksalt) and synthetic NaCl by exposure to ionizing radiation is well known.[1] Irradiation induced increases in yield stress and indentation hardness are caused by point defects and defect aggregates which interfere with the motion of dislocations. The details of the hardening process, however, are not clear. Nadeau[2] has shown that the increases in yield stress he observed in a number of alkali halides were introduced at the same rate as F-centers, but comparisons between additively colored and irradiated material indicated that the interstitial halogen was involved in the hardening center and not the charged vacancy. Other than the indicated role of the interstitial, nothing more can be said with definiteness about the center as has been concluded in the review by Clark and Crawford.[1]

The aggregation of F-centers can occur when they become mobile (∿150-200°C in NaCl). In this temperature range Na colloids can form due to the aggregation of F-centers. Hughes and Jain[3] have reviewed the literature regarding colloids in alkali halides and conclude that more work is needed to establish the effect, if any, of colloid formation on hardness or yield stress. The temperature region for colloid growth is rather narrow, i.e., from about 150-250°C. Above 250°C, colloids break up rapidly. The influence of the irradiation temperature

*This article sponsored by the U. S. Department of Energy under Contract DE-AC04-76-DP00789.
†A U. S. Department of Energy facility.

on colloid formation is of importance in nuclear waste repository studies, especially those involving bedded salt deposits. Lidiard[4] has recently examined the influence of colloids in NaCl on stored energy. The present paper gives indentation hardness results for irradiated (^{60}Co γ-irradiated and ion implanted) natural and synthetic NaCl for irradiation temperatures less than 95°C and relates these measurements to colloid formation.

EXPERIMENTAL

Rocksalt samples were obtained from the bedded salt deposits associated with the Waste Isolation Pilot Plant (WIPP) in southeastern New Mexico. The samples came from an approximate depth of 2430 ft. Pure synthetic material was purchased from Harshaw Chemical and crystals doped with 1.1 wt % $CaCl_2$ were grown by R. J. Baughman of Sandia National Laboratories.

Hardness measurements were made using both a Kentron and a Sheffield hardness tester. Loading varied from 5 to 50 g and the indentations were made with either a Knoop or a Vickers point, depending upon whether surface or volume changes were being investigated. The irradiations were made with both a ^{60}Co source and an Accelerators, Inc. ion implanter. In one instance, 2 MeV H ions were implanted using a High Voltage Engineering tandem accelerator. Optical density changes were made using a Cary 14 spectrophotometer.

RESULTS AND DISCUSSION

Figure 1 shows the measured microhardness (Vickers, 25 g load) as a function of γ-ray dose in rads. The data show typical hardness values of about 18 kg/mm^2, as in unirradiated material, which do not change appreciably in the fluence range from 10^5 to 10^8 rads. Above 10^8 rads there is a fairly rapid increase to an apparent saturation level of about 31 kg/mm^2. The F-center concentration, on the other hand, increases rapidly in the low dose region and becomes very difficult to measure above about 5 x 10^8 for the sample thicknesses used (about 0.1 cm). As discussed in the Introduction, however, the hardness is not related to the halogen vacancy concentration but only to the interstitial component. Typical optical absorption features are shown in Fig. 2. The F-center absorption is located at about 4630A and the band at about 7250A is due to the F_2 (M)-center. Colloidal Na produces absorption in the spectral range of 5600-6000A depending on colloid radius; larger colloid radii are associated with longer wavelength absorption. The F_3(R)-center absorption, ascribed to three associated F-centers, if present, is also found near 5900A. Neither band is evident in Fig. 2. The concentration of F-centers is on the order of 1 x 10^{17}/cm^3 for this level of absorbed dose.

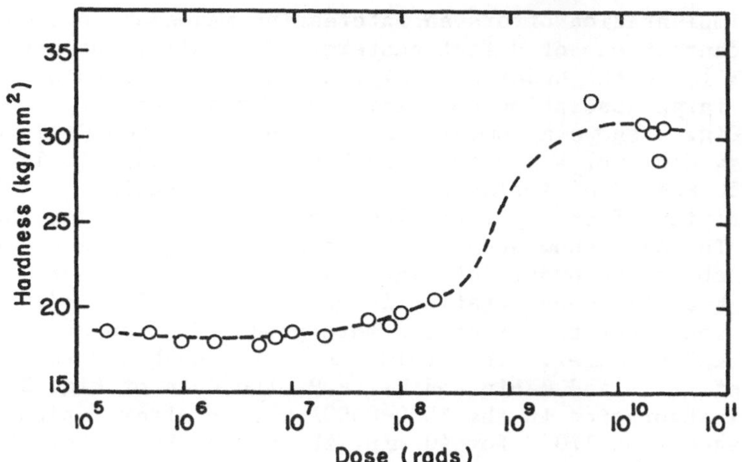

Figure 1. Hardness (Vickers) (kg/mm^2) vs. ^{60}Co gamma ray dose (rads) for New Mexico WIPP rocksalt at 25 g loading for irradiation temperatures less than 95°C.

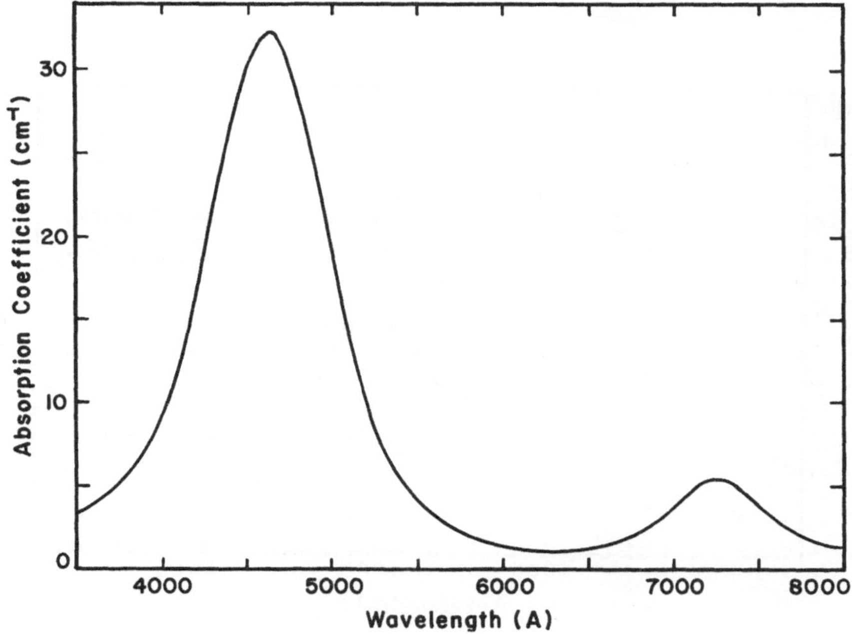

Figure 2. Optical absorption coefficient (cm^{-1}) vs. wavelength (A) for New Mexico WIPP rocksalt after ^{60}Co gamma irradiation to 2.5 x 10^7 rads.

Ion implantation offers an interesting means of introducing large concentrations of defect centers. Typically, the ion penetration depth is on the order of a micron and one can obtain, as a consequence, large absorption coefficients at reasonable optical density levels. Fig. 3 shows a comparison between New Mexico WIPP rocksalt and Harshaw NaCl for a 1 x 10^{16} 250 keV H/cm^2 implant followed by a 5 x 10^{16} 50 keV H/cm^2 implant. One expects to obtain a fairly uniform deposition of energy into electronic processes under these conditions. The data show absorption coefficients for the F-centers which are about two orders of magnitude greater than shown in Fig. 2, i.e., the F-center concentration is on the order of 10^{19}/cm^3. Calculations show that the F-centers were produced at an expenditure of about 2.5 keV/F-center. This value is in reasonable agreement with the data of Ritz[5] and Rabin and Klick.[6] The data of Fig. 3 show additional absorption in the 5500-6000A region after implantation. After an anneal at 170°C for 40 min. the F- and F_2-center absorption is removed and is replaced by the absorption centered about 5700A. This colloidal absorption disappears after 4 hrs. annealing time at 170°C.

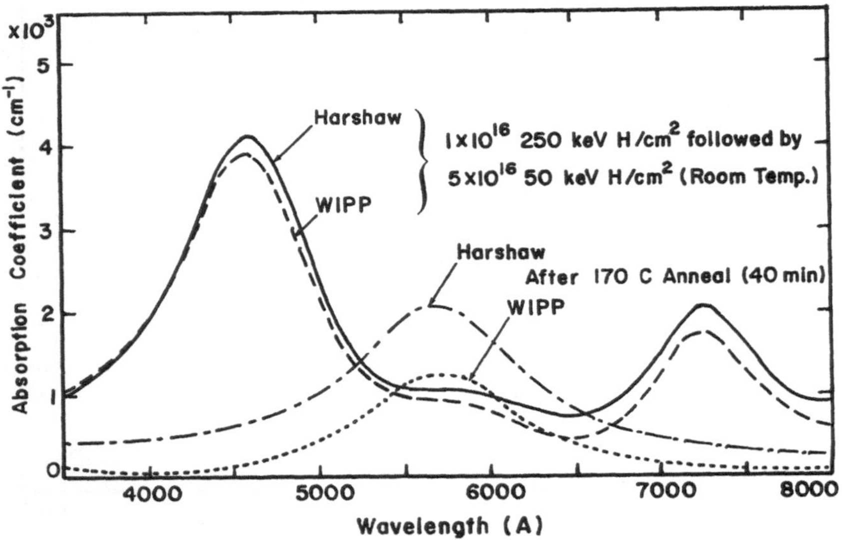

Figure 3. Optical absorption coefficient (cm^{-1}) vs. wavelength (A) for New Mexico WIPP rocksalt and Harshaw synthetic NaCl after implantation with 1 x 10^{16} 250 keV H/cm^2 followed by 5 x 10^{16} 50 keV H/cm^2 and annealing at 170°C for 40 min.

It has not been possible to obtain unequivocal indentation hardness results in the case of low energy ion implantation. In order to obtain hardness values only in the implanted region, the shallow Knoop indenter with 5 g loading was used. Values of 3-8 kg/mm^2 were typically obtained for both the implanted and unimplanted sides of the sample. This observation suggests that exposure to the atmosphere resulted in surface absorption of water vapor to a depth exceeding the penetration or implantation depth. The data of Fig. 1 do not reflect this effect because, presumably, the reacted layer thickness is small with respect to the penetration depth (~ 10 μm) and the entire volume of the sample has been uniformly irradiated. An attempt was made to assess the influence of colloidal centers on hardness by implanting a Harshaw sample with 1×10^{16} 2 MeV H/cm^2. At this energy, the penetration depth is expected to be ~ 50 microns. The sample developed a strong colloidal band when annealed at 170°C for 80 min. Knoop hardness measurements at 5 g load showed about 3 kg/mm^2 on the unimplanted side and about 7 kg/mm^2 on the implanted surface, again within the uncertainty of the measurements with the shallow Knoop indenter. However, using the Vickers indenter (pyramidal) with 25 g load, as in Fig. 1, the hardness value for the unimplanted side was about 19 kg/mm^2 while the implanted surface gave a value of about 43 kg/mm^2. This result indicates that colloidal centers do play a significant role in the hardening process and show the value of using ion implantation in such studies.

Measurements on unirradiated $NaCl:CaCl_2$ synthetic material yielded hardness values on the order of 40 kg/mm^2 in agreement with the results of Reddy et al.[7] Because this result is not typical of the data obtained on the WIPP and Harshaw materials, it is concluded that much of the Ca impurity is probably excluded from the crystalline rocksalt during the growth process and precipitates at grain boundaries.

CONCLUSIONS

Both rocksalt and synthetic NaCl increase in indentation hardness (by a factor of about 1.7) when exposed to γ-ray doses in excess of 10^{10} rads at temperatures less than 95°C. The large optical densities produced make it very difficult to assess the importance of colloidal Na upon the change in hardness. However, this difficulty has been circumvented by using ion implantation to get high concentrations of defects in a thin surface layer. Annealing of such implanted layers produces colloidal Na. Increases in hardness by a factor of about 2.1 have been found by this technique when only the colloidal Na band is present in the optical absorption. It is important to note that the colloidal band disappears in a matter of hours at 170°C and thus the changes in hardness we have noted will depend critically upon the interface temperature in the repository, the accumulated dose, and the time of exposure at a given temperature.

REFERENCES

1. See, e.g., J. H. Crawford, Jr., Advances in Physics 17, 93 (1968).
2. J. S. Nadeau, J. Appl. Phys. 34, 2248 (1963).
3. A. E. Hughes and S. C. Jain, Advances in Physics 28, 717 (1979).
4. A. B. Lidiard, Phil. Mag. A 39, 647 (1979).
5. V. H. Ritz, Phys. Rev. A133, 1452 (1964).
6. H. Rabin and C. C. Klick, Phys. Rev. 117, 1005 (1960).
7. N. K. Reddy, M. L. Rao, and V. H. Babu, Ind. J. Pure and Appl. Phys. 17, 806 (1979).

RADIATION EFFECTS IN CRYSTALLINE HIGH-LEVEL NUCLEAR WASTE SOLIDS*

W. J. Weber, J. W. Wald, and W. J. Gray

Pacific Northwest Laboratory

Richland, Washington 99352

INTRODUCTION

Glass, cement, and crystalline-ceramic waste forms are being considered as potential solid forms for incorporation of nuclear wastes. The solidified waste will be subjected to high doses of many different radiations which may measurably alter physical properties and/or affect the durability of the solid waste form. As a result, the long-term stability of these waste forms is a subject of continuing research. The present paper defines the general radiation damage problem and summarizes experimental results from studies of the effects of alpha decay, alpha bombardment, and transmutations on crystalline waste forms, related single-phase compounds, and some glass waste forms.

SOURCES OF RADIATION DAMAGE

The principle sources of radiation in high-level nuclear waste forms are alpha decay of the actinide elements and beta decay of the fission products. Alpha decay produces two particles: a high energy alpha particle (~4 to 6 MeV) and a recoil nucleus (~0.1 MeV). Nearly all the energy of the recoil nucleus is lost through elastic collisions producing several thousand atomic displacements. In the case of the alpha particle, most of its energy is dissipated in ionization processes, but sufficient energy is lost through elastic collisions to produce several hundred displacements. Beta decay produces high energy beta particles and gamma rays which interact with the solid by ionization processes and produce very

*Work supported by the U.S. Department of Energy under Contract DE-AC06-76RLO 1830.

few direct atomic displacements. It is generally assumed that the primary source for radiation damage in nuclear waste forms results from displacement damage (i.e., alpha decay), but beta decay does result in significant transmutations (e.g., Cs→Ba and Sr→Zr) which may have a detrimental effect on waste form stability.

In complex crystalline waste forms where the crystal size is small (in the range of 0.1 to 10 μm), the alpha particle, with a range of approximately 20 μm, effectively bombards the entire solid including the nonactinide-containing phases, while the damage due to the recoil nucleus, because of its short range (~0.01 μm), is confined to the phase in which the actinide was chemically incorporated. Consequently, alpha decay actually results in two related but different damage mechanisms: alpha-recoil damage (i.e., damage resulting from the synergistic effect of both alpha particle and recoil nucleus) in the actinide host phases and alpha particle damage in the nonactinide-containing phases.

ALPHA-RECOIL DAMAGE

The long-term effects of alpha-recoil damage in nuclear waste forms are most effectively simulated in the laboratory by incorporating a short-lived actinide isotope, such as Cm-244, in either a simulated waste form or the proposed actinide host phase (or phases). Several multiphase waste forms and actinide host phases have been studied or are currently being studied at the Pacific Northwest Laboratory (PNL) by this method. These materials are summarized in Table 1 along with the maximum volume expansion observed. The actinide host phases which are listed represent some of the crystalline phases or structure types found containing actinides in the crystalline waste forms.

Changes in density, stored energy, and lattice parameters are generally monitored as a function of accummulated alpha-decay events. It has been found that these property changes follow an exponential dose dependence that can be expressed by the following general relationship:

$$\frac{\Delta x}{x_o} = A[1 - \exp(-BD)] \tag{1}$$

where x is the property measured, A and B are material dependent constants, and D is the dose. In general, the changes in property measurements of the waste forms tend to saturate at a dose of about 5×10^{18} α-decays/cm^3. X-ray diffraction (XRD) measurements have been used to determine both lattice parameter changes and any tendency toward amorphization. The fluorite structure, PuO_2, is found to be stable with respect to amorphization and undergoes only a small lattice parameter increase (~0.32%). Several crystalline phases in the simulated waste forms and the pure compound,

$Ca_2Nd_8(SiO_4)_6O_2$, are found to become x-ray amorphous with increasing dose. These crystalline phases and the results are summarized in Table 2. Since the distribution and concentration of the actinide dopant in the crystalline phases is not accurately known in the case of the simulated waste forms, the amorphization dose for a given structure type varies from waste form to waste form. It is known, however, that the actinides tend to concentrate in the crystalline phases listed in Table 2; therefore, the amorphization dose given is a minimum dose that should be corrected by a concentration factor. This is borne out by the relatively large dose that appears to be required for complete amorphization of the pure compound, $Ca_2Nd_8(SiO_4)_6O_2$.

Table 1. Crystalline Materials Studied at PNL by Actinide Doping and the Maximum Volume Expansion Observed

Material	Swelling $\Delta V/V$ (%)	Dose (α-decays/cm^3)
Simulated Waste Forms		
Partially Devitrified Glass[(a)]	1.0	5×10^{18} (f)
Celsian Glass Ceramic[(b)]	0.5	3×10^{18} (f)
Supercalcine, SPC-2[(c)]	1.4	5×10^{18} (f)
Portland Cement, Type II[(d)]	<0.1	5×10^{17}
Actinide Host Phases		
$Ca_2Nd_8(SiO_4)_6O_2$	6	6×10^{18}
PuO_2[(e)]	1.0	5×10^{18} (f)
$CaZrTi_2O_7$[(b)]	Data Not Available	
$Gd_2Ti_2O_7$[(b)]	Data Not Available	

(a) PNL-77-260 waste glass, cooled at 6 K/h (Weber et al., 1979).
(b) Studied in cooperation with the Hahn-Meitner Institute (Berlin, West Germany).
(c) Rusin, Gray, and Wald (1979).
(d) Contains 10 wt% simulated PW-9 calcine.
(e) Chikalla and Turcotte (1973).
(f) Dose at saturation.

Table 2. Crystalline Phases Observed to Become X-ray Amorphous

Parent Solid Form	Crystalline Phase	Volume Change To Crystalline Phase (%)	Amorphization Dose[(e)] (α-decays/cm^3)
Partially Devitrified Glass	$Ca_3Gd_7(SiO_4)_5(PO_4)O_2$ [(a)]	3[(c)]	1.5×10^{18}
Partially Devirified Glass	$Gd_2Ti_2O_7$ [(b)]	Not Measurable	$>5 \times 10^{18}$
Celsian Glass Ceramic	$Gd_2Ti_2O_7$ [(b)]	1[(c)]	3.2×10^{18}
Supercalcine SPC-2	$Ca_2Nd_8(SiO_4)_6O_2$ [(a)]	4[(c)]	4.2×10^{18}
Pure Compound	$Ca_2Nd_8(SiO_4)_6O_2$ [(a)]	>6[(d)]	$>6 \times 10^{18}$ [(f)]

(a) Apatite structure type
(b) Pyrochlore structure type
(c) Determined from XRD measurements
(d) Determined from density measurements
(e) Volume averaged dose to complete sample
(f) This compound has not yet reached a complete x-ray amorphous state

There are also studies reported in the literature of pure actinide or actinide-containing compounds which contribute to the understanding of alpha-recoil damage in nuclear waste forms. These compounds include other fluorite-related oxides (Fuger, 1976), zircon (Holland and Gottfried, 1955), and a perovskite structure, $CmAlO_3$ (Mosley, 1971). The data from these literature sources, along with the data on PuO_2, the apatite structure in supercalcine, and the pyrochlore structure in the celsian glass ceramic, are summarized in Figure 1. All of the data fit the exponential relationship given by Eq. 1. (The doses for the apatite and pyrochlore were estimated by assuming that 50% of the curium dopant concentrated in the apatite phase in supercalcine, and that 100% of the curium concentrated in the pyrochlore phase in the celsian glass ceramic.) The results show that the fluorite structures (which are stable) and the pyrochlore structure saturate at low lattice expansions relative to the silicates (apatite and zircon) and the aluminate (perovskite).

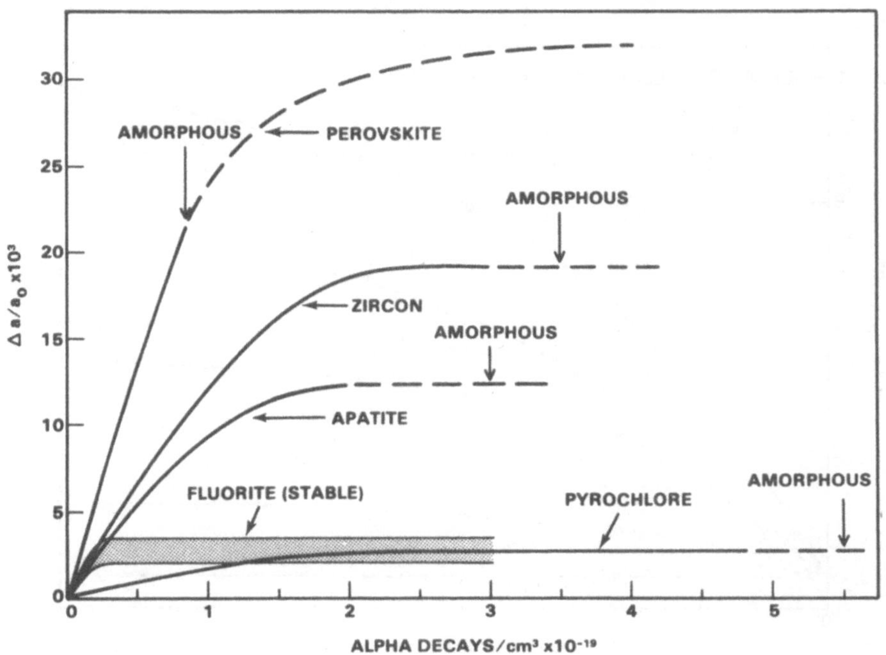

Fig. 1. Lattice expansions as a function of internal alpha-recoil dose.

ALPHA DAMAGE

Several crystal structures have been irradiated at room temperature with alpha particles emitted from a thick $^{238}PuO_2$ source (E_{max} = 5.5 MeV) in order to simulate the alpha particle flux incident on the nonactinide-containing phases in nuclear waste solids. The relative changes in lattice parameters were measured as a function of alpha dose by x-ray diffraction techniques. The results are shown in Figure 2. The observed damage ingrowth also follows the same exponential dose dependence as observed in alpha-recoil damage. The data show that nonactinide-containing phases can undergo significant lattice expansions as a result of alpha bombardment from adjacent actinide host phases. In fact, the results for UO_2 show a significantly greater lattice parameter increase (factor of two) than expected from the review of Fuger shown in Figure 1. This is discussed in greater detail in a separate paper by Weber (submitted to <u>J. Nucl. Mater.</u>). From the data available, it appears that alpha doses of 5×10^{16} α/cm^2 may be expected to cause changes approaching saturation. None of the materials evaluated in the present study has shown any tendency toward amorphization. Alpha bombardment studies of more complex structures (apatite, hollandite, etc.) are in progress.

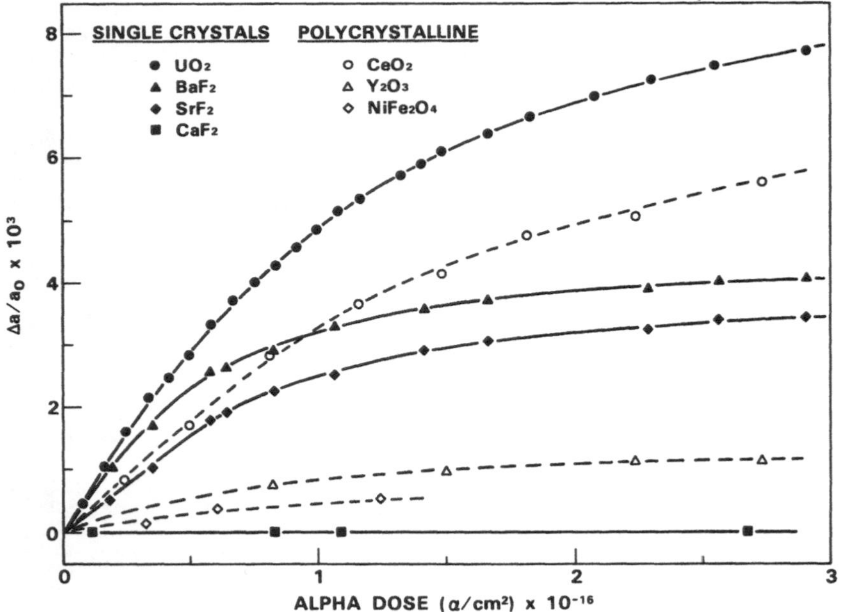

Fig. 2. Lattice expansions as a function of external alpha-recoil dose.

TRANSMUTATION EFFECTS

The effects of transmutations, due to beta decay of the fission products in high-level nuclear waste forms, are currently being investigated. Several nuclear waste forms (glasses and supercalcine) and some single-phase materials listed in Table 3 have been neutron irradiated to convert about 12% of the natural Cs-133 to Cs-134 which then transmutes to Ba-134 with a 2.06 year half life. Immediately following irradiation, the materials were annealed to remove neutron damage. About 3% of the total Cs has transmuted to Ba at the present time. Changes in density, leach rates, x-ray diffraction parameters and microstructure, as determined by optical and scanning electron microscopy, are being determined at intervals of 6 to 12 months.

Table 3. Density Data From Transmutation Study

Sample		Density (g/cm^3)	
No.	Type	June, 1979	February, 1980
3	standard	3.209	3.203
5	standard	2.689	2.690
H-1	simulated waste glass	1.993	1.990
H-2	high Cs glass[(a)]	2.704	2.727
H-3	high Cs glass	2.835	2.857
H-4	high silica glass	2.362	2.360
H-5	high alumina glass	2.315	2.310
H-7	pollucite	2.970	2.970
H-8	supercalcine	4.207	4.208

(a) This glass is doped with approximately 2 wt% $^{244}CmO_2$.

Data obtained during the second round of post-irradiation examinations have shown little if any change in any of the parameters being studied. For example, the data given in Table 3 shows that the densities of samples H-2 and H-3 have changed with time but these data must be viewed with some skepticism unless and until subsequent measurements, 6 to 12 months from now, shows that the trend continues. Similarly, Cs leach rates as determined by 7-day static tests at 90°C have shown little change. No difference in microstructure in any of the materials was detected. X-ray diffraction data obtained to date have been of poor quality and, therefore, inconclusive.

SUMMARY

The results lead to the following conclusions:

- Both alpha-recoil damage in actinide host phases and alpha damage to all other phases must be considered since significant structural changes may occur from either source.
- The ingrowth of damage follows exponential behavior for both alpha-recoil damage and alpha damage, leading to saturation effects in most materials.
- Preliminary results show no significant effect of transmutations on waste form stability.

REFERENCES

Chikalla, T. D. and Turcotte, R. P., 1973, Self-Radiation Damage Ingrowth in $^{238}PuO_2$, Rad. Effects, 19:93.

Fuger, J., 1976, Effects of Radioactivity in Actinide Chemistry, in: "Inorganic Chemistry, Series 2, Vol. 7," K. W. Bagnall, ed., University Park Press, Baltimore.

Holland, H. D. and Gottfried, D., 1955, The Effect of Nuclear Radiation on the Structure of Zircon, Acta Cryst., 8:291.

Mosley, W. C., 1971, Self-Radiation Damage in Curium-244 Oxide and Aluminate, J. Amer. Cer. Soc., 54:475.

Rusin, J. M., Gray, W. J., and Wald, J. W., 1979, "Multibarrier Waste Forms Part II: Characterization and Evaluation," PNL-2668-2, Pacific Northwest Laboratory, Richland, Washington.

Weber, W. J., Turcotte, R. P., Bunnell, L. R., Roberts, F. P. and Westsik, J. H., 1979, Radiation Effects in Vitreous and Devitrified Simulated Waste Glass, in: "Ceramics in Nuclear Waste Management," T. D. Chikalla and J. E. Mendel, eds., National Technical Information Service, Springfield, Virginia.

RADIATION DAMAGE EFFECTS ON THE LEACH RESISTANCE OF GLASSES AND MINERALS: IMPLICATIONS FOR RADIOACTIVE WASTE STORAGE

J. C. Dran, M. Maurette, J. C. Petit, B. Vassent

Laboratoire René Bernas, C.N.R.S.

91406 Orsay, France

INTRODUCTION

Most experts in nuclear waste management agree that a safe long-term isolation of high-level waste (HLW) from the biosphere could be achieved by incorporating such wastes into a solid matrix, which would then be disposed in a deep seated (∿1000 m) geological repository. However, there is still no consensus either on the preferred geological formation or on the best solid form. Salt domes, granites, shales, etc., are currently investigated by the nuclear countries as potential disposal sites. On the other hand, borosilicate glasses have been the most favored matrices for more than 20 years, but a number of material scientists still question their long-term stability against corrosion by groundwaters. In particular, glasses could be subjected to hydrothermal alteration[1] or could undergo uncontrolled thermally induced devitrification,[2] which would probably decrease their durability. Because of such drawbacks, crystalline solid forms such as Synroc B[2] or super-calcines[3] are now considered by several authors as potential advantageous alternates to borosilicate glasses.

In previous work[4] intended to study the particular effect of α-recoils emitted during the decay of actinides on the leach resistance of radwaste glasses, we developed a simulation technique based on an external implantation of low energy (∿1 keV/amu) lead ions (see below). We showed that the durability of such glasses could be drastically reduced above a critical dose of α-recoils (α-recoil aging). On the other hand, the existence of metamict minerals, which have retained their U, Th and daughter-products over millions of years although exposed to severe environmental conditions,[5] has been invoked as indicative of the long-term stability of actinide bearing crystalline matrices. However, no laboratory experiments have been conducted so far to quantitatively evaluate the sensitivity of such matrices to α-recoil aging. We thus decided to compare the respective stability of glasses and minerals against radiation damage in well-defined environmental conditions. For this purpose we recently extended our ion implantation approach to both borosilicate glasses of a wider range of compositions and a great variety of minerals.

PREVIOUS EVIDENCES ABOUT THE COMPARATIVE DURABILITY OF GLASSES AND MINERALS

One of the most difficult issues in the disposal of HLW is to evaluate the long-term durability (>100 years) of the proposed storage matrices, which could be partly related to their sensitivity to α-recoil aging. In the particular case of borosilicate glasses, their durability can only be inferred from laboratory simulations such as soxhlet tests. The specific effect of α-recoil aging on their leach resistance is usually assessed by incorporating short-lived α-emitters in order to accumulate in a short time a dose of α-recoils equivalent to that expected over a time-scale of several thousand years in the real radwaste glasses. These tests have generally shown no pronounced effect of radiation on leachability of the waste form for the relatively mild leachants employed. Moreover, the extrapolations of data obtained with natural glasses such as tektites, obsidians, etc., to radwaste glasses are particularly difficult for two main reasons: (1) The chemical composition of these two types of materials is very different. In particular, the generally high durability of natural glasses in various terrestrial environments (a few million years) is directly related to their very high content in network-forming elements ($SiO_2 + Al_2O_3$, $\gtrsim$85 wt %). In contrast, radwaste glasses usually contain $\lesssim$50% of such elements, because of technological constraints, and thus have much higher leach rates in groundwaters (by a factor of 10^3). (2) Natural glasses usually incorporate very small amounts ($\sim$1 ppm) of actinides and the total dose of α-recoils accumulated even in the oldest samples ($\gtrsim 10^8$ years) is negligible with respect to that expected in radwaste glasses ($\sim 10^{18}$ g^{-1}) after a few thousand years of disposal.

On the other hand, the constituent minerals of ceramic waste forms have natural analogues from which long-term stability assessments can be derived. As first pointed out by Ringwood,[5] these analogues have strictly the same structure and chemical composition and furthermore they contain high concentrations of actinides. For example, zirconolite, one of the major actinide bearing constituents of Synroc B, can be currently found in nature with concentrations of UO_2 and ThO_2 as high as 1.5 and 6.6 wt %, respectively. In addition, samples of a wide range of ages and actinide contents are currently investigated in which the accumulated α-recoil dose can be as high as 10^{20} g^{-1}. Even for such high levels of irradiation doses which lead to complete metamictization, these natural zirconolites do not show any marked evidence of alteration and/or loss of incorporated actinides or daughter-products in spite of severe environmental conditions. These favorable observations have led Ringwood to optimistic predictions about the long-term durability of Synroc B. However, such conclusions have been recently criticized. Nestbitt et al.[6] questioned the stability of Synroc B constituent minerals in natural waters, on the basis of both thermodynamic data and leach tests, which suggest in particular that these minerals are not thermodynamically compatible with siliceous rocks of candidate disposal formations. Moreover, we believe that it is very difficult to trace back the variations of the environmental conditions during the history of metamict minerals. As no specific laboratory leach tests have been performed to date on α-recoil aged actinide bearing minerals, we extended our ion implantation approach for investigating the effect of α-recoil aging on the leach resistance of this class of minerals.

THE STEP-HEIGHT METHOD

In this method which has been extensively described elsewhere,[4] a polished section of the material under evaluation is covered with an electron microscope grid and then exposed to a beam of 1 keV/amu lead ions, thus delineating a periodic succession of irradiated and unirradiated areas. The irradiated areas, which extend to a depth of ∿500 Å, should be similar to an internal ultrathin slice of the real storage material, where waste elements are homogeneously dispersed on an atomic scale. By varying the incident lead ion fluence, Φ_i, from 10^{10} up to 10^{14} ion·cm^{-2}, a range of equivalent storage periods, Ts, can be continuously simulated. The absolute value of Ts depends on several parameters such as the residual concentration of actinides and the cooling time of the nitric liquid wastes. In the case of PWR wastes originating from a pure UO_2 fuel, it is expected that the residual actinide concentration in both glasses and the actinide bearing minerals of ceramics would be roughly similar (0.5 wt %). In this case an internal fluence of α-recoils of 2×10^{18} g^{-1} corresponds to a value of Ts of a few thousand years. After a given chemical etching, the etched surface is then probed with a diamond stylus device. If the irradiation increases the etching rate of the material, a periodic succession of very shallow holes is formed. The depth of the hole is detected as a step-height, Δ, with a diamond stylus device. The values of Δ, which can be inferred with an accuracy of ∿10 Å by averaging the measurements for a large number of irradiated areas, directly scale the increase in the chemical reactivity of the irradiated area, $\Delta V = V^* - Vo = \Delta/t$, where t is the etching time anf V^* and Vo the etching rates of the irradiated and unirradiated materials, respectively. A recent improvement of this step-height technique makes possible the direct measurement of Vo and thus the evaluation of V^*. In this procedure, the grid-covered polished section is coated with a very adhesive gold film by ion sputtering, which thus produces a succession of protected and unprotected areas. When the sample is immersed in a reactant, the noncoated areas only are etched and the depth of the resulting shallow holes is measured with the diamond stylus device.

This method presents two main advantages: (1) very low etching rates, $\Delta/t \sim 5 \times 10^{-9}$ $g \cdot cm^{-2} \cdot d^{-1}$ can be easily measured within a reasonable time (∿100 days); and (2) for glasses and minerals, which develop good surfaces upon cleavage or polishing, this method applies nicely. Then its remarkable versatility should allow the investigation of variations of V* and Vo with various parameters (incident ion fluence, nature of matrix, leaching conditions), which define specific storage conditions in natural environments.

One could, however, argue about the following limitations of the methods: (1) We only investigate the specific effects of α-recoil aging. (2) We look at the earliest phase of etching (∿days), which removes a very thin (∿500 Å) layer from the radwaste materials. Consequently the results of such an early etching are possibly not relevant to the much longer ($\gtrsim$years) etching, which can possibly occur in the natural repository. This is a real limitation of the method, which very unfortunately also applies to all other methods as yet proposed (including the actinide loading method) as soon as the value of the etching rate reaches a range of values $\lesssim 10^{-6}$ $g \cdot cm^{-2} \cdot d^{-1}$.[2] The anisotropic lead ion irradiations used in our work is possibly responsible for the build-up of strong stress fields, which would be responsible for the increased etching rate of the irradiated material. As the real radwaste matrices are isotropically irradiated by α-recoils, the results of our ion implantation experiments would possibly lead to misleading conclusions. This argument can be rejected for several distinct reasons which will be more conveniently discussed elsewhere.[7] We simply note here that all <u>insulator</u> minerals so far investigated, except those (corundum) where nuclear tracks do not register as "etchable" tracks, show a drastic increase in their etching rate at the same critical fluence ($\sim 5 \times 10^{12}$ cm^{-2}) of lead ions. This unique characteristic is not compatible with the assumption of "stress fields," which would build up very differently in amorphous materials such as glasses and in a highly anisotropic mineral such as mica. On the other hand, this unique characteristic is well interpreted by a "planar" nuclear track model for the shallow layer of implanted material. This model attributes the increased etching rate to the radiation-damaged islands produced by the incident ions, which are roughly spherical in shape, and not highly anisotropic like the long tracks due to high energy ions such as fission fragments.

RESULTS AND DISCUSSION

Various glasses and minerals have been tested with the step-height method, including simulated radwaste glasses and constituent minerals of ceramics (see Tables 1 and 2). Our results clearly demonstrate the effect of α-recoil damage on the leach resistance of radwaste insulator matrices, that has not been observed in previous work. Indeed the leach rate of most irradiated materials shows a sudden increase when the fluence of the incident lead ions reaches a critical value, $\Phi c \sim 5 \times 10^{12}$ ions·cm^{-2}. Then for higher ion fluences, $\Delta V = V^* - Vo$, slightly increases and reaches a saturation value when $\Phi = \Phi s \sim 100\Phi c$. In addition, for the few materials for which all relevant data are available the value of Φs is similar to the amorphizing ion fluence.

By assuming a roughly spherical shape for the tiny islands of radiation damage produced by low energy lead nuclei and α-recoils, the critical lead ion fluence, $\Phi c \sim 5 \times 10^{12}$ cm^{-2}, can be expressed as an equivalent dose of α-decays ($\sim 2 \times 10^{18}$ α-decays·g^{-1}). This equivalent dose in turn should be accumulated over a critical storage time, Tc, of a few thousand years, for a glass loaded with a 0.5 wt % concentration of actinides expected from a "typical" PWR cycle. By assuming that the values of the etching rate amplification factor, $K = V^*/Vo$, measured in the earliest phase of etching (∿days) do not drastically vary with both the etching time and the etching conditions, we thus previously suggested[4] that the durability of radwaste glasses could be drastically reduced by a factor $K \gtrsim 20$, when $Ts \gtrsim Tc$. However, the much larger survey of materials now performed (11 glasses and 14 minerals), as well as preliminary results pertaining to the effects of the etching conditions on the values of K, stresses the great complexity of α-recoil aging. This was not sufficiently assessed in our earlier work, which had a very limited sampling of three glasses and did not include real radwaste glasses. In fact, the ΔV values measured for glasses (Table 1) etched in a NaCl brine at 100°C now spread over a 4 orders of magnitude range, and such huge variations cannot as yet be correlated to compositional differences between glasses. Indeed the Vo scaling, which is qualitatively understood in terms of such compositional differences (i.e., concentration of oxide additives such as Al_2O_3, Fe_2O_3, CaO, etc.) does not match the ΔV scaling. But even worse, the K-ratio, which also shows a large variation from ∿1 to 53 in the NaCl brine, clearly depends on the etching conditions. For example, a preliminary investigation, so far conducted with the BONI glass, indicates that K decreases from a value of ∿53 in the NaCl brine at 100°C to a low value of ∿3 in distilled water at the same temperature. In addition, for other glasses, the K-ratio reaches different values in strong etchants. Consequently, our results relevant to a NaCl brine as well as those of the widely used Soxhlet test can hardly be extrapolated to other etching conditions, and this important characteristic of α-recoil aging was

Table 1. Sensitivity of Various Glasses to α-Recoil Aging

Glass[a]	Composition (wt %)					Vo[b]	ΔV[b]	K
	SiO_2	Al_2O_3	B_2O_3	Na_2O	Other oxides	($g \cdot cm^{-2} \cdot d^{-1}$)	($g \cdot cm^{-2} \cdot d^{-1}$)	
4681	56		29	10	Li_2O:5	$\lesssim 2.3 \times 10^{-3}$	9.5×10^{-3}	>5
4680	38		15	6	ZnO:30 K_2O:6 CaO:2 BaO:2	2.9×10^{-4}	4.5×10^{-3}	16
4678	64		12	24		$\sim 1 \times 10^{-4}$	1.9×10^{-3}	20
ASA 15 N° 7	67	5		16	SWO:12[c]	4.6×10^{-5}	1.2×10^{-3}	27
BONI	48	5	15	18.5	SWO:13.5	1.5×10^{-5}	8.2×10^{-4}	53
4679		25	15	21	P_2O_5:38	3.3×10^{-5}	4.3×10^{-4}	14
Soda lime	75	1		11	CaO:9 MgO:2	$\lesssim 5 \times 10^{-6}$	1.4×10^{-4}	>29
SON 69 14	54		13.5	13	CaO:5, SWO:19.5	9×10^{-4}	7×10^{-5}	1
4677	50		20	12	Fe_2O_3:17		5.8×10^{-5}	
PIVER 40-16	50	13	16	16	SWO:5		2.4×10^{-5}	
Pure silica	100						1.4×10^{-7}	

[a]ASA 15 N° 7, PIVER 40-16 and SON 69 14 15 C_3 are simulated radwaste glasses prepared at Marcoule, the last one being the only one relevant to PWR wastes. BONI is a laboratory simulated radwaste glass prepared at ISPRA. Glasses 4677 to 4681 have been manufactured by Centre de Recherches St Gobain, Aubervilliers (France) for investigating the influence of glass composition on α-recoil aging.

[b]The Vo and ΔV values refer to both a lead ion fluence of 10^{13} cm^{-2} and to our standard NaCl solution (250 $g \cdot l^{-1}$, T ∿100°C, P ∿1 bar).

[c]SWO: simulated mixture of waste oxides.

Table 2. Sensitivity of Various Minerals to α-Recoil Aging

Mineral[a]	Formula	Etched track detector	Ceramic constituent	ΔV_2^b (g $cm^{-2} \cdot d^{-1}$)	Radiation effect with strong etchants
Labradorite	$(Na_2Ca)Al_2Si_2O_8$	yes		6×10^{-6}	
Quartz	SiO_2	yes		6×10^{-6}	
Baddeleyite	ZrO_2	n.r[c]	SC[d]	$\sim 3 \times 10^{-6}$	
Pollucite	$CsAlSi_2O_2$	yes	SC	$\sim 3 \times 10^{-6}$	
Olivine	$(Mg,Fe)SiO_4$	yes		$\sim 3 \times 10^{-7}$	
Uraninite	UO_2	no		10^{-6} to 10^{-7}	
Monazite	$(Ce,La,Th)PO_4$	yes	SC	$\lesssim 10^{-7}$	
Muscovite	$KAl_2(AlSi_3O_{10})(OH)_2$	yes		0	yes
Perovskite	$CaTiO_3$	n.r	SR[e]	0	yes
Davidite	$Fe_6Ti_{15}O_{36}$	n.r	SC	0	yes
Hollandite	$BaAl_2Ti_6O_{16}$	n.r	SR	0	yes
Ilmenite	$FeTiO_3$	no		0	no
Thorianite	ThO_2	no	SC	0	no
Corundum	Al_2O_3	no		0	no

[a]The first column identifies minerals which are solid state track detectors.

[b]The ΔV values refer to a lead ion fluence of 10^{13} cm^{-2}. The leaching was performed in a standard solution of NaCl (250 $g \cdot l^{-1}$, T ∿100°C, P ∿1 bar). Strong etchants (H_2SO_4, H_3PO_4, HF, NaOH) were also used for the most leach resistant minerals.

[c]n.r: not reported. [d]SC: super calcine. [e]SR: Synroc B.

clearly overlooked in our earlier study.[4] Moreover, the lack of any measurable increase in etching rate, noted recently at Marcoule for a type of glass (SAN) not listed in Table 1, in which the accumulated dose of α-decays (4×10^{18} g^{-1}) clearly exceeded our critical value of $\sim 2 \times 10^{18}$ g^{-1}, could also be possibly related to a smaller value of K in the tap water used at Marcoule.

As a result of this complex effect of the etching conditions on the K-ratio, it is difficult to pinpoint a choice of "best" radwaste glasses in Table 1. The only safe statement which can be made as yet is that in the NaCl brine at 100°C one of the glasses listed in Table 1, and which is a simulated radwaste glass, does show a low value of K. However, we have still to demonstrate that this still holds true at slightly higher lead ion fluences and for other etching conditions, such as those expected in a confined environment (high-pH waters) before considering this glass as fairly resistant to α-recoil aging. Obviously, more work is needed to sort out the most stable glass composition with respect to α-recoil aging, among the possibilities offered by the present nuclear waste vitrification technology.

In Table 2 the most sensitive minerals clearly show measurable values of ΔV even in the mildest etchant, namely the sodium brine used as our standard glass etchant. Minerals found in the group of intermediate sensitivity only show measurable ΔV values in strong etchants, such as HF and NaOH. The most resistant minerals included in the third group of sensitivity are not preferentially etched out after irradiation in the strongest etchants, even though the corresponding Vo values are not negligible. Their high resistance to lead ion damage probably reflects a much higher value of the critical ion fluence Φc, beyond which a measurable ΔV value is observed. It is noteworthy to realize that most constituent minerals of ceramics belong to the groups of intermediate and low sensitivities to α-recoil damage. However, the leach rate of ceramics could be overdominated by the physico-chemistry of grain boundaries, which have a very large specific area in the case of Synroc B prepared by cold pressing. In this case, leach rate measurements of the constituent minerals of Synroc B and/or of individual metamict minerals considered as their valid analogues, would be hardly relevant to the long-term durability of Synroc B.

CONCLUSION

The present work is only relevant to α-recoil aging, which can affect nearly all radwaste matrices. This peculiar effect, which should be triggered in most radwaste matrices when the internal dose of α-decays reaches an equivalent value of $\sim 2 \times 10^{18}$ α-decays·g^{-1}, could be of particular concern for the durability of radwaste glasses. However, the formation of the "giant" α-recoil track could be markedly delayed in glasses by further decreasing their actinide contents below 0.5 wt %, and one of the glasses reported in Table 1 (SON type) already shows favorable low values of ΔV and $K = V^*/Vo$ in the NaCl brine at 100°C. If these values do not drastically increase in high-pH waters expected in confined environments, then borosilicate glasses would present major advantages as storage matrices, as their technology has proved to be economical on an industrial scale. On the other hand, it is urgent to demonstrate whether the better leach resistance of crystalline materials both before and after α-recoil aging, is not counterbalanced either by the prohibitive cost of their large scale production or by the still unexplored physicochemistry of grain boundaries.

ACKNOWLEDGMENTS

This work greatly benefited from both the active help of J. Chaumont, R. Klapisch, M. Lagache and G. Vidal, and the superb technical skill of F. Lalu. We are also grateful to Museum d'Histoire Naturelle and Ecole des Mines de Paris for supplying mineralogical samples.

REFERENCES

1. G. J. McCarthy, W. B. White, R. Roy, B. E. Scheetz, S. Komarneni, D. K. Smith and D. M. Roy, "Interaction Between Nuclear Waste and Surrounding Rock," Nature, 273: 216 (1978).
2. A. E. Ringwood, S. E. Kesson, N. G. Ware, W. Hibberson and A. Major, "Immobilisation of High-Level Nuclear Reactor Wastes in Synroc," Nature, 278: 219 (1979).
3. G. J. McCarthy, "High-Level Waste Ceramics: Materials Considerations and Product Characterization," Nucl. Technol. 32: 92 (1977).
4. J. C. Dran, M. Maurette and J. C. Petit, "Radioactive Waste Storage Materials: Their Alpha-Recoil Aging," Science, 209: 1518 (1980).
5. A. E. Ringwood, V. Oversby and W. Sinclair, "Effects of Radiation Damage on Synroc," Scientific Basis for Nuclear Waste Management, vol. 2, C. J. M. Northrup, Ed., Plenum Press, New York (1980).
6. H. W. Nesbitt, G. M. Brancroft, W. S. Fyfe, S. Karkhanis, P. Melling and A. Nishijima, "On the Thermodynamic and Kinetic Stability of Perovskite in Natural Waters," Workshop on Alternate Nuclear Waste Forms and Interactions in Geologic Media, Gatlinburg, Tenn. (1980).
7. J. C. Dran, Y. Langevin, M. Maurette, J. C. Petit, and B. Vassent, "A Step-Height Microetching Method for Investigating Glass-Water Systems," submitted to J. Appl. Phys. (1980).

CESIUM MIGRATION THROUGH SOLID CORES OF MAGENTA DOLOMITE*

A. W. Lynch and R. G. Dosch

Sandia National Laboratories**

Albuquerque, New Mexico 87185

ABSTRACT

Column-flow experiments have been conducted on Magenta dolomite rock taken from a potential site for radioactive waste disposal (WIPP). The results indicate qualitative agreement with results from batch sorption coefficient (K_d) measurements. In one experiment deionized water pre-equilibrated with crushed rock containing 0.1 μCi/ml ^{137}Cs flowed through a column with an average velocity of 1.1 cm/day for 18 months. The average ^{137}Cs penetration was 0.3 cm into the column indicating a sorption coefficient of about 120 ml/g. Batch sorption experiments gave a value of 650 ml/g. When the water was changed to a pre-equilibrated groundwater and the Cs concentration was increased, both column and batch experiments gave values for the sorption coefficient of approximately 19 ml/g. Transmission electron microscopy results showed the Magenta samples to be primarily composed of dolomite and calcium sulfate and that a minor montmorillonite clay phase may control the retention of Cs on these rocks.

INTRODUCTION

The currently favored method for radioactive waste disposal is burial in deep geologic formations based on the assumption that the geologic environment will provide permanent isolation. To evaluate

---

* This work was supported by the U. S. Department of Energy under contract #DE-AC04-76-DP00789.

** A U. S. Department of Energy facility.

this assumption, experiments are performed to determine the potential for interaction between radiochemicals and the rocks obtained from the proposed site. This report describes three column-flow migration experiments that were performed in support of the Safety Assessment Program for the Waste Isolation Pilot Plant (WIPP) in southeastern New Mexico.

For the WIPP studies it is assumed that a breach in the rock salt repository allows groundwater to contact and leach the waste form. One potential path to the biosphere is a dolomite aquifer located in the Rustler Formation which lies above the salt beds[1] proposed to host the repository. The dolomite samples used in the column-flow studies were obtained from the Magenta member of the Rustler Formation.

Cesium was used to represent a leached fission product; it does not form precipitates or hydrolysis products in WIPP waters, and it is one of the most soluble, mobile and hazardous species in high level waste. The liquid phase in these experiments was either deionized water or WIPP groundwater simulant; both solutions were pre-equilibrated with crushed dolomite prior to use. These solutions were chosen because of the magnitude of batch sorption coefficient values (K_d) determined in separate experiments. The three experiments reported here differed in the composition of the water and in the method used to add the ^{137}Cs tracer to the system.

The objectives of the experiment were to: 1) determine the concentration profile of ^{137}Cs in the porous rock, and 2) compare the sorption coefficients (K_d's) calculated from column-flow data with K_d's measured for the same solid-liquid systems using batch-equilibration techniques.

ROCK SAMPLES

The dolomite samples were obtained from a core from AEC #8 (749 ft. horizon) drilled into the Rustler Magenta Formation near the proposed WIPP site. The rock was composed of alternating layers (spaced from ~0.1 to ~5 mm apart) of dolomite, $CaMg(CO_3)_2$, and anhydrite, $CaSO_4$, or gypsum, $CaSO_4 \cdot 2H_2O$.[2] Large variations in composition are known to exist with different locations within the Magenta member.[3] This rock was interspersed with montmorillonite clays.

The rock columns used in these experiments were lithologically identical, monolithic cylinders, machined from the cores so that the bedding planes were oriented parallel to the direction of flow. The density of the rock was 2.37 g/cc (measured by water displacement). The porosity (13.8%) used was an average value determined by measuring the volume of water absorbed under vacuum, values ranged from 12-16 percent, reflecting the inhomogeneity of the rock.

LIQUID PHASE

The liquid phase was either rock-equilibrated deionized water[4] or rock-equilibrated groundwater simulant.[5] All liquid phases were equilibrated before use by contacting 50 grams of crushed rock per liter of solution for 24 hours and filtering before use.

The groundwater simulant contained 5% by volume salt brine B[5] and 0.5 ppm stable Cs. Both the salt brine and the Cs were added specifically to decrease the sorption coefficient so that significant migration of Cs into the column could be observed in a shorter time frame (months). The appropriate concentrations were established in separate experiments.

EXPERIMENTAL PROCEDURE

Argon gas (≤10 psi) was used to pressurize the feed solution and provide constant fluid flow rates from 0.1 to 3 cm/day, a range typical of groundwater flow in the area.[6] The edges of the rock cores were sealed with silicone rubber to prevent leaks between the sample and the apparatus. Flow was continuous and aliquots (1 to 5 ml) of the effluent were analyzed for ^{137}Cs activity on a Searle Model 1185 gamma counter with a 3 in. diameter NaI well detector.

A description of the three experiments follows.

Column I

The rock column measured 10 cm in diameter and 2.5 cm in thickness. The liquid was pre-equilibrated deionized water doped with 0.1 μCi/ml (~10^{-3} ppm) ^{137}Cs. The column flow averaged 1.1 cm/day and was maintained for approximately 18 months. A total of 0.5 millicurie of ^{137}Cs was deposited on the column. The batch sorption coefficient measured for this system was 650 ml/g.

Column 2

This column experiment was conducted to determine the effect of 1) increased Cs concentration and 2) increased ionic strength on ^{137}Cs migration by using the groundwater simulant with 5% salt brine and 0.5 ppm ^{133}Cs. The core measured 5 cm in diameter and 2.6 cm in thickness. ^{137}Cs concentration in the groundwater was 0.1 μCi/ml (~10^{-3} ppm). The flow rate was 1.6 cm/day. The solution composition was altered in order to significantly decrease the ^{137}Cs breakthrough time by reducing the K_d of the system to ~19 ml/g.

Column 3

This column duplicated column 2 except for the flow rate and the method used to add the ^{137}Cs. Instead of adding it to the bulk solution, 1 ml of 0.4 millicurie/ml of ^{137}Cs solution was added directly to the rock surface and allowed to penetrate (~16 hours) before adding the pre-equilibrated groundwater simulant (including the 5% salt brine and 0.5 ppm ^{133}Cs). The column measured 5 cm in diameter and 1.7 cm in thickness. The flow rate used to elute the ^{137}Cs was 2.8 cm/day. The batch sorption coefficient corresponding to this system was also taken as 19 ml/g.

RETARDATION FACTOR CALCULATIONS

The retardation factor from column data is defined as the ratio of the velocity of the radionuclide to the velocity of the water. The velocity of the water was determined by measuring effluent volume for a specified time interval. Three methods were used to determine radionuclide velocity:

(1) Column 1 - depth of ^{137}Cs infiltration determined by a scan of the column's cross section
(2) Column 2 - half maximum from the breakthrough curve, and
(3) Column 3 - maximum on elution curve.

The well-known equation[7]

$$\frac{V_n}{V_w} = \frac{1}{1 + \frac{\rho}{\theta} K_d} , \qquad (1)$$

where V_n = velocity of radionuclide, V_w = velocity of water carrier, θ = porosity, ρ = density, and K_d = sorption coefficient (porous flow assumed) was used to back calculate K_d from the column-flow data; i.e., by measuring fluid flow, nuclide migration and thereby approximating V_n and V_w.
NOTE: There are several sources of error that must be considered when comparing results from batch-equilibration and column-flow measurements: 1) rock heterogeneity (i.e., composition, porosity), 2) formation heterogeneity[3]. 3) compositional changes to the rock and water due to dissolution which may occur during the experiment, and 4) the effect of solid/liquid ratio. Sorption coefficients vary with changes in solid/liquid ratio with other systems[8] as well as with the Cs-magenta system (unpublished data).

RESULTS

Column 1

A trace of ^{137}Cs activity (~0.1 counts per minute per milliliter or ~2 x 10^{-9} ppm) was detected in effluent samples after ~56 void volumes or 1.4 liters of solution had passed through the dolomite sample. The activity in the effluent gradually decreased to ~1 cpm/ml

(~2 x 10^{-8} ppm) after ~88 void volumes. Increasing the flow rate from 0.8 to 1.6 cm/day caused an increase in effluent activity to ~5 cpm/ml (~1 x 10^{-7} ppm) after ~160 void volumes. (Count rates of 5 x 10^4 cpm/ml were typically measured for solutions containing 0.1 μCi/ml ^{137}Cs.) Apparently a small amount of ^{137}Cs penetrated the entire length of rock. In addition a small amount of ^{137}Cs activity (~100 cpm/g) was found in a monolith cut from the bottom of the column (possible contamination from handling).

After 18 months the sample was removed from the apparatus and a cross section was taken to measure ^{137}Cs infiltration. An autoradiograph made from the cross section (Fig. 1) shows different concentrations in different bedding planes as well as an apparent maximum depth of penetration (4 mm). The ^{137}Cs progressed with a sharp, well-defined wave front which is expected in a system having a favorable adsorption isotherm and a high K_d.

The curve shown in Fig. 2 was obtained by moving the cross-sectioned rock across a NaI detector mounted behind a 0.05 cm slit. The average ^{137}Cs penetration (1/2 maximum) was measured at 0.3 cm. A K_d of ~120 ml/g was calculated (Eq. 1) using an average ^{137}Cs velocity of 5.4 x 10^{-4} cm/day (0.3 cm in 18 months) and an average water velocity of 1.1 cm/day.

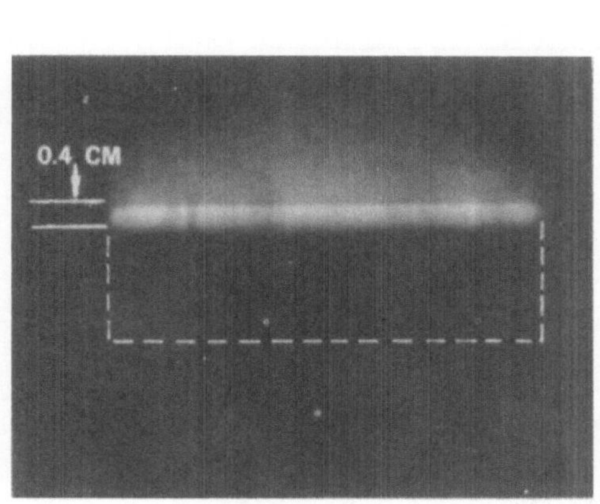

Fig. 1. ^{137}Cs penetration in column 1.

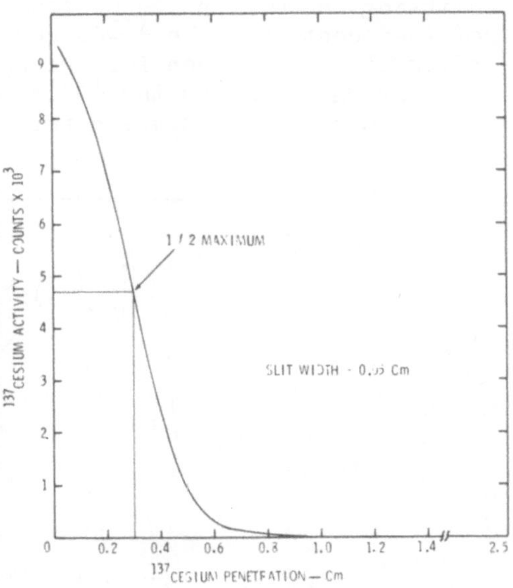

Fig. 2. ^{137}Cs activity in cross section of column 1.

Column 2

Measurable ^{137}Cs activity was observed after ~5 void volumes or 50 ml had passed through the column. The data collected from effluent samples are represented in Fig. 3. The experiment was discontinued after the ^{137}Cs activity in the effluent reached about 85% of the activity in the feed solution.

An autoradiograph made from a cross section of column 2 (Fig. 4) shows complete infiltration of ^{137}Cs. The sorption coefficient calculated from the half maximum value from the breakthrough curve (~200 void volumes) for column 2 is the same, within estimated experimental error, as the corresponding batch value (19 ml/g). ^{137}Cs migrated faster in column 2 than in column 1, as predicted, due to the increased Cs concentration and/or ionic strength.

Column 3

The elution curve (Fig. 3) shows that 550 void volumes (~2.8 liters) were required to elute most of the ^{137}Cs from column 3. A small amount remained on the column (indicated by the long tail on the elution curve). The appearance of ^{137}Cs activity in the initial effluent samples may be an indication of the effect of high localized Cs concentration on migration. A second effect of Cs concentration is displayed by the plateau occurring between void volumes 75 to 100 which is apparently the result of inadvertently omitting the 0.5 ppm ^{133}Cs from the feed solution. The next addition of feed contained the ^{133}Cs resulting in an increase of ^{137}Cs in the effluent. The reason for the spikes is currently unknown. An autoradiograph made from the ^{137}Cs remaining on the sample (Fig. 5) shows complete penetration through the sample.

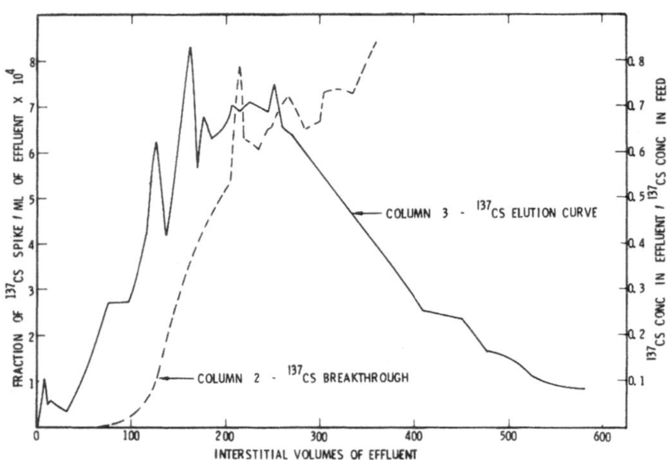

Fig. 3. ^{137}Cs migration through magenta columns 2 and 3.

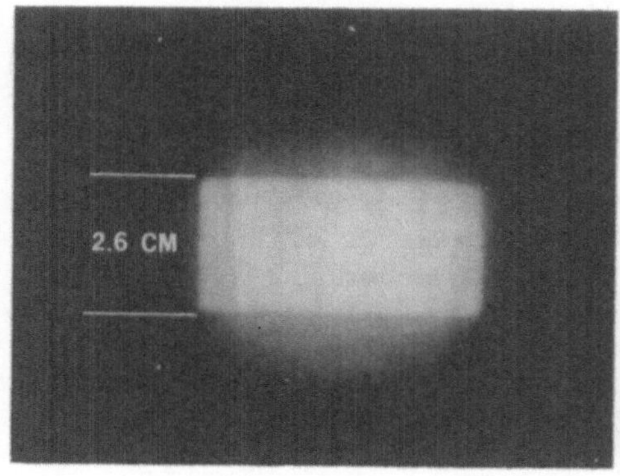

Fig. 4. ^{137}Cs penetration in column 2.

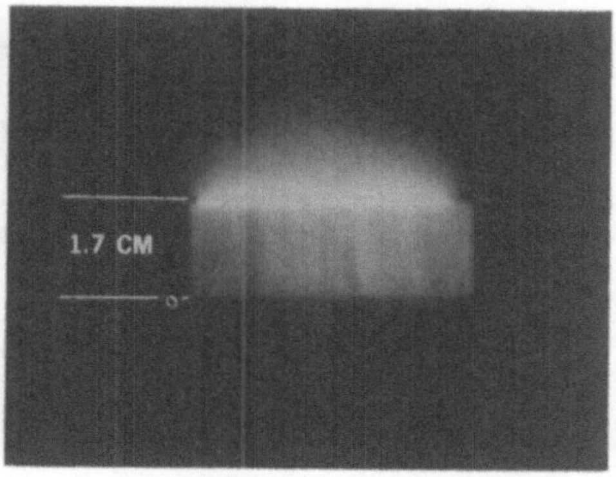

Fig. 5. ^{137}Cs penetration in column 3.

The sorption coefficient calculated (maximum on elution curve ~225 void volumes, initial spike ignored) from the column data is the same, within estimated experimental error, as the corresponding batch value (19 ml/g).

TRANSMISSION ELECTRON MICROSCOPY

Transmission electron microscopy techniques were used in an attempt to determine how Cs was being sorbed on the Magenta dolomite. A sample of the rock was soaked in 1 molar $^{133}CsCl$ solution. Fragments of this sample were compared to fragments from an untreated sample. The results indicate that ^{133}Cs is preferentially sorbed on montmorillonite clay particles.[2] Assuming that the clay's affinity for Cs does not decrease with decreasing Cs concentration, the result could indicate that a minor clay constituent in Magenta is providing the principal sites for ^{137}Cs sorption.

CONCLUSIONS

The objective of these experiments was to compare results from two experimental techniques typically used for predicting radionuclide migration in geologic media. Several conclusions obtained from the experimental results are summarized below.

There was qualitative agreement between batch K_d and column-flow results in all systems studied. Conditions which led to decreasing batch K_d's resulted in increasing velocity of ^{137}Cs in the column. For example, simultaneously increasing Cs concentration and ionic strength caused the batch K_d's to decrease and caused the increase in ^{137}Cs migration through columns 2 and 3.

Quantitatively, the predicted average ^{137}Cs penetration in column 1 (K_d = 650 ml/g) was 0.05 cm. The average ^{137}Cs penetration indicated on the profile obtained from the cross section of column 1 was 0.3 cm. The small amount of ^{137}Cs that appears to have penetrated the entire length of column 1 suggests that a second mechanism such as anomalous equilibria at low concentrations, reaction species (silicates) or contaminants (organics) may be a factor in Cs migration and should be investigated further. In columns 2 and 3 where elution curves were actually obtained, the average ^{137}Cs velocities observed were within estimated experimental error of those predicted from the batch K_d (19 lm/g).

Montmorillonite clay particles found in the Magenta were shown to sorb Cs from 1 molar CsCl solution and may control ^{137}Cs sorption at the tracer level in this rock.

REFERENCES

1. D. W. Powers, S. J. Lambert, S. E. Shaffer, L. R. Hill and W. D. Weart, "Geological Characterization Report, Waste Isolation Pilot Plant, Southeastern New Mexico," SAND78-1596 (1978).*
2. A. W. Lynch, R. G. Dosch, and C. R. Hills, "Migration of Cesium-137 Through a Solid Core of Magenta Dolomite Taken from the Rustler Formation in Southeastern New Mexico," SAND80-1259 (1980).*
3. Basic Data Reports, SAND79-0267 through 0291 (in preparation).
4. M. G. Seitz, P. G. Ricket, S. M. Fried, A. M. Friedman, and J. Steindler, "Studies of Nuclear Waste Migration in Geologic Media," Argonne Report ANL-79-30 (1978).
5. R. G. Dosch and A. W. Lynch, "Interaction of Radionuclides Associated with the Waste Isolation Pilot Plant Site in New Mexico," SAND78-0297 (1978).*
6. S. J. Lambert, J. W. Mercer, "Hydrologic Investigations of the Los Medanos Area, Southeastern New Mexico," SAND77-1401 (1977).*
7. F. Holfferich, "Ion Exchange," McGraw-Hill, New York (1962), p. 402.
8. R. G. Dosch "Assessment of Potential Radionuclide Transport in Site-Specific Geologic Formations," SAND79-2468 (1980).*

---

* Available from National Technical Information Service, U. S. Department of Commerce, 5285 Port Royal Rd., Springfield, VA, 22161.

SORPTION STUDIES OF $H^{14}CO_3^-$ ON SOME GEOLOGIC MEDIA AND CONCRETE

B. Allard, B. Torstenfelt, and K. Andersson

Department of Nuclear Chemistry
Chalmers University of Technology
S-412 96 GÖTEBORG, SWEDEN

INTRODUCTION

Carbon (as ^{14}C, half-life 5730 y) might give a significant contribution to the long-term biological hazards from medium-level radioactive wastes. In the evaluation of the feasibility of underground storage of radioactive wastes, it is generally assumed that inorganic carbon (as carbonate or hydrogen carbonate) would have a retention factor (i.e., water velocity/nuclide velocity) of 1 in groundwater,[1] although no sorption measurements have actually been performed. In the present study the sorption of hydrogen carbonate on some geologic media, as well as on concrete and cement paste, has been studied.

THE CARBONATE SYSTEM IN GROUNDWATERS

Most groundwaters have fairly high concentrations of carbonate/hydrogen carbonate (in the millimolar range[2]), partly from exchanges and contact with the biosphere and the atmosphere. In granitic groundwaters a typical concentration range for total carbonate is 1-6 m$\underline{M}$ (60-100 mg/l)(ref. 3) although deep groundwaters might have substantially lower carbonate contents.[4]

The only metal ions in the groundwater that would form precipitates or significant concentrations of soluble carbonate complexes are calcium and magnesium with the following formation constants (K_1) and solubility products (K_s) at zero ionic strength:[5]

$$Ca^{2+} + HCO_3^- \rightleftharpoons CaHCO_3^+ \qquad \log K_1 = 1.0$$
$$Ca^{2+} + CO_3^{2-} \rightleftharpoons CaCO_3 \qquad \log K_1 = 3.15$$
$$Ca^{2+} + CO_3^{2-} \rightleftharpoons CaCO_3(s) \qquad \log K_s = -8.35$$
$$Mg^{2+} + HCO_3^- \rightleftharpoons MgHCO_3^+ \qquad \log K_1 = 0.95$$
$$Mg^{2+} + CO_3^{2-} \rightleftharpoons MgCO_3 \qquad \log K_1 = 2.88$$
$$Mg^{2+} + CO_3^{2-} \rightleftharpoons MgCO_3(s) \qquad \log K_s = -7.46$$

The concentration ranges for common constituents in granitic groundwaters are given in Table 1 (ref. 3) as well as the composition of an artificial groundwater[6] used in this study. The carbonate equilibria in this water, as illustrated in Fig. 1, is slightly undersaturated with respect to $CaCO_3(s)$, which is common in natural waters. This means that an increase in pH or an increase of the calcium concentration might have substantial influence on the complexation and sorption of carbonate/hydrogen carbonate.

Table 1. Composition of granitic groundwater

Specie	Probable concentration mg/l	Artificial groundwater[a] mg/l
Ca^{2+}	25 - 50	18
Mg^{2+}	5 - 20	4.3
Na^+	10 - 100	65
K^+	1 - 5	3.9
Fe^{2+}	0.5 - 15	-
Mn^{2+}	0.1 - 0.5	-
HCO_3^-	60 - 400	123
Cl^-	5 - 50	70
SO_4^{2-}	1 - 15	9.6
NO_3^-	0.1 - 0.5	-
HPO_4^{2-}	0.01 - 0.1	-
F^-	0.5 - 2	-
SiO_2(tot)	5 - 30	12
HS^-	0 - 1	-
pH	7.2 - 8.5	8.0 - 8.2

[a]Initial concentrations

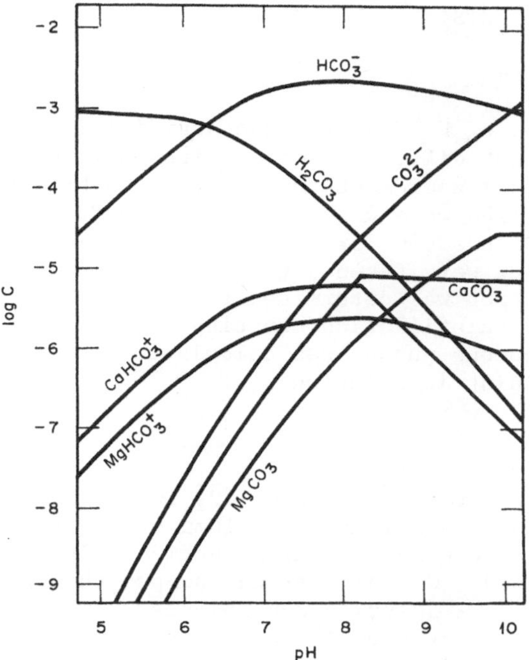

Fig. 1. Carbonate equilibria in the artificial groundwater (log concentration vs pH).

The following experimental conditions were chosen:

Solid phase:	Granite (Stripa);[a] 0.09-0.25 mm Na-montmorillonite Bentonite/quartz (10/90)[b] Sandy moraine;[c] not sieved Clayish moraine;[c] not sieved Calcite; 0.063-0.090 mm Concrete;[d] 0.063-0.090 mm Cement paste;[d] 0.063-0.090 mm
Liquid phase:	Artificial groundwater; compare Table 1
Solid/liquid ratio:	20 g/l
Radionuclide:	^{14}C as $H^{14}CO_3^-$; trace concentration
Contact time:	0.5 h to 6 months; compare Table 2
Temperature:	Ambient

[a] Characterization given in Ref. 8.
[b] Compare ref. 9.
[c] Characterization given in Ref. 10.
[d] Characterization given in Ref. 11.

SORPTION MEASUREMENTS

Batch Measurements

The sorption of $H^{14}CO_3^-$ from the artificial groundwater given in Table 1 on eight different solids with varying calcium content was studied. The measurements were performed by a batch technique.[6,7] The solids were washed with a groundwater solution and then contacted with a similar solution containing $H^{14}CO_3^-$. Aliquots of the water were taken after various contact times and the $H^{14}CO_3^-$ concentration was determined from the ^{14}CO activity obtained by liquid scintillation counting. The distribution between the solid and aqueous phases was calculated from the $H^{14}CO_3^-$ concentration remaining in solution.

Column Measurements

A study of the retention of $H^{14}CO_3^-$ in a column was also performed for some of the solids. A column of the type used for HPLC together with a HPLC pump, sample injection valve, and fraction collector were used. Groundwater was pumped through the column and a pulse of $H^{14}CO_3^-$ solution was injected. The samples from the fraction collector were counted for ^{14}C activity in the same way as in the batch experiments.

To determine the column volume in the system, a pulse of ^{22}Na was injected in a separate run. This was assumed to pass the column with little retention. Similar measurements using ^{131}I are in progress.

The retention of the $H^{14}CO_3^-$ compared to ^{22}Na was evaluated from the concentration profile in the outlet of the column by means of momentum analysis.[12]

The following experimental conditions were chosen:

Solid phase:	Sandy moraine; compare batch experiments Calcite; compare batch experiments
Liquid phase:	Artificial groundwater; compare Table 1
Column diameter:	4 mm
Column height:	130 mm
Porosity of solid in column:	0.09 (sandy moraine), 0.24 (calcite)
Flow rate:	0.01-0.08 ml/min; compare Table 3
Fraction size:	0.25 ml

Radionuclides: ^{22}Na; trace concentration
$H^{14}CO_3^-$; trace concentration

Injected radionuclide volume: 0.05 ml

Temperature: Ambient

RESULTS AND DISCUSSION

Batch Measurements

A summary of measured distribution coefficients (K_d, mol/kg solid per mol/m^3 liquid) at different contact times is given in Table 2. The sorption is low compared with what is observed for Cs or for the actinides. Values for $K_d > 0$ are, however, obtained for all the studied solids except for granite and Na-montmorillonite.

Table 2. Distribution coefficient (K_d) vs contact time

Solid	$K_d \times 10^3$ (m^3/kg) at contact times							
	0.5 h	2 h	6 h	24 h	3 d	1 w	5 w	6 m
Stripa granite					0	0	0	0
Na-montmorillonite					0	0	0	0
Bentonite/quartz					2.8	8.6	7.6	7.8
Sandy moraine	1.0	0.2	0.7	0.5	0.8	1.1	2.6	2.2
Clayish moraine	1.9	0.7	1.3	1.4	1.3	2.0	3.0	2.3
Calcite	1.0	1.0	1.5	1.7	3.7	3.5	9.0	83
Concrete					5.3	1600	$>10^4$	
Cement paste					7.4	1600	$>10^4$	

The sorption is enhanced by increasing calcium content of the solid. This may be due to several mechanisms. The groundwater is almost saturated with respect to $CaCO_3$. A small change in composition of the water caused by weathering of calcium-containing minerals or ion-exchange (sodium-calcium), giving an increased calcium concentration in solution or an increase of pH might cause a precipitation, where $^{14}CO_3^{2-}$ is precipitated together with inactive carbonate. In the case of cement paste and concrete, the carbonate will most certainly be precipitated. The difference between Na-montmorillonite and bentonite/quartz may be explained by the calcium content of the bentonite used (MX-80 with 18 mmoles Ca^{2+}/100 g) allowing a sodium-calcium exchange.

Another mechanism might be exchange between carbonate in solution and in the solid. The slow increase of distribution coefficients with time for calcite may be explained by this mechanism, which would have very slow kinetics.

A third mechanism might be a diffusion of $H^{14}CO_3^-$ into microfissures of the particles. This seems not to be the case for calcite, since the observed increase in the distribution coefficient vs time is not what would be expected for a pure diffusion process [K_d increases with $(time)^{-0.7}$ instead of $(time)^{-0.5}$ for a diffusion process].

Column Measurements

Column studies were performed for calcite and sandy moraine only. (The other solids, showing a $K_d \neq 0$ had too low permeabilities for this kind of column experiments.) The measured retention factors (R) and the corresponding distribution coefficients are given in Table 3. The distribution coefficients have been calculated from $R = (1 + K_d\rho_p(1-\varepsilon)/\varepsilon)$, where ρ_p is the density of the solid (kg/m^3) and ε the porosity in the column. An example of concentration profiles obtained in the column experiments is given in Fig. 2.

Table 3. Retention factors from column experiments

Solid	Flow rate ml/min	R	$K_d\rho_p$	$K_d x10^3$ m^3/kg
Sandy moraine	0.01	2.7	0.27	0.1
	0.03	1.1	0.01	0
	0.05	1	0	0
	0.08	1	0	0
Calcite	0.01	3.2	0.68	0.3
	0.03	2.5	0.46	0.2
	0.05	2.3	0.40	0.2
	0.08	2.2	0.37	0.1

A retention factor of 2 to 3 is obtained for calcite, indicating that at least this retention may be expected in a natural groundwater system if calcite is present.

CONCLUSIONS

The investigation has shown that a significant retention of $H^{14}CO_3^-$ in a groundwater system may be expected in the presence of

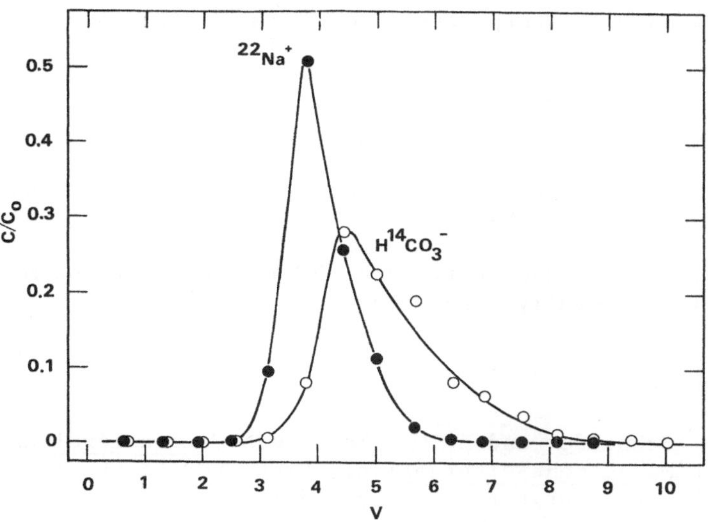

Fig. 2. Concentration profiles obtained in column experiments. Flow rate 0.05 ml/min, V = column volumes/porosity.

calcite- and calcium-containing materials. Higher distribution coefficients were obtained in the batch experiments than in the column experiments. Thus a lower flow rate would probably give a higher retention than what was observed in the present column experiments. In a potential waste repository the water flow rate would be low and calcite would be a common fracture mineral.

Concrete and bentonite/quartz have been suggested as backfill materials in a repository.[9] Concrete may be expected to retain a carbonate leakage completely, as long as the material is intact.[11] Also a bentonite/quartz mixture would give a retention by at least a factor of 10.

The ^{14}C exchange for inactive carbon is also of importance when using the ^{14}C method for groundwater dating. In the use of mathematical models for prediction of the carbon evolution in a groundwater system, a quantitative knowledge of possible sinks and sources of carbonate is required. The sorption of ^{14}C on solid surfaces constitutes a sink that will, if not considered in the age determination, lead to too high estimated ages. Both precipitation and weathering may change the $^{14}C/^{12}C$ ratio in the undisturbed groundwater.

Thus, the role of calcium-containing media, particularly calcite, in the repository/bedrock system must be considered in assessing ^{14}C transport times from the repository to the biosphere as well as in the interpretation of ^{14}C age determinations of groundwater.

ACKNOWLEDGEMENTS

This work was supported by the Swedish Research Council for Radioactive Waste (PRAV).

The skillful laboratory work of Ms. Wanda May and Ms. Lena Eliasson is gratefully acknowledged.

REFERENCES

1. K. J. Schneider and A. M. Platt (eds), High-level Radioactive Waste Management Alternatives, BNWL-1900, Battelle Pacific Northwest Laboratories, Richland (1974).
2. W. Stumm and J. J. Morgan, Aquatic Chemistry, Wiley Interscience, New York (1970).
3. G. Jacks, Groundwater Chemistry at Depth in Granites and Gneisses, KBS Technical Report 88, Kärnbränslesäkerhet, Stockholm (1978).
4. P. Fritz, J. F. Baker, and J. E. Gale, Geochemistry and Isotope Hydrology in the Stripa Granite, Swedish-American Cooperative Program on Radioactive Waste Storage in Mineral Caverns in Crystalline Rock, Technical Information 12, LBL-8285, SAC-12, UC-70, Berkeley (1979).
5. L. G. Sillen and A. E. Martell, Stability Constants of Metal-ion Complexes, Special Publication 17, The Chemical Society, London (1964).
6. B. Allard and G. W. Beall, "Sorption of Americium on Geologic Media," J. Environ. Sci. Health, 6:507-18 (1979).
7. B. Allard, K. Andersson, and B. Torstenfelt, to be published.
8. B. Torstenfelt, K. Andersson, and B. Allard, Sorption of Sr and Cs on Rocks and Minerals. Part I, KBS Technical Report, in press.
9. Handling of Spent Nuclear Fuel and Final Storage of Vitrified High-level Reprocessing Waste, Kärnbränslesäkerhet, Stockholm (1977).
10. K. Andersson, B. Torstenfelt, and B. Allard, Sorption of Some Radionuclides on Sandy and Clayish Moraine and Comparison with Granite and Montmorillonite, CTH, (1980) (draft).
11. K. Andersson, B. Torstenfelt, and B. Allard, Sorption and Diffusion of Cs and I in Concrete and Cement Paste, KBS Technical Report, in press.
12. M. Kubin, "Beitrag zur Theorie der Chromatographie," Coll. Czech. Chem. Comm., 30:1104-18 (1965).

SOME DIFFICULTIES IN INTERPRETING IN-SITU TRACER TESTS

Ivars Neretnieks

Department of Chemical Engineering
Royal Institute of Technology
S 100 44 Stockholm 70

SUMMARY

Two mechanisms for the spreading of a tracer pulse are discussed. It is demonstrated that in the case of stratified blow as well as when there is diffusion into the rock matrix, the normally used Fickian-dispersion model is not applicable.

BACKGROUND

By dispersion in a very broad sense, we mean the spreading of a species carried by a fluid as the fluid moves along a flow path. Bear (1969) gives a comprehensive treatment on hydrodynamic dispersion theories. The most advanced models treat the spreading process by modelling more of less randomly oriented pores combined with some assumptions on how velocities in the channels vary as well as how distribution at channel divisions and mixing at channel intersections occur. An early, but fairly advanced, such treatment is found in de Josselin de Jong's paper (1958).

The common basis for practically all these treatments is that the spreading is described by one parameter — the standard deviation σ_x (or variance σ_x^2) of a pulse as it spreads with distance. Fig. 1 shows the standard deviation of a pulse which is spreading as it moves along a flow path.

If the spreading were a random process such as molecular diffusion, a dispersion coefficient D_L analogous to the diffusion coefficient could be determined from $D_L = \sigma_x^2/2\bar{t}$. This is called Fickian dispersion.

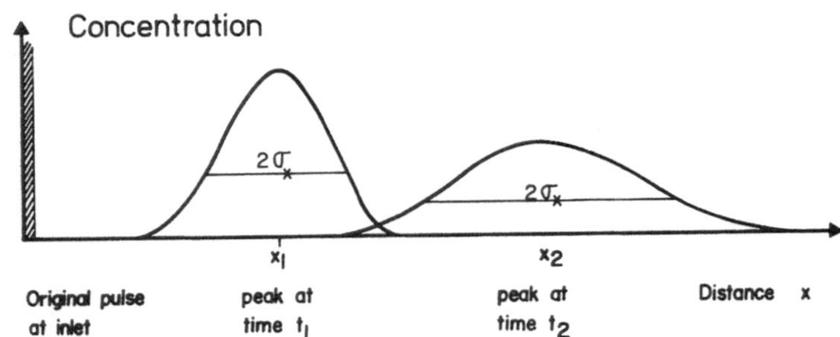

Fig. 1. The concentration profile and the standard deviation of a tracer at different times.

The dispersion coefficient is proportional to the velocity and the particle size, but is independent of distance; $D_L \propto U_f d_p$. This also implies that the variance is proportional to the distance $\sigma_x^2 \propto x$.

Recently, Schwarz (1977) by computer simulation has shown that the uneven distribution of resistances in a porous medium may not lead to a variance which increases proportionally to the distance travelled. Neretnieks (1979) showed that in a medium where severe channelling occurs — parallel, unconnected channels with velocity differences between channels — the standard deviation σ_x is proportional to the distance travelled — $\sigma_x \propto x$ — instead of $\sigma_x^2 \propto x$ as in the Fickian dispersion case above. Matheron and Marsily (1979) arrived at the same conclusion.

Lallemand-Barrès and Peaudecerf (1978) gives a compilation of field data. Their Fig. 2 is a plot of D_L/U_f versus the distance between injection point and observation point. Although the spread in data is large, there is a definite increase in D_L/U_f with distance which ranges from 3 m to 5000 m. The swarm of points may roughly be approximated by $D_L/U_f = \text{const}\cdot x$. With const = 0.1, 90% of the points lie within the bounds $U_f x/D_L = 1 - 30$.

The hypothesis that the dimensionless quantity $U_f x/D_L$, which is commonly called the Peclet number, is constant, is much better than the hypothesis that D_L/U_f is constant. The first spans over 1.5 orders of magnitude whereas the second spans over more than 2.5 orders of magnitude. Field data thus do not indicate that D_L/U_f is constant.

A mechanism of "dispersion" which is seldom treated in the hydraulic literature, is the effect of interaction with the solid material. Here only one such aspect will be discussed — that of diffusion of the species into the rock matrix. In the chemical engineering literature (Kučera 1965, Perry 1973) this effect is not treated as dispersion but is modelled independently and thus separated from the hydrodynamic dispersion. For many cases involving diffusion into the solids, the shape of the breakthrough front is more determined by this mechanism than by hydrodynamic dispersion. It has been shown (Neretnieks 1980) that these effects may have a considerable influence on the breakthrough curve for flow in fissured granitic bedrock.

STRATIFIED FLOW

The variance and dispersion for flow in a set of parallel fissures with varying sizes is derived below. At the inlet end of the channels a tracer is introduced, and at a place downstream the fluid from all channels is collected and mixed. The concentration in this place is measured over time. The residence time distribution and its mean and variance can be determined from the observed concentration-time curve. If the variance of the tracer pulse at the inlet is known, the dispersion coefficient can be determined from (Levenspiel 1972)

$$\frac{\sigma^2_{t,out} - \sigma^2_{t,in}}{\bar{t}^2} = 2\,\frac{D_L}{\bar{U}x} \tag{1}$$

Determination of the variance of a breakthrough curve from parallel fissure flow

The fissure width distribution is $f(\delta)$. In a fissure of width δ, the flow rate $Q(\delta)$ is proportional to the fissure width to the third power for laminar flow

$$Q(\delta) = k_1 \delta^3 l \tag{2}$$

where l is the breadth of the fissure. The velocity is proportional to the fissure width squared:

$$U = k_1 \delta^2 \tag{3}$$

The residence time in fissures with width δ over a given distance x is:

$$t = \frac{x}{k_1 \delta^2} = \frac{k_2}{\delta^2} \tag{4}$$

If a step with concentration C_0 is introduced at the inlet end of the set of fissures, the front will travel the distance x in time t in fissures with width δ. The fissures with residence times less than t will carry tracer. This is shown in Fig. 2.

The concentration obtained at the outlet end, at a time t when the effluent from all fissures is mixed, is given in eq. (5).

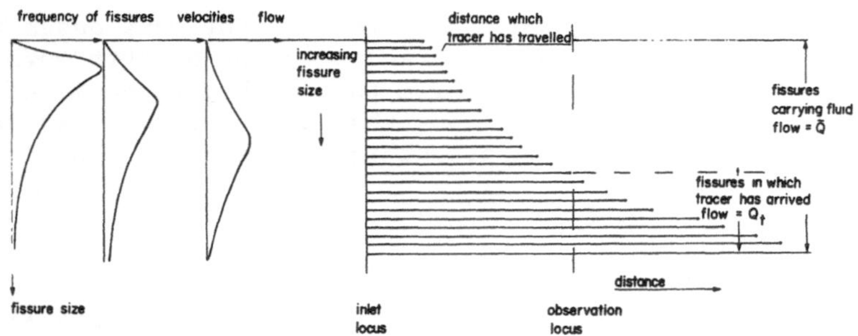

Fig. 2. Tracer flow in a medium with parallel fissures of different size.

$$\frac{C(t)}{C_o} = \frac{\int_{\delta}^{\infty} f(\delta)Q(\delta)d\delta}{\int_{0}^{\infty} f(\delta)Q(\delta)d\delta} = \frac{Q_t}{\overline{Q}} \tag{5}$$

t is the residence time in fissure $\delta(t)$. The above expression says that the flow Q_t from the tracer carrying fissures with widths $\delta(t) \geq \delta$, is diluted by the total flow of water Q from all fissures.

In general for individual channels with concentration breakthrough curves $C(\delta,t)$ at the outlet, the concentration of the mixed effluent from all channels is

$$\frac{C(t)}{C_o} = \frac{\int_{0}^{\infty} f(\delta)Q(\delta)C(\delta,t)d\delta}{\int_{0}^{\infty} f(\delta)Q(\delta)d\delta} \tag{5a}$$

The result for a Dirac concentration pulse at the inlet i.e. $C_o dt = 1$ and $dt \rightarrow 0$ can be obtained by differentiation of equation (5) with respect to time.

$$\frac{C(t)}{C_o dt} = \frac{1}{2}k_1 \frac{f(\delta)Q(\delta)\delta^3}{x\overline{Q}} \tag{6}$$

where

$$\overline{Q} = \int_{0}^{\infty} f(\delta)Q(\delta)d\delta \tag{7}$$

The mean residence time for a response curve C(t) from a Dirac pulse is obtained from

$$\overline{t} = \frac{1}{C_o dt} \int_{0}^{\infty} C(t)t\,dt \tag{8}$$

and the variance of the residence time or second statistical central moment from

$$\sigma_t^2 = \frac{1}{C_o dt} \int_0^\infty C(t)(t-\bar{t})^2 dt \tag{9}$$

By inserting eqs. (6), (7), (2), and (4) into (8) and (9), we obtain

$$\bar{t} = \frac{xl}{\bar{Q}} \int_0^\infty f(\delta)\delta d\delta = \frac{x}{k_1} \frac{\int_0^\infty f(\delta)\delta d\delta}{\int_0^\infty f(\delta)\delta^3 d\delta} \tag{10}$$

and

$$\frac{\sigma_t^2}{\bar{t}^2} = \frac{\int_0^\infty f(\delta)\delta^3 d\delta \cdot \int_0^\infty \frac{f(\delta)}{\delta} d\delta}{\left[\int_0^\infty f(\delta)\delta d\delta\right]^2} - 1 \tag{11}$$

equation (11) shows that $\sigma_t^2/\bar{t}^2$ is independent of the distance. For the Dirac pulse at the inlet the variance is 0, and thus equation (1) directly gives

$$D_L = \frac{\sigma_t^2}{\bar{t}^2} \cdot \frac{1}{2} \bar{U}x \; .$$

Thus $D_L \propto x$ for stratified flow.

Snow (1970) obtained the fissure width distributions $f(\delta)$ for various consolidated rocks including granites. Snow used data from water injection tests in boreholes and from direct measurements of fissure widths. He found the distribution to be log normal

$$f(\frac{\delta}{\mu}) = \frac{1}{A} \cdot e^{-\frac{1}{2}\left(\frac{\log \delta/\mu}{\sigma_1}\right)^2} \tag{12}$$

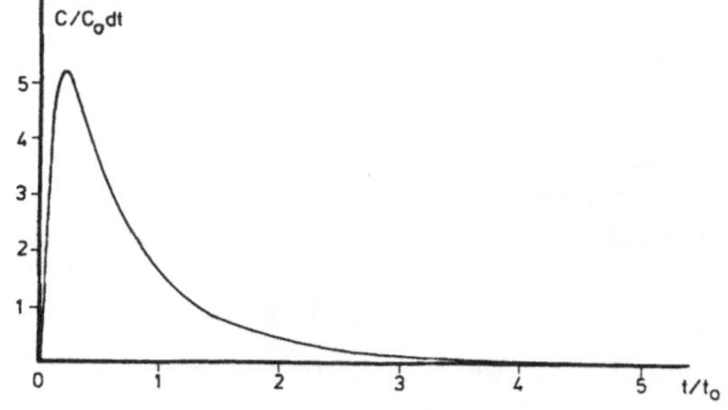

Fig. 3. Pulse response for a medium with parallel fissures.

The standard deviations σ_1 range from 0.057 to 0.394. The mean of σ_1 in Snow's investigation is 0.22. Fig. 3 shows the response to a Dirac pulse as determined by eq. (6) for $\sigma_1 = 0.22$. For this case $(\sigma_t/\bar{t})^2 = 1.82$, and $\sigma_t/\bar{t} = 1.35$, and $D_L/\bar{U} = 0.91\,x$.

The parallel fissure flow model thus predicts that the "dispersion coefficient" is proportional to the distance between the injection point and the observation point. It is also proportional to the velocity.

The recent compilation of dispersion coefficients obtained from field measurements by Lallemand-Barrès and Peaudecerf (1978), show this tendency. Although their data are very scattered, the relation predicted above falls within their data.

MATRIX DIFFUSION

Matrix diffusion has been treated recently by Neretnieks (1980) and is only summarized here.

When water which contains a tracer flows in a fissure, the tracer will migrate into the porous structure of the rock by diffusion. The flowing water will thus be depleted of the tracer. For a case where the distance between the fissures is very large and where the tracer thus does not penetrate more than a fraction of the distance between fissures, the transport can be mathematically described by the following expressions.

Diffusion in the rock is given by:

$$\frac{\partial C_p}{\partial t} = D_a \frac{\partial^2 C_p}{\partial z^2} \tag{13}$$

where $D_a = \frac{D_e}{K}$

For flow and sorption from the water in the fissure we have:

$$\frac{\partial C_f}{\partial t} + U_f \frac{\partial C_f}{\partial x} = \frac{2D_e}{\delta} \cdot \left.\frac{\partial C_p}{\partial z}\right|_{z=0} \tag{14}$$

For a system which is initially free of tracer and where the tracer concentration suddenly is increased to C_o at the inlet of the fissure $(x = 0)$, the initial and boundary conditions are:

IC	$C_p = C_f = 0$	$t = 0$, all x and z	(15)
BC1	$C_p = 0$	when $t > 0$ for $z \to \infty$	(16)
BC2	$C_f = C_o$	at $x = 0$ for $t > 0$	(17)

The solution is available in the literature (Carslaw & Jaeger 2nd ed. p. 396).

For the fluid in the fissure and for $t > x/U_f = t_w$, the following expression results:

$$\frac{C_f}{C_o} = \text{erfc}\left(\frac{G}{\sqrt{t - t_w}}\right) \tag{18}$$

where $G = \dfrac{2D_e t_w}{\delta\sqrt{D_a}}$.

For a unit pulse at the inlet i.e. $C_o\,dt = 1$ and $dt \to 0$, the response at the outlet is obtained by differentiation of equation (18) with respect to time, for $t > t_w$ we have

$$\frac{C_f}{C_o dt} = \frac{G}{(t - t_w)^{3/2}\pi^{1/2}} \cdot e^{-\frac{G^2}{t-t_w}} \tag{19}$$

The same result is of course obtained by using the Dirac pulse as boundary condition BC2 instead of eq. (17) when solving eqs. (13) and (14).

With the aid of eqs. (8) and (9) it could be attempted to find the mean residence time and variance of this residence time distribution. We find, however, that

$$\frac{C_f}{C_o dt} \propto \frac{1}{t^{3/2}} \quad \text{for } t \to \infty \tag{20}$$

and eqs. (8) and (9) thus are unbounded. This means that there is neither a mean residence time nor a variance if there is any diffusion into the porous walls and if the penetration depth is small compared to the distance to the next fissure. As most known rocks are more or less porous, it is in principle impossible to determine a dispersion coefficient by determining first and second moments of a response curve. In some cases, the dispersion coefficients in the literature may possibly have been obtained only because the integration of eqs. (8) and (9) were discontinued due to limited detection capability of the tracer at the tail of the pulse.

An example is used to demonstrate the importance of the detection limit. The case chosen is the following: There is steady flow in a fracture where a tracer pulse is introduced at one point and the response is measured at another point 22 m downstream. The gradient over the fracture is 0.11 m/m, the rock matrix has an effective diffusivity $D_e = 10^{-12}$ m^2/s, an apparent diffusivity $D_a = 2 \cdot 10^{-10}$ m^2/s. The water transport time is 10 hours. The D_e and D_a values are chosen to describe the diffusion of solved ions in the water in the

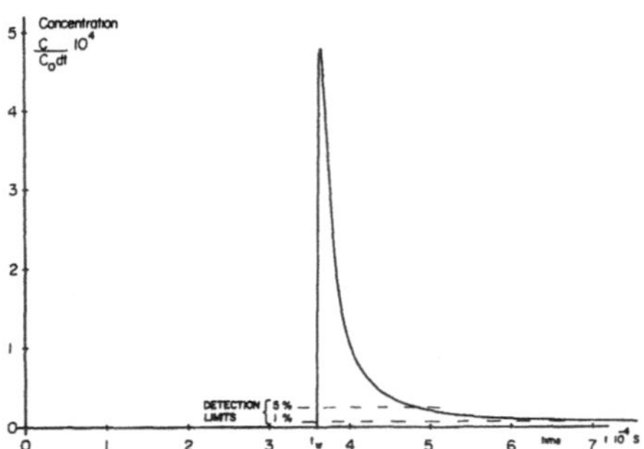

Fig. 4. Pulse response for a single fissure with porous walls.

pores in the rock matrix. There are no sorption effects, and the rock porosity is 0.5% which is a good value for a granite. Further details on diffusivities are given by Neretnieks (1980). For laminar flow of water in a parallel fissure at ambient temperature, the fissure width δ would be 0.082 mm.

The species will diffuse into the rock matrix with a penetration depth of about 10 mm during a 10-h contact time. The condition that the tracer should penetrate only a fraction of the block size of the rock is well fulfilled for crystalline rocks such as granites and gneisses, where the distances between fissures are usually considerably more than 22 mm.

The response at the outlet is shown in Fig. 4. It has been calculated using eq. (19). Two detection limits, 1% and 5%, are indicated by the dashed lines. The figure shows the long tail which gives an infinite mean residence time and variance. Table 1, case 1, shows the mean residence time $\bar{t}$, the relative standard deviation $\sigma_t/\bar{t}$, the dispersion coefficient D_L, and evaluated Peclet number Pec for the detection limits 0.2, 1, and 5%. It is clear that the detection limit will have a dominating influence on the results.

Table 1. The influence of the detection limit on dispersion

	case 1				case 2		
detection limit %	0	0.2	1	5	0	1	5
$\bar{t}$ h	∞	13.0	11.7	10.9	∞	33.3	27.1
$\sigma_t/\bar{t}$	∞	0.39	0.18	0.078	∞	0.5	0.24
$D_L \cdot 10^4\ m^2/s$	∞	7.9	1.84	0.33	∞	5.06	1.31
Pec	0	13	62	330	0	8.0	36

Even in this case when the peak is fairly narrow, the long tail will give a high dispersion coefficient, although there is no hydrodynamic dispersion. An increasing water transport time will quickly aggravate the problem. Table 1, case 2, is the same as case 1, but with twice the residence time, 20 h instead of 10 h. For 1% detection level, the Peclet number is 8.

DISCUSSIONS AND CONCLUSIONS

It is doubtful whether the Fickian dispersion description of a pulse spreading is applicable for sparsely fissured rock where flow channels may be poorly connected and flow may be stratified. For long contact times, diffusion into the rock matrix may become the dominating mechanism for pulse spreading. Neither of these mechanisms can be described by using a constant "dispersion coefficient."

NOTATION

C	concentration in the liquid	mol/m^3
C_o	initial concentration in the liquid	mol/m^3
C_f	concentration in the liquid in a fissure	mol/m^3
C_p	concentration in the liquid in a pore	mol/m^3
d_p	particle diameter	m
D_a	apparent diffusivity, $D_a = D_e/K$	m^2/s
D_e	effective diffusivity	m^2/s
D_L	dispersion coefficient	m^2/s
K	volume equilibrium constant	m^3/m^3
l	fissure length perpendicular to flow direction (breadth)	m
Pec	Peclet number $\bar{U} \cdot x/D_L$	-
Q	flow rate	m^3/s
$\bar{Q}$	mean flow rate	m^3/s
Q_t	flow rate carrying tracer	m^3/s
t	time	s
$\bar{t}$	mean time	s
t_w	water residence time	s
dt	time for tracer injection	s
$\bar{U}$	mean velocity	m/s
U_f	velocity in a fissure	m/s
x	distance in flow direction	m
z	distance into rock matrix	m
δ	fissure width	m
μ	mean fissure width	m
σ_l	standard deviation in the logarithm of fissure widths	-
σ_t	standard deviation in the residence time distribution	s
σ_x	standard deviation in the spreading of a tracer pulse	m

REFERENCES

Bear, J. 1969. Hydrodynamic dispersion. In Flow through porous media, ed. R. J. M. de Wiest. p. 109. Academic Press, New York.

de Josselin de Jong, G. 1958. Longitudinal and transverse diffusion in granular deposits. Trans. Am. Geophys. Union 39:67.

Kučera, E. 1965. Contribution to the theory of chromatography. Linear nonequilibrium elution chromatography. J. Chromatogr. 19:237.

Lallemand-Barrè, A., and Peaudecerf, P. 1978. Recherche des relations entre la valeur de la dispersivitè macroscopique d'un milieu aquifere, ses autres caracteristiques et les conditions de mesure. Bull. Bur. Rech. Geol. Minieres (Fr.) Sect. 3, 4:277.

Levenspiel, O. 1972. Chemical Reaction Engineering, 2nd ed. Wiley, New York.

Matheron, G., and de Marsily, G. 1979. Is transport in porous media always diffusive? A counter example. Private communication.

Neretnieks, I. 1979. Analysis of some tracer runs in granite rock using a fissure model. In Scientific basis for nuclear waste management, vol. 1, ed. G. J. McCarthy. p. 411. Plenum Press, New York.

Neretnieks, I. 1980. Diffusion in the rock matrix - an important factor in radionuclide retardation? J. Geophys. Res. (in press).

Perry, H., and Chilton, C. H. 1973. Chemical Engineer's Handbook. 5th ed. McGraw-Hill.

Schwartz, F. W. 1977. Macroscopic dispersion in porous media: the controlling factors. Water Resour. Res. 13:743.

Snow, D. T. 1970. The frequency and apertures of fractures in rock. Int. J. Rock Mech. Min. Sci. 7:23-40.

RADIONUCLIDE TRANSPORT AND RETARDATION IN TUFF*

E. N. Vine, B. P. Bayhurst, W. R. Daniels,
S. J. DeVilliers, B. R. Erdal, F. O. Lawrence,
and K. Wolfsberg

Los Alamos Scientific Laboratory
Los Alamos, NM 87545

INTRODUCTION

The suitability of tuff at the Nevada Test Site (NTS) for the isolation of radioactive waste is being investigated as part of the Nevada Nuclear Waste Storage Investigations. Tuff is a geological term applied to pyroclastic rocks composed of particles fragmented and ejected during volcanic eruptions. Such deposits are complex and may exhibit a wide range of properties, depending on their cooling and alteration history (see, for example, Ref. 1).

The migration of radionuclides from a deep geologic nuclear waste repository would likely be the result of transport by groundwater. Retardation due to interactions with the surrounding geologic media should be a significant factor in minimizing such transport. Many lithologic types of tuff contain highly sorptive minerals. In addition, long hydrologic flow paths are typical of the NTS. These are both important reasons for the consideration of tuff as a geologic medium for isolation of radioactive waste.

An understanding of the mechanisms of transport and sorption-desorption is essential for prediction of the behavior of radionuclides during the time required for decay to safe levels and, thus, for demonstration of the effectiveness of tuff, or any potential medium. Sorptive properties of tuff are being studied with both static (batch measurements) and dynamic methods (crushed and solid rock columns). Studies were made on tuff core samples from two drillholes at the NTS: J-13 (Jackass Flats)[2] and UE25a-1 (Yucca Mountain).[3] Water from well J-13 was pretreated at least two weeks with crushed tuff from the core being studied prior to use in the experiments.

---

*Work supported by the U. S. Department of Energy.

The distribution coefficient, K_d, is commonly used to describe the partition of a radionuclide between solid and aqueous phases. K_d is defined as the concentration of a species per gram solid phase divided by its concentration per milliliter in the liquid phase at equilibrium. The term sorption ratio, R_d, which does not imply equilibrium but is otherwise identical to K_d is used here. In column experiments the relative velocity of a radionuclide with respect to the groundwater velocity is measured. This retardation factor, R_f, is related to the sorption ratio R_d, or K_d, by the simple expression $R_f = (\rho/\varepsilon)R_d + 1$, where ρ is the density of the rock column and ε is the porosity.

The sorption ratios calculated from flow measurements can be compared with data obtained by batch techniques. Because the batch experiments are simple and fast, it is feasible to measure the influence on R_d of a large number of parameters. It is hoped that such comparisons will lead to an understanding of the relationship between behavior in a dynamic laboratory experiment (and, ultimately, behavior in the field) and the many available batch R_d data.

SORPTION PROPERTIES: BATCH MEASUREMENTS

Several parameters were studied with batch experiments. Details of the batch technique and tracer preparation used have been reported.[4,5,6,7] Particle size had the least effect on R_d values; little or no variation was obtained among fractions of <106 μm, 106-150 μm, 355-500 μm and 106-500 μm. Measurements were also made at two temperatures and sorption ratios at 70°C were generally greater than at 22°C. Strontium, cesium, and barium sorption ratios increased by factors of approximately 1.5 to 4, 1 to 2.8, and 2.5 to 5.6, respectively. Americium values changed very little. Increases in R_d with sorption time were often observed, although the changes were generally small.

The effects of atmosphere on sorption behavior were investigated by comparing the results of studies performed in a pure nitrogen atmosphere having ≦0.2 ppm oxygen and ≦20 ppm carbon dioxide present with similar measurements made under natural atmospheric conditions on the same materials. The controlled-atmosphere studies are not truly representative of the conditions to be found in deep geologic systems since essentially no carbon dioxide was present and bicarbonate may have been lost from solution. However, the pH values when the rock was present in solution were similar to those observed under atmospheric conditions, indicating that the rocks themselves may have supplied some bicarbonate ion buffering. The effects of atmosphere are summarized in Table 1. The sorption ratios for technetium were most affected and were higher when measured under controlled atmosphere conditions, as was the sorption of plutonium. For tuff samples YM-22 and YM-54 the sorption ratios of uranium were not affected by atmosphere; however, they were somewhat higher in a nitrogen atmosphere for YM-38 (zeolitized tuff). U(VI) was apparently not reduced to U(IV); only on zeolitized tuff YM-38 did the sorption ratio of uranium increase in the nitrogen atmosphere.

TABLE 1

Comparison of Sorption Ratios[a,b] (mℓ/g) Measured under Atmospheric and Controlled-Atmosphere Conditions

	YM-22		YM-38		YM-54	
	Air	Controlled	Air	Controlled	Air	Controlled
^{237}Pu	140(27)	220(21)	250(35)	800(11)	84(20)	120(24)
^{241}Am	4000(30)	1400(15)	5500(20)	5600(17)	590(34)	1000(21)
^{237}U	1.9(36)	0.48(140)	5.3(14)	14.7(7.1)	1.5(43)	1.48(43)
^{137}Cs	384(1.1)	243(1.8)	7870(2.6)	8515(4.2)	238(1.3)	246(1.8)
^{85}Sr	52.2(1.1)	56.5(1.9)	9880(2.0)	7900(3.7)	89(1.2)	84.9(1.7)
^{133}Ba	914(1.2)	210(1.6)	40500(1.9)	36300(11.9)	601(1.1)	426(1.2)
^{141}Ce	1348(2.1)	672(3.6)	675(3.7)	592(6.3)	141(1.8)	127(2.9)
^{152}Eu	1378(2.0)	751(2.8)	1670(3.8)	935(6.1)	443(2.0)	383(2.5)
^{95m}Tc	0.3[1.2]	2.1[18]	0.2[c]	2.8[120]		2.7[79]

[a]Standard deviations of replicate measurements expressed in percent are given in parentheses.

[b]Desorption values are given in brackets.

[c]Value obtained using two similar, zeolitized tuffs.

The U(VI) may remain strongly complexed, probably by carbonate, in all of these groundwaters. As expected, strontium, cesium, and barium were least affected by the presence or absence of oxygen and carbon dioxide. Although cerium and europium were also unaffected on tuffs, sorption of both those elements increased by a factor of 10 or more in a nitrogen atmosphere on granite and argillite.[7] We have been able to observe apparently negative E_h values only on some tuff-water systems; none on granite or argillite systems.

Groundwater composition may also influence the sorption ratio for many radionuclides. The compositions of the two solutions used were selected to represent extremes for media being studied. While the dependence of the sorption ratio on specific major ion concentrations could not be determined since the concentrations of several ions were simultaneously varied, overall effects were distinguished. The approximate initial compositions of the two solutions prepared and the amounts of the corresponding elements in water from well J-13 are given in Table 2. Results are also given for sorption times of ~60 days along with average sorption ratios (at ~40 days) of the same tuff samples and radionuclides in pretreated J-13 water for comparison. Water from well J-13 is intermediate in concentration compared to solutions I and II, and, in general, R_d values with J-13 water are also intermediate. The results indicate the importance of determining both the cation and anion compositions of the solutions used in making R_d measurements.

In addition to providing information on the influence of numerous variables, batch measurements are also providing relative sorption data on a wide variety of lithologic types of tuff. Although minor components in a rock sample can certainly play a major role in sorption, there is a fairly good correlation between sorption and major phases, as determined by x-ray diffraction, in the tuffs studied to date. Average sorption ratios are shown in Table 3,

TABLE 2

The Effect of Groundwater Composition on Sorption Ratios (mℓ/g)

Sample	Water[a]	Cs	Sr	Ba	Ce	Eu
JA-18	I	15200(7.6)	2700(7.5)	119000(12.9)	1470(11.3)	1700(10.7)
	II	8440(6.0)	4850(3.0)	82200(7.6)	≧34200	15400(10.6)
	J-13	13900	16200	86000	2900	570
JA-32	I	143(3.7)	85.0(2.0)	65.6(2.1)	≧54700	4200(7.6)
	II	73.0(3.3)	17.7(4.4)	182(1.8)	≧45900	≧61000
	J-13	130	63	479	---	---
JA-37	I	1390(3.6)	385(1.8)	882(1.9)	≧48400	≧25600
	II	757(7.8)	149(2.4)	339(1.8)	≧33000	50900(10.0)
	J-13	763	303	898	---	---

[a]Water I, pH 8.17. Constituents (mg/ℓ): Na(10), K(5), Ca(10), Mg(2), SO_4(5), Cl(4.5)
Water II, pH 8.62. Constituents (mg/ℓ): Na(50), K(5), Ca(50), Mg(20), SO_4(70), Cl(15)
J-13 Water, pH 8.32. Constituents (mg/ℓ): Na(50), K(5), Ca(13), Mg(2), SO_4(20), Cl(7.6)

along with the approximate percentages of major phases (J. R. Smyth, Los Alamos Scientific Laboratory, personal communication, 1980). The devitrified tuff YM-54 has among the lowest sorption ratios of the samples studied. Cores JA-26 and JA-28 contain analcime in addition to quartz and feldspar, yet sorption ratios on those tuffs have R_d values very similar to those for YM-54. Analcime apparently does not behave as the zeolite clinoptilolite (and heulandite); sorption ratios for strontium, cesium, and barium on tuffs YM-42, YM-49, YM-38, and YM-48, all with clinoptilolite (or heulandite), are at least one to two orders of magnitude larger than analcime-containing cores. Cerium and europium sorption, however, does not seem to be dependent on zeolite content. Ranges of sorption values obtained on samples containing quartz and feldspar (with and without analcime), samples containing glass (with and without clay) and samples containing clinoptilolite (heulandite) are summarized in Table 4. Intermediate values for strontium, cesium, and barium were obtained on glassy cores. Additional samples now being studied include tuffs containing primarily cristobalite and feldspar, mordenite and clinoptilolite, glass, or montmorillonite. As more samples are studied, a more detailed correlation may be possible.

SORPTION PROPERTIES: COLUMN MEASUREMENTS

Crushed Rock Columns

The migration of radionuclides through crushed-rock-core columns (35-106 μm) was measured. Batch sorption ratios measured on the same cores provide a comparison with the column results. A description of the columns and apparatus used was published elsewhere.[7] Groundwater from well J-13 pretreated with crushed tuff was

TABLE 3

Sorption Ratio vs. Mineralogy

Tuff	Major Phases, ≥20%	R_d (mℓ/g)				
		Sr	Cs	Ba	Ce	Eu
YM-54	quartz, feldspar	90	250	620	140	500
YM-22	quartz, feldspar	53	340	980	1400	1400
JA-26[a]	quartz, (feldspar)[b], analcime	35	487	209	385	118
JA-28[a]	quartz, (feldspar)[b], analcime	111	1230	720	1350	1170
YM-42	quartz, feldspar, clinoptilolite	2780	15300	42600	36400	49700
YM-49	clinoptilolite	2700	29000	33000	550	1200
JA-18	glass, clinoptilolite	~20000	16000	~70000	2600	1400
YM-38	feldspar, cristobalite, clinoptilolite	12000	86000	66000	830	2300
YM-5	glass	268	7640	1011	23700	19900
JA-8[a]	glass, clay	305	1963	358		1710
YM-48	glass, feldspar, (clinoptilolite)[b]	1800	17000	15000	1900	2500

[a]Preliminary results, measurements still in progress.
[b]Abundance approximately 15%.

TABLE 4

Ranges of Sorption Ratios (mℓ/g)

MAJOR PHASES	Sr	Cs	Ba	Ce	Eu
quartz, feldspar (+/- analcine)	35-100	250-1200	210-980	140-1400	500-1400
glass (+/- clay)	270-300	2000-7600	360-1000	24000	1700-20000
clinoptilolite (heulandite)	1800-28000	8600-17000	1500-130000	550-36400	1200-49700

used, and generally the columns were loaded with small (~10 μℓ) spikes of the pretreated water containing the tracer(s). Flow rates generally ranged from 0.041-0.082 mℓ/h (30-60 m/y). The free column volumes (used to calculate the effective column porosity) were determined with both HTO and $^{131}I^-$, which gave identical results.

The calculated column sorption ratios for the isotopes ^{85}Sr, ^{137}Cs, and ^{133}Ba are given in Table 5 with data from batch sorption measurements for comparison. Several kinds of elution behavior were observed: symmetric peaks where fifty percent of the activity was eluted at the peak, asymmetric peaks, and "no peaks" - but instead a slow, usually uniform elution. With the exception of a J-18 column, the elution curves of ^{85}Sr were symmetric, and the column R_d values were one to three times lower than the corresponding batch R_d value. The JA-18 batch R_d value for ^{85}Sr was >10,000 mℓ/g. Strontium on a JA-18 column was eluted at a rate of ~0.07%/day for ~72 days, then a small, sharp peak was observed. Elution of JA-18 was continued, and the slow, uniform "leaking" resumed. JA-18 is a highly glassy tuff, and the slow elution may be due to a gradual dissolution of the glass. This, however, would

TABLE 5

A Comparison of Batch and Column R_d Values (mℓ/g)

Column	^{85}Sr Batch	^{85}Sr Column	^{137}Cs Batch	^{137}Cs Column	^{133}Ba Batch	^{133}Ba Column
YM-22 (core)	50	20				
YM-22 (crushed)	50	30	287	122	899	355
YM-38			8600	>"21900"		
YM-54	84	44	247	97	620	124
JA-18	16000	381(9%)[a]	6600-15000	(b)	4800	(b)
JA-32	56	42				
JA-37	300	106	740	>560[b]		

[a]Slow, "peakless" elution.
[b]In progress.

not explain the weak, sharp peak which seems to indicate that more than one "sorption" mechanism exists. Columns of JA-18 and YM-38 tuff were also loaded with ^{137}Cs, and the same slow "leaking", without a sharp peak, was observed. The slow elution of ^{85}Sr and ^{137}Cs might be the result of exchange of the sorbed radioactive species with stable isotopes in the waters used, which contain ~10^{-9} M cesium and ~6 x 10^{-7} M strontium.

Three JA-32 columns were loaded with ^{85}Sr and flow rates from 0.082 mℓ/h to 18 mℓ/h were used. The sorption ratio from a fast-flow column, run with ^{85}Sr added to the groundwater, was identical to that obtained from a slow-flow column loaded with a spike. However, an increase of 50% in R_d was obtained on another spike-loaded column run at a fast flow rate. When two granite columns were loaded with spikes of activity and run at 0.04 mℓ/h and 4.98 mℓ/h, a considerable increase in R_d was also observed at the faster flow rate.

Cesium column data are given in Table 5. The most straightforward results are from tuff columns of YM-54 (3 columns) and YM-22 (1 column), where the batch sorption ratios are fairly small and values from desorption and sorption experiments are approximately equal. Elution of the activity occurred in a peak. The three YM-54 columns were run at two ^{137}Cs concentrations, 10^{-6} M and 10^{-9} M; the R_d value calculated for cesium was not affected by the cesium ion concentration. On the JA-37 column there was a gradual elution of ^{137}Cs. In another column, ^{137}Cs was loaded onto YM-38 and run at 4-5 mℓ/h. Fifty percent of the total ^{137}Cs was eluted in 7680 mℓ, in slowly increasing amounts. A sharp peak was never observed. A "column R_d" value of 21900 mℓ/g could only be estimated from the volume required to elute 50% of the activity. This value is well above that obtained in batch measurements (8600 mℓ/g) and is probably due in part to the fast flow rate used. It is also indicative of the "complications" revealed by flow experiments.

Other isotopes studied were ^{237}U and ^{133}Ba. The columns loaded with ^{133}Ba showed peaks, followed in some cases by a gradual elution of activity. The R_d values are ~2 to 5 times lower than the corresponding batch R_d values. An R_d value of 0.72 mℓ/g for uranium was obtained on tuff YM-54. The uranium peak was quite asymmetric, and the activity eluted per mℓ slowly decreased. Again, the marked asymmetry could be an effect of the complicated "sorption" illustrated by the large difference between batch sorption (1.5 mℓ/g) and desorption (11 mℓ/g) ratios.

Whole Core Columns

Because studies using crushed rock involve newly exposed mineral surfaces, experiments with intact rock cores were also undertaken. Such measurements should provide a better understanding of transport of radionuclides through rock, either by porous or fracture flow. In addition, they are an intermediate step in the extrapolation from batch and crushed-rock column studies to the field.

An "elution" curve was obtained on one tuff core YM-22, which was 15.9 mm high and 25.4 mm in diameter. The apparatus has been described elsewhere.[7] Approximately 40% of the ^{85}Sr loaded was eluted at a fairly constant, "peakless," rate in one year. The estimated sorption ratio of ~20 mℓ/g is less than the value obtained from batch measurements and similar to the value obtained on a column of YM-22. Other intact (and fractured) rock columns now being eluted should indicate whether this is a general trend.

SORPTION PROPERTIES: CIRCULATING SYSTEM MEASUREMENTS

Sorption ratios measured using a batch technique have frequently been observed[6,7] to increase slowly with contact time. A series of measurements were initiated to determine whether this effect is due to "weathering," and self-grinding which occur during the shaking operation used in the batch technique. A circulating system was developed in which a 355-500 μm fraction of crushed tuff, placed in a 1.0 cm diameter x 5.0 cm long polycarbonate tube, was contacted with groundwater continuously circulated through the crushed tuff in a closed loop. The tuff was held in place with polyethylene bed supports and polyproplyene Luer fittings at each end. Three tuff samples of ~4.5 g each were used: JA-37, YM-22, and YM-54. Results from groundwater tagged with ^{85}Sr, ^{137}Cs, ^{133}Ba are given in Table 6. In general, results from the circulating system are comparable to those from crushed rock columns and lower than those from batch measurements.

CONCLUSIONS

Batch measurements provide an understanding of which experimental variables are important. For example, sorption ratios vary little with particle size (and surface area); however, groundwater composition and rock composition are quite important. A general correlation has been identified between mineralogy (major phases) and degree of sorption for strontium, cesium, and barium. Although these are approximate, a more detailed analysis may be possible as more samples are studied and the data base increased.

Data from crushed tuff columns indicate that, except in simple cases where sorption coefficients are relatively low, and ion-exchange equilibria not only exist but are the dominant mechanism for removal of radioisotopes from solution, the simple relation between the sorption ratio R_d (or K_d) and the relative velocity of radionuclides with respect to groundwater velocity may be insufficient to permit accurate modeling of the retardation of radionuclides. Additional work on whole core columns and larger blocks of intact material is required to better understand radionuclide sorption and transport through rock.

TABLE 6

Sorption Ratios on Crushed Tuffs in mℓ/g

Sample	Batch			Circulating System			Columns		
	Sr	Cs	Ba	Sr	Cs	Ba	Sr	Cs	Ba
YM-22	51	264	945	28	423	129	30	122	355
YM-54	84	247	653	47	131	137	44	97	124
JA-37	283	627	747	396	1820	886	106	---	---

References

1. C. S. Ross and R. L. Smith, "Ash-Flow Tuffs: Their Origin, Geologic Relations, and Identification," U. S. Geol. Soc. Prof. Paper 366 (1961).

2. G. H. Heiken and M. L. Bevier, "Petrology of Tuff Units from the J-13 Drill Site, Jackass Flats, Nevada," Los Alamos Scientific Laboratory report LA-7563-MS (1979).

3. M. L. Sykes, G. H. Heiken, and J. R. Smyth, "Mineralogy and Petrology of Tuff Units from the UE25A-1 Drill Site, Yucca Mountain, Nevada," Los Alamos Scientific Laboratory report LA-8139-MS (1979).

4. B. R. Erdal et al., "Sorption-Desorption Studies on Argillite," Los Alamos Scientific Laboratory report LA-7455-MS (1979).

5. B. R. Erdal et al., "Sorption-Desorption Studies on Granite," Los Alamos Scientific Laboratory report LA-7456-MS (1979).

6. K. Wolfsberg et al., "Sorption-Desorption Studies on Tuff. I. Initial Studies with Samples from the J-13 Drill Site, Jackass Flats, Nevada," Los Alamos Scientific Laboratory report LA-7480-MS (1979).

7. E. N. Vine et al., "Sorption-Desorption Studies on Tuff. II. A Continuation of Studies with Samples from Jackass Flats, Nevada and Initial Studies with Samples from Yucca Mountain, Nevada," Los Alamos Scientific Laboratory report LA-8110-MS (1980).

TEMPERATURE AND pH EFFECTS ON SORPTION PROPERTIES OF SUBSEABED CLAY*

B. T. Kenna

Sandia National Laboratories**

Albuquerque, New Mexico 87185

ABSTRACT

The effect of time (1-14 days), temperature (20-60 C) and pH (2-10) on sorption of Cs, Sr, Eu and Ba onto subseabed smectite clay was studied through batch equilibrium techniques. Sorption equilibrium was reached by all elements within four days. All except Cs exhibited a pH dependence. The equilibrium sorption distribution coefficient (K_d) for Cs was inversely dependent on temperature (Cs, 1180/20 C, 340/60 C), whereas the others showed little or no dependence (Sr, 130/20 C, 170/60 C), (Eu, 1.0×10^7/20C, 1.1×10^7/60C) and (Ba, 7.0×10^4/20 C, 7.0×10^4/60 C). This would indicate that for the conditions used, the subseabed smectite can be considered as a viable geological release barrier.

INTRODUCTION

A waste package, i.e. nuclear waste contained within a canister, emplaced in the subseabed could produce elevated temperatures and therefore, pH changes in the area immediately surrounding the waste package.[1] These changes could possibly alter the sorption properties of the subseabed clay which would directly effect the adequacy of the clay to serve as a barrier to radionuclide migration away from the waste form.[2-4] Therefore, the possible influence of temperature and pH must be evaluated before appropriate modeling studies can be completed. The present study addressed the sorption of Cs, Sr, Eu

* This work was supported by the U. S. Department of Energy under contract DE-AC04-76-DP00789.

** A U. S. Department of Energy facility.

and Ba by subseabed "smectite" clay (composed of quartz, illite, clinoptilolite and montmorillonite[5,6,7]) at two temperatures (20 C and 60 C) in the pH range of 2-10. Erickson[8] has studied the sorption properties of these elements by smectite clay at the nominal temperature of undisturbed seabed sediment (4 C).

The sorption properties of materials are generally represented mathematically by the equilibrium sorption distribution coefficient ($K_{d,i}$) for each of the species involved, viz.

$$K_d = C_{s,i}/C_{\ell,i} \quad ,$$

where $C_{s,i}$ is the equilibrium solid-phase concentration (mMol/gm) of species $\underline{i}$ and $C_{\ell,i}$ is the equilibrium solution-phase concentration (mMol/mL) of species $\underline{i}$. The general technique of batch equilibration can be readily used to determine $K_{d,i}$ values for specific experimental conditions.

EXPERIMENTAL

Batch equilibration under controlled temperature and pH conditions was used to study the sorption behavior of smectite clay for Cs, Sr, Eu and Ba. The procedure is delineated below.

The smectite clay was provided by G. R. Heath, University of Rhode Island from core LL44-CPC-2, collected in the Pacific Ocean October 11, 1976, at 30°20.9'N, 157°57.85'W, water depth 5821 meters, and was representative of the smectite-rich region which occurs in the sediment at depths below 10 meters. This material contains 5-6 w/o leachable iron and manganese in the form of hydrous oxides and the minerals, illite, clinoptilolite (a zeolite) and montmorillonite.

Duplicate samples for each temperature and/or pH were prepared by suspending 0.25 gm of the clay in 50 mL of 0.68 N NaCl solution (approximate salinity of sea water). Each sample pair had 1 μCi/mL of ^{137}Cs, ^{85}Sr, ^{154}Eu, or ^{133}Ba and a known amount of the corresponding stable element added, and the pH was adjusted to a preselected value with HCl or NaOH. They were then placed in a constant temperature shaker bath and maintained at temperature with agitation for various time periods. A second suite of sample pairs also were made in which the pH of the 0.68 N NaCl-smectite solution was permitted to remain at its initial value of 7.5-7.8.

After various time periods, a sample pair was removed from the constant temperature bath, the solutions filtered through 0.2 μ filters and a 5 mL aliquot taken for counting. Duplicate standards for each pH were prepared by adding a known spike of radionuclide plus stable element to 50 mL of 0.68 N NaCl and adjusting the pH to a preselected value. These standards were subjected to the same treatment as the other samples.

RESULTS AND DISCUSSION

Figures 1 and 2 summarize the effect of time, temperature, and pH on the sorption of Cs, Sr, Eu and Ba by subseabed smectite clay. For Cs, the sorption equilibrium appears to be established in 3-4 days at 20 C and sooner at 60 C with pH contributing little effect. This is in accord with previous work.[8,9] Ion exchange has been suggested as a primary sorption mechanism for Cs,[8] and the inverse relationship between sorption and temperature (Fig. 1) might be consistent with this, but further work is required before a definite statement can be made.

In the case of Sr, sorption equilibrium is established in 1-2 days at both 20 C and 60 C. At both temperatures, a marked increase in K_d accompanies an increase in pH. This is not due to precipitation of strontium since the initial strontium concentration in the samples (ca. 10^{-6} mMol/mL) was well below the solubility limits of sparingly soluble compounds such as $Sr(OH)_2$ (10^{-3} mMol/mL) and $SrCO_3$ (10^{-5} mMol/mL). A possible explanation for the increase is that, in basic solutions, the apparent capacity of the smectite clay increases in terms of Sr^{+2} sorption. This could be caused either by the Sr^{+2} reacting with OH^- to form species such as $Sr(OH)^+$, thereby halving the capacity demand compared to Sr^{+2}, or by sites, which are normally not available for Sr^{+2} sorption, reacting with OH^- to produce additional sorption sites, e.g., Eqs. 1 and 2. In either case, Sr^{+2} sorption would increase with increasing pH because of a change in sorption mechanism. A similar phenomenon and explanation has been reported previously[10] for titanate materials.

$$>Si=O=OH^- \longrightarrow >Si<^{O^-}_{OH} \quad (1)$$

$$>Si<^{O^-}_{OH} + Sr^{+2} \longrightarrow \left[>Si<^{O}_{OH} >Sr \right] \quad (2)$$

Sorption equilibrium for both Eu and Ba appear to be established within 2-3 days, and temperature does not appear to have a significant effect. However, the sorption of both seems to be pH dependent, and the changes in slope of the log K_d vs. pH curves may indicate a change in sorption mechanism, perhaps due to processes similar to those suggested for Sr^{+2} or to polymeric complex formation. Sorption K_d values obtained for samples exposed at 20 C for 8 days and then after the samples were exposed at 60 C for 8 days showed the sorption processes to be insensitive to temperature and possibly

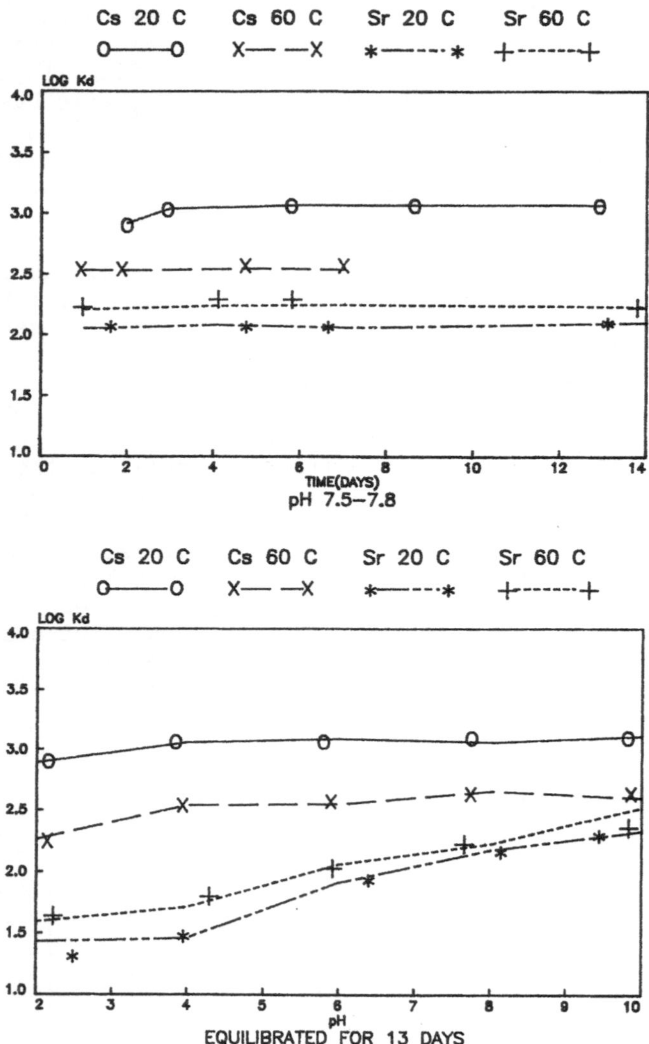

Figure 1. Effect of time, temperature, and pH on sorption of Cs and Sr by subseabed smectite clay.

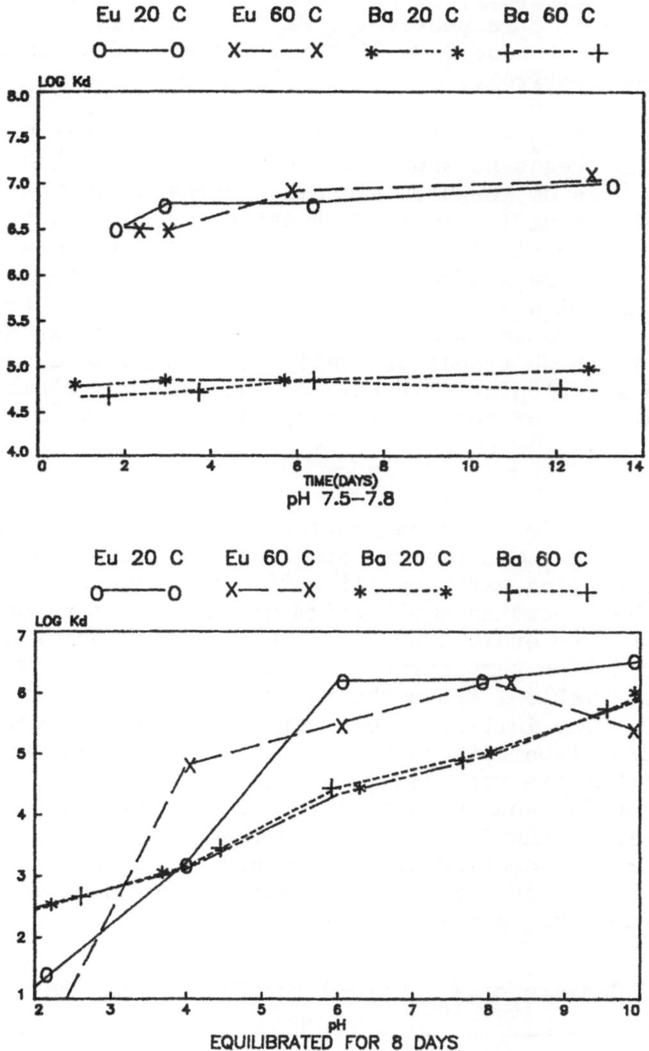

Figure 2. Effect of time, temperature, and pH on sorption of Eu and Ba by subseabed smectite clay.

irreversible. Similar results were obtained for samples exposed to 60 C for 8 days and then exposed to 20 C for 8 days. The log K_d values determined were: Eu 20 C initial and 60 C final, log K_d = 7.0 and 6.8, respectively; Eu 60 C initial and 20 C final, log K_d = 6.9 and 6.8, respectively; Ba 20 C initial and 60 C final, log K_d = 4.9 in both cases; Ba 60 C initial and 20 C final, log K_d = 4.9 and 5.1, respectively.

The change of K_d as a function of temperature for Cs, Sr, Eu and Ba exhibits trends as shown in Figure 3 (data at 4 C determined previously[8]). Cesium sorption steadily decreases as the temperature increases. Strontium is similar between 4 and 20 C, but then exhibits a possible slight increase as temperature increases. The sorption of Eu and Ba apparently is much greater at temperatures above 20 C than below 20 C. The explanation for these trends is not clear at the present time. It is sufficient to say at the moment that, if these trends are valid, then subseabed smectite can be considered a geological barrier to radionuclide release in general, regardless of the pH, at temperatures below 60 C.

SUMMARY

The radionuclide sorption properties of a smectite subseabed clay for Cs, Sr, Eu, and Ba were studied as functions of time, pH, and temperature using batch equilibration techniques. All of the elements reached sorption equilibrium within four days and Cs was the only one not to exhibit pH dependence. Sorption of Cs is inversely dependent on temperature whereas the others show little or no dependence. Table I lists the K_d for the elements indicated at 20 C and 60 C. The initial concentrations are also denoted to allow intercomparison with other work. The values appear to be consistent with those reported in the literature for Cs,[3,9,11] Sr,[3,9,11,12] Eu,[3,9] and Ba.[3,9,11,12] Ion exchange may be the primary sorption mechanism for Cs, but there appear to be other mechanisms for the sorption of Ba, Sr, and Eu on subseabed sediments. Significant pH effects were measured for the sorption of the latter elements on subseabed sediments.

Table I
Summary of K_d (mL/gm) for Cs, Sr, Eu, Ba

T (°C)	Cs	Sr	Eu	Ba
20	1180 ± 24	126 ± 4	$(1 \pm 0.2) \times 10^7$	$(9.3 \pm 0.7) \times 10^4$
60	343 ± 10	170 ± 21	$(1.1 \pm 0.3) \times 10^7$	$(5.5 \pm 0.3) \times 10^4$
*	3×10^{-7}	7×10^{-7}	5×10^{-6}	1.7×10^{-7}

* Initial concentration in mMol/mL; solution pH 7.5 - 7.8.

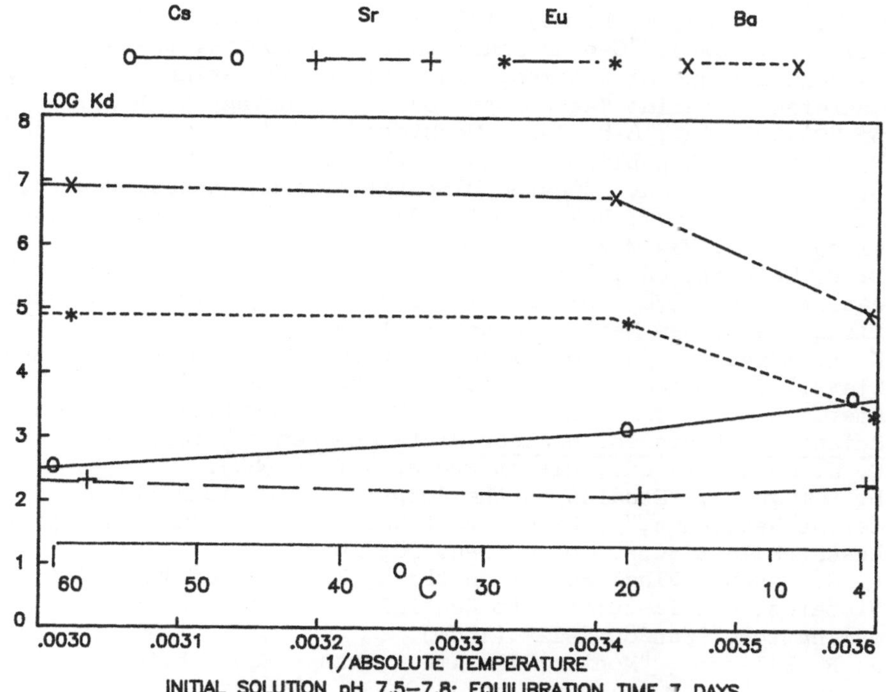

Figure 3. Effect of temperature on sorption of Cs, Sr, Eu, and Ba by subseabed smectite clay,

ACKNOWLEDGEMENT

The author expresses his appreciation to M. K. Gordon for laboratory assistance and to K. L. Erickson for helpful discussions.

REFERENCES

1. J. L. Bischoff and W. E. Seyfried, "Hydrothermal Chemistry of Seawater from 25 to 350 C," Am. Jour. Sci., 278, 838-860 (1978).
2. O. P. Mehra and M. L. Jackson, "Specific Surface Determination by Glycerol Sorption for Vermiculite and Montmorillonite," Surf. Sci. Soc. Am. Proc., 23, 351 (1959).
3. G. R. Heath, G. Epstein and R. A. Prince, in "Seabed Disposal Program Annual Report - Part II," Appendix E, D. M. Talbert, ed., SAND77-1270* Oct., 1977.

4. K. L. Erickson, W. E. Seyfried, G. R. Heath, J. L. Krumhansl, and V. T. Bowen, "Geochemical Studies Supporting the U. S. Subseabed Disposal Program," in 16th Annual Marine Technology Conference, Marine Technology Society, Washington, D.C., pp 500-509, Oct. 6-8, 1980, Washington, D.C.
5. B. T. Kenna, Unpublished results (1980).
6. D. B. Talbert, ed., "Seabed Disposal Program, Annual Report," Jan. to Dec. 1977, Vol. I & II, SAND78-1359, Sandia Laboratories, Albuquerque, New Mexico, January, 1979.*
7. D. B. Talbert, ed., "Seabed Disposal Program, Annual Report," Jan. to Dec. 1978, Vol. I & II, SAND79-1618, Sandia Laboratories, Albuquerque, New Mexico, October, 1979.*
8. K. L. Erickson, "Radionuclide Sorption Studies on Abyssal Red Clays," Symposium on Radioactive Waste in Geologic Storage, Amer. Chem. Soc. meeting, Miami, Florida, Sept. 11-15, 1978; Scientific Basis for Nuclear Waste Management, Vol. II, C. J. Northrup, ed., Plenum Press, N.Y. (1980).
9. H. Hamaguchi, "Studies on the Sorption of Radionuclides on Marine Sediments," IAEA Contract No. 881R2/RB, Japan Analytical Research Institute, Tokyo, Dec. 28, 1963.
10. R. G. Dosch, "Final Report on the Application of Titanates, Niobates, and Tantalates to Neutralized Defense Waste Decontamination," Sand80-1212* (1980).
11. J. R. Elliason, "Montmorillonite Exchange Equilibria with Sr-Na-Cs," Amer. Mineral., 51, 324-335 (1966).
12. R. J. Lewis and H. C. Thomas, "Adsorption Studies on Clay Minerals: VIII Consistency Test of Exchange in the Systems Na-Cs-Ba Montmorillonite," J. Phys. Chem., 67, 1781-1783 (1963).

* Available from National Technical Information Service, U. S. Department of Commerce, 5285 Port Royal Rd., Springfield, VA, 22161, U.S.A.

$SrCl_2$ SOLUBILITY IN COMPLEX BRINES

M. A. Clynne,[1] I-Ming Chou,[2] and J. L. Haas, Jr.[2]

U.S. Geological Survey
[1]Menlo Park, CA 94025
[2]Reston, VA 22092

INTRODUCTION

A large percentage of the heat generated by decay of fission products in spent fuel aged more than 5 years in high-level waste is due to ^{90}Sr and ^{137}Cs, which are significant heat producers for about 100 years.

Most scientists and engineers agree that many proposed waste-canister materials will suffer corrosion in salt. Estimates vary widely, but few expect more than a decade or two of integrity for unprotected canisters. During this interval the highest container surface temperatures, the greatest thermal gradients in their vicinity and most of the heating of the salt in the repository workings will take place. Substantial plastic flow of the salt likely will occur during this interval. The rate and magnitude of these processes will affect and be affected by the amount of fluid in the heated volume.

Fluids can reach the waste canisters in several ways; by decrepitation, by decomposition of hydrous minerals, by migration of fluid inclusions along thermal gradients, by release due to rupture during plastic flow, by desorption from salt backfill, or by adsorption from air.

Two chemical groups of fluids are commonly found in salt, NaCl brines, and high $MgCl_2$ and/or $CaCl_2$ bitterns. Waste canisters made of stainless steel, copper, or Zircaloy coming in contact with brines or bitterns will be susceptible to stress crack corrosion and likely will be breached soon after contact (Jenks, 1979).

In the presence of alkali and/or alkaline-earth chloride-bearing brines, strontium and cesium radionuclides are leachable from spent fuel and other commonly considered waste forms as ionized chlorides. Table 1 gives some experimental data of McCarthy et al., 1979, on the extractability of Sr and Cs from simulated spent fuel, supercalcine ceramic, and borosilicate glass. In each case significant Sr and Cs were leached after only 30 days. For longer times and the higher temperatures of actual waste these values must be considered low. If the fluid volume is large enough, it is probable that essentially all the Sr and Cs contained in the waste is leachable.

Brines containing dissolved ^{90}Sr and ^{137}Cs possess their own heat source. Thus, their migration could disturb the designed thermal evolution of a repository. In order to assess the problem better, a knowledge of the solubility of Sr and Cs in complex brines is necessary. This paper reviews known $SrCl_2$ solubility relationships in brines and bitterns and presents new experimental data for the system $SrCl_2$-NaCl-H_2O.

SYSTEM $SrCl_2$-H_2O

The pattern of the solubility properties of $SrCl_2$ is shown by the simple binary system $SrCl_2$-H_2O (Fig. 1). In this system the $SrCl_2$ content is 31.94 wt % at 0°C (Clynne and Potter, 1979a) and increases to 62 wt % at 200°C (Menzies, 1936) and 81 wt % at 412°C (Benrath, 1941). Breaks in the slope of the solubility curve (decreases in the rate of solubility increase with increasing temperature caused by changes in the hydration state of the solid $SrCl_2$ phase) occur at 61.45°C, 46.09 wt % $SrCl_2$ (Clynne and Potter, 1979a), where

$$SrCl_2 \cdot 6H_2O = SrCl_2 \cdot 2H_2O + 4H_2O$$

at 134.4°C, 56.1 wt % $SrCl_2$ (Menzies, 1936) where

$$SrCl_2 \cdot 2H_2O = SrCl_2 \cdot H_2O + H_2O$$

and at 320°C, 78.5 wt % $SrCl_2$ (Benrath, 1941) where

$$SrCl_2 \cdot H_2O = SrCl_2 + H_2O.$$

SYSTEM $SrCl_2$-NaCl-H_2O

In the system $SrCl_2$-NaCl-H_2O, we have determined the solubility of halite (NaCl), $SrCl_2 \cdot 6H_2O$, and $SrCl_2 \cdot 2H_2O$ from 18°C to 115°C, using the technique of Potter and Clynne, 1978. The NaCl content was fixed at 0, 5, 10, 15 and 20 wt %. These data,

Table 1. Leaching of Sr and Cs from Various Simulated Waste Forms in Bittern (from McCarthy et al., 1979)

Waste form	Percent leached	
	Sr	Cs
Spent fuel	30	100
Supercalcine ceramic	90	40
Borosilicate glass	50	50

Test conditions: sealed gold capsules, 300 bars, 30 days, borosilicate glass and supercalcine ceramic 300°C, spent fuel 200°C, bittern NBT-6 10 wt % $CaCl_2$, 10 wt % $MgCl_2$, 5 wt % NaCl, 5 wt % KCl

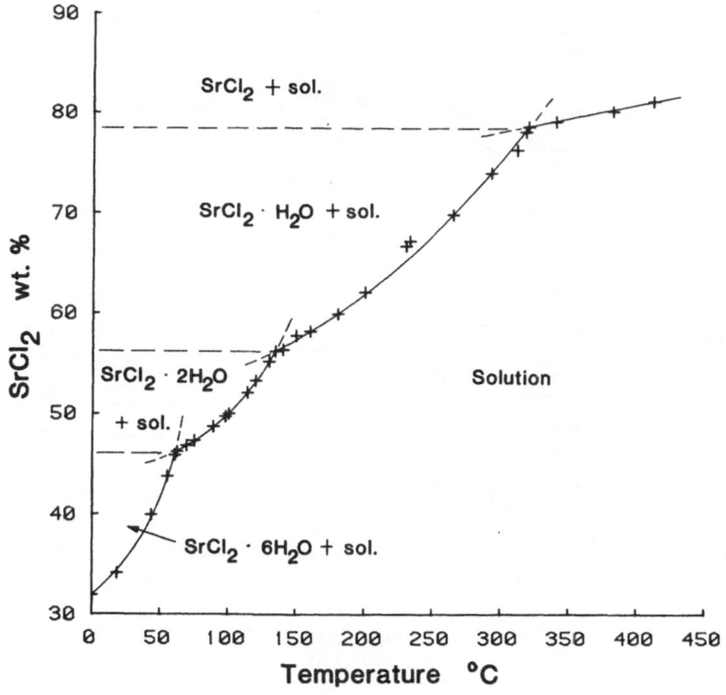

Fig. 1. The system $SrCl_2$-H_2O from 0° to 420°C.

together with those of Assarson, 1953, were fitted with equations using polynomial regression techniques. The relationships are given in Fig. 2.

The presence of NaCl in solution reduces the $SrCl_2$ solubility by an amount dependent on the NaCl concentration. Along the univariant curves where halite, a hydrate of $SrCl_2$, and solution coexist, the solubility of $SrCl_2$ hydrates is only slightly lower than in the $SrCl_2$-H_2O system, and the NaCl content is only a few weight percent. For example, at 100°C the solubility of $SrCl_2 \cdot 2H_2O$ in the $SrCl_2$-H_2O system is 49.87 wt %; when that solution is saturated with NaCl, the $SrCl_2 \cdot 2H_2O$ solubility is reduced to 48.0 wt %. The NaCl content is 3.3 wt %; thus the total dissolved solids content is 51.3 wt %. The higher total dissolved solid content and hence lower activity of H_2O lowers the temperature of dehydration of $SrCl_2 \cdot 6H_2O$ to $SrCl_2 \cdot 2H_2O$. In this case the temperature of dehydration is lowered only slightly from 61.45°C to 60.3°C; however, larger effects can be anticipated for the more complex systems to be discussed.

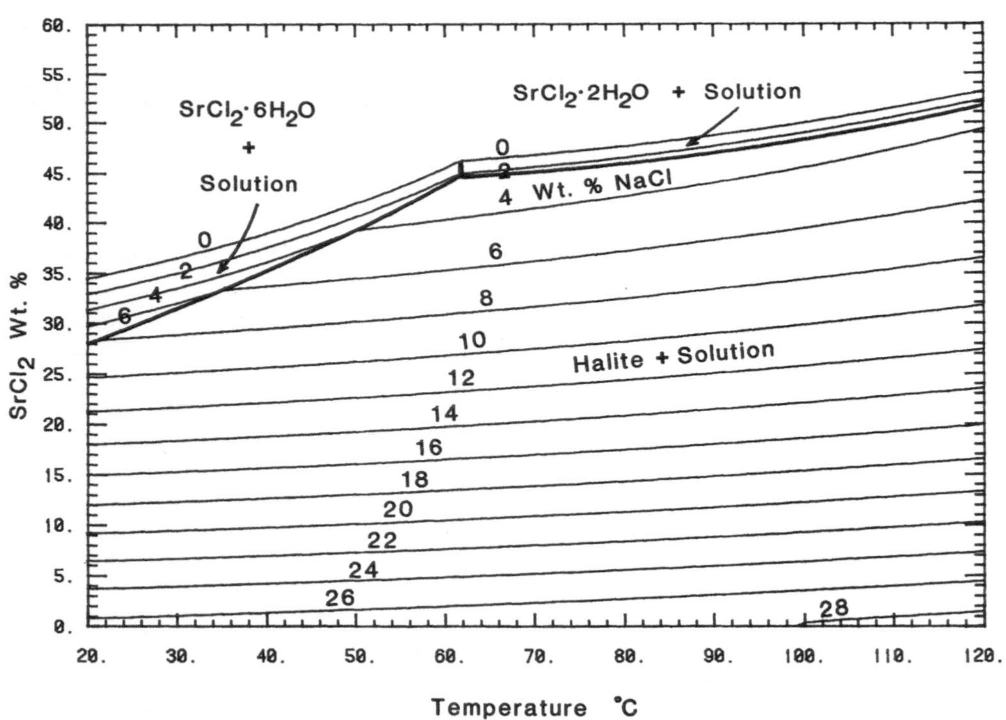

Fig. 2. The system $SrCl_2$-NaCl-H_2O from 20° to 120°C.

SYSTEMS WITH KCl

The effect in the $SrCl_2$-KCl-H_2O system is similar with the exception that the total dissolved solid content is higher (Table 2). At 100°C a solution saturated in KCl and $SrCl_2 \cdot 2H_2O$ contains 46.7 wt % $SrCl_2$ and 10.8 wt % KCl for a total of 57.5 wt % dissolved solids.

In the quaternary system $SrCl_2$-NaCl-KCl-H_2O, at similar temperatures, the solubility of each component is depressed slightly compared to the individual ternary systems; however, the total dissolved solid content is still higher. For example, at 100°C, the solution saturated with $SrCl_2 \cdot 2H_2O$, NaCl, and KCl contains 45.0 wt % $SrCl_2$, 10.3 wt % KCl, and 2.8 wt % NaCl, for a total of 58.1 wt % dissolved solids (Table 2).

Table 2. $SrCl_2$ Solubility in Brines and Bitterns (concentration in wt %)

T (°C)	$SrCl_2$	$CaCl_2$	$MgCl_2$	NaCl	KCl	H_2O	Ref.
100	48.0			3.3		48.7	1
100	46.7				10.8	42.5	2
100	45.0			2.8	10.3	41.9	2
100	2.7	59.1				38.2	3
80.5	1.7		39.1			59.2	4
74.5	2.0	56.0			3.5	38.5	5
100	2.5	57.2		1.0	3.0	36.3	5
93	3.0	39.0	9.9	0.3	6.2	41.6	6

1 This study.
2 Assarsson, 1953.
3 Assarsson and Balder, 1953.
4 Assarsson and Balder, 1954b.
5 Assarsson and Balder, 1954a.
6 Assarsson and Balder, 1955.

SYSTEMS WITH $CaCl_2$ AND $MgCl_2$

The solubility of $SrCl_2$ in bittern brines rich in $CaCl_2$ and/or $MgCl_2$ is considerably different. Both are more soluble than $SrCl_2$. Bitterns saturated in either component have $SrCl_2$ solubilities up to about 3.0 wt % $SrCl_2$ at the temperatures indicated in Table 2. In actual repository conditions, however, $SrCl_2$ solubility could be considerably higher. The equilibria in Table 2 are for systems saturated in all components; whereas, natural bitterns in rock salt are normally saturated only with NaCl, i.e., the bitterns with various amounts of dissolved salts are in equilibrium only with halite. Such solutions can leach Sr and precipitate Na as halite simultaneously. For the given temperature of the site, the composition of the solution would change toward increased $SrCl_2$ content and decreased NaCl (or KCl if present) content until saturation is achieved in a Sr-bearing compound such as one of the $SrCl_2$ hydrates.

The equilibria in Table 2 are for temperatures of 100°C and below; some postulated repository temperatures are higher than 100°C, and canister temperatures will almost certainly be higher. Solubilities of $SrCl_2$ in complex systems at temperatures higher than 100°C have not been measured. The increase in $SrCl_2$ solubility could be large if more soluble components, i.e., $MgCl_2$ and/or $CaCl_2$,are not available to the solution. In addition, bitterns containing dissolved ^{90}Sr would be self-heating, thereby driving themselves away from saturation and allowing increased dissolution of $SrCl_2$.

The situation in an actual repository would be a dynamic one in which the system would be continually changing. Making a quantitative model of $SrCl_2$ solubility would be difficult because of ^{90}Sr leached from the waste and heating the solution, precipitation of NaCl in cooler zones, and unsaturated bittern continuously arriving at the canister site and being added to the system.

SO_4^{2-}-BEARING SYSTEMS

The equilibria in Table 2 are for the indicated metal-chloride systems. Natural NaCl brines and bitterns have low SO_4^{2-} content, controlled by the amount of Ca present. $SrSO_4$ is much less soluble than $SrCl_2$ and exhibits retrograde solubility (solubility decreasing with increasing temperature), an important consideration for self-heating systems. In the system $SrSO_4$-H_2O, the $SrSO_4$ content is 0.0003 wt % at 200°C, 0.00012 wt % at 300°C, and only 0.00002 wt % at 395°C (Booth and Bidwell, 1950). The solubility of Sr in complex Cl^- and SO_4^{2-}-bearing systems is unexplored; however, by analogy to the behavior of $CaSO_4$ in complex systems, the increase in $SrSO_4$ solubility because of the presence of additional components is probable. At 200°C the solubility of $CaSO_4$ in H_2O is 0.0008 wt %; in NBT-6 (a bittern containing 10 wt % $CaCl_2$, 10 wt % $MgCl_2$, 5 wt % NaCl, and 5 wt % KCl) $CaSO_4$ solubility is 0.027 wt % or 34 times the pure H_2O value; and in saturated NaCl brine $CaSO_4$ solubility is 0.27 wt % or 338 times the pure H_2O value (Clynne and Potter, 1979b). Applying these factors to the $SrSO_4$ solubility in H_2O at 200°C gives a solubility of about 0.1 wt % $SrSO_4$ in the NaCl saturated brine (a worst case because pure NaCl brines are rare) and about 0.01 wt % in the bittern brine. The solubility of $SrSO_4$ in either medium decreases as temperature is increased. By analogy to the $CaSO_4$ systems, at 300°C $SrSO_4$ solubility would be about 40% of the 200°C value, and at 400°C it would be less than 10% of the 200° value.

To reduce Sr solubility to the level discussed above, SO_4^{2-} content exceeding the combined contents of Ca, other cations that form insoluble sulfates, and potentially leachable Sr will be needed in the canister environment. The necessary excess can be provided by the addition to the canister overpack of a SO_4^{2-}-bearing phase soluble at repository temperatures, such as Na_2SO_4 (Schroeder et al., 1935), that will reduce Sr solubility without adding an additional cation to the system.

REFERENCES

Assarsson, G. O., 1953, Equilibria in aqueous systems containing Sr^{2+}, K^+, Na^+ and Cl^-, J. Phys. Chem. 57:207-210.

Assarsson, G. O., and Balder, A., 1953, Equilibria between 18 and 114° in the aqueous ternary system containing Ca^{2+}, Sr^{2+} and Cl^-, J. Phys. Chem. 57:717-720.

Assarsson, G. O., and Balder, A., 1954a, Equilibria in the aqueous systems containing Ca^{2+}, Sr^{2+}, K^+, Na^+ and Cl^- between 18 and 114°, J. Phys. Chem. 58:253-255.

Assarsson, G. O., and Balder, A., 1954b, Equilibria between 18 and 100° in the aqueous ternary system containing Sr^{2+}, Mg^{2+} and Cl^-, J. Phys. Chem. 58:416.

Assarsson, G. O., and Balder, A., 1955, The poly-component aqueous systems containing the chlorides of Ca^{2+}, Mg^{2+}, Sr^{2+}, K^+ and Na^+ between 18 and 93°, J. Phys. Chem. 59:631-633.

Benrath, A., 1941, The solubility of salts and salt mixtures in water at temperatures above 100°C III, Zeit. Anorg. Allgem. Chem. 247:147-160.

Booth, H. S., and Bidwell, R. M., 1950, Solubilities of salts in water at high temperatures, J. Am. Chem. Soc. 72:2567-75.

Clynne, M. A., and Potter, R. W., II, 1979a, Solubility of some alkali and alkaline earth chlorides in water at moderate temperatures, J. Chem. Eng. Data 24:338-340.

Clynne, M. A., and Potter, R. W., II, 1979b, P-T-X relations of anhydrite and brine and their implications for the suitability of anhydrite as a nuclear waste repository medium, pp. 323-28 in: "Scientific Basis for Nuclear Waste Management, Vol. 1," G. J. McCarthy, ed., Plenum, New York.

Jenks, G. H., 1979, Effects of gaseous radioactive nuclides on the design and operation of repositories for spent LWR fuel in rock salt, ORNL-5578.

McCarthy, G. J., Komarneni, S., Scheetz, B. E., and White, W. B., 1979, Hydrothermal reactivity of simulated nuclear waste forms and water-catalysed waste-rock interactions, pp. 329-44 in: "Scientific Basis for Nuclear Waste Management, Vol. 1," G. J. McCarthy, ed., Plenum, New York.

Menzies, A. W. C., 1936, A method of solubility measurement - solubilities in the system $SrCl_2$-H_2O from 20° to 200°C, J. Am. Chem. Soc. 58:934-7.

Potter, R. W., II, and Clynne, M. A., 1978, The solubility of highly soluble salts in aqueous media: I. NaCl, KCl, $CaCl_2$, Na_2SO_4, and K_2SO_4 solubilities to 100°C, U.S. Geol. Survey J. Research 6:701-4.

Schroeder, W. C., Gabriel, A., and Partridge, E. P., 1935, Solubility equilibria of sodium sulfate at temperatures of 150 to 350°C. I. Effect of sodium hydroxide and sodium chloride, J. Am. Chem. Soc. 57:1539-46.

GEOLOGIC STABILITY OF MONAZITE AND ITS BEARING ON THE IMMOBILIZATION OF ACTINIDE WASTES

R. J. Floran,† M. M. Abraham,* L. A. Boatner,* and M. Rappaz*

Oak Ridge National Laboratory **
Oak Ridge, Tennessee 37830

INTRODUCTION

A common approach used to evaluate crystalline waste forms for high-level waste disposal is to examine the long-term stability of their natural analogs in geological/geochemical environments. Synthetic analogs of the monazite group of minerals [(Ce,La,Ca,Th,U)(P,Si)O_4] represent one of the most promising ceramic host phases for the isolation of actinide-bearing wastes. The most important properties of monazite that are relevant to actinide waste disposal are: a) the similar geochemical behavior of lanthanides and actinides which allows energetically favorable ionic substitutions of radioactive species within the crystal structure, and b) the reportedly high resistance of natural monazites to alteration and nuclear radiation damage.[1] Synthetic monazite is an important constituent of high-temperature supercalcine ceramics[2] as well as low-temperature cements.[3] If spent nuclear fuel is reprocessed, the actinide and lanthanide fractions could conceivably be separated from the remaining high-level waste and accommodated within a synthetic monazite phase.[4]

In view of the increased attention recently accorded monazite, it was felt that a general review regarding its behavior in natural environments would be useful. This paper examines the geologic evidence for monazite stability during weathering (i.e. at ambient T and P); under hydrothermal conditions (at elevated T and P); and with regard to radiation damage (metamictization). Recommendations are made for further research based on the available data.

---

†Environmental Sciences Division, *Solid State Division
**Operated by Union Carbide Corporation for the U.S. Department of Energy under contract W-7405-eng-26.

OCCURRENCE AND ALTERATION OF NATURAL MONAZITES

Monazite [$(Ce,La)PO_4$], the major member of the monazite group of minerals, is widely distributed and occurs in a variety of igneous, metamorphic and sedimentary terrains. It occurs as an accessory phase in granitic and gneissic rocks; as large crystals in pegmatites; in vein deposits of hydrothermal origin; and as a heavy detrital mineral in sedimentary deposits and unconsolidated beach sands. The tendency of monazite to be concentrated along with other relatively insoluble heavy minerals in placer deposits is a reflection of its chemical and physical inertness. Monazite also forms earthy fine-grained masses pseudomorphous after other phases in oxidized weathering environments.[5]

A recent review of minerals considered as candidate hosts for radionuclides concluded that, while monazite is extremely stable, it alters to members of the rhabdophanite group of minerals:[4] rhabdophane $(Ce,Y,La)(PO_4,CO_3)\cdot H_2O$, brockite $(Ca,Th,Ce)(PO_4,CO_3)\cdot H_2O$, and grayite $(Th,Pb,Ca)PO_4\cdot H_2O$. Despite this claim there are no well-documented descriptions of monazite altering to these hydrated phosphates. The rhabdophanite group and the related phosphate ningyoite [$CaU(PO_4)_2\cdot 1.5H_2O$] typically occur as rare, fine grained secondary phases in supergene deposits and as weathering products in oxidizing environments; only occasionally are they associated with monazite (Table 1). In most occurrences the rhabdophanite minerals appear to have precipitated from circulating, oxidized solutions. In other instances these hydrous phosphates are found replacing other phases[6] such as apatite;[7] have formed coatings on minerals other than monazite;[7-9] or are associated with fine-grained earthy monazite of secondary origin.[10] There is no petrochemical evidence to suggest that these hydrated phosphates formed from monazite-group minerals. On the contrary, earthy monazite may have formed by the dehydration of rhabdophane.[10] This observation is consistent with laboratory experiments demonstrating that the crystal structures of the rhabdophanite group readily transform to the monazite structure on heating to 800-900°C.[6,8,11,12] The transformation of rhabdophane to monazite has also been observed at much lower temperatures ($\leq$ 100°C) if sufficient time is allowed for the reaction to take place.[13]

When reported, the alteration of monazite is typically described as a surface incrustation of fine-grained white, yellowish or reddish-brown material of unknown nature[14,15] often tentatively identified as rhabdophane. In decomposed granitic rocks that have undergone intense *in situ* chemical weathering (saprolitization), this surficial alteration tends to be lost by abrasion and is usually absent from detrital monazite. However, some monazites from saprolite have been observed to alter internally along cleavage traces and in patches to an iron-rich non-hydrous amorphous material; the degree of alteration, from greenish yellow to opaque brown, is directly correlated with an increased development of multiple cleavages and an opacity of up to 75%.[16]

Alteration of monazite affects its U-Th-Pb isotopic systematics used in geochronology. U-Pb and Th-Pb ages tend to be discordant due to loss of Pb or U/Th, indicating that open system conditions prevailed at some time in the past. However, monazite has less tendency than zircon to lose its radiogenic Pb.[17] The discordance in some (perhaps most) dated monazite samples may be due to surficial weathering processes. A recent study by Williams and Silver[18] of the U-Th-Pb systems in coexisting zircon, monazite and xenotime revealed that U can be selectively lost relative to Th in monazites from weathered granite. These observations have important implications for the mobility of actinide elements of variable valence state, since U/Th partitioning in nature is usually ascribed to oxidation of U^{4+}.

RADIATION DAMAGE

Radiation-induced changes caused by alpha (α) particle emission and recoil nuclei can severely affect the long-term physical and chemical integrity of a waste-form. In natural minerals radiation damage often leads to metamictization. Disordering of the crystal structure is invariably accompanied by changes in physical properties, e.g. a decrease in density, increase in volume, etc. Some minerals, however, are more resistant to radiation damage than others. Approximately 18 radioactive minerals have been reported to occur only in the crystalline state; in rare instances some of these minerals including monazite, huttonite ($ThSiO_4$), and xenotime (YPO_4) have also been described as being partially metamict.[4] As a result, there is considerable controversy as to whether these latter minerals actually undergo metamictization. Partly metamict monazites have been reported only from Indian beach sands;[19,20] it was concluded that these monazites were metamict on the basis of diffuse x-ray diffraction patterns which became noticeably sharper after heat treatment. Karkhanavala and Shankar[19] noted that the birefringence of their samples greatly increased upon heating despite the fine grain size of the particles. However, these studies are difficult to interpret because of a lack of detailed sample descriptions, especially compositional data which would indicate whether the samples were altered. Such data are important because the rate at which metamictization occurs is believed to be due to the inherent stability of the crystal structure and α-particle flux plus other poorly known factors such as hydrothermal alteration and surficial weathering.

The stability of the monazite structure is demonstrated by the extreme rarity of reported partially metamict occurrences, not only of monazite but also of the isostructural mineral huttonite.[21] Totally metamict or fully amorphous examples of these minerals have never been reported. A puzzling observation worth noting is an apparent discrepancy in susceptibility to metamictization by minerals with very similar crystal structures. This is well illustrated by comparing the monoclinic/tetragonal polymorphs of $ThSiO_4$ (huttonite/thorite) with their rare-earth orthophosphate counterparts (monazite/xenotime). Thorite, like zircon ($ZrSiO_4$) is often found in either a partially or fully metamict state. In contrast, xenotime (YPO_4) is rarely if ever metamict despite the fact that it is isostructural with both thorite and zircon (Table 2). The reason for this differential response toward metamictization is not known.

Very large α-particle doses have not noticeably affected the crystallinity of monazites containing high concentrations of U and Th [e.g. 36% ThO_2 + 6% U_3O_8;[22] 16% U_3O_8 + 11% ThO_2[23]]. Synthetic crystals of $LaPO_4$ doped with various transuranic isotopes (^{246}Cm, ^{244}Cm, ^{241}Am, ^{242}Pu, ^{237}Np) have also retained their crystallinity although color-center formation was observed in the $LaPO_4$:^{241}Am and $LaPO_4$:$^{246,244}Cm$ systems due to internal radiation effects.[24] The concentration of americium and curium in the $LaPO_4$ single crystals was 0.2 wt% and 0.25 wt% respectively. No radiation effects were observed in $LaPO_4$ crystals containing $\sim$ 5 wt% ^{242}Pu.

These observations are consistent with the conclusion that factors other than structural stability and α-particle flux (e.g. primary or secondary alteration) may be responsible for the diffuse diffraction patterns exhibited by some monazites and attributed previously to metamictization. Taylor and Ewing[21] suggested that hydration may be a major factor responsible for the metamictization of thorite. Natural thorites with up to 70 mol% water have been identified but, interestingly, water of hydration has not been observed in xenotime (Table 2).

An alternative hypothesis was presented by Ueda[25] who interpreted similar x-ray data as being due to the growth of randomly oriented monazite crystallites formed during intense heating accompanying α bombardment. He proposed a model, consistent with experimental data, suggesting that monazite and xenotime are resistant to metamictization because they undergo a localized disordering process akin to melting followed by recrystallization to the original phase. In contrast, other radioactive minerals that often occur in a fully amorphous state (e.g. zircon, allanite) decompose to other phases. In the case of zircon the ultimate products of metamictization are complex but are often aggregates of crystalline ZrO_2 and amorphous SiO_2.[26]

CONCLUSIONS

Based on the available geologic data it appears that additional critical information needs to be gathered and evaluated before synthetic monazite-type minerals can be selected as host phases for actinide wastes. Research needs include a better understanding of:

- the leaching mechanism(s) involved in monazite decomposition and how these mechanisms affect the mobility of rare-earth elements as well as the U-Th-Pb systematics.
- the relative stabilities of the monazite and rhabdophanite groups of minerals as a function of kinetic and thermodynamic parameters, e.g., pH and Eh. The genetic relationship between these mineral groups (which also include the synthetic rhabdophane-like minerals, $AcPO_4 \cdot 0.5\ H_2O$ and $PuPO_4 \cdot 0.5\ H_2O$) needs to be established in the presence of weathering solutions and hydrothermal fluids.
- the physical and chemical conditions under which monazite can (?) become metamict and the potential role that alteration (including the possible formation of hydrated phosphates) may play in metamictization. Detailed investigations of a large suite of monazites are needed, as well as a careful reexamination of the partially metamict monazites reported in the literature. A comparative study of monazite and related minerals (e.g. Table 2) that examines the critical role of crystal structure and hydration in the metamictization process might shed considerable light on this problem.

Table 1. Geologic Occurrence of Rhabdophanite Minerals with Monazite.

Mineral/ Reference	Geologic Setting	Associated Phosphates	Remarks
Rhabophane			
(27)	granite pegmatite	monazite	no evidence of alteration from monazite
(28)	supergene mineral in oxidized fault zone	monazite	no evidence of alteration from monazite
(10)	weathered pegmatite	monazite apatite	monazite occurs as pseudomorphs after apatite and (?) rhabdophane
(29)	nepheline syenite	monazite steenstrupine	no evidence of alteration from monazite
(30)	weathered granite	monazite, xenotime, apatite	--
Brockite			
(31)	veins near alkalic rocks; carbonatites	monazite	no evidence of formation from monazite

Table 2. Some Properties of Monazite and Related Minerals.

Mineral	Ideal Formula	Metamictization Total/Partial	Hydrated Form
Monazite[a]	$(La,Ce)PO_4$	no/?	no
Cheralite	$(Ca_{.5}Th_{.5})PO_4$	no/?	no
Huttonite	$ThSiO_4$	no/?	no
Xenotime	YPO_4	no/?	no
Thorite	$ThSiO_4$	yes/yes	yes
Zircon	$ZrSiO_4$	yes/yes	yes

[a]Dashed line separates isostructural phases crystallizing in the monoclinic system (above) from phases crystallizing in the tetragonal system (below).

REFERENCES

1. R. J. Floran, M. Rappaz, M. M. Abraham, and L. A. Boatner, Hot and cold pressing of $LaPO_4$-based nuclear waste forms, presented at the "Workshop on Alternate Nuclear Waste Forms and Interactions in Geologic Media", Gatlinburg, Tennessee (1980).
2. G. J. McCarthy, J. G. Pepin, and D. D. Davis, Crystal chemistry and phase relations in the synthetic minerals of ceramic waste forms: I. Fluorite and monazite structure phases, in: "Scientific Bases for Nuclear Waste Management," Vol. II, C. J. Northrup, ed., Plenum, N. Y. (in press).
3. D. M. Roy, B. E. Sheetz, L. D. Wakeley, and S. D. Atkinson, Low-temperature ceramic radioactive waste form characterization of supercalcine-based monazite-cement composite, D.O.E./ET/41900-1 (1980).
4. Draft Environmental Impact Statement, Management of Commercially Generated Radioactive Waste, D.O.E./EIS-0046-D, Washington (1979).
5. H. J. Rose, L. V. Blade, and M. Ross, Earthy monazite at Magnet Cove, Arkansas, Amer. Mineral. 43: 995 (1958).
6. F. G. Fisher and R. Meyrowitz, Brockite, a new calcium thorium phosphate from the Wet Mountains, Colorado, Amer. Mineral. 47: 1346 (1962).
7. R. P. Geitgey, Mineralogy of a deeply weathered perrierite-bearing pegmatite in Amherst County, Virginia, M.S. Thesis, Univ. Va. (1967).
8. T. Muto, R. Meyrowitz, A. M. Pommer, and T. Murano, Ningyoite, a new uranous phosphate mineral from Japan, Amer. Mineral. 44: 633 (1959).
9. I. Haapala, On the granitic pegmatites in the Peräseinäjok.-Alaus area, South Pohjanmaa, Finland Bull. Comm. Géol. Finlande 224 (1966).

10. R. S. Michell, S. M. Swanson, and J. K. Crowley, Mineralogy of a deeply weathered perrierite-bearing pegmatite, Bedford County, Virginia, Southeast. Geol. 18: 37 (1976).
11. R. S. Mitchell, Rhabdophane from the champion pegmatite, Amelia County, Virginia, Amer. Mineral. 50: 231 (1965).
12. J. R. Dooley and J. C. Hathaway, Two occurrences of thorium-bearing minerals [grayite] with rhabdophane-like structure, U.S. Geol. Surv. Prof. Paper 424-C: 339 (1961).
13. M. K. Carron, C. R. Naeser, H. J. Rose, and F. A. Hildebrand, Fractional precipitation of rare earths with phosphoric acid, U.S. Geol. Surv. Bull. 1036-N (1958).
14. C. Frondel, Systematic mineralogy of uranium and thorium, U.S. Geol. Surv. Bull. 1064 (1958).
15. J. B. Mertie, Monazite in the granitic rocks of the southeastern Atlantic States - an example of the use of heavy minerals in geologic exploration, U.S. Geol. Surv. Prof. Paper 1094 (1979).
16. M. W. Malloy, A comparative study of ten monazites, Amer. Mineral. 44: 510 (1959).
17. P. Pasteels, A comparison of methods in geochronology, Earth Sci. Rev. 4: 5 (1968).
18. I. S. Williams and L. T. Silver, The U-Th-Pb system compared in coexisting zircon, monazite and xenotime (abst.), EOS, 60: 411 (1979).
19. M. D. Karkhanavala and J. Shankar, An x-ray study of natural monazite: 1, Proc. Indian Acad. Sci. A40: 67 (1954).
20. K. M. Ghouse, Refinement of the crystal structure of heat-treated monazite crystal, Ind. Jour. Pure Appl. Phys. 6: 1 (1968).
21. M. Taylor and R. C. Ewing, The crystal structures of the $ThSiO_4$ polymorphs: Huttonite and thorite, Acta Cryst. B34: 1074 (1978).
22. J. J. Finney and N. N. Rao, The crystal structure of cheralite, Amer. Mineral. 52: 13 (1967).
23. C. M. Gramaccioli and T. V. Segalstad, A uranium- and thorium-rich monazite from a south-alpine pegmatite at Piona, Italy, Amer. Mineral. 63: 757 (1978)
24. L. A. Boatner, G. W. Beall, M. M. Abraham, C. B. Finch, R. J. Floran, P. G. Huray, and M. Rappaz, Lanthanide orthophosphates for the primary immobilization of actinide wastes, International Symp. Management Alpha-Contaminated Wastes, IAEA-SM-246/73 (1980).
25. T. Ueda, Studies on the metamictization of radioactive minerals, Mem. Coll. Sci. Univ. Kyoto XXIV: 82 (1957).
26. M. von Stackelberg and K. Chudoba, Dichte und Struktur des Zirkons (II), Zeit. Krist. 97: 252 (1937).
27. V. I. Pavlishin, Rhabdophanite - A new mineral of granite chambered pegmatites, Akad. Nauk Ukr. RSR, Dopov. Ser. B, No. 7: 598 (1969).

28. J. W. Adams, Rhabdophane from a rare-earth occurrence, Valley County, Idaho, U.S. Geol. Surv. Prof. Paper 600-B: B48 (1968).
29. H. Sorensen, Low-grade uranium deposits in agpaitic nepheline syenites, South Greenland, IAEA-PL-391/22: 151 (1970).
30. V. V. Burkov and Y. K. Podporina, Rare earths in the weathering crusts of granitoids, Akad. Nauk SSSR Dokl. 177: 691 (1967).
31. M. H. Staatz, Thorium veins in the United States (abstr.), Min. Engin. 25: 58 (1973).

DEVELOPMENT OF STRUCTURAL ENGINEERED BARRIERS FOR THE LONG-TERM CONTAINMENT OF NUCLEAR WASTE

R. E. Westerman

Pacific Northwest Laboratory
Battelle Memorial Institute
Richland, Washington 99352

INTRODUCTION

The concept for high-level nuclear waste disposal currently receiving the most attention in the United States involves deep emplacement of waste in continental geologic formations. Though the geologic medium is viewed as the ultimate barrier protecting the biosphere from excessive concentrations of radioactive materials, there is a recognition of the desirability of the added protection that would be provided by a localized system of engineered barriers that would last for ∿1000 yr. Such a barrier system would offer primary protection to the biosphere during the early disposal period, in which the geologic formation is potentially most vulnerable to failure because of the high thermal loadings from the waste. The toxicity of the waste is also highest during this period.

A task directed toward the development of licensable engineered barrier systems for the long-term containment of high-level nuclear waste under conditions of deep geologic disposal has been underway at the Pacific Northwest Laboratory (PNL) since January 1979. The work has been funded by the U.S. Department of Energy (DOE), Division of Waste Products, under the auspices of the PNL High-Level Waste Immobilization Program (HLWIP). Metallic, ceramic, and polymeric materials are being investigated by means of mechanical property and corrosion screening studies to determine the suitability of these materials for use as long-lived structural barriers, i.e., canister, overpack, and hole sleeve, in nuclear waste packages.

APPROACH

The development of a licensable package for containing high-level nuclear waste requires cooperation between the materials specialists responsible for the barrier elements, the groups responsible for the waste form development, and the groups responsible for characterization of the repository environment. The present program represents only a part of the total effort required, but it in itself must be guided by a comprehensive plan because of the complexities inherent in materials selection and their testing under repository-relevant conditions. The elements of the plan are presented below in approximate correspondence to the logical order of accomplishment:

- Establishment of Package Longevity Criteria--In the present task, longevity has been set at 1,000 yr. It is possible that this longevity requirement eventually will be more closely matched to the fission product heat liberation period, which essentially ends in $\sim$400 yr.

- Selection of Promising Candidate Barrier Materials--Metallic, ceramic, polymeric, and graphitic materials are all potentially useful in barrier system applications. Candidate materials from each of these broad material classes have been selected for corrosion and mechanical testing relevant to expected barrier service conditions.

- Development of Preliminary Packaging Concepts--A group of preliminary packaging concepts utilizing materials from the major material classes mentioned above have been developed for purposes of discussion. Three of these concepts, illustrating use of a broad range of materials, are presented in Figures 1, 2, and 3. The oxygen fugacity in the environment is considered a major factor in the development of package concepts, primarily because of its profound effect on the stability of metals. It is expected to vary from a high level, while the repository is open, to a low level, after the repository has been sealed and the effects of radiation on the groundwater have diminished.

  No detailed engineering analyses have been performed on any of the concepts presented; therefore, the range of applicability of materials potentially sensitive to heat and radiation, such as polymers and concretes, has yet to be defined.

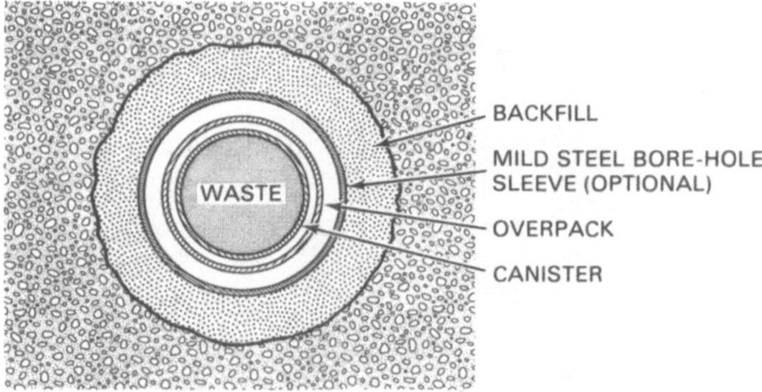

- Number of engineered barriers, exclusive of waste form: 3 (canister, overpack, backfill)
- Materials for overpack: Ti-base, Zr-base alloys; Ni-Cr, Fe-Cr alloys; ceramics; graphite
- Materials for canister: Cu, Cu-base alloys; Fe-base, Ni-base alloys

Fig. 1. Engineered barrier system compatible with both oxidizing and anoxic environments. Overpack resistant to oxidizing conditions, canister resistant to anoxic conditions

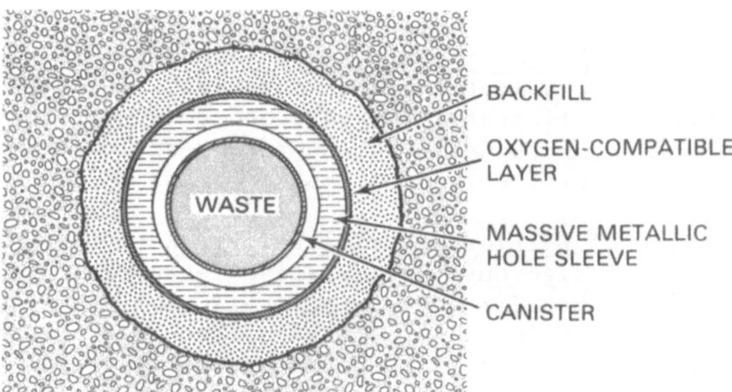

- Number of engineered barriers, exclusive of waste form: 4 (canister, hole sleeve, hole sleeve protective layer, backfill)
- Materials for canister: Cu, Cu-base alloys; Fe-base, Ni-base alloys
- Materials for hole sleeve: Cast iron; Cu, Cu-base alloys
- Materials for oxygen-compatible layer: Ti-base, Zr-base alloys, polymer

Fig. 2. Engineered barrier system compatible with both oxidizing and anoxic environments, featuring massive metallic hole sleeve. In principle, hole sleeve could be overpack

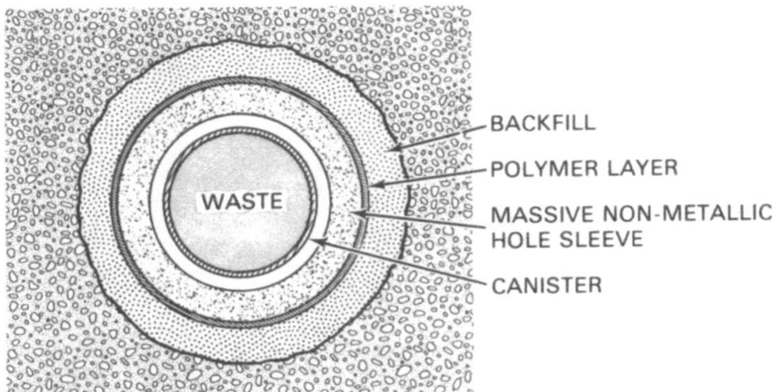

- Number of engineered barriers, exclusive of waste form: 4 (canister, concrete, polymer, backfill)
- Materials for canister: Cu, Cu-base alloys; Fe-base, Ni-base alloys
- Materials for hole sleeve: Concrete; polymer concrete; ceramics;graphite

Fig. 3. Engineered barrier system compatible with both oxidizing and anoxic environments, featuring massive non-metallic hole sleeve

- Performance of Screening Studies to Identify Most Promising Materials--The candidate barrier materials, selected initially on the basis of literature surveys and consultations with other investigators, are undergoing laboratory-scale corrosion and mechanical property screening tests.

- Anticipation of Barrier Material Failure Mechanisms and Performance of Rigorous, Conservative Tests--Materials appearing most useful to barrier system applications after completion of the screening studies will be tested in a rigorous fashion to reveal any susceptibility to degradation under severe service simulation conditions. The severity of the test(s) will be relied upon, in general, to reveal unanticipated modes of material degradation.

- Final Designs of Site-Specific Barrier Systems Based on Applicable Materials, Engineering Analysis of Systems, and Cost-Benefit Assessments--The most efficient solution to the waste disposal problem will require a careful analysis of all engineering factors associated with the repository as well as the economic aspects of all package components. It is part of the operating philosophy of the present program that at least two long-lived barriers, exclusive of the waste form and the geologic medium, will be designed into each package system. Each of the barriers will offer a high statistical probability of lasting for the full, desired package lifetime. Common-cause failure will be anticipated and avoided. This approach provides assurance of barrier redundancy.

- Prototype Waste Packages Ultimately Validated by In-Situ Studies--It is anticipated that the prototype waste packages (including, by definition, the appropriate barrier systems) will be validated by means of in-situ tests wholly relevant to the specific disposal sites for which the packages are intended. The in-situ tests must duplicate the most severe conditions expected in each environment, including the effects of temperature, pressure, stress, and radiation. It is possible that these in-situ studies could be best accomplished in "pseudo" sites, i.e., large pressure vessels containing site-relevant environments.

At the present time the barrier program at PNL is primarily involved in conducting screening studies, though some in-depth investigations are underway, notably in the mechanical testing area. The actual status of the subtask work elements, and the results obtained to date, are presented briefly in the following section of this report. More detailed accounts of the technical results of the program may be found in documents recently published[1,2].

RESULTS

Mechanical Property Evaluation, Metallic Materials

Titanium alloys Grade 2 (commercial purity) and Grade 12 (nominally 0.8 wt% Ni, 0.3 wt% Mo) were chosen for thorough mechanical property characterization, on the basis of a literature review; preliminary corrosion data; and the work of other contemporary researchers, such as those at PNL (Geothermal Energy Program) and Sandia Laboratories (Seabed and WIPP Programs).

Several different techniques for material characterization are being used in this task, including tensile tests; static stress corrosion tests; static and dynamic fracture toughness tests; and fatigue-crack-growth rate tests, with and without a simulated groundwater environment. The first four techniques offer the potential of supplying parameters useful in barrier design; the last technique (crack-growth tests) is useful in assessing the extent, if any, of environmentally induced crack propagation.

The fracture toughness of Grade 2 titanium was found to be substantially higher than that of Grade 12. Grade 2 is extremely tough;

in room temperature air, a critical crack length of 12.5 in.(32 cm) in length, at a stress level of 20,000 psi (138 MPa), can be tolerated without initiating fracture. In Grade 12 the critical crack length is about 1.2 in.(3 cm) under the same conditions.

The fatigue-crack-growth rates obtained at frequencies of 10 to 0.1 Hz in a simulated Hanford groundwater at 90 to 94°C show no increase over those found in laboratory air.

Corrosion Evaluation, Metallic Materials

Corrosion samples of 300 and 400 series stainless steels, Inconels, Hastelloy C-276, titanium, Zircaloy, copper-nickel and cast irons were prepared as U-bend specimens with simulated welds. These samples, electrically isolated, have been exposed in Inconel autoclaves at 250°C for as long as six months to simulated Hanford basalt groundwater and a Waste Isolation Pilot Plant (WIPP) brine.

The titanium alloys show outstanding corrosion resistance in both the Hanford groundwater and the brine media. Inconel, Incoloy, Hastelloy C-276, and Zircaloy-2 also showed very low corrosion rates, on the basis of weight change data, in both media. Cast irons were attacked in the Hanford groundwater, but appear usable in reasonable thicknesses (a few centimeters) over time periods of $\sim$1000 years. Cast irons, copper and a 70-30 copper-nickel alloy were severely attacked in the brine environment.

Ceramic Materials

Screening experiments were conducted to evaluate a number of candidate ceramic materials. The principal method used to screen the candidates was to subject samples of each material to a series of leaching tests and determine their relative resistances to attack by the leach solutions. A total of 14 ceramic materials, including two grades of alumina, plus graphite and basalt, were evaluated under both static and flow conditions using three different leach solutions: demineralized water, a synthetic Hanford groundwater, and a synthetic WIPP brine solution.

In general, increases in leach rate were associated with increases in test temperature, water purity, and solution flow rate.

Based on all of the results obtained, five of the materials, namely graphite, the two grades of alumina, TiO_2, and ZrO_2, exhibited greater resistance to attack by the leach solutions than the other materials tested.

Polymeric Materials

Initially, eighteen generic types of polymers were selected for evaluation. These included samples from all three of the major polymeric materials categories: thermoplastics, thermosets, and elastomers.

Samples were exposed to deionized water in autoclaves at 150°C, 200°C, or 250°C (temperature depending upon material properties in these initial studies). Samples were also exposed to ionizing radiation (gamma) to a total dose of 5 X 10^8 rads.

In testing at 150°C a number of polymers retain their properties: polyphenylene oxide, polymethylpentene, polysulfones, and glass-reinforced polyimides. After seven days at 200°C, only furan, EPDM rubbers, filled polyphenylene sulfide, metal-filled epoxy, and fluoropolymer samples remained intact and retained good mechanical properties. Only EPDM rubber, polyphenylene sulfide, furan and fluoropolymer materials survived the 250°C test. Polymers that appeared little affected by gamma doses of 5 X 10^8 rads were polyphenylene sulfide, polyimide, furan and metal-filled epoxy. Both EPDM and fluoropolymer samples, although suffering considerable loss in elongation (due to increased crosslinking) still retained good tensile strengths after radiation exposure. Impact measurements confirm the findings of the tensile measurements, i.e., loss of tensile ductility is associated with a decrease in impact energy absorption.

PREDICTION OF BARRIER LONGEVITY

The development of barrier elements for a licensable waste package must ultimately include presentation of proof that the proposed barrier element will indeed demonstrate, with a high degree of confidence, the required useful lifetime. (In the present program, "useful" is defined as complete freedom from penetration.)

Some potential barrier failure modes are amenable to prediction by examination of relatively short-term data; others are not. In the case of metals, for example, the failure modes most likely to end the life of a barrier are 1) uniform corrosion; 2) non-uniform corrosion (pitting, intergranular attack, stress-corrosion cracking); and 3) hydrogen embrittlement (here considered separate from stress-corrosion cracking, though they are, in many cases, related). Of these, uniform corrosion can probably be estimated in a satisfactory manner by knowledge of the rate law involved and the activation energy of the critical rate control step. An assumption must be made, of course, that the operative mechanism will not change with time; but we believe that a conservative approach to obtaining and interpreting the laboratory data can result in predictions of adequate confidence. Similar reasoning can be applied to any process of degradation where the basic mechanism is known and rate laws/activation energies can be determined, such as microstructural changes brought about by diffusion.

In the case of a degradation mechanism with unpredictable kinetics, such as stress corrosion cracking, a different approach is taken. The materials that become serious candidates after general screening tests are completed will be evaluated in terms of basic fracture toughness characteristics, K_{Iscc} or J_{Iscc}, in the relevant repository medium. Knowledge of K_{Iscc}/J_{Iscc} will yield allowable stress/flaw size values for use by the barrier designer. This approach attempts to utilize barrier materials in such a way that the conditions that would cause stress corrosion cracking would never exist.

The most conservative environmental fracture toughness values are obtained by means of corrosion fatigue studies, emphasized in the present program.

Other materials degradation phenomena that are not amenable to avoidance by either an understanding of the kinetics or control of design and environmental parameters may have to be treated on a "go - no go" basis, i.e., a material exhibiting the susceptibility in rigorous testing may have to be eliminated totally from consideration as a barrier material. Some non-uniform corrosion phenomena, such as severe pitting or crevice corrosion, might exemplify this condition. Of course, good engineering judgment must always be exercised in any predictive evaluation, and the results of rigorous tests must be always viewed in the proper rational context. Lack of confidence in individual barriers in an engineered barrier system may be circumvented, of course, by increased barrier redundancy.

REFERENCES

1. Fullam, H. T., Use of Ceramic Materials in Waste Package Systems for Geologic Disposal of Nuclear Wastes, PNL-3447, Pacific Northwest Laboratory, Richland, Washington, 1980.

2. Westerman, R. E., Investigation of Metallic, Ceramic, and Polymeric Materials for Engineered Barrier Applications in Nuclear Waste Packages, PNL-3484, Pacific Northwest Laboratory, Richland, Washington, October, 1980.

EVALUATION OF METALLIC MATERIALS FOR USE IN ENGINEERED BARRIER SYSTEMS

S. G. Pitman, B. Griggs, and R. P. Elmore

Pacific Northwest Laboratory, Operated by Battelle Memorial Institute, P.O. Box 999, Richland, Washington, 99352

INTRODUCTION

A program is underway at PNL to select and evaluate candidate materials for engineered barrier applications in long-lived nuclear waste packaging systems. The effort is funded by the Division of Waste Products, U.S. Department of Energy.

This presentation deals with corrosion and enviro-mechanical tests of metallic materials. The program includes corrosion, tensile, impact, crevice corrosion, stress-corrosion cracking, static and dynamic fracture toughness, fatigue-crack-growth rate, and environmental fatigue-crack-growth rate tests.

A parallel testing approach was taken, in which thorough testing was begun on two promising materials at the same time that a screening program was in progress. Titanium grades 2 and 12 were chosen for thorough enviro-mechanical testing on the basis of a literature review and preliminary corrosion data. Grade 2 is commercial-purity titanium; grade 12, containing nominally 0.8 wt% Ni and 0.3 wt% Mo, is noted for its resistance to pitting and crevice corrosion at elevated temperatures.

A complete description of the test program and results is beyond the scope of this presentation, so an attempt has been made to describe the test methods and to summarize the more important results.

EXPERIMENTAL METHODS

Corrosion Tests

The corrosion tests performed to date have been of a screening nature designed to eliminate materials from consideration that do not exhibit potential for long life in a repository environment. Samples of 1.9 x 13.7 cm were bent into U-bends (ASTM-G#30-72) prior to creating a simulated weld bead using a tungsten-inert gas welding torch. After bending, the samples were cleaned by etching in HNO_3 or an HNO_3/HF mixture, and were finally rinsed in water and methanol. The samples were insulated from the stressing bolts by means of alumina washers.

Two general water chemistries (Table 1) were used in the present study: a Hanford basaltic ground water and a NaCl-$MgCl_2$ brine. The Hanford ground water (HGW) formulation used in the tests was based on analyses of several deep aquifers that exist in the Hanford Reservation. The simulated ground water passed upward through a 25-cm deep crushed basalt rock layer, over the corrosion specimens, and out of the autoclave through a cooler and back-pressure regulator (Fig. 1). The autoclave was held at 5.52 MPa to maintain the liquid phase at 250°C. Pumping rates were 300 ml/h initially and were reduced to 30 ml/h for the second series, with a slight change in water composition. The storage tanks were sparged with nitrogen for oxygen control.

TABLE 1. Chemistry of Leach Solutions

	Synthetic Hanford Ground Water					
	August 1979 to May 1980		May 1980 to July 1980		Brine Solution	
Ion	Make-up mg/l	Effluent Analyses[a] mg/l	Make-up mg/l	Effluent Analyses[a] mg/l	Ions	Make-up
Al	0	10	0	6	Mg^{++}	34.98 g/L
Na	139	140	111	127	Na^+	41.34
Mg	0.5	0.05	0.5	0.05	K^+	29.95
Ca	1.6	0.1	2.5	0.1	Ca^{++}	0.60
K	13	55	13	21	Li^+	0.020
CO_3 (total)	167	210	167	240	Rb^+	0.020
SiO_2	36	210	36	331	Sr^{++}	0.0053
Cl^-	52	50	52	58	Cs^+	0.0008
F^-	8	8	8	8	Cl	191.3
SO_4	0.8	--	0.8	5	SO_4	3.51
O_2	1	--	0.05	--	BO_3	1.24
pH	9.5	8.5	9.0	7.7	Br^-	0.39
Conductivity, Mho	475	55	630	600	I^-	0.010
					Fe^{++}	0.002
					pH	9.0

(a) Sample was cooled before sampling. The water passed through a bed of crushed basalt at 250°C in the autoclave, which increased the Al and Si, and decreased pH, Ca and Mg.

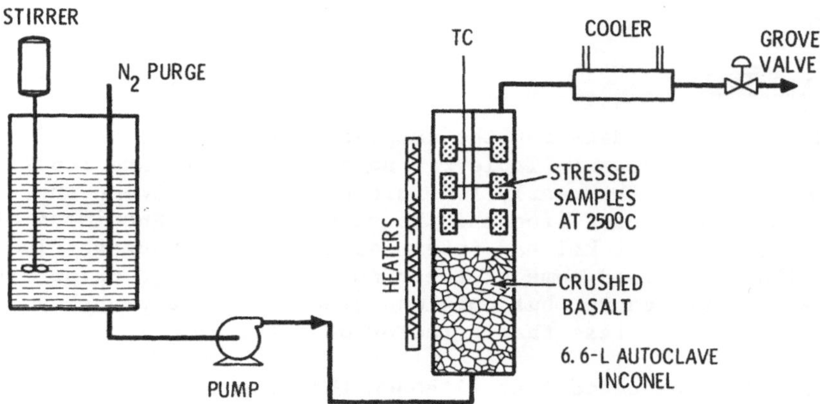

Fig. 1. Autoclave System

The brine test was run nearly static and thus essentially without oxygen. Samples were held in a 1-L Inconel-625 autoclave and electrically insulated from each other by Teflon spacers.

Enviro-Mechanical Tests

Tensile, impact, and fracture toughness tests were conducted between 20 and 250°C, using round tensile specimens, Charpy V-notch specimens, and compact tension specimens, respectively. Tests of the fatigue-crack-growth rate in air were done using compact tension specimens from plate material and center-cracked tension specimens from sheet material. Fatigue-crack-growth rate tests were performed at 90°C in a flowing simulated Hanford ground-water solution (Table 1) using an environmental chamber to surround the cracked portion of the specimen.

Static fracture toughness tests were performed on 1.3-cm plates of titanium grades 2 and 12 at room temperature, using compact tension specimens. Linear-elastic fracture toughness testing was not applicable because of the degree of deformation preceeding fracture, so an elastic-plastic (J-integral) approach was used. This approach measures crack extension as the specimen is loaded, to determine the energy required to cause crack initiation. This energy value is used to calculate J_{Ic}, which is relatable to fracture toughness K_{Ic}. Both single-specimen and multiple-specimen J-integral tests were used.

RESULTS

Corrosion Tests

Weight change data for the samples in the basalt ground water tests are shown in Table 2. Some conclusions are readily apparent, such as the nearly 100-fold difference between the weight changes of cast iron and the other alloys. Specimens of nickel and copper-nickel had intermediate weight changes. The other alloys retained some of their corrosion product oxide after the cleaning procedure, but complete cleaning would probably result in a change less than a factor of five.

It should be noted that although the weight change data of cast irons shown in Table 2 appear high, an extrapolation of these data to 1000 yr would lead to an expected loss of only about 2.5 cm of material, if the corrosion were uniform. The specimens were highly pitted, however, indicating much higher local rates of penetration.

Examination of Table 2 shows the interactions of materials and an aggressive brine postulated to be possible in an accident scenario for salt storage of waste. Once again, there was little middle ground. The contrast of cast iron, copper, copper-nickel, and nickel to stainless steel, Inconel or titanium is as much as 1000-fold. Inconel 600, Inconel 625, Incoloy 800, Hastelloy C-276, titanium grades 2 and 12, and Zircaloy-2 all had a very low corrosion rate. The contrast between the Zircaloy-2 in the HGW and in the brine is possibly caused by the 8 ppm of fluoride in the HGW. The austenitic stainless steels all performed well from a weight change viewpoint, but would probably stress-corrosion-crack if oxygen were present. The 400 series stainless steels lost much more weight and completely cracked through in the heat-affected zone near the weld. This cracking is probably a hydrogen-induced cracking in the marten-site layer produced by the welding heat. These screening studies are to eliminate the obvious metals which are inappropriate as waste barriers. The surviving metals will be more rigorously tested in a facility, illustrated in Fig. 2, which more completely simulates the geologic respository conditions. The simultaneous influence of temperature (up to 250°C), pressure (600 psi), environmental chemistry (groundwater plus fractured rocks), and radiation (up to 10^6 R/hr) will enable improvements on corrosion rates measured during the screening tests.

Enviro-Mechanical Tests

The static and dynamic yield strength and fracture toughness of grades 2 and 12 titanium plate are given in Table 3. Each

Table 2. Corrosion of Samples in Simulated Repository Environments at 250°C[a]

Material	Basalt Ground Water: Weight Change First 3 Mo, mg/dm^2		Basalt Ground Water: Weight Change Second 3 Mo, mg/dm^2		$MgCl_2$-NaCl Brine Weight Change		Remarks
	Before Cleaning	After Cleaning[b]	Before Cleaning	After Cleaning[b]	mg/dm^2[c]	Time Days	
304 L S/S	+11	-0.9	+4.8	-1.5	-5	72	Bright HAZ-Black[d]
304 S/S	+14	-3.3	+7.4	-0.9	-5	72	Bright HAZ-Black
316 S/S	+6	-5.9	--	-1.6	-5	72	Bright HAZ-Black
321 S/S	+14	-2.2	+4.4	-2.2	-5	72	Bright HAZ-Black
405 S/S	+15	-3.3	+2.7	-3.1	-351	72	Broken, in HAZ-Black
410 S/S	+28	-5.1	+18.0	-4.5	-354	72	Broken, in HAZ-Black
Inconel-600	+13	-3.3	+7.6	-4.2	12	58	
Inconel-625	+43	-0.9	+8.0	-1.8	-2	72	Track of tarnish
Incoloy-800	+24	0.0	+20.0	-1.3	-9	58	Light tarnish
Hastelloy-C-276	+14	-1.8	+14.0	-3.5	-6	58	
Nickel-200	+245	+168.0[e]	+20.0	-345.0	-486	58	
Titanium-Grade 2	+32	+0.4	+6.3	+1.8	+1.0	58	Light straw tarnish
Titanium-Grade 12	+17	+0.2	+5.2	-0.7	+0.9[f]	58	Light straw tarnish
Titanium-6AL-4V			+39.0	+0.2	+0.2	58	Light black tarnish
Zircaloy-2	+80	+23.0	+3.3	-11.0	+8.0[f]	58	Black
Cast-Iron 180-7	+90	-415.0	+277.0	-442.0			
Cast-Iron 22-8	+225	-345.0	+263.0	-476.0	-7713	20	Cleaned to metal
Cast-Iron 142-12	+345	-200.0	+284.0	-512.0			
Cast-Iron 166-3	+135	-180.0	+235.0	-422.0			
Cast-Iron 136-4	+32	-411.0	+224.0	-530.0			
Copper			+168.0	-171.0	-3858	38	Spalling
Copper-Nickel 70-30	+58	-89.0	+268.0	-580.0	-996	20	Spalling

(a) See Table 1 for composition.
(b) Samples of Inconel, stainless steel, titanium, Zircaloy, Incoloy and Hastelloy only had the deposited film removed, not the corrosion product. Samples of copper, copper-nickel and cast-iron had all deposited and most of the corrosion product film removed. The same specimens were used for both 3-mo. tests.
(c) Weight change after cleaning in Na_2 EDTA solution with ultrasonics to remove deposited scale. Samples of copper, copper-nickel and nickel were cleaned in 18% HCl (room temperature) 3 min. Cast-iron cleaned in diammonium citrate plus phenylthiourea (85°C).
(d) HAZ: heat-affected zone of weld.
(e) Residual deposited film apparent.
(f) Weight change, second period mg/dm^2 to titanium grade 12 (0 for 72 days) and for Zircaloy-2 (+10 for 72 days).

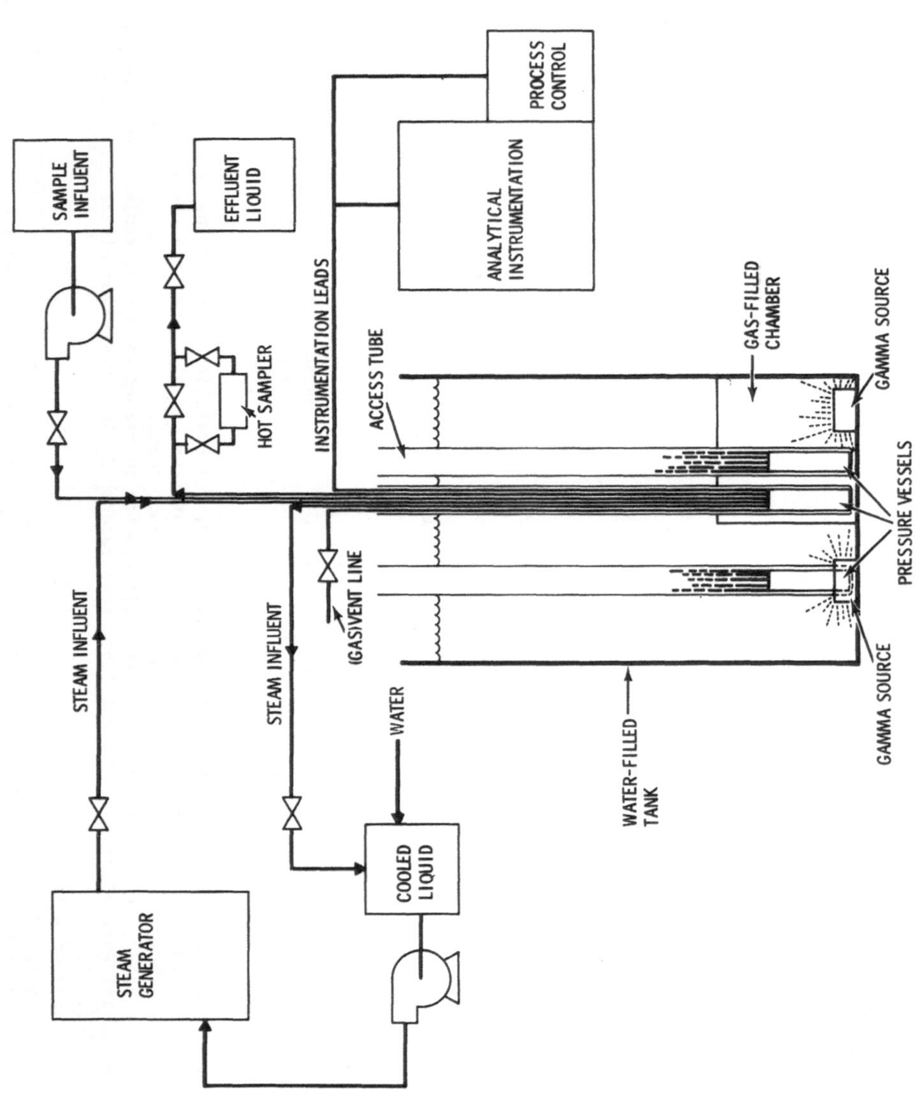

Fig. 2. Radiation Test Facility

Table 3. Yield Strength and Fracture Toughness of Grades 2 and 12 Titanium

	Temperature °C	Orientation	σ_{ys} (a) MPa	σ_{yd} (b) MPa	$K_{Ic}(J)$ (c) $MPa\sqrt{m}$	K_{Id} (*) or K_Q (d) $MPa\sqrt{m}$
Grade 2	25	TL	333	641	146	158
		LT	292	792	136	93
	250	TL	136	448	---	181
		LT	120(e)	427	---	153
Grade 12	25	TL	476	806	94(e)	60
		LT	340	703	38	33*
	250	TL	299	586	118	115
		LT	194	531	120(e)	35*

(a) Static yield strength, from round tensile specimens.
(b) Dynamic yield strength, fom Charpy V-notch specimens.
(c) Static fracture toughness from j-integral test.
(d) Dynamic fracture toughness from pre-cracked, instrumented Charpy specimens.
(e) Estimate.

material is stronger in the orientation transverse to the rolling direction. Grade 12 is more anisotropic, but it is not presently known whether this difference results from processing or from alloying content. The higher strength of grade 12 at elevated temperatures may be advantageous in a load-carrying barrier element. The strength and toughness of both alloys are sensitive to strain rate, as may be expected. One discrepancy is the low dynamic toughness of the LT orientation (cracking plane perpendicular to the rolling direction) of grade 12 titanium at 250°C. This may suggest a sharp toughness transition curve, which is shifted toward higher temperatures by high strain rates. Further testing will be needed to explain this anomaly.

Results of the fatigue-crack-growth rate tests of titanium plate in air, using compact tension specimens, are summarized in Table 4. The crack-growth rate of grade 2 titanium was lower than that of grade 12 titanium at all stress intensities studied. At $\Delta K = 20$ MPa $\sqrt{m}$, the TL orientation (cracking plane parallel to the rolling direction) of each material had lower crack-growth rates. From extrapolation of the data to high stress intensities ($\Delta K = 40$ MPa $\sqrt{m}$), it appears that the LT orientation will have a lower crack-growth rate. However, there may be a knee in the curve which would change behavior at high crack-growth rates.

Table 4. Summary of Fatigue-Crack-Growth Rates of Titanium Plate in Air, 20°C

$$\frac{da}{dn} = C \quad \Delta K^m, \text{ in./cycle}$$

	Titanium-2		Titanium-12	
	TL	LT	TL	LT
C	2.9×10^{-9}	1.2×10^{-8}	1.8×10^{-10}	8.9×10^{-10}
M	2.9	2.4	4.0	3.6

residually-stressed environmental tests of grades 2 and 12 titanium cannot be presented in detail here. No evidence of environmental degradation was found in any of the tests. Fatigue-crack-growth rates at various frequencies were no higher in groundwater solution than in air at the same temperature (90°C). There was no evidence of environmentally assisted crack growth or crevice corrosion at 250°C for a 3-mo test duration; however, more conclusive tests need to be run to verify the structural integrity of these materials when exposed to aqueous environments and radiation fields for extended times. One environmental interaction was the pickup of up to 10 ppm of hydrogen by grade 12 titanium, when coupled with cast iron and exposed in the ground-water solution. Grade 2 titanium showed no hydrogen uptake in similar tests.

CONCLUSIONS

1. Inconel 600, Inconel 625, Incoloy 800, Hastelloy, C-276, and grades 2 and 12 titanium all had excellent corrosion resistance in both postulated repository environments tested. Further work will be required to evaluate the pertinent enviro-mechanical properties of these materials.

2. The mechanical properties of grade 2 titanium are better than those of grade 12 titanium, except the tensile and yield strengths. These properies include fatigue-crack-growth rate, environmental fatigue-crack-growth rate, fracture toughness, impact toughness, and dynamic fracture toughness.

3. There is no evidence in the current data to indicate that the simulated repository environment is aggressive to grade 2 or grade 12 titanium. This includes data from corrosion-fatigue, crevice corrosion, wedge-loaded cracked specimens, and residual-stress specimens.

EVALUATION OF CERAMIC AND POLYMERIC MATERIALS FOR USE IN ENGINEERED BARRIER SYSTEMS

H.T. Fullam and W.E. Skiens

Battelle
Pacific Northwest Laboratory
P.O. Box 999, Richland, WA 99352

A program is underway at the Pacific Northwest Laboratory (PNL) to develop long-lived engineered barrier systems for use in the geologic disposal of nuclear wastes. The work is funded by the Division of Waste Products of the Department of Energy. One task of the program consists of a study to evaluate the potential use of nonmetallic materials as elements in engineered barrier systems. The purpose of the nonmetallic barrier elements would be to isolate the primary metallic waste canister and waste form from the geologic repository environment for the time required to allow the major heat-producing radioisotopes to decay to a negligible level--approximately 1,000 years. The work with nonmetallics is limited to two classes of materials: polymeric materials and ceramic materials. The initial objective of the work was to screen a large number of materials of each class and select the best materials for detailed evaluation.

CERAMIC MATERIALS

An initial screening of the ceramic materials was accomplished by subjecting specimens of the materials to a series of leaching tests carried out under various temperature and pressure conditions and determining their relative resistances to attack by the leach solutions. Complete details of the screening studies are given in PNL-3447 (Fullam, 1980) and are summarized below.

Experimental

The ceramic materials evaluated in the screening studies were Al_2O_3 (99%), Al_2O_3 (99.8%), mullite ($2Al_2O_3 \cdot SiO_2$), vitreous silica (SiO_2), $BaTiO_3$, $CaTiO_3$, $CaZrO_3$, $CaTiSiO_5$, TiO_2, $ZrSiO_4$, basalt, Pyroceram 9617®, and Marcor code 9658 machinable glass ceramic®. One grade of graphite (Toyotanso IG-11®) was also evaluated.

Five different leach tests were used to screen the ceramic materials

- A static leach test at 100°C.
- A static leach test at 150°C.
- A static leach test at 250°C.
- A dynamic leach test at 90°C.
- A dynamic leach test at 250°C.

Three leach solutions were used in the screening tests: demineralized water, a synthetic Hanford groundwater, and a synthetic NaCl brine solution.

Specimens of each material were subjected to each of the five leach tests. The leach rate of each specimen was determined by monitoring the specimen weight change and the leach solution composition as a function of time. After testing the ceramic specimens were subjected to ceramographic examination.

Results and Discussions

The primary objective of this study was to screen a large number of materials and determine which ones were most resistant to attack by the leach solutions. An initial comparison of the materials can be obtained by determining their leach rates under the various test conditions. Leach rates can be misleading, however. Localized attack can result in more serious attack than is evident from leach rate measurements; and alteration reactions can result in extensive attack that may not be apparent from either the solution analyses or the specimen weight change measurements.

® Corning Glass Works, Corning, New York.

® Toyo-Tanso Company, Ltd., Osaka, Japan.

All of the materials tested except two exhibited their highest leach rates in the dynamic test at 250°C; the two exceptions were graphite and TiO_2, which had slightly higher leach rates in some of the static tests at 250°C. The results obtained in the dynamic test at 250°C are summarized in Table 1. The leach rates shown are the average rates for the total exposure time. Many of the materials suffered alteration reactions to a considerable depth. The actual reactions involved were not determined, but with most of the materials they appeared to involve hydration reactions. The two grades of alumina tested suffered some localized attack which consisted of the leaching of impurities from grain boundaries. The cylindrical TiO_2 specimen exhibited radial cracks to a considerable depth, but it appears that the cracking was probably due to problems in fabricating the specimens rather than attack by the leach solution.

Overall, leach rates of the materials increased with temperature as expected, although in most cases the increases were not as great as one might anticipate. For example, the leach rate of TiO_2 in flowing demineralized water was only about twice as great at 250°C as at 90°C. In the static tests at 150°C and 250°C, leach rates were generally highest in the demineralized water and lowest in the brine solution. Alteration reactions and localized attack were not serious problems in the tests at 150°C and lower.

In Table 2 the materials are ranked in order of decreasing leach resistance for each of the leach tests. Since the leach rates varied with time in many of the tests, it is difficult to compare the materials accurately, and the rankings shown in Table 2 should only be considered as approximations. A cursory examination of Table 2 shows that five materials--namely graphite, the two grades of alumina, TiO_2 and ZrO_2--exhibit the greatest resistance to leaching in most of the tests. The data also show that none of the five materials are susceptible to alteration reactions at temperatures up to 250°C. Some of the other materials tested exhibited a high leach resistance in one or two tests, but on an overall basis they are much inferior to the five materials mentioned. In addition, most of the materials having relatively low leach resistances are also susceptible to alteration reactions at temperatures above 150°C.

Based on all of the results obtained, graphite appears to be the most leach resistant of the materials tested with the two grades of alumina being the best of the ceramic materals. Titanium dioxide and ZrO_2 are the most leach resistant of the remaining materials. Taking into account other facts, such as availability, cost, mechanical and physical properties, etc., as well as leach resistance, graphite and alumina are the preferred materials for detailed evaluation as elements in engineered barrier systems.

Table 1. Attack of Ceramic Materials Exposed to Flowing Demineralized Water at 250°C

Material	Total exposure d	Average leach rate[a] g/cm^2-d	Depth of material affected, μm: Localized attack	Depth of material affected, μm: Alteration reactions
Graphite	43	7.6×10^{-6}	0	0
TiO_2	45	6.7×10^{-6}	*b*	0
ZrO_2	43	1.4×10^{-4}	0	0
Al_2O_3 (99.8%)	43	1.3×10^{-4}	25	0
Al_2O_3 (99%)	43	1.8×10^{-4}	50	0
Vitreous silica	43	5.0×10^{-4}	0	0
$CaZrO_3$	45	4.4×10^{-4}		360
$CaTiSiO_5$	45	7.0×10^{-4}		200
$BaTiO_3$	44	5.1×10^{-4}		50
$CaTiO_3$	44	7.2×10^{-4}		100
Mullite	43	1.3×10^{-3}		>800
$ZrSiO_4$	45	1.1×10^{-3}		100
$BaZrO_3$	45	1.4×10^{-3}		150
Basalt	45	3.0×10^{-3}		550
Marcor 9658	42	3.5×10^{-3}		>850
Pyroceram 9617	43	8.3×10^{-3}		>800

[a] Calculated from the geometric surface areas of the specimens.

[b] Specimens exhibited cracks to a considerable depth — see text.

POLYMERIC MATERIALS

Polymeric materials have been particularly outstanding in their ability to protect metals and other materials in a wide range of corrosive environments. Polymers have received little attention as components of containment systems for nuclear wastes, possibly because it was felt that they could withstand neither the environment nor radioactive conditions in this application. Since most synthetic polymers of interest in this program have been developed within the past 30 years, information related to long-term environmental stability is unavailable. However, after initial evaluation, a number of polymer systems appeared to have the potential stability for long term use under the radiation and environmental conditions expected in waste repositories (Westerman et al., 1979).

Candidate materials from all three general classes of polymers--thermoplastics, thermosets, and elastomers--were considered in the selection for experimental study in this program. The maximum storage conditions were assumed as 1) time--1000 years, 2) radiation dose--10^{10} rads, 3) environment--contact with water or brine, and 4) temperature--250°C. These conditions of temperature and radiation levels were not assumed as absolute cut-off points, such that promising materials were excluded by these maxima.

Table 2. Relative Resistance of Ceramic Materials to Leaching Under Various Test Conditions — in Order of Decreasing Resistance

Static leach test at 100°C, demineralized water	Dynamic leach test at 90°C, demineralized water	Dynamic leach test at 250°C, demineralized water	Static leach test at 150°C			Static leach test at 250°C		
			Demineralized water	Hanford ground water	Brine	Demineralized water	Hanford ground water	Brine
Graphite	Graphite	Graphite	Graphite	Al_2O_3 (99.8%)	Graphite[a]	Al_2O_3 (99.8%)	Al_2O_3 (99.8%)	Graphite[a]
Al_2O_3 (99.8%)	Al_2O_3 (99.8%)	TiO_2	Al_2O_3 (99.8%)	Graphite	TiO_2[a]	Graphite	Graphite	Al_2O_3 (99.8%)
Al_2O_3 (99%)	Mullite	Al_2O_3 (99.8%)	Al_2O_3 (99%)	Al_2O_3 (99%)	Al_2O_3 (99.8%)	Al_2O_3 (99%)	ZrO_2	ZrO_2
Pyroceram 9617	Al_2O_3 (99%)	ZrO_2	Mullite	Mullite	Al_2O_3 (99%)	TiO_2	Al_2O_3 (99%)	Al_2O_3 (99%)
Vitreous silica	TiO_2	Al_2O_3 (99%)	Vitreous silica	ZrO_2	Mullite	$CaTiO_3$	TiO_2	TiO_2
TiO_2	ZrO_2	Vitreous silica	ZrO_2	Pyroceram 9617	Vitreous silica	ZrO_2	$CaTiO_3$	$CaTiO_3$
Basalt	Pyroceram 9617	$CaZrO_3$	Pyroceram 9617	$CaZrO_3$	Pyroceram 9617	Vitreous silica	$CaTiSiO_5$	Mullite
$CaTiO_3$	$CaTiO_3$	$CaTiSiO_5$	TiO_2	$ZrSiO_4$	ZrO_2	Basalt	$BaZrO_3$	$BaZrO_3$
ZrO_2	$CaZrO_2$	$BaTiO_3$	$CaTiSiO_5$	$CaTiO_3$	$CaTiO_3$	$CaTiSiO_5$	$CaZrO_3$	Vitreous silica
$ZrSiO_4$	Basalt	$CaTiO_3$	$ZrSiO_5$	TiO_2	$ZrSiO_4$	Mullite	Mullite	Pyroceram 9617
Marcor	Vitreous silica	Mullite	$CaTiO_3$	$CaTiSiO_5$	Marcor	$CaZrO_3$	Basalt	$ZrSiO_4$
$BaTiO_3$	Marcor	$ZrSiO_4$	Basalt	Vitreous silica	$BaTiO_3$	$ZrSiO_4$	Vitreous silica	Marcor
$BaZrO_3$	$ZrSiO_4$	$BaZrO_3$	Marcor	$BaTiO_3$	$BaZrO_3$	Pyroceram 9617	Pyroceram 9617	$BaTiO_3$
$CaTiSiO_5$	$BaTiO_3$	Basalt	$CaZrO_3$	Basalt		$BaZrO_3$	$ZrSiO_4$	
Mullite	$CaTiSiO_5$	Marcor	$BaZrO_3$	Marcor		$BaTiO_3$	Marcor	
$CaZrO_3$	$BaZrO_3$	Pyroceram 9716	$BaTiO_3$	$BaZrO_3$		Marcor	$BaTiO_3$	

[a]Some specimens exhibited weight gains due to incomplete removal of brine prior to drying.

Containment of Nuclear Wastes by Polymeric Materials

Polymers that have been considered for this application have been selected from commercially available materials; however, no economic or processing restrictions have been put on materials in this preliminary study. All generic families of polymers have been considered, and after study of available information concerning their response to the expected environment, a number were selected for initial evaluation. Polymers in this application are not expected to need high structural strength, and change in properties due to radiation (crosslinking or scission) can occur in many cases without loss of function.

Experimental

Samples of the selected polymers in the form of tensile or deflection bars were obtained from manufacturers or prepared in our laboratories from resins. As an initial screening step, weighed and measured samples were placed in an autoclave under the following conditions: temperature - 150°, 200°, or 250°C (depending upon known sample properties); time - 168 hours; autoclave contained deionized water. Samples that remained intact through this test were evaluated by measuring tensile strength and elongation and compared with control samples. Polymers appearing promising at lower temperatures were evaluated at a higher temperature. Promising materials were also tested by exposing samples to gamma radiation (^{60}Co source). Samples were irradiated to a dose of $5x10^8$ rads and tested for tensile and elongation. Additionally, promising materials were tested for impact strength.

Results and Discussion

Table 3 lists 14 different generic polymers with a total of 29 differing formulations tested in autoclaves (1 l/hr water flow) with surviving specimens evaluated for tensile strength and elongation. Selected samples were further exposed to gamma radiation and submitted to tensile and elongation measurements--shown in Table 4. Those materials that appeared promising through both autoclave and radiation testing were also evaluated by impact tests--Table 5.

Although all of the polymers tested are little affected by exposure to ground waters or brine at ambient temperatures, exposure to water or vapor at 150-250°C is a much more severe exposure and Table 3 indicates that polymers such as the polyimides which are very resistant to temperature and radiation in the ranges studied do not stand up well in a hot aqueous environment. Materials that appeared most promising after autoclave testing were EPDM rubbers, polyphenylene sulfide, poly (ethylene-tetrafluoroethylene) copolymer, and polyfurfuryl alcohol. Although each of these materials had

Table 3. Polymers Autoclaved in Flowing Deionized Water at Various Temperatures (exposure — 168 hours)

Sample	Tensile strength psi[a]	Elongation %[a]	Comments
	150°C Autoclave		
ABS (2 types)			swollen, distorted
Polyethylene (hi-density, 2 types)			softened, distorted
Polyphenylene oxide	9,300 (10,200)	16 (62)	appear good
(2 types)	9,200 (7,000)	11 (21)	elongation decreased
Polyimide (glass-filled)	6,900 (8,200)	1 (1)	essentially unchanged
Polysulfone	12,200 (9,000)	6 (90)	elongation decrease
Polyphenylene sulfone	11,500 (9,000)	21 (18)	unchanged
Polyamide-imide (filled)			embrittled, broken
Polyether sulfone	13,400 (12,100)	6 (12)	elongation drop
Polymethylpentene	3,000 (3,400)	22 (25)	no change in properties
Epoxy (3 types)			all crazed or broken
	200°C Autoclave		
Polyphenylene oxide (2 types)			softened, distorted
Polysulfone			melted, flowed
Polyphenylene sulfone			slightly distorted
Polyimide 3 types (glass or aramid fiber-filled)			all degraded, embrittled
Polyarylene			swollen, embrittled
Polyphenylene sulfide	1,900 (15,700)	0.8 (1.4)	embrittled, lost tensile
(2 types)	3,000 (8,000)	0.6 (0.8)	and elongation
Silicone rubber			crumbled to powder
Poly (ethylene-tetrafluoroethylene)	5,200 (5,800)	250 (225)	no change
Metal-filled epoxy			brittle, broken
	250°C Autoclave		
Chlorosulfonated rubber			distorted, embrittled
EPDM rubber (ethylene-	700	150 (>370)	significant drop in
propylene-4	300	110 (>370)	elongation
formulations)	1,900	100 (190)	looks good
	500	110 (>360)	properties decrease
Polyphenylene-sulfide	3,100 (15,700)	0.1 (1.4)	tensile and elongation
(3 types)	1,400 (8,000)	0.2 (0.8)	both drop significantly
	2,300 (10,300)	0.1 (1.2)	
Polyimide (glass or aramid fiber-filled)			crumbled, untestable
Poly (ethylene-tetrafluoroethylene)	2,700 (5,800)	19 (225)	tensile drop, large elongation decrease
Silicone rubber (foam)			swollen, degraded
Metal-filled epoxy			embrittled, broken
Polyfurfuryl alcohol	3,000 (3,300)	13 (25)	elongation drop

[a]( ) control sample values.

significant property changes after the 7-day exposure at 250°C, further testing and evaluation followed by reformulation may provide materials with acceptable properties.

The radiation exposure of samples was carried out to only 5 x 10^8 rads initially to expedite testing. This provided an initial evaluation of radiation effects. Data in Table 4 indicate that, while this radiation dose had little effect on polyfurfuryl alcohol and polyphenylene sulfide, very significant decreases in elongation occurred with fluorocarbon copolymer and EPDM rubbers, as might be expected with the high polyethylene content in both these materials. Impact data in Table 5 verifies the tensile strength and elongation data seen in the previous table.

Table 4. Evaluation of Polymers Exposed to Gamma Radiation (total dose 5 x 10^8 rads — 50°C in air)

Sample	Tensile strength psi[a]	Elongation %[a]	Comments
Polymethyl pentene	1,200 (3,400)	3 (25)	degraded, sticky
Polyphenylene oxide	3,000 (10,150)	1 (62)	lost tensile-elong.
Polyphenylene sulfone	8,600 (9,000)	37 (18)	little property change
Polysulfone	2,600 (9,000)	18 (90)	lost tensile and elong.
Polyphenylene sulfide	18,400 (15,700)	6 (1.4)	tensile unchanged
(2 types-filled)	9,000 (10,300)	3 (1.2)	increase in elongation
Polyimide	8,100 (8,200)	1 (1)	no change
EPDM rubber	700	11 (>370)	little change in tensile
(3 formulations)	1,100	3 (190)	decreased elongation
	600	8 (>360)	
Polyethersulfone	12,000 (12,000)	2 (12)	decrease in elongation
Metal-filled epoxy	744 (1,100)	<1 (<1)	brittle breaks easily
Poly (ethylene-tetrafluoroethylene)	4,900 (3,800)	3 (225)	large decrease in elongation
Silicone rubber			degraded, torn

[a]( ) control values.

Table 5. Impact Strength Tests on Selected Polymers (Notched Izod — samples exposed to 5 x 10^8 rads gamma at 50°C)

Sample	Impact strength lb/in.[a]	Maximum load lb/in.	Energy absorbed ft-lb/in.[a]	Fracture mode
Polymethylpentene	0.05 (0.4	65 (240)	0.05 (0.5)	Elastic
Polyphenylene sulfide				
(radiation exposure)	1.3 (1.1)	820 (790)	1.4 (1.8)	Elastic
(autoclave exposure)	1.3 (1.1)	740 (790)	1.5 (1.8)	Elastic
Poly (ethylene-tetrafluoroethylene)	0.9 (no break)	250 (280)	0.8 (1.3)	Elastic
Polyfurfuryl alcohol	0.13 (0.15)	130 (150)	0.12 (0.14)	Elastic

[a]( ) control values.

Based on these tests some polymeric materials have promise as components for barrier systems in nuclear waste containment. However, since these studies were carried out under isolated conditions of temperature, environment, and radiation, they do not indicate any synergistic effects which might occur as the materials were tested simultaneously for all conditions. Equipment for simultaneous evaluation of all of these conditions is being developed for use in the near future.

REFERENCES

H. T. Fullam, 1980, Use of Ceramic Materials in Waste-Package Systems for Geologic Disposal of Nuclear Wastes, PNL-3447, Pacific Northwest Laboratory, Richland, Washington.

R. E. Westerman et al., 1979, Preliminary Conceptual Designs for Advanced Packages for the Geological Disposal of Spent Fuel, PNL-2990, Pacific Northwest Laboratory, Richland, Washington.

TECHNICAL BARRIERS AGAINST A CONTACT OF GROUNDWATER WITH DISPOSED RADIOACTIVE WASTE

Ernst-Peter Uerpmann

Gesellschaft für Strahlen- und Umweltforschung
Theodor-Heuss-Str. 4
D-3300 Braunschweig
Federal Republic of Germany

The Federal Republic of Germany considers salt formations to be the most suitable geologic media for the final disposal of radioactive wastes. Safety analyses of deep salt disposal showed that groundwater transport is the most important long-term (>1000 y) transport mechanism by which the nuclides could be spread.

In spite of the past successes of hydrologic controls in salt mining, critics of disposal in salt dwell on the solubility of salt and the potential danger of water intrusion into the repository. They cite examples of those potash mines that have flooded and argue against the final disposal in salt formations. Because of this, I have investigated the previous cases and found that there have been seven causes of flooding in salt mines:[1]

1. weakened shaft lining,
2. insufficient physical separation from other flooded mines,
3. thin salt layers in the upper part of the dome,
4. mistakes in mining metamorphic kainite salt,
5. inflows from underlying parts of the dome,
6. inflows from salt/caprock interface or dissolution fronts in bedded salt, and
7. inflows from the "main-anhydrite" depositional sequence.

Proper site selection avoids the first six of these causes. Only the "main-anhydrite"* need be considered as a potential brine-bearing layer. But even here, if a cautious investigative program is

*The "main anhydrite" is a formation specific to the third of four depositional cycles of the late permian Zechstein salt. It is relevant only to internal dome structures of north Germany.

performed, the discovered cracks could be backfilled with suitable materials and this potential danger minimized. Despite these facts, the Reactor Safety Commission (RSK) recommended that water inrush into the final repository still be considered in the safety assessment. They assumed that a water inrush through the shaft by an unidentified accident (for example, sabotage) could happen and that the final repository should be protected against this occurrence.

In examining the protecting barriers, the first consideration is the chemical state and the packaging of the waste. There are a number of different fixation methods and materials for the different radioactive waste types, several of which will be described below.

One of the advantages of some of the organic and inorganic binding agents used to fix low and intermediate level radioactive waste is that the materials mix readily with the waste. For example, gypsum or cementitious grouts are used to solidify radioactive waste such as slurries, filtering materials, concentrates, or ion exchange beads. These binders show a rather high radiation resistance and are good shielding materials. They can be handled in a simple manner without heating.

With time, these inorganic materials are corroded by saline water, especially if the water contains alkaline or alkaline earth chlorides and sulfates, as will be the case in rock salt repositories. This particular disadvantage can be counteracted by using a coating material. The process should be easily adapted as a simple addition to an existing solidification process. A 200-L drum is first provided with a bottom layer of saltwater-resistant resin and then fitted with a prefabricated insert of foamed polystyrene. The drum is then filled to within about 5 cm from the top with the waste-binder slurry. After the waste has solidified, a measured amount of polyester resin liquid monomer plus hardener is poured over the waste by remote handling. The resin liquid dissolves the polystyrene insert and after polymerizing, replaces it by a coating of polyester resin.[2] Thus, another 1- to 2-cm-thick chemical-resistive barrier is provided which acts to prohibit leaching and may even hinder the release of radioactive gases from the barrels (Fig. 1). Polyester resins show a sufficient radiation resistance up to 5×10^8 to 2×10^9 rad.[3]

High-level waste originates from the first extraction cycle of a reprocessing plant and contains approximately 99% of the radioactivity that requires permanent disposal. This waste is heated by its radioactive decay. Therefore, a heat-resistant matrix such as glass or ceramic may be more appropriate as a fixation material.

Another release barrier is suggested by the different types of underground waste disposal facilities. Depending on the surface

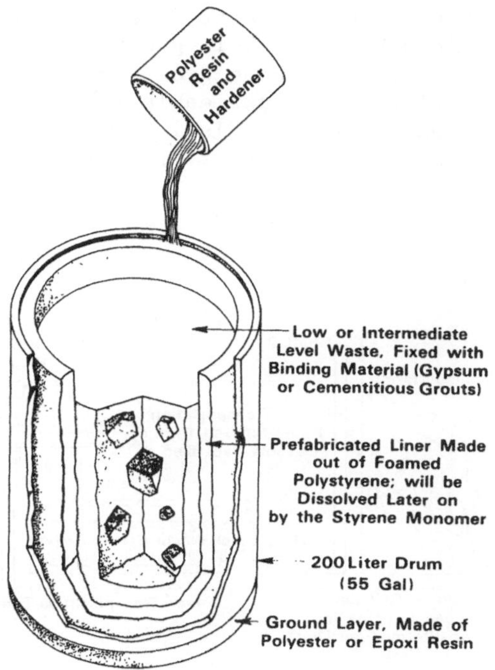

Fig. 1. Coating low- or intermediate-level radioactive waste with polymers.

dose-rate of the waste package, it may be stored in either chambers, caverns, galleries, or boreholes. The entrances leading to those disposal areas have to be backfilled after the operations are completed. The backfilling can take the form of walls or dams, forming bulkheads which isolate sections of the underground repository. The backfill supports the overburden, absorbs migrating radionuclides, and blocks diffusion-convection transport by reducing the free cross-section (to zero when intact) of the underground tunnels.

Swelling agents such as silica gel, methyl cellulose, or polysaccharides may be added to the backfill materials to increase the viscosity of brines passing through the material, thus decreasing the convection currents. If there are enough such gels present the convection currents within the brines could be avoided. In this case, the transport phenomena for radionuclides would depend on slow diffusion processes.

Another barrier can be provided by the design of the repository. For example, in the concept shown in Fig. 2, the storage tunnel level is elevated above the main transportation level. In case of water inrush through the shaft, a gas bubble forms which will be at hydrostatic pressure. This gas bubble could be enlarged by pumping compressed air from above ground into these chambers. Since rock salt formations are impermeable to gases or liquids, the bubble of compressed gas within levels 1-3, the upper transportation level and storage rooms will be a barrier to the transport of radioactive ions.

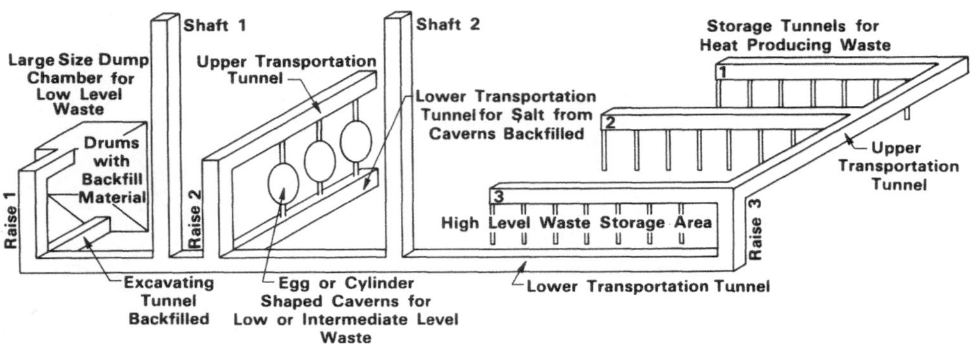

Fig. 2. Underground Repository in salt formation with a special design against flooding.

In conclusion, there may be no way of assuring with absolute certainty that water cannot contact the waste in a geological time frame. One must analyze scenarios which consider the possibility of the repository filling with water during or after storage even though there is only a very slight probability of such an event occurring.

If a water break-in must be assumed, it would probably take place through a shaft, penetrating covering layers with groundwater. The shaft is generally considered the most vulnerable portion of a final repository, since here the water has its shortest path through the geological formations. For that reason, the greatest attention should be focused on a shaft design that is permanently and completely watertight.

In some current concepts, connections with the level of the final repository are made horizontally and directly with air shafts via a drift at the same elevation. If water should break in, such water would flow to the lowest point in the shaft and flow from there into the horizontally branching bottom drifts of the repository. This direct horizontal connection at the bottom of the repository would favor propagation of the radioactivity due to convection.

Figure 2 shows a type of mine layout which is applicable to all different disposal concepts for radioactive waste, whether they utilize tunnels, dumping chambers, caverns, or emplacement boreholes. However, this may complicate the operational phase of the repository since the haulage ways become longer and the ventilation system may be more complex. Nevertheless, it provides an effective passive countermeasure against the flooding of the disposal tunnels. Also, it allows time for active postaccident countermeasures such as backfilling with slurries and grouts.

If several barriers are placed in series, the escape of disposed nuclides requires a long time and the nuclides decay to harmless levels. Therefore, radioactive waste disposal within specially designed disposal sites in deep geologic formations is a sufficiently safe way to isolate radioactive waste.

REFERENCES

1. E.-P. Uerpmann, "Hydrogeologische Fragen bei der Endlagerung Abfalle," GSF-Bericht T 106, 1980.

2. E.-P. Uerpmann, Method for Encasing Waste Barrels in a Leakproof Closed Sheath, U.S. Patent 4,222,889, Sept. 16, 1980.

3. H. Schönbacher, U. Van de Voorde, G. Kruska, and K. M. Oesterle, "Verhalten von Farbanstrichen im Strahlenfeld von Kernreaktoren und Hochenergie-Beschleunigern und bei Kontamination durch Radionuklide Kerntechnik," Bd. 19, S. 209-217, 1977.

COMPOSITE BACKFILL MATERIALS FOR RADIOACTIVE WASTE ISOLATION BY DEEP BURIAL IN SALT*

E. J. Nowak

Sandia National Laboratories**

Albuquerque, New Mexico 87185

ABSTRACT

Bentonite and hectorite were found to sorb Pu(IV) and Am(III) from concentrated brines with distribution coefficients K_d >3000 ml/g. The permeability of bentonite to brine was less than 1 microdarcy at a confining pressure of 18 MPa, the expected lithostatic pressure at the 800-m level in a salt repository. Getters for sorption of TcO_4^- (charcoal, K_d ∿300 ml/g), I^- (charcoal, $K_d \geqq$ 30 ml/g), Cs (mordenite, $K_d \geqq$ 30 ml/g), and Sr (sodium titanate, $K_d \geqq$ ∿100 ml/g) from brines were identified. Their sorption properties are presented. Thermal conductivity results (>0.5 W/mK) and evidence for bentonite stability in brines at hydrothermal conditions are also given. It is shown by calculated estimates that a 3-ft-thick mixture of bentonite with other getter materials could retain Pu, Am, and TcO_4^- for >10^4 years and I^- for >10^3 years. Another tailored mixture could retain Cs for ∿600 years, Sr for ∿700 years, TcO_4^- for ∿4000 years, and I^- for ∿400 years. The backfill can offer a significant contribution to the isolation capability of a waste package system.

INTRODUCTION

The probable effectiveness of backfill material with low permeability to groundwater or brine, acting as a barrier to radionuclide migration, has been shown by calculation of radionuclide migration rates.[1-3] Further, the backfill can be designed to form a hydrologic barrier to the inflow of groundwater or brine, to

---

* This work was supported by the U. S. Department of Energy under contract DE-AC04-76-DP00789.

** A U. S. Department of Energy facility.

chemically modify or buffer brines for minimum corrosivity, to form a mechanical stress buffer between the waste containers and the repository rock, and to provide adequate heat transfer from high level waste.[1,4-6]

Sorption of radionuclides from concentrated brines that may intrude into a waste repository constructed in salt is likely to be a complex process. Large decreases in the amount sorbed with increasing competing ion concentration have been demonstrated for sorption of trivalent actinides, typical fission products, and rare earths.[7,8] Large and complex effects of changing pH, Eh, and other composition variables on sorption and sorption mechanisms have also been reported.[2,9-16]

This complexity and specificity of sorption requires that site- and material-specific studies be done to develop backfill materials for testing in the Waste Isolation Pilot Plant (WIPP), a proposed deep geologic radioactive waste repository in bedded salt. Previous studies[1,2] had shown that the use of swelling clay materials bentonite and hectorite, alone and in combination with sand or crushed salt, result in a backfill mixture with low permeability, adequate thermal conductivity, support strength for a waste container, other favorable chemical and physical properties, and the potential for adequate sorption of transuranics. Consequently, these and other potential getter materials were chosen for more detailed studies. The results and calculated estimates of effectiveness for mixtures of getters as backfills are presented.

MATERIALS

Brines were synthesized to represent water in contact with either the potash-containing horizons in the vicinity of the proposed WIPP site, brine A, or water in contact with the halite of the proposed WIPP horizons, brine B. Brine A contains 1.8 $\underline{M}$ NaCl, 0.8 $\underline{M}$ KCl, and 1.4 $\underline{M}$ MgCl as major constituents, along with 0.01 $\underline{M}$ bicarbonate and 0.02 $\underline{M}$ borate as potential pH buffers. 5 $\underline{M}$ NaCl is the single major constituent of brine B. Both brines contain 0.02 to 0.03 $\underline{M}$ Ca^{++} and 10^{-5} to 10^{-4} $\underline{M}$ Cs^{+} and Sr^{++}. Detailed composition data for these brines are given elsewhere.[16]

Samples of commercially available smectite swelling clay materials bentonite and hectorite, sand, and other potential getter materials in this study are also described in detail elsewhere.[16,19]

SORPTION MEASUREMENTS

Sorption measurements were carried out in duplicate by the batchwise contacting of a tracer-containing brine with granular or powdered solid sorbent material. Interphase contact and mixing were enhanced by continuous agitation. Initial values of pH = 6.5

for brine A and pH = 7.5 for brine B were attained by additions of aqueous HCl or NaOH. Those values were measured for a wide range (0.3:1 to 20:1) of brine:solids ratios using bentonite, hectorite and clay-sand mixtures.[16] pH was not controlled during agitation. After a period of agitation, the final batch sample pH was measured and an aliquot was centrifuged at 12,000 G for 10 minutes. Supernatant brine was drawn off and filtered through a 0.8 μm pore size Micropore filter before analysis by radiation counting. Blank samples containing brine but no solids were included in each experiment. Detailed procedures have been documented elsewhere.[16]

Results are presented as duplicate sample averages of the solid to liquid phase concentration ratios, K_d, calculated from the differences between sample and blank sample liquid phase concentrations.[16] Repeatability was within a factor of 2 for duplicates, and values of K_d calculated from the known quantity of added radiotracer agreed within a factor of 2 with values calculated from the liquid phase concentrations for blank samples. Details of these calculations are documented elsewhere.[16]

Results for the sorption of Pu(IV) and Am(III) from brines A and B on bentonite and bentonite-sand mixtures are summarized in Table 1. The average K_d value in brine A was 2900 ml/g for bentonite and 520 ml/g for the mixture of 10 wt % bentonite with sand.

Table 1. Batch Sorption of ^{238}Pu(IV) and ^{243}Am(III) on Bentonite and Bentonite-Sand Mixtures

Initial tracer concentrations, C_o = ~3x10^{-8} $\underline{M}$ Pu(IV)
~5x10^{-7} $\underline{M}$ Am(III)
Final tracer concentrations, C = ~10^{-11} to ~10^{-9} $\underline{M}$
Batch contact time = 3 weeks
Bentonite & Sand = 10 wt % bentonite and 90 wt % sand

Species	Solid Sorbent	Brine	K_d, ml/g	pH†
Pu(IV)	bentonite alone	A	2900	6.7
"	bentonite & sand	A	520	6.7
"	bentonite in sand*	A	5200*	6.7
"	bentonite alone	B	30000	7.1
"	bentonite & sand	B	3600	7.1
"	bentonite in sand*	B	36000	7.1
Am(III)	bentonite alone	B	9000**	7.3
"	bentonite & sand	B	9500	7.3

* Value of K_d calculated from data for bentonite & sand but based on mass of bentonite only.

** Values for duplicate samples were 4100 and 14,000 ml/g.

† Final pH reading.

If the sand were an inert diluent, the K_d value for the bentonite alone in the clay-sand mixture would be 10-fold larger, or 5200 ml/g. That value is within a factor of 2 of the measured K_d for bentonite (2900 ml/g). Precision within a factor of 2 was expected, as discussed previously. Therefore, sand was considered to be an inert diluent for Pu(IV) sorption. In brine B, the average K_d was equal to 30,000 ml/g for bentonite at pH = 7.1, and again sand acted as an inert diluent. The larger K_d value is attributable to the higher pH value [11-13,16] which may affect solution chemistry and include the possibility of precipitation. For Am(III) in brine B, K_d was equal to ~9000 ml/g for both 100% bentonite and 10 wt % bentonite in sand. This large apparent sorption effect by sand is difficult to explain with ion exchange alone and suggests a more complex mechanism. Invoking chemical similarities among the actinides, the K_d value for Am(III) in brine A was estimated to be no more than an order of magnitude smaller than for brine B, as was the case for Pu(IV). Then, K_d was estimated to be ≥900 ml/g for sorption of Am(III) by bentonite from brine A. These empirical concentration ratios (K_d) were used to calculate approximate breakthrough times for the estimation of backfill effectiveness.

Because neither bentonite nor hectorite alone or in combination with sand adequately sorb TcO_4^- or I^- from brines A and B,[16] other getters for these species were sought. Tetrahedrite, chalcopyrite, biotite, pyrite, and galena were all found to be ineffective, with $K_d \leq 1$ ml/g for the removal of TcO_4^- from these brines. Bournonite had K_d = ~10 ml/g. Following a suggestion by R. G. Dosch and A. W. Lynch[10,17] (personal communications) and results of Akatsu et al.,[18] several charcoals were tested. The results for the best sorbers are summarized in Table 2. Activated charcoal #1 had K_d values of ~400 ml/g in brine A and ~300 ml/g in brine B for TcO_4^-, and 35 ml/g for I^- in brine A. Further screening and mechanistic studies are needed.

Bentonite and hectorite are also inadequate sorbers for the cationic fission products Cs^+ and Sr^{++} in brines A and B. A summary of some screening and mechanistic results[19] for the most potent getters is given in Table 3. Sodium titanate,[20] a synthetic inorganic ion exchanger was found to have the largest K_d values for Sr^{++} in both brines. A synthetic mordenite (zeolon) had the largest K_d for Cs^+ in brine A and nearly the largest K_d for the same species in brine B. The larger K_d values for sorption from 1% brine A in deionized water illustrate a large competing ion effect, since separate experiments with modified brine A showed invarient K_d values of ~27 ml/g for both 10^{-5}M and 10^{-8}M Cs^+ on mordenite and ~130 ml/g for both 6×10^{-5}M and 6×10^{-11}M Sr^{++} on sodium titanate. Other experiments verified that a constant batch composition was reached during a contact time of 1 week. These results support the validity of a linear equilibrium sorption model for Cs^+ and Sr^{++}.

Table 2. Batch Sorption of $^{99}TcO_4^-$ and $^{129}I^-$ on Charcoals

Initial tracer concentrations, C_o = ~5×10^{-5} $\underline{M}$ TcO_4^-
~3×10^{-4} $\underline{M}$ I^-
Final tracer concentrations, C = 2×10^{-5} to 2×10^{-4} $\underline{M}$
Batch contact time = 1 week

Species	Solid Sorbent	Brine	K_d,ml/g	pH
TcO_4^-	Charcoal #1	A	380	6.3 - 6.5
"	"	B	300	6.6
I^-	Charcoal #1	A	35	6.6
"	Charcoal #2	A	53	6.7
"	Charcoal #3	A	18	6.5

Charcoal #1 = activated "Nuchar." See reference 16.
Charcoal #2 = activated cocoanut charcoal.
Charcoal #3 = activated "Darco" G-60.
Characterization of these charcoals is under way.

Table 3. Batch Sorption of $^{137}Cs^+$ and $^{85}Sr^{++}$ Getters

Initial and final concentrations were in the range of ~10^{-5} to ~10^{-11} $\underline{M}$. Separate experiments showed K_d invariant within that range.

Species	Solid Sorbent	Brine	K_d,ml/g	pH
Sr^{++}	sodium titanate	A	125	6.7
"	"	1%A*	100,000	7.2
"	"	B	540	7.2 - 7.5
Cs^+	synthetic mordenite	A	27	6.7
"	"	1%A*	12,000	7.3
"	"	B	74	6.6
"	illite	B	115	5.6

sodium titanate = synthetic inorganic exchanger.[20]
synthetic mordenite = Norton Zeolon 900.[16]
illite = illitic clay material from Silver Hills, Montana.[19]
* brine A diluted with deionized water.

PHYSICAL PROPERTIES AND STABILITY

Hydraulic permeability and thermal conductivity are important properties which influence the effectiveness of a backfill material. At a confining pressure of 18 MPa (lithostatic pressure at the 800 m level of the proposed WIPP) the permeability of a 10 wt % bentonite-in-sand mixture to brine B was 0.4 microdarcy. Bentonite had a permeability of 0.1 microdarcy to water at the same confining pressure. These data will be fully documented elsewhere.[21] Permeabilities in the microdarcy range are low enough to prevent significant convective transport of radionuclides relative to transport by molecular diffusion.[1] Thermal conductivities at zero confining pressure were found to be 0.57 W/mK for 30 wt % bentonite in sand, 0.59 W/mK for 30 wt % hectorite in sand, and 0.72 W/mK for 10 wt % bentonite in crushed WIPP salt. These values compare favorably with the value of 0.49 W/mK measured for crushed salt.

Autoclave experiments to measure hydrothermal reactions of backfill materials with brines have been initiated to help predict the long term stability of backfills. After bentonite was contacted with brine B at 250°C for 11 days, its montmorillonite was better crystallized, and initially present chlorites and illites were undetectable by x-ray diffraction analysis. These initial results favor the long term maintenance of bentonite's essential properties in brine B at elevated temperatures. More extensive and detailed experiments are needed.

EFFECTIVENESS

A tailored mixture of bentonite, charcoal, synthetic mordenite, and sodium titanate[20] could be an effective backfill material. Estimates of backfill effectiveness in terms of radionuclide retention times (times to breakthrough at 1% of entering concentration) can be calculated[1] for retarded diffusion through bentonite- and getter-containing backfills having the low permeability and K_d values presented above. For example, as shown in Table 4, a three-feet-thick backfill composed of 30 wt % bentonite and 70 wt % charcoal #1 could retain Pu, Am, and TcO_4^- for more than 10^4 years and I^- for more than 10^3 years in brine A. Designed for retention of Cs and Sr as well, a backfill composed of 30 wt % bentonite, 40 wt % synthetic mordenite, 10 wt % sodium titanate, and 20 wt % charcoal #1 could retain Cs for ~600 years, Sr for ~700 years, Pu and Am for more than 10^4 years as above, TcO_4^- for ~4000 years, and I^- for ~400 years. These retention times can be significant contributions to the radionuclide isolation capability of a waste package system. Details of our experimental program to further the development of backfill materials are given elsewhere.[22]

Table 4. Calculated Retention Times for a Three-Feet-Thick Backfill Layer Assuming Retardation by Linear Sorption from Brine A.

Bulk Density = 2 g/cm^3 Effective Porosity = 0.1
Migration is by molecular diffusion and breakthrough is defined at 1% of entering concentration.

Backfill Composition	Retention Times, Years			
30% bentonite, 70% charcoal #1	Pu:	5×10^4	TcO_4^-:	1×10^4
	Am:	2×10^4	I^-:	1×10^3
30% bentonite, 40% mordenite, 20% charcoal #1, 10% sodium titanate	Pu:	5×10^4		
	Am:	2×10^4	TcO_4^-:	4×10^3
	Cs:	6×10^2	I^-:	4×10^2
	Sr:	7×10^2		

Method of calculation has been documented in reference 1.

REFERENCES

1. E. J. Nowak, "The Backfill Barrier as a Component in a Multiple Barrier Nuclear Waste Isolation System," SAND79-1109 (1980), NTIC.*
2. E. J. Nowak, "The backfill as an engineered barrier for nuclear waste management," in Scientific Basis for Nuclear Waste Management, Vol.2, C. J. Northrup, ed., Plenum Press, New York, (1980).
3. I. Neretnieks, "Transport of Oxidants and Radionuclides Through a Clay Barrier," KBS Report 79, Stockholm, (1978).
4. A. Jacobson and R. Pusch, "Deposition of High-Level Radioactive Waste Products in Bore-holes with Buffer Substance," KBS Report 03, Stockholm (1977).
5. KBS, "The KBS Annual Report 1979," KBS Report 79-28, Stockholm (1980).
6. W. E. Coons, E. L. Moore, M. J. Smith, and J. D. Kaser, "The Functions of an Engineered Barrier System for a Nuclear Waste Repository in Basalt," Rockwell Hanford Operations Informal Report RHO-BWI-LD-23 (1980).
7. G. W. Beall, B. H. Ketelle, R. G. Haire, and G. D. O'Kelley, "Sorption behavior of trivalent actinides and rare earths on clay minerals," in Radioactive Waste in Geologic Storage, Sherman Fried, ed., ACS Symposium Series 100, American Chemical Society, Washington, D.C. (1979).
8. S.-Y. Shiao, P. Rafferty, R. E. Meyer, and W. J. Rogers, "Ion exchange equilibria between montmorillonite and solutions of moderate-to-high ionic strength," ibid., (1979).

* National Technical Information Center (NTIC), U.S.Dept. of Commerce, 5285 Port Royal Rd., Springfield, VA, 22161,U.S.A.

9. R. G. Dosch and A. W. Lynch, "Interactions of Radionuclides with Geomedia Associated with the Waste Isolation Pilot Plant (WIPP) Site in New Mexico," SAND78-0297 (1978), NTIC*
10. R. G. Dosch and A. W. Lynch, "Interaction of Radionuclides with Argillite from the Eleana Formation on the Nevada Test Site," SAND78-0893 (1979), NTIC.*
11. B. Allard, G. W. Beall, T. Krajewski, and J. R. Peterson, "The sorption of americium on major rock-forming minerals," Trans. Amer. Nuclear Soc., 32:167 (1979).
12. G. W. Beall and B. Allard, "Chemical factors controlling actinide sorption in the environment," ibid., 32:164 (1979).
13. T. Krajewski, G. W. Beall, B. Allard, and J. R. Peterson, "Mineral-contributed anion effects on the retention of trivalent actinides in the environment," ibid., 32:168 (1979).
14. D. Rai, R. J. Serne, and J. L. Swanson, "Solution species of $^{239}Pu(V)$ in the environment," in Proceedings of the Task 4 Waste Isolation Safety Assessment Program Second Contractor Information Meeting, Vol. 1, Battelle Pacific Northwest Laboratories Report PNL-SA-7352, (1978).
15. G. W. Beall, B. Allard, J. Krajewski, and G. D. Kelly, "Chemical reactions in the bedrock-groundwater system of importance for the sorption of actinides," in: Scientific Basis for Nuclear Waste Management, Vol. 2, C. J. Northrup, ed., Plenum Press, New York (1980).
16. E. J. Nowak, "Radionuclide Sorption and Migration Studies of Getters for Backfill Barriers," SAND79-1110 (1980), NTIC.*
17. A. W. Lynch and R. G. Dosch, "Interaction of Radionuclides with Geomedia from the Nevada Test Site," in: Scientific Basis for Nuclear Waste Management, Vol. 2, ed. C. J. Northrup, Plenum Press, New York (1980).
18. E. Akatsu, R. Ono, K. Tsukuechi, and H. Uchiyama, "Radiochemical study of adsorption behavior of inorganic ions on zirconium phosphate, silica gel, and charcoal," J. Nuclear Science and Technology, 2:141 (1965).
19. C. D. Winslow, "The Sorption of Cesium and Strontium from Concentrated Brines by Backfill Barrier Materials," SAND80-2046, in preparation, NTIC.*
20. A. W. Lynch, R. G. Dosch, B. T. Kenna, J. K. Johnstone, and E. J. Nowak, "The Sandia Solidification Process: A Broad Range Aqueous Waste Solidification Method," in:Management of Radioactive Wastes from the Nuclear Fuel Cycle, Vol. 1, International Atomic Energy Agency, Vienna (1976).
21. E. Peterson and C. Cooley, "Initial Permeability Measurements on Backfill Materials for the Waste Isolation Pilot Plant - Progress Report," SAND80-7095, in preparation, NTIC.*
22. M. A. Molecke and E. J. Nowak,"Applicability of a Backfill-Getter for Radioactive Waste Isolation: Chemical and Physical Functions," Sandia National Laboratories, Preprint SAND80-0838C (1980).

* National Technical Information Center (NTIC), U.S. Dept. of Commerce, 5285 Port Royal Rd., Springfield, VA, 22161, U.S.A.

HIGHLY COMPACTED BENTONITE - A SELF-HEALING SUBSTANCE FOR NUCLEAR WASTE ISOLATION

Roland Pusch

Div. Soil Mechanics
University of Luleå
S- 951 87 Luleå

Anders Bergström

Nuclear Fuel Safety Project
S-102 48 Stockholm

INTRODUCTION

The ability of bentonites to serve as effective barriers in the migration of water and solutes through waste repositories in rock has been conclusively demonstrated by experimental evidence[1,2]. It is required, in this context, that there is a perfect contact between an artificially applied clay layer barrier and the confining rock, and that the layer does not contain inhomogeneities which can serve as passages. This implies a swelling and self-healing potential of the clay, i.e. an ability to generate internal readjustment of the interparticle distance and water distributions. The matter is discussed in this article.

MICROSTRUCTURAL PROPERTIES

Bentonite is the name of natural smectite-rich clays which have been formed by transformation of volcanic ash in the sea or in estuaries or lakes. The crystal structure of the dominant clay mineral montmorillonite may be either one of the two types shown in Fig. 1. Most outcrops form very stiff clay layers or shales, some of which are being mined in Wyoming and South Dakota in the U.S., and at various sites all over the world. The Wyoming clays have been processed and used in various industries for many years and they are also the source of the commercially available bentonite powder MX-80 (American Colloid Co), which is the KBS* reference material and which will be discussed here. The cation adsorption sites of the montmorillonite crystallites, which have a high ion exchange capacity

*Nuclear Fuel Safety Project, organized by the Swedish nuclear utilities.

(~80 mE/100 g of solid substance), are mainly occupied by sodium but calcium and magnesium adsorbants are probably present as well. The processing of the Wyoming bentonite involves air drying of excavated material, and some additional drying through heating, as well as a grinding procedure. This yields a granular character with aggregate sizes mainly ranging from a few microns to approximately a millimeter.

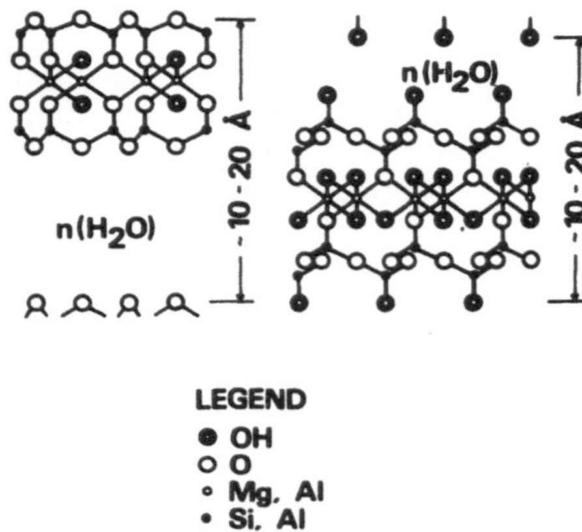

Fig. 1. Two possible models of the montmorillonite crystal lattice. Left picture according to Hofman, Endel & Wilm, right picture according to Edelmen & Favajee.

The aggregate size distribution and low water content of the powder make it possible to compact it to a high density. Blocks of highly compacted Na bentonite form the clay barrier constituents of the Swedish KBS concept for isolating highly radioactive waste products from the biosphere (cf. Fig. 2).

The microstructure of granular bentonite powders, compacted at their natural water contents, is illustrated by Fig. 3. It shows a dominant anisotropy of the aggregates, which are characterized by a very small interparticle and interlamellar spacing in the original, "air-dry" condition.

At moderate pressures the overall particle orientation is largely random, while at very high pressures (>50 MPa) the aggregates are forced together which yields considerable contact areas and aggregate deformation, as well as a tendency of the particles to be oriented perpendicularly to the applied pressure.

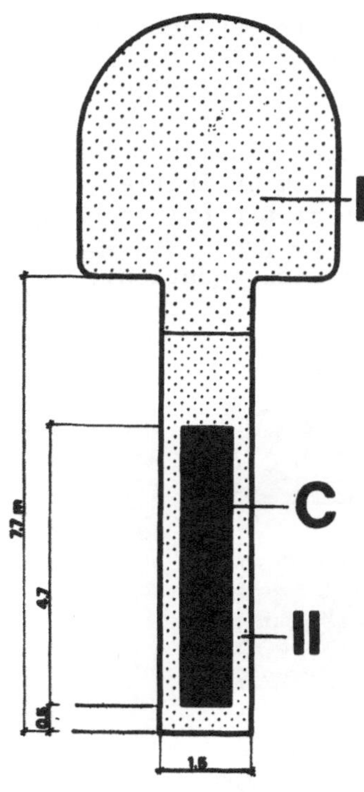

Fig. 2. Schematic section through tunnel and deposition hole with canister (C), highly compacted bentonite (II), and tunnel backfill (I).

Fig. 3. Schematic microstructure of compacted granular "air-dry" bentonite.

WATER UPTAKE AND SWELLING

Blocks of highly compact air-dry bentonite to be used in clay barriers in deposition holes will have an initial bulk density of 2-2.1 t/m^3, a water content (ratio of the mass of water to that of the solid mass) of about 10%, and an approximate degree of water saturation of 60% ($S_r = 0.6$). After the application of such blocks (cf. Fig. 2), water from the surrounding rock will successively be taken up by the swelling clay, which eventually will be water-saturated.

A large amount of energy is released on adsorption of the first 1-3 water molecule layers on dry clay particle surfaces, which is partly due to the hydration of exchangeable cations and partly to the mineral surface hydration. This "affinity" to water yields a negative pore pressure which is of the same order of magnitude as the swelling pressure but with the opposite sign. It can actually be measured if the block is unloaded and thus free to absorb water and swell. The strong dependence of the "suction" potential on the thickness of the adsorbed water film, and thus on the interparticle or interlamellar distance, produces potential gradients which generate water migration and a tendency to create homogeneous conditions within the clay elements.

Figure 4 illustrates, very schematically, the successive hydration and occupation of water molecule sites in the moistening and expansion processes. Keeping in mind the "steric" character of near-surface interstitial water, the migration and formation of water layers might well be termed "water crystallization". For the special case of one-dimensional water uptake from an initial, partly saturated state to complete saturation of a confined bentonite element, the rate of water migration can be predicted with reasonable accuracy by applying a simple diffusion equation.

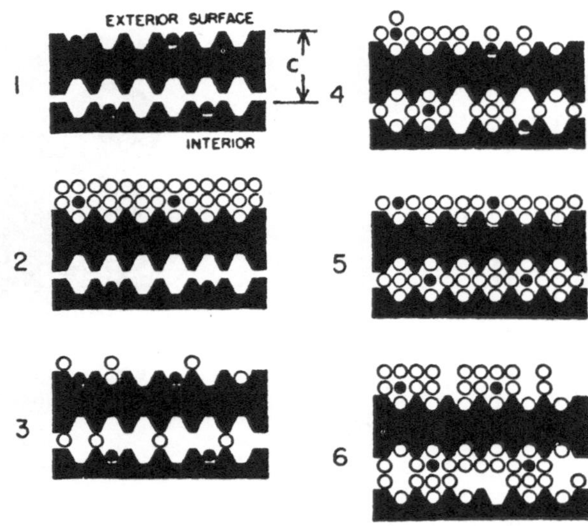

Fig. 4. Schematic representation of the hydration and interlayer expansion processes of montmorillonite. Black geometrical pattern = crystal lattice. (after Barshad). Black o = exchangeable cation. White o = water molecules. 1, Anhydrous and contracted stage; 2, the initially hydrated stage; 3, the initially expanded stage; 4 and 5, advanced stages of hydration after the initial expansion; 6, beginning of the second stage of expansion.

The equation is of the form:

$$\frac{\partial w}{\partial t} = D \frac{\partial}{\partial x} \left(\frac{\partial w}{\partial x}\right) \tag{1}$$

where w = water content
x = direction of migration
t = time
D = diffusion coefficient ($\sim 10^{-9}$ m^2/s) (2)

The water uptake is associated with microstructural changes. The major, initial process is an expansion of the aggregates by which the larger interaggregate voids will be filled by a clay gel. Eventually, a condition of considerable isotropy and homogeneity with respect to particle orientation and interparticle distance will be reached, at least at moderate and high densities. The average particle spacing, without referring specifically to inter- or intraparticle (interlamellar) distances is schematically illustrated by Fig. 5.

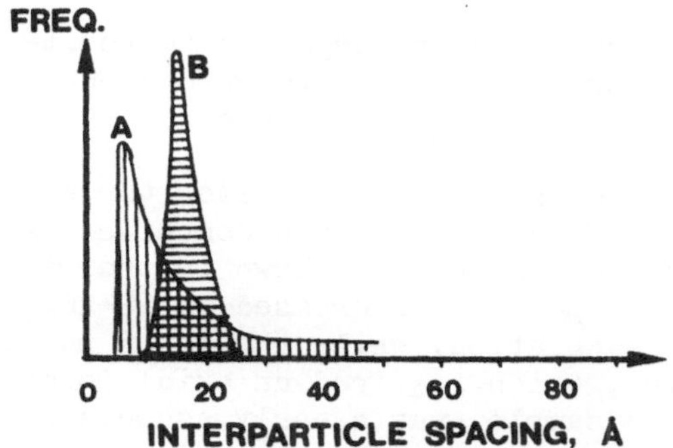

Fig. 5. Schematic distributions of the interparticle spacing. A) original spectrum for air-dry bentonite. B) narrow peak for homogeneous, swollen sample with high density.

EXPERIMENTAL

The ability of Na bentonite to swell and become homogeneous has been investigated in tests where precompacted samples were allowed to take up water and swell to fill a certain space. The MX-80 clay material used in the experiments is characterized by a minus 2 µm content of approximately 85%, and a montmorillonite content of about 80-90% of this fraction. The water used for saturation and swelling was an artificial "brackish" ground water.

A pilot test was made on a precompacted MX-80 sample with a void ratio of approximately 0.9, an initial water content of about 10%, and a theoretical bulk density of about 1.9 t/m^3 after saturation. The diameter was 43 mm and the total height of the three-piece column 65 mm, while the inner diameter and height of the steel container were 49 mm and 70 mm, respectively (Fig. 6). Water

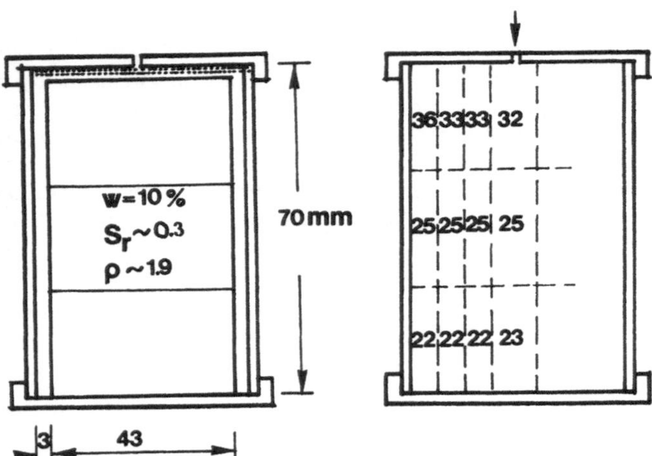

Fig. 6. Swelling test with precompacted bentonite. The right picture shows the average water contents of symmetrically situated elements after 3 weeks.

was taken up from one end of the container through a paper filter. Three weeks after the test start, the container was opened and the water distribution determined. As shown in the figure the water content of the upper part had increased to 32-36%, corresponding to approximately 50% saturation, while the lower parts were less saturated. The swelling, which required an axial force of approximately 2 kN to extrude the sample, was clearly associated with a strong ability of the clay to become homogeneous as manifested for instance by the fact that the joints between the original blocks disappeared completely. Although additional time was certainly required to yield a completely homogeneous state, it would finally have been obtained.

The same conclusion was drawn from a test series performed by means of swelling pressure odometer in which the bentonite samples were laterally confined. Bentonite powder was compacted by applying a pressure of 50 MPa in 3 tests which yielded 20 mm high samples with a water content of about 10%, and a bulk density of 2-2.1 t/m^3. The samples were then allowed to swell during the water uptake to a pre-determined volume and bulk density. At that stage, the samples were also axially confined and the swelling pressure exerted in the vertical direction was measured at regular intervals. When the pressure had become practically constant, the samples were extruded and the water content determined. The results are given in Fig. 7, which also shows the vertical extension of the space which the individual samples had to fill up through swelling. Water entered the space from below as well as from above, and the time to reach the

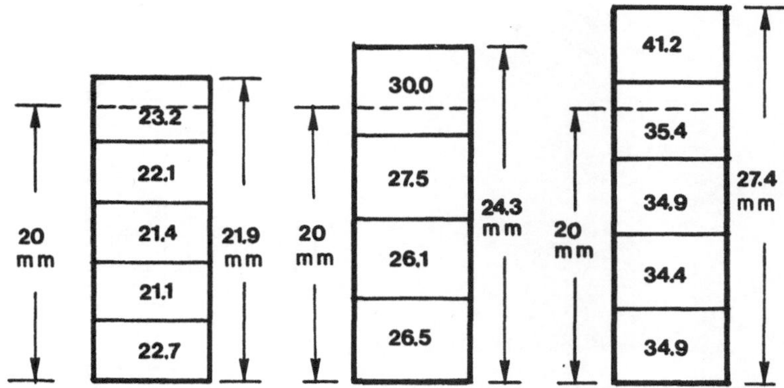

Fig. 7. Water content distributions in three swelling tests. Time of test for the shorter samples was 3 weeks, while the longest sample required 7 weeks.

condition of constant swelling pressure, including the preceeding swelling, was found to be 3-7 weeks. The high degree of homogeneity after the relatively short periods for self-adjusting processes, is remarkable.

The main beneficial property of homogeneous Na bentonite for sealing purposes is the extremely low permeability. Current laboratory research shows that the coefficient of permeability ranges between 10^{-12} and $2 \cdot 10^{-14}$ m/s when the bulk density is in the interval 1.7-2.1 t/m^3. This means that the water flow through repositories which contain such barriers is negligible when only regional hydraulic gradients prevail. The swelling pressure ranges between 0.3 and 20 MPa in the same density interval. The considerable pressure even at fairly low densities illustrates the ability to create a perfect contact with confining rock or structures. It also offers the required self-healing potential, i.e. the ability to redistribute water and particles to yield homogeneous conditions.

PRACTICAL APPLICATIONS

Apart from the very important main case shown in Fig. 2, where a practically homogeneous clay barrier is required to isolate the canister from the rock, there are at least two more applications of bentonite sealing in repositories.[3] Boreholes in rock can be sealed by filling them with highly compacted blocks of bentonite in perforated metal pipes in the holes. The bentonite takes up water, swells through the perforations, and eventually forms a

Fig. 8. Extruded perforated pipe with swelled bentonite. Part of the gel removed to show the perforation.

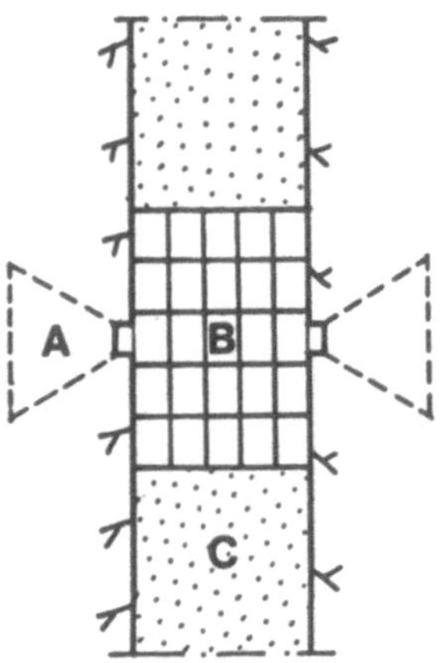

Fig. 9. Bentonite block sealing (B). A, grouted rock. C, ordinary backfill.

homogeneous mass which fills the hole completely and in which the pipes will be embedded (Fig. 8). Also, shafts and tunnels can be sealed by applying zones of highly compacted blocks of bentonite. These "packers" are separated by ordinary, less dense backfill (Fig. 9).

CONCLUSIONS

Granular Na bentonite acquires considerable microstructural homogeneity through water and particle redistribution processes. This yields a very low permeability and thus excellent barrier functions in repositories. This self-healing property also means that large bentonite volumes tend to reach a homogeneous condition. Thus, local voids or inhomogeneities produced, for instance, by minor displacement of the surrounding rock will be healed. The swelling potential also means that a perfect contact is established between the clay barrier and the rock.

REFERENCES

1. I. Neretnieks, "Transport of oxidants and radionucleids through a clay barrier", KBS Technical Report No. 79 (1978).
2. R. Pusch, "Highly compacted sodium bentonite for isolating rock-deposited radioactive waste products". Nuclear Technology Vol 45 Sep. (1979).
3. R. Pusch & A. Bergström, "Highly compacted bentonite for borehole and shaft plugging". Workshop on borehole and shaft plugging, OECD/NEA, Columbus Ohio 7-9 May 1980 (In press).

THE BELL CANYON TEST AND RESULTS

Charles L. Christensen and Thomas O. Hunter

Experimental Programs Division 4512
Sandia National Laboratories
Albuquerque, NM 87185

INTRODUCTION

The Borehole Plugging Program (BHP) at Sandia National Laboratories is sponsored by the Office of Nuclear Waste Isolation (ONWI) under contract to the U. S. Department of Energy (DOE). Its purposes are to identify issues associated with sealing boreholes and shafts, to establish a data base from which to assess the importance of these issues, and to develop sealing criteria, materials, and demonstrative tests for the Waste Isolation Pilot Plant (WIPP).[1,2]

The Bell Canyon Test (BCT)[3] (Figure 1), in which a 2-m-long grout plug was emplaced 1370 m below the surface in a 20-cm-diameter borehole and exposed to a 12.4 MPa (1800 psi) pressure differential, was conducted to evaluate in situ the state of the art in borehole plugs and to identify and resolve problems encountered in evaluating a "typical" plug installation in anhydrite.

GROUT DEVELOPMENT

The most advanced candidate material currently available for plugging boreholes is high-quality cement grout and is the focus for the long-range development and testing program at the U.S. Army Corps of Engineers, Waterways Experiment Station (WES), Vicksburg, MS.[2]

Desirable properties of plugging grouts are: low permeability, low porosity, high bond strength, expansivity,

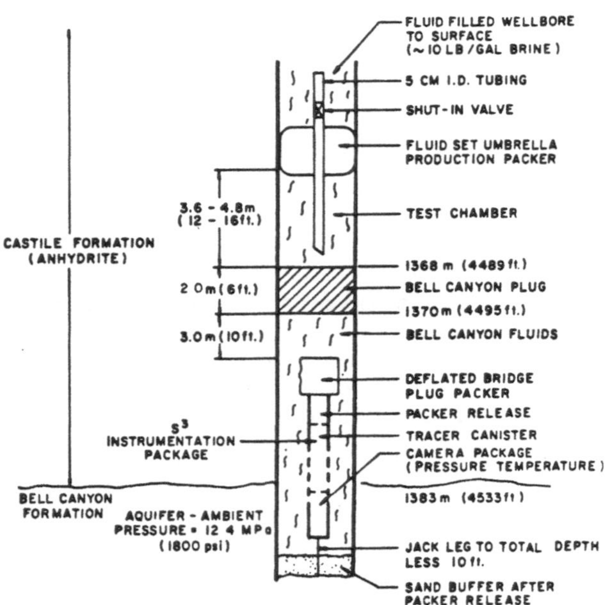

Fig. 1. As-built BCT configuration during monitoring operation (not to scale)

homogeneity, easy pumpability, and stability. Lower water-to-cement (w/c) ratios, are correlated with lower permeability and porosity and higher density and strength. Reduction of the w/c ratio has been achieved by controlling the coarseness of the cement grind, inclusion of super-plasticizers and turbulence-inducing compounds, use of retarders, and slurry temperature control.

Previous laboratory experience at WES suggested a grout mixture of Class H cement and a proprietary additive provided by Dowell Division of the Dow Chemical Company. WES formulated 24 candidate mixtures and tested these for time of efflux, workability, density, and strength under accelerated curing. Dowell further developed this mixture (Table 1), designated BCT-1F, by including a powder dispersant.

After the borehole was cored to intercept the Bell Canyon aquifer, the plug location was selected in the basal Castile anhydrite. Anhydrite core was tested for compatibility with BCT-1F by WES, PSU, Oak Ridge National Laboratory (ORNL), and Dowell.[4,5,6,7,8,9]

Permeability of various grout/anhydrite samples ranged from 10^{-3} to 10^{-6} darcies.[4,6,9] BCT-1F grout samples consistently had permeabilities less than 10^{-6} darcies. Grout-filled anhydrite cores leaked at the interface and some samples, particularly those cured at ambient temperatures of about 22°C (72°F), had white powder at the interface. The push-out bond strength was in excess of 2.5 MPa (360 psi).[4,8]

An alternate fresh-water grout, designated BCT-1FF, was also formulated for further testing (Table 1). The w/c ratio was increased to an acceptable viscosity and pumpability.

The fresh-water grout (BCT-1FF) had a higher strength and greater expansion than the salt grout (BCT-1F), and push-out bond strengths were equal to or greater than those of the salt-water grout. Fresh-water grout/rock sample permeability ranged from 1 to 15 microdarcies, the grout adhered more tightly to the rock, and there was no observable leakage at the interface.[4,6] The BCT-1FF mixture was selected for the BCT plug.

Results of studies to date have been published and investigations are continuing at Sandia, WES, and PSU.[4,5,6,7,8,9]

Table 1
Grout Ingredients and Properties

Ingredients (Wt%)	BCT-1F	BCT-1FF
Class H Cement	50.1	52.2
Expansive Additive	6.7	7.0
Fly Ash	16.9	17.6
Salt (NaCl)	6.5	-
Dispersant	0.2	0.2
Defoamer	0.02	0.02
Water	19.5	23.0
Properties		
Water-to-cement ratio	0.26:1.0	0.30:1.0
Fluid Density, g/cm^3(lb/gal)	2.04(17.0)	1.98(17.0)

INSTRUMENTATION

Plug performance was evaluated by monitoring the upward migration of tracer gas and brine through the plugged region.

A packer below the plug was deflated at a preset time after the grout had been placed by dump bailer and cured for 12 days (Figure 1) allowing the aquifer brine to reach the plug. Tracer gas (SF_6) was released below the plug by the instrument package and allowed for assessment of effective permeability through the plugged region.[10,11]

TEST RESULTS

The field tests are summarized from the work of Peterson and Christensen.[11] Three types of tests were performed to evaluate the isolation characteristics of the 2.0-m-long plug. The various testing configurations employed are illustrated in Figure 2. Plug performance was inferred from fluid flow into the test chamber immediately above the plug. This influx could come from either below the plug (Bell Canyon) or the wellbore fluids above the packer. Conclusions about plug performance rely on the ability to distinguish between these sources.

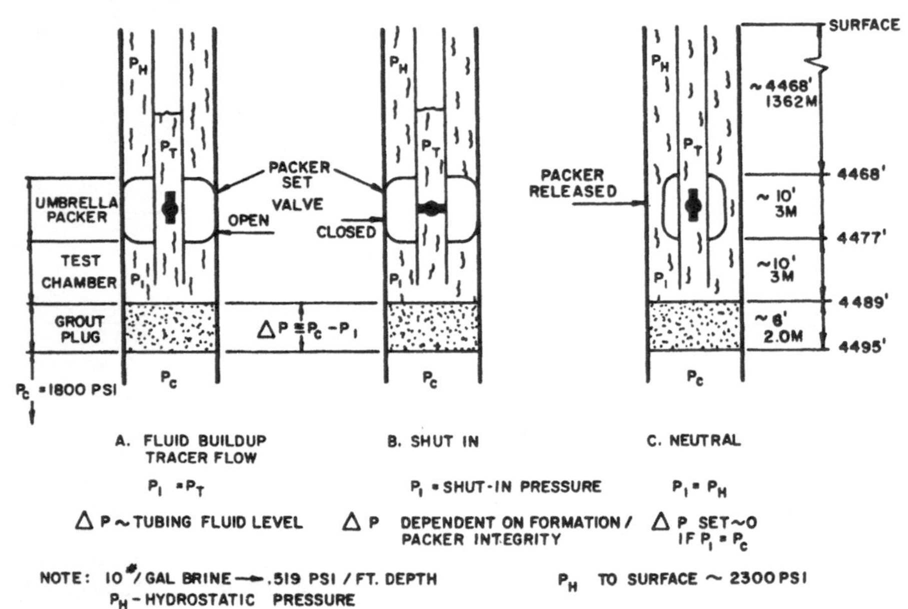

Fig. 2 BCT during monitor operations

Fluid Build-up Tests

Flow into the test chamber was derived from changes in fluid level above the plug determined by measured pressure changes. Data from four of these tests are shown in Figure 3 and summarized in Table 2. In all tests a pressure differential of 12.4 MPa (1800 psi) was maintained across the plug.

The October 9-10 data definitely indicate the lowest flow rates and are consistent with shut-in test and tracer transit data of the plug/formation system performance. The results from the last two tests are likely due to leakage into the test chamber. Fluid build-up tests are most susceptible to leakage in the packer/tubing assembly. This effect is aggravated by normal wear in the tubing joints and is considered to be the cause of the apparent increased leakage rates observed with later tests. These data do not indicate plug degradation since the shut-in tests and tracer transit tests remained consistent throughout the testing period exceeding three months.

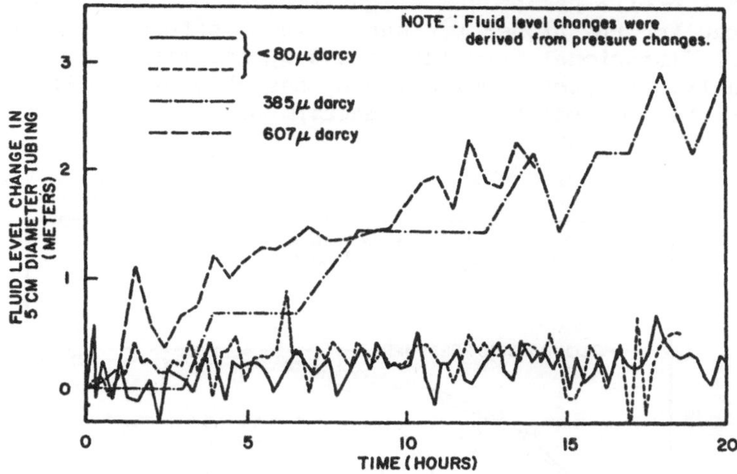

Fig. 3. Fluid build-up test data obtained during plug performance evaluation studies

Table 2
Summary of Fluid Build-up Test Results

Test Initiation Date	Test Duration (Hours)	Measured Test Region Inflow Rate (liters/day)	Permeability-Area Product, kA (10^{-10} cm^4)
10/09/79	18	0.32	0.9 (27 darcies)*
10/10/79	18	0.67	1.8 (57 darcies)
10/19/79	144	4.6	12.0 (385 darcies)
12/12/79	16	7.1	20.0 (607 darcies)

* parentheses indicate the permeability for a flow path with "cross-section" area equal to the wellbore.

Shut-in Tests

Figure 4 shows data from three shut-in tests. These data can be represented well by a one-dimensional porous-flow model using those values of the permeability-area product (kA) shown. The shut-in tests were consistent, and an effective plug/formation system permeability of approximately 50 microdarcies can be determined from these kA values (approximately 2×10^{-10} cm^4) and assuming a cross-section area equal to that of the wellbore.

During the December 19 shut-in test (Figure 4), the time lapse between shut-in and the initial pressure increase is believed to be caused by an air bubble trapped in the tubing below the umbrella packer. A slight flow from the annulus region can account for the higher pressure. (Both effects were modeled with a one-dimensional formulation and qualitatively explained.) Subsequently, the packer-valve-tubing assembly was modified to eliminate the possibility of air entrapment.

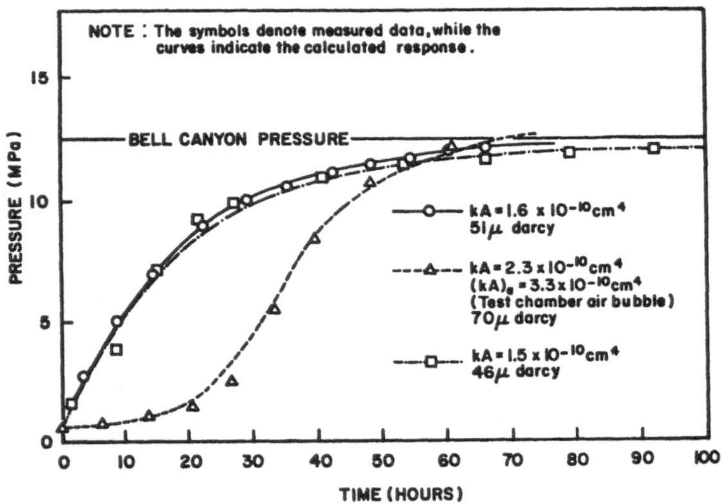

Fig. 4. Shut-in test data obtained during plug performance evaluation studies

There is no evidence of flow from the high pressure annulus above the packer into the test region during the November 2 and January 18 shut-in tests. This strongly suggests that, at least in the area where the umbrella packer is positioned, wellbore damage is slight. Assuming a damage area cross-section equal to that of the wellbore, these shut-in tests indicate a formation permeability of less than 5 microdarcies.[11]

Tracer Flow Tests

Tracer arrival times measured during the test series indicated approximately a 36-hr interval between gas release and detection at the top of the plug. This time interval has been modeled both as porous flow through the plug and crack flow at the interface between the plug and host rock.

Under the crack flow assumption, a crack width of 2×10^{-5} cm is required to match the arrival times. However, approximately 4000 of these cracks with length equal to the plug circumference (63 cm) running the full height of the plug (2m) would be required to match the volumetric flow observed during the fluid build-up and shut-in tests. Laboratory testing on grout/rock samples do not show this type of fracture density.

Alternatively, using a porous flow assumption, the tracer arrival time data corresponds to a permeability-to-porosity ratio (k/ϕ) of 3.3×10^{-11} cm^2. For reasonable plug porosities (0.02) This corresponds to a permeability of 66 microdarcies. While this is consistent with shut-in test data, the plug and formation permeabilities determined in the laboratory are both less than this. Thus, we infer that the flow occurs at a porous microstructure at the grout/rock interface. The width of this zone cannot be specified since the flow area is not uniquely determined from the permeability area product (kA) obtained in the Darcy flow model.

CONCLUSIONS

The program to develop grouts provides materials that are effective for sealing boreholes. Brine and fresh-water grouts, with acceptable physical properties in the fluid and hardened states, have been developed.

The field data, taken together with laboratory data, suggest that the predominant flow into the test region occurs through the cement plug/borehole interface region, with lesser contributions occurring through the wellbore damage zone, the plug core, and the surrounding undisturbed anhydrite bed.

The 2.0-m-long by 20-cm-diameter grout plug, installed in anhydrite at a depth of 1370 m in the AEC-7 borehole, limits flow from the high pressure (12.4 MPa) Bell Canyon aquifer to 0.6 liters/day. If the flow path crosssection area is assumed equal to that of the wellbore, the corresponding permeability would be

50 microdarcies and the porosity would be 0.015. However, field tests and laboratory tests together suggest that most of the flow is through a porous microstructure interface zone. Shut-in test data indicate the formation permeability surrounding the umbrella packer positioned approximately 4 m above the plug is at least one order of magnitude less than that through the plug/formation system.

If more precision in measuring plug performance is required, current oil-field testing tools will not be adequate. New techniques and equipment will be required to measure plug-formation systems with effective permeabilities around 1 microdarcy.

REFERENCES

1. Christensen, C. L. and Hunter, T. O., "Waste Isolation Pilot Plant (WIPP), Borehole Plugging Program Description," January 1979. SAND79-0640, Sandia National Laboratories (SNL), Albuquerque, NM, August 1979.
2. Gulick, C. W., Jr., "Borehole Plugging--Materials Development Program," SAND78-0715, SNL, Albuquerque, NM, June 1978.
3. Christensen, C. L., "Test Plan, Bell Canyon Test, WIPP Experimental Program, Borehole Plugging," SAND79-0739, SNL, Albuquerque, NM June 1979.
4. Gulick, C. W. Jr. et al., "Bell Canyon Test Cement Development Report," SAND80-0358C, SNL, Albuquerque, NM, May 1980.
5. Grutzek, M. W. et al., "Modified Cement-Based Borehole Plugging Materials: Properties and Potential Longevity," Materials Research Laboratory, the PSU, University Park, PA.
6. C. W. Gulick, "Bell Canyon Test (BCT) Cement Grout Development Report," (DRAFT). SAND80-1928, SNL, Albuquerque, NM.
7. Boa, J. A., Jr., "Borehole Plugging Program (Waste Disposal), Report 1, Initial Investigations and Preliminary Data," U.S. Army WES, Misc. Paper 6-78-1, SAND77-7005 January 1978.
8. Gulick, C. W., et al., "Borehole Plugging Materials Development Program, Report No. 2," SAND79-1514, SNL, Albuquerque, NM 87185, February 1980.
9. Moore, J. G., M. T. Morgan, E. W. McDaniel, H. B. Grune, and G. A. West, "Cement Technology for Plugging Boreholes in Radioactive Waste Repository Sites: Progress Report for the Period October 1, 1978, to September 30, 1979," ORNL 5610, Oak Ridge National Laboratory, Oak Ridge, TN, 37830.
10. Cook, C. W., C. B. Kinabrew, P. Lagus, and R. Broce, "Bell Canyon Test (BCT) Instrumentation Development," SAND80-0408C, SNL, Albuquerque, NM 87185, May 1980.
11. Peterson, E. W. and C. L. Christensen, "Analysis of Bell Canyon Test Results," SAND80-7044C, Albuquerque, NM, May 1980.

[M]ODELLING OF ROCK MASSES FOR RADIOACTIVE WASTE REPOSITORIES IN HARD ROCK

Ove Stephansson
Per Jonasson
Dept. of Rock Mechanics
University of Luleå
Luleå, Sweden

Tommy Groth
Dept. of Soil and Rock Mechanics
Royal Institute of Technology
Stockholm, Sweden

INTRODUCTION

A repository for radioactive waste must be so located and designed as to be protected against (i) natural changes of first order (faulting, glaciation, meteorite impact), (ii) human activity (war, sabotage, mining), and (iii) degradation of engineering and natural barriers (corrosion of canisters, penetration of bentonite buffer, groundwater transport). In order to protect the repository from first order faulting, we suggest that it be located in a large block of a jointed rock mass surrounded by faults or shear zones. Any large scale movements will then appear along pre-existing faults and the repository in the center of the large block will be protected.

Stress concentration leading to displacements, groundwater flow and retention of certain waste substances prompt the following requirements if the jointed rock mass is to be suitable for a repository. They are:

- a low frequency of joints favors a low rate of flow of groundwater,
- a large number of joints and joint surfaces favor the restraint of certain waste substances if they should seep into groundwater, and
- a moderately jointed rock mass is preferable in terms of rock mechanics as it prevents the accumulation of mechanical energy.

This clearly demonstrates that the discovery of a suitable site for a repository is a first order optimization problem where various aspects have to be considered. Mathematical modelling thus becomes an important tool when analyzing the effects of these aspects.

FINITE ELEMENT MODELLING OF STORAGE HOLE

A program system for analysis of the stability of two-dimensional structures in jointed rock has been used. When failure occurs in any of the elements in the model the stability of the structures is disturbed. If this is the case, an equilibrium-iteration process starts which will show whether equilibrium can be restored or not. This is performed through redistribution of the excessive stresses to the surrounding joints.

The program system is composed of the reprocessor BEMESH, the main processor BEFEM, and the post processor POSTPROG, Stephansson et al. (1980).

Model geometry

A correct analysis of a storage tunnel and storage hole must be executed by three-dimensional modelling. In this paper we present a two-dimensional model of a horizontal cross-section of the storage hole. As a result of geometrical symmetry only one quarter of the storage hole need be analyzed with the finite element technique. Figure 1 depicts a part of the model analyzed, which consists of 1802 nodes and 1687 elements of which 438 are joint elements.

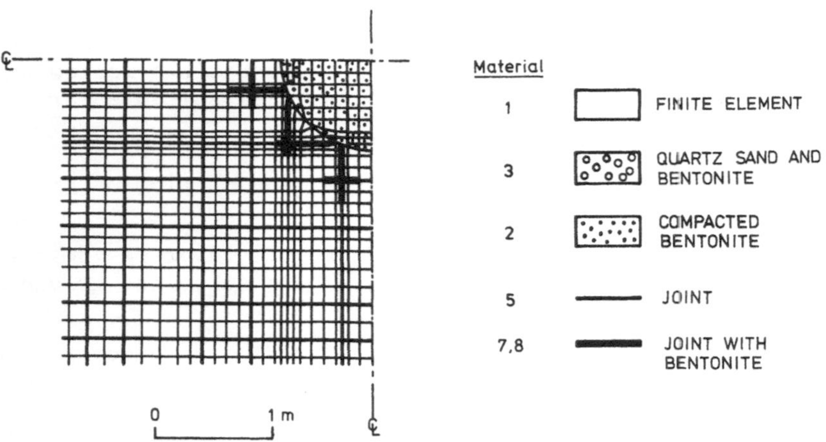

Fig. 1. Element mesh of horizontal section of storage hole.

Material properties

The rock mass in the vicinity of the openings is a discontinuum consisting of blocks and joints. Towards the boundary of the models the rock mass is continuous, homogeneous isotropic and linearly elastic. The parameters are chosen to correspond with a typical granitic terrain, and in particular the rock properties at the Stripa Mine, Sweden. The elastic Young's modulus of homogeneous, nonfractured rock samples of Stripa Granite is thereby found to be about 60 GPa, whereas the stress-deformation characteristic of the jointed rock mass with an average joint spacing of 0.5 m approaches 40 GPa. Young's modulii of compacted bentonite in the storage hole is 0.3 GPa.

The modelling of joints in the finite-element discretization is in accordance with the joint model by Barton. Their model gives a relationship between the peak value of the shear strength as a function of normal stress. The relationship between normal stress and displacement of a joint is hyperbolic. The joint properties for the finite-element discretization are given in Stephansson et al. (1980).

Sequences of loading

Modelling of a storage tunnel and storage hole for spent fuel was performed for six sequences of loading (Fig. 2). Prior to excavation of tunnel and storage hole the rock mass is in equilibrium under the virgin stresses. In the model of the horizontal section of the storage hole the ratios for horizontal stresses, σ_x'/σ_y', 10/3.2, 7/10, 10/10 and 20/20 MPa where applied, where σ_x' is the virgin stress along the shortest extent of the model. The value 20 MPa in the two principal directions is in accordance with the state of horizontal stress at 500 m depth in the Scandinavian countries. The vertical component of the state of stress, σ_v, is a result of the weight of the overburden and the horizontal components range from 0.5 σ_v to 3 σ_v.

<u>The second sequence</u> of modelling is the excavation of tunnel and storage hole. Excavation causes displacement of the periphery of the openings, and the new state of stress formed in the model is stored and used as initial stresses for each of the following sequence

<u>Sequence 3</u> takes into consideration the loading due to swelling of the compacted bentonite in the storage hole. The loading will exert pressure perpendicular to the walls with a value of 10 MPa.

<u>In the 4th sequence</u> the effect of thermal loading is analyzed. The maximum thermal stresses in the center of a single-level repository, which are σ_z = 2 MPa and $\sigma_r = \sigma_\theta$ = 12 MPa are added to the

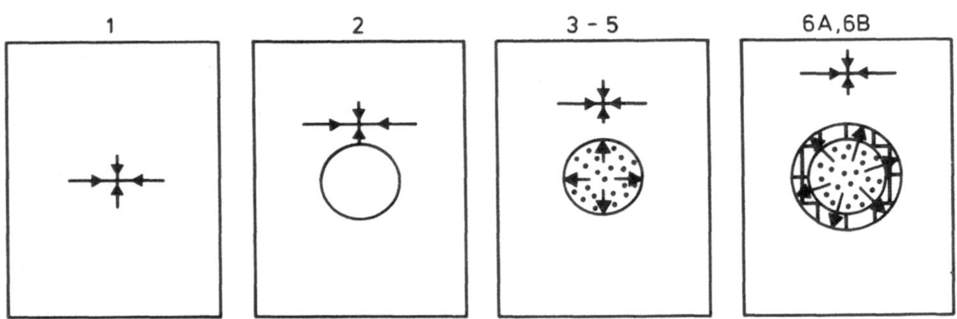

Fig. 2. Sequences of loading in a jointed rock mass for final storage of spent fuel. (1) application of virgin stress to the rock mass; (2) excavation of tunnel and storage hole; (3) loading due to swelling of compacted bentonite in the hole and swelling of quartz-sand and bentonite mixture in the tunnel; (4) thermal loading from the heat generation of the canisters; (5) combination of thermal loading and swelling; (6A) thermal loading and swelling in compacted bentonite, quartz-sand mixture and injected joints in the vicinity of the storage hole. Thermal stresses in joints (6B) the same as sequence 6A but without thermal stresses in joints.

stresses of the rock mass after excavation. This method of superposition of stresses is strictly valid in the realm of linear elasticity.

Sequence 5 of the analysis involves superposition of pressure loading due to swelling bentonite and thermal loading due to heat generation of the spent fuel. The two loadings are superimposed on the stresses of the rock mass after excavation.

The last sequence of modelling, 6A and 6B, are storage with thermal loading, swelling pressure in the bentonite and 0.5 m penetration of bentonite into the joints of the rock mass around the deposition hole. Penetration of bentonite causes a swelling pressure of 3 MPa in the outer 0.2 m of the joint and 1 MPa for the interval 0.2-0.5 m. The pressure is exerted perpendicular to the surfaces of the joints. Sequence 6A gives the situation when thermal stresses are superimposed on solid blocks and joints whereas thermal loadings are omitted for the joints in the analysis for sequence 6B. Hence, sequence 6A simulates a situation with filled joints capable of transmitting thermal loadings, whereas sequence 6B is more likely to simulate joints without filling material, except for the bentonite which penetrates the outermost parts of the joints in the periphery of the storage hole.

RESULTS

In our model of a jointed rock mass for a deposition hole in hard rock we chose an orthogonal joint pattern with an average interval of 40 cm between the joints. Each block of solid rock consists of 24 solid elements surrounded by the same number of joint elements, cf. Fig. 3, and the joint pattern was chosen to reveal wedge shaped block at the periphery. The virgin stresses applied to the model are of immediate concern. We chose virgin stresses which represent expected and extreme values for hard rock in Scandinavia.

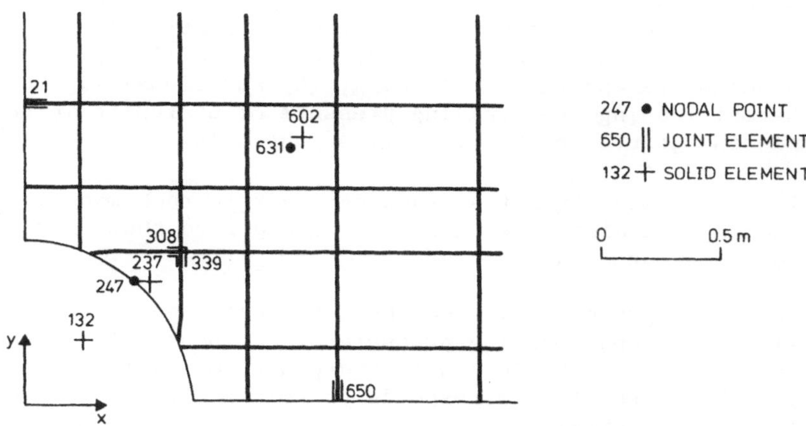

Fig. 3. Joint pattern in the vicinity of the deposition hole and location of nodal points, joint elements, and solid elements mentioned in Table 1.

Typical results of displacement of the structure for the various sequences of loading are shown in Fig. 4A-C, for the virgin stresses $\sigma_{xx} = \sigma_{yy} = 20$ MPa. Excavation of the hole will cause displacement of the blocks, e.g. the wedge-shaped block at the periphery, Fig. 4A. High contact forces at the corners of some of the blocks will lead to indentation and overlapping of the displacement for the contingent blocks. Rotation of blocks is another common feature. Since the structure is loaded by thermal stresses and swelling pressure, corresponding to sequence 5 of the loading scheme, the joints are closed by the excess of radial and tangential compressive stresses, Fig. 4B. The swelling pressure will cause a positive displacement of the nodal points as indicated in Table 1 for sequence 3, while the thermal loading will cause a negative displacement as indicated for sequence 4. Note that displacements are given as cumulative from excavation of the hole in sequence 2.

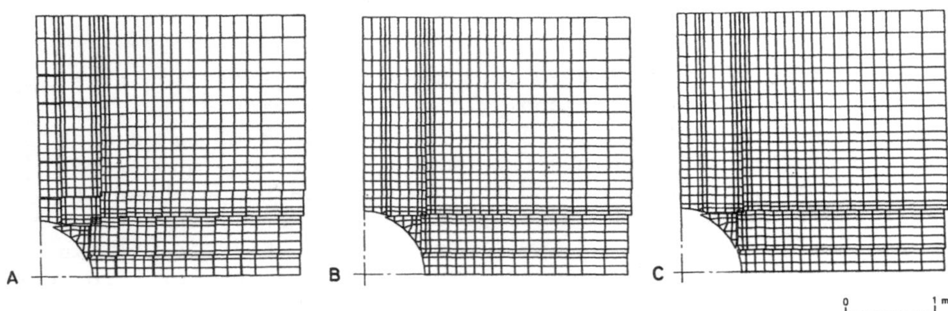

Fig. 4. Deformation of a deposition hole in a jointed rock mass, subjected to virgin stresses $\sigma_{xx} = \sigma_{yy} = 20$ MPa. One quarter of the deposition hole is shown. (A) after excavation of the hole, sequence 2; (B) additional displacement due to swelling pressure of compacted bentonite and thermal loading; (C) additional displacement due to thermal loading and swelling pressure of bentonite in deposition hole and joints in the vicinity of the hole.

The final displacement from the stage of a virgin rock mass is found by adding the displacement from sequence 2 to those for the sequence of interest.

The computer graph in Fig. 4C shows the deformed structure due to thermal loading and swelling pressure of the bentonite in the deposition hole and in the joints to a depth of 0.5 m from the periphery. This situation corresponds to sequence 6A of the loading scheme, where thermal stresses are assumed to cause thermal expansion in the joint fillings.

The normal displacements across the joints are of particular interest in this study as they will affect the groundwater flow in the vicinity of a deposition hole. Displacements of four joint elements in the structure of Fig. 4 are presented in Table 1 for various loading sequences and virgin stresses. A positive value indicates an opening and a negative closure of the joints. Excavation of the hole for a state of stress where $\sigma_{xx} = \sigma_{yy} = 20$ MPa causes an opening of joint 339 and 308 of 0.26 and 0.34 mm respectively, cf. Table 1. Thermal loading and swelling of bentonite in the outermost 0.5 m of the joints (sequence 6) will open them. Note how a loading sequence without thermal stresses in the joints (sequence 6B) will lead to a closure of the joints surrounding the wedgeshaped block. Normal stresses, σ_x and σ_y, of solid elements in the rock mass and in the highly compact bentonite are illustrated in Table 1. The position of the elements in the structure is depicted in Fig. 3. The virgin stresses, σ_{xx}/σ_{yy}, are doubled from 10/10 to 20/20 MPa. The normal stresses in the solid elements after excavation (sequence 2) are also nearly doubled. The detachment of the wedge at the periphery causes a release of the stresses, cf. element

Table 1 Result of FEM modelling of deposition hole

Virgin stress σ_{xx}/σ_{yy} [MPa]	Seq	Displacement of nodal points x/y [mm]		Normal displacement in joint elements [mm]		Normal stresses in solid elements σ_x/σ_y [MPa]		
		247	631	339 308	650 21	132	237	602
10/3.2	2	-1.69 -0.66	-0.05 -0.01	0.189 0.336	0.188 0.097	* *	- 3.1 - 2.5	-10.8 - 2.9
7/10	2	-1.22 -1.11	-0.07 -0.00	0.248 0.244	0.147 0.188	* *	- 3.8 - 3.8	- 7.5 - 9.6
10/10	2	-1.51 -1.09	-0.07 -0.01	0.259 0.271	0.164 0.177	* *	- 4.6 - 4.3	-10.9 - 9.5
20/20	2	-2.05 -1.39	-0.13 -0.02	0.262 0.337	0.158 0.176	* *	- 9.6 - 7.6	-22.5 -18.9
20/20	3	1.14 0.69	0.05 0.01	-0.159 -0.211	-0.050 -0.062	- 9.6 - 9.7	-21.8 -17.5	-21.3 -19.5
20/20	4	-2.04 -1.26	-0.04 -0.01	0.151 0.515	0.072 0.105	- 0.4 - 0.2	- 2.8 - 5.0	-35.4 -30.0
10/3.2	5	-0.50 -0.13	+0.00 -0.00	0.003 0.001	0.004 0.004	-10.1 -10.1	-14.9 -13.0	-22.6 -14.8
7/10	5	0.44 -0.16	0.06 0.01	0.002 0.000	0.006 0.005	- 9.8 -10.0	- 9.9 -12.1	- 8.6 -22.3
10/10	5	-0.52 -0.17	-0.00 -0.00	0.000 0.001	0.005 0.005	-10.1 -10.1	-15.9 -14.3	-22.8 -21.4
20/20	5	-0.83 -0.26	+0.00 +0.00	0.001 0.000	0.005 0.006	-10.2 -10.1	-19.8 -16.9	-34.2 -30.7
10/3.2	6A	-0.67 -0.27	+0.00 -0.00	0.115 0.128	0.005 0.005	-10.2 -10.1	-15.0 -14.8	-22.6 -14.9
7/10	6A	0.41 -0.30	0.12 0.01	0.093 0.139	0.006 0.004	- 9.9 -10.0	-13.5 -14.4	- 8.6 -22.4
10/10	6A	-0.68 -0.32	0.00 -0.00	0.098 0.138	0.005 0.005	-10.2 -10.1	-15.8 -15.6	-22.8 -21.4
20/20	6A	-1.03 -0.42	+0.00 +0.00	0.116 0.146	0.007 0.007	-10.3 -10.1	-17.6 -17.1	-34.2 -30.7
20/20	6B	0.96 0.85	-0.02 -0.01	-0.842 -0.672	-0.052 -0.063	- 9.8 - 9.7	-44.5 -43.4	-34.0 -29.7

237. With the onset of the loading from swelling bentonite, sequence 3, the normal stresses in the vicinity of the opening alter for element 237, whereas the stresses one diameter out from the opening, element 602, are almost unaffected. The effect of only thermal loading, sequence 4, for the same virgin stress 20/20 MPa is an increase of normal compressive stresses for most of the solid elements. The further load imposed by the penetration of bentonite into the joints, sequence 6A, gives a minor increase in the state of normal stress.

CONCLUSIONS

1. The models studied of a deposition hole are overall structurally stable for the applied virgin stresses.
2. The maximum compressive and tensile stresses in the solid block of the rock mass never exceed uniaxial compressive and tensile strength of granite.
3. Excavation of the deposition hole will cause displacement of blocks inward to the opening. The maximum displacement is found to be about 2 mm and the average aperture of joints is one-tenth of a millimeter.
4. Loading the rock mass with thermal stresses from the waste, swelling pressure from bentonite in the deposition hole and the joints at its periphery will cause an additional maximum displacement of about 1 mm, and aperture of joints of the order of tenths to hundredths of a millimeter.
5. Superposition of thermal loading and swelling pressure of bentonite causes an increase of compressive stresses in the structure. Further increase of stress due to injection of bentonite into joints is minor.
6. A uniform increase of virgin stresses causes a corresponding increase in the state of stress in the solid blocks of the rock mass.
7. Modelling of joints results in nonlinear displacements as a function of applied stresses.
8. After excavation of the hole normal and shear stresses are low for the joints in its vicinity. Modelling of thermal loading and swelling of bentonite cause the normal compressive stress to approach the value of the applied swelling pressure. The shear stresses transmitted by the joints are found to be very small.
9. States of stress in the model is almost unaltered as thermal expansion of joints are excluded from the modelling.

REFERENCES

O. Stephansson, P. Jonasson, and T. Groth, 1980, Modelling of rock mass deformation for radioactive waste repositories in hard rock. KBS, Technical Report 80-02. KBS Project, Box 5864, S-102 48 Stockholm, Sweden.

A COMBINED FRACTURE/POROUS MEDIA MODEL FOR CONTAMINANT TRANSPORT OF RADIOACTIVE IONS

H. E. Nuttall

Department of Chemical and Nuclear Engineering
The University of New Mexico
Albuquerque, New Mexico 87131

A. K. Ray

Department of Chemical Engineering
University of Kentucky
Lexington, Kentucky 40506

ABSTRACT

A model, with an analytical solution, was developed to study the effects of combined fracture and porous media flow on the migration rate of radioactive nuclides. The objective is to determine the degree of enhanced migration due to a thin fracture within the otherwise homogeneous porous medium. The geological waste disposal zone is treated as an infinite two-dimensional medium consisting of a narrow channel surrounded by a flowing aquifier. The top and bottom of the zone are assumed to be bounded by impermeable clays. The two-dimensional convective-dispersion equations, with a line source have been solved giving analytical expressions for the contaminant concentration, flux and total discharge rate downstream of the source. The total discharge rate of nuclides downstream of the pulse source have been calculated and compared with the results for ion migration in a homogeneous porous medium.

NOMENCLATURE

Roman Letters

$A = \dfrac{mu_p^2}{hD_p^2C_{ref}}$ dimensionless parameter in Eq. (8)

C_c'	concentration in the fractured channel, kg/m^3
C_c	concentration in the fractured channel averaged over the width, kg/m^3
C_p	concentration in the porous section, kg/m
C_{ref}	reference concentration, kg/m^3
d	half-width of the fractured channel, m
$d^* = du_p/D_p$	dimensionless half-width of the fractured channel
D_c	dispersion coefficient in fractured channel, m^2/yr
D_p	dispersion coefficient in porous section, m^2/yr
h	height of source and thickness of seam
h^*	dimensionless height, $u_p h/D$
H'	heavyside unit step function
J_c^*	dimensionless flux at crack
J_p^*	dimensionless flux in porous medium
K_d	adsorption coefficient, m^3/kg
m	mass of contaminant released, kg
$N_D = D_c R/D_p$	dimensionless parameter
$N_u = u_c R/u_p$	dimensionless parameter
$N_\lambda = \lambda R D_p/u_p^2$	dimensionless parameter
Q	total discharge rate for homogeneous porous media, kg/yr
Q_c	discharge rate in crack, kg/yr
Q_p	discharge rate in porous media, kg/yr
Q_t	total discharge rate in fractured porous media, kg/yr
R	retardation factor
t	time, yr

$t_{1/2}$	half-life of radionuclides
t^*	dimensionless time
u_c	velocity of underground water flow in the fractured channel, m/yr
u_p	Darcy's velocity in the porous section, m/yr
x	axial coordinate, m
x_1	axial location of source
x^*	dimensionless axial coordinate
y	lateral coordinate
y_1	lateral location of source
y^*	dimensionless lateral coordinate
z	dimensionless variable of integration defined in Eq. (13)

Greek Letters

δ	Dirac delta function
ε	porosity
λ	decay constant, $\frac{0.693}{t_{1/2}}$, y^{-1}
σ	variable of integration in Eq. (10)
ρ_B	bulk density, kg/m^3
ϕ	dimensionless concentration

Superscripts

*	dimensionless parameter

Subscripts

c	fractured channel
p	porous section

INTRODUCTION

In radioactive waste modeling, problems of ion transport in either a porous medium or a fracture have been treated; however, the problem of bulk flow and ion transport in both the fracture and the porous medium is new. This paper describes the development of a combined fracture/porous medium contaminant transport model. The model, which was solved analytically, is based upon the two-dimensional, convective-dispersion equation with a line source located along the fracture.

The problem of ion migration in a homogeneous porous medium under various initial conditions was presented by Homsy[1] in his study of contaminant transport within a coal seam. And Erickson,[2] in a one-dimensional analysis of jointed media, modeled the problem of flow through a slit with diffusion of the ion species into the porous but nonflowing surrounding matrix. The combined problem of ion migration in both a fracture and the surrounding porous media was presented by Nuttall and Ray;[3] however, that model did not treat radioactive decay of the ion species. The current model extends the previous treatment to the migration of radioactive ions, and permits an analytical analysis of ion tranport for a waste deposit site that has been breached by a fracture.

Nelson[4] recently pointed out that a quantitative description of the rate of contaminant reaching a specified boundary is needed to evaluate the consequences of ground water contamination. The model described in this paper provides a worst case analysis as is often needed for an environmental assessment.

MODEL

The analytical model developed in this section describes the convective transport of decaying radionuclides through a fractured porous medium. The two-dimensional "x-y" region of uniform thickness used to formulate the problem is illustrated in Fig. 1. The problem is formulated for an instantaneous line source lying at an arbitrary location within the fracture. This simulation represents a worst possible case analysis. The problem is symmetrical about the y axis whose origin is located at the center of the fracture. The fracture width is 2d. The fluid is moving in the x direction and x and y dispersion coefficients are assumed to be equal for problem simplification though it is recognized that axial dispersion is greater and that they are not equal. This simplification does not alter the basic results and conclusions.

Under the assumption of uniform convective velocity, u_c, in the crack and u_p, in the porous section and for constant physical

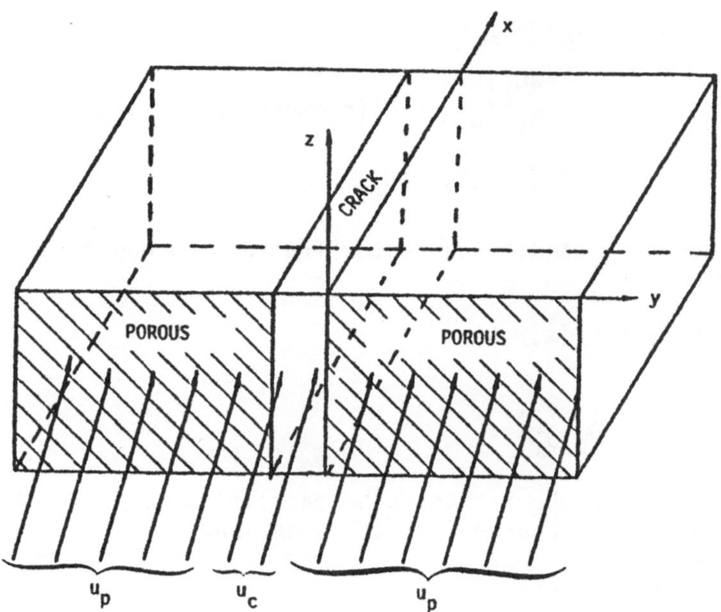

Fig. 1. Schematic diagram of a coal seam with fracture.

properties, the partial differential equations describing the concentration profile in the configuration shown in Fig. 1 may be written as:

for the fractured channel $(0 \leq y \leq d)$

$$\frac{\partial C_c'}{\partial t} = D_c \frac{\partial^2 C_c'}{\partial x^2} + D_c \frac{\partial^2 C_c'}{\partial y^2} - u_c \frac{\partial C_c'}{\partial x}$$

$$- \lambda C_c' + \frac{m}{h} \delta(x - x_1)\, \delta(y - y_1)\, \delta(t - 0) \ , \tag{1}$$

for the porous section $(d \leq y \leq \infty)$

$$R \frac{\partial C_p}{\partial t} = D_p \frac{\partial^2 C_p}{\partial x^2} + D_p \frac{\partial^2 C_p}{\partial y^2} - u_p \frac{\partial C_p}{\partial x} - R\lambda C_p \tag{2}$$

with boundary conditions

$$C_c'(x,y,0) = C_p(x,y,0) = 0 \tag{3a}$$

$$C_c'(\pm\infty,y,t) = C_p(\pm\infty,y,t) = 0 \tag{3b}$$

$$\left.\frac{\partial C_c'}{\partial y}\right|_{y=0} = 0 \text{ (symmetry condition)} \tag{3c}$$

$$C_p(x,\infty,t) = 0 \; . \tag{3d}$$

The boundary conditions at the fracture porous region interface ($y = d$) are

$$C_c'(x,y,t) = C_p(x,y,t) \; , \tag{4a}$$

$$\dot{D}_c \frac{\partial C_c'}{\partial y} = D_p \frac{\partial C_p}{\partial y} \; . \tag{4b}$$

Here we have taken the x-axis to be parallel to the direction of flow, and R is the retardation factor defined as

$$R = 1 + K_d \rho_B / \varepsilon \; . \tag{5}$$

The last term in the right-hand side of Eq. (1) describes the release of radionuclides of mass, m, instantaneously at $t = 0$ uniformly along a line of height h at (x_1,y_1).

The exact solution of the above coupled partial differential equations is very complicated. But the problem can be somewhat simplified under the assumption that the thickness of the fractured channel is small. In this situation the concentration gradient in the y direction across the channel can be neglected and the partial differential equation describing the concentration profile of radionuclide in the channel reduces to

$$\frac{\partial C_c}{\partial t} = D_c \frac{\partial^2 C_c}{\partial x^2} - u_c \frac{\partial C_c}{\partial x} + \frac{D_p}{d} \left(\frac{\partial C_p}{\partial y}\right)_{y=d} - \lambda C_c$$

$$+ \frac{m}{2hd} \, \delta(x - x_1) \, \delta(t - 0) \; , \tag{6}$$

and the boundary conditions given by Eqs. (4a) and (4b) reduce to

$$C_c(x,t) = C_p(x,d,t) \; . \tag{7}$$

Equation (6) is obtained by making a material balance on the channel.

Now the problem is rendered dimensionless by introducing the following variables:

$$\phi_c = C_c/C_{ref} , \qquad x^* = u_p x/D_p ,$$

$$\phi_p = C_p/C_{ref} , \qquad y^* = u_p y/D_p ,$$

$$t^* = u_p^2 t/RD_p ,$$

With the above dimensionless variables, Eqs. (7) and (2) transform to

$$\frac{\partial \phi_c}{\partial t^*} = N_D \frac{\partial^2 \phi_c}{\partial x^{*2}} - N_u \frac{\partial \phi_c}{\partial x^*} + \frac{R}{d^*}\left(\frac{\partial \phi_c}{\partial y^*}\right)_{y^*=d^*}$$

$$- N_\lambda \phi_c + \frac{A}{2d^*} \delta(x^* - x_1^*) \ \delta(t^* - 0) \qquad (8)$$

and

$$\frac{\partial \phi_p}{\partial t^*} = \frac{\partial^2 \phi_p}{\partial x^{*2}} + \frac{\partial^2 \phi_p}{\partial y^{*2}} - \frac{\partial \phi_p}{\partial x^*} - N_\lambda \phi_p , \qquad (9)$$

where

$$N_\lambda = \lambda R D_p/u_p^2 , \qquad d^* = d u_p/D_p ,$$

$$N_D = \frac{D_c R}{D_p} , \qquad A = \frac{m u_p^2}{h D_p^2 C_{ref}} ,$$

$$N_u = u_c R/u_p ,$$

Equations (8) and (9) are solved by the Fourier-Laplace transform technique.[5] The following solutions are obtained

$$\phi_c(x^*,t^*) = \frac{AR}{8\pi} \exp(-N_\lambda t^*)$$

$$\times \int_1^\infty \frac{(\sigma - 1)\ H[t^* - d^*(\sigma - 1)]}{[t^* - d^*(\sigma - 1)]^{3/2}\ [t^* + d^*(N_D - 1)(\sigma - 1)]^{1/2}}$$

$$\times \exp\left\{- \frac{R^2(\sigma - 1)^2}{4[t^* - d^*(\sigma - 1)]}\right.$$

$$\left. - \frac{[x^* - x_1^* - d^*(\sigma - 1)(N_u - 1) - t^*]^2}{4[t^* + d^*(\sigma - 1)(N_D - 1)]}\right\} d\sigma \quad (10)$$

and

$$\phi_p(x^*,y^*,t^*) = \frac{A}{8\pi} \exp(-N_\lambda t^*)$$

$$\times \int_1^\infty \frac{[y^* - d^* + R(\sigma - 1)]\ H[t - d^*(\sigma - 1)]}{[t^* - d^*(\sigma - 1)]^{3/4}\ [t^* + d^*(\sigma - 1)(N_D - 1)]^{1/2}}$$

$$\times \exp\left\{- \frac{[y^* - d^* + R(\sigma - 1)]^2}{4[t^* - d^*(\sigma - 1)]}\right.$$

$$\left. - \frac{[x^* - x_1^* - d^*(\sigma - 1)(N_u - 1) - t^*]^2}{4[t^* + d^*(\sigma - 1)(N_D - 1)]}\right\} d\sigma\ . \quad (11)$$

For $d^* = 0$, which means that there is no channel present, Eq. (11) can be integrated to give

$$\phi(x^*,y^*,t^*) = \frac{A}{4\pi R t^*} \exp\left[- \frac{(x - x_1^* - t^*)^2 + y^{*2}}{4t^*} - N_\lambda t^*\right], \quad (12)$$

which is the same solution obtained by solving the partial differential equation describing concentration profile in a homogeneous porous media.

For $d^* \neq 0$, Eqs. (10) and (11) can be numerically integrated. A convenient form for numerical integration can be developed by means of the following transformation

$$z = 1 + \frac{d^*(\sigma - 1)(N_D - 1)}{t^*} . \tag{13}$$

Substitution of the above transformation in Eqs. (10) and (11) produces the following

$$\phi_c = \frac{AR \exp(-N_\lambda t^*)}{8\pi d^{*2}(N_D - 1)^{1/2}} \int_1^{N_D} \frac{(z - 1)}{(N_D - z)^{3/2} z^{1/2}}$$

$$\times \exp\left\{- \frac{R^2 t^*(z - 1)^2}{4d^{*2}(N_D - 1)(N_D - z)}\right.$$

$$\left. - \frac{[(x^* - x_1^*)(N_D - 1) - t^*(N_u z - N_u - z + N_D)]^2}{4(N_D - 1)^2 zt^*}\right\} dz \tag{14}$$

and

$$\phi_p = \frac{A \exp(-N_\lambda t^*)}{8\pi t^* d^{*2}(N_D - 1)^{1/2}}$$

$$\times \int_1^{N_D} \frac{[d^*(y^* - d^*)(N_D - 1) + R(z - 1)t^*}{(N_D - z)^{3/2} z^{1/2}}$$

$$\times \exp\left\{- \frac{[d^*(y^* - d^*)(N_D - 1) + R(z - 1)t^*]^2}{4d^{*2}t^*(N_D - 1)(N_D - z)}\right.$$

$$\left. - \frac{[(N_D - 1)(x^* - x_1^*) - t^*(N_u z - N_u - z + N_D)]^2}{4(N_D - 1)^2 zt^*}\right\} dz . \tag{15}$$

The dimensionless flux, J_c^*, in the fracture is

$$J_c^* = -\frac{\partial \phi_c}{\partial x^*} = \frac{AR\ \exp(-N_\lambda t^*)}{16\pi d^{*2} t^*(N_D - 1)^{3/2}}$$

$$\times \int_1^{N_D} [(x^* - x_1^*)(N_D - 1) - t^*(N_u z - N_u - z + N_D)]$$

$$\times \frac{(z - 1)}{[(N_D - z)z]^{3/2}} \exp\left\{- \frac{R^2 t^*(z - 1)^2}{4d^{*2}(N_D - 1)(N_D - z)}\right.$$

$$\left.- \frac{[(N_D - 1)(x^* - x_1^*) - t^*(N_u z - N_u - z + N_D)]^2}{4(N_D - 1)^2\ zt^*}\right\} dz\ . \quad (16)$$

The dimensionless flux, J_p^*, in the porous section is

$$J_p^* = -\frac{\partial \phi_p}{\partial x^*} = -\frac{A\ \exp(-N_\lambda t^*)}{16\pi t^{*2} d^{*2}(N_D - 1)^{3/2}}$$

$$\times \int_1^{N_D} \frac{[(N_D - 1)(x^* - x^*_1) - t^*(N_u z - N_u - z + N_D)]}{[(N_D - z)z]^{3/2}}$$

$$\times [(N_D - 1)d^*(y^* - d^*) + R(z - 1)t^*]$$

$$\times \exp\left\{- \frac{[(N_D - 1)d^*(y^* - d^*) + R(z - 1)t^*]^2}{4d^{*2} t^*(N\ \ - 1)(N\ \ - z)}\right.$$

$$\left.- \frac{[(N_D - 1)(x^* - x_1^*) - t^*(N_u z - N_u - z + N_D)]^2}{4(N_D - 1)^2\ zt^*}\right\} dz\ . \quad (17)$$

Similarly, the dimensionless flux in the homogeneous porous media is given by

$$J^* = -\frac{\partial\phi}{\partial x^*} = \frac{A(x^* - x_1^* - t^*)}{8\pi R t^{*2}}$$

$$\times \exp - \frac{(x^* - x_1^* - t^*)^2 + y^{*2}}{4t^*} - N_\lambda t^* \quad . \qquad (18)$$

The dimensionless discharge rate, Q_c^*, in the channel is

$$Q_c^* = \frac{2h^*}{R}\int_0^{d^*} [N_D J_c^* + N_u \phi_c]\, dy^* = \frac{Ah^* \exp(-N_\lambda t^*)}{8\pi d^* t^* (N_D - 1)^{3/2}}$$

$$\times \int_1^{N_D} \frac{[(x^*-x_1^*)(N_D-1)N_D + t^*(N_D N_u z + N_D N_u + N_D z - 2N_u z - N_D^2)]}{[(N_D - z)]^{3/2}}$$

$$\times (z - 1) \exp\left\{- \frac{R^2 t^* (z - 1)^2}{4d^{*2} (N_D - 1)(N_D - z)}\right.$$

$$\left. - \frac{[(x^* - x_1^*)(N_D - 1) - t^*(N_u z - N_u - z + N_D)]^2}{4(N_D - 1)^2 z t^*}\right\} dz \ . \quad (19)$$

The dimensionless discharge rate, Q_p^*, through the porous section is

$$Q_p^* = 2h^* \int_{d^*}^{\infty} [J_p^* + \phi_p]\, dy^* = \frac{Ah^* \exp(-N_\lambda t^*)}{4\pi t^* d^* (N_D - 1)^{3/2}}$$

$$\times \int_1^{N_D} \frac{[(N_D - 1)(x^* - x_1^*) - t^*(N_u z - 2N_D z + z - N_u + N_D)]}{(N_D - z)^{1/2}\, z^{3/2}}$$

$$\times \exp\left\{- \frac{R^2 (z - 1)^2\, t^*}{4d^{*2} (N_D - 1)(N_D - z)}\right.$$

(NOTE: Eq. (20) continued on next page)

$$- \frac{[(x^* - x_1^*)(N_D - 1) - t^*(N_u z - N_u - z + N_D)]}{4(N_D - 1)^2 \, zt^*} \Bigg\} dz \; . \qquad (20)$$

The total discharge rate, Q_t^*, through a fractured porous media is

$$Q_t^* = Q_c^* + Q_p^* \; .$$

The dimensionless discharge rate, Q^*, for homogeneous porous media is given by

$$Q^* = 2h^* \int_0^\infty (J^* + \phi) \, dy^*$$

$$= \frac{Ah^*(x^* - x_1^* + t^*)}{4R\sqrt{\pi t^{*3}}} \exp\left[- \frac{(x^* - x_1^* - t^*)^2}{4t^*} - N_\lambda t^*\right] . \qquad (21)$$

To obtain dimensional discharge rate the non-dimensional discharge rate should be multiplied by

$$D_p^2 \, C_{ref}/u_p \; .$$

A test study is presented in the following section which illustrates the adverse effect of a fracture.

CASE STUDY

A test case is presented to demonstrate the adverse effect of a fracture on the migration rate of radioactive nuclides. The downstream discharge rates were calculated for both the homogeneous medium case and for the combined fracture/porous medium case. An instantaneous line source was used to simulate a leaky canister. The ion line source is located within the fracture. This is a worst case formulation of the problem and helps to illustrate the adverse effect of a fracture on ion migration rates. The study parameters are shown in Table 1.

TABLE 1

Study Parameters

d = 0.00079 m (1/32 in.) half-width of the fractured channel

h = 6.1 m height of source and thickness of coal seam

$D_p = D_c$ = 0.25 m^2/yr dispersion coefficient in porous medium and crack

C_{ref} = 2.1 kg/^{3} reference concentration

m = 1066 kg mass of contaminant released

u_p = 40 m/yr Darcy velocity in the porous medium

u_c = 2.46 x 10^5 m/yr velocity within the crack where $u_c/u_p = (\varepsilon/k)(d^2/3)$

$\varepsilon = 0.01$ and $k = 3.4 \times 10^{-13}$ m^2

R = 68.0

λ = 0.0 decay constant

The discharge rate of the ions at a point 100 meters downstream of the source was evaluated for transport in the crack and the porous medium case as well. The resulting curves are presented in Fig. 2. The results clearly show the early discharge of ions from the crack followed by a later discharge of those ions that had diffused from the fracture into the porous medium and were eventually swept past the 100 meter point. If the porous medium were fracture free the resulting discharge occurs much later in time as illustrated in curve 1. The effect of radioactive decay enlarges the difference in the two cases with greater reactivity release in the fractured medium case. This is of course due to the retarding nature of the homogeneous porous medium and thus permitting greater time for radioactive decay.

ACKNOWLEDGEMENT

The authors express their appreciation to Sandia National Laboratory and the U.S. Department of Energy for supporting this work.

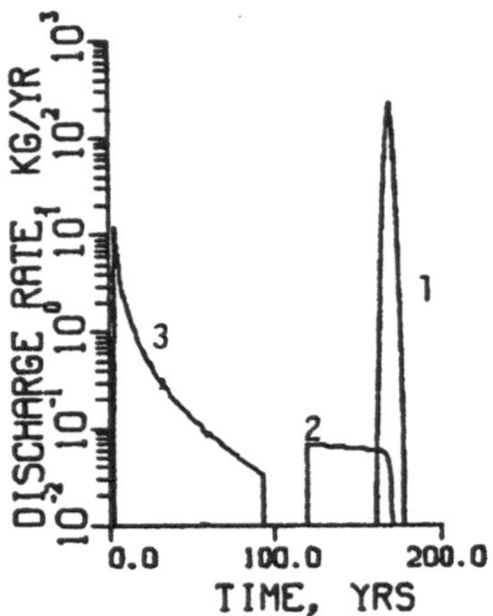

Fig. 2. Discharge rate at 100 meters: 1. homogeneous case; 2. porous medium in crack case; 3. crack discharge.

REFERENCES

1. Homsy, Robert V., Two-Dimensional Transient Dispersion and Adsorption in Porous Media, Lawrence Livermore Laboratory.

2. Erickson, Kenneth L, "Preliminary Rate Expressions for Analysis of Radionuclide Migration Resulting from Fluid Flow Through Jointed Media," paper presented at the Annual Materials Research Society Meeting in Boston, MA, November 1979.

3. Nuttall, H. E. and A. K. Ray, "A Combined Fracture/Porous Media Model for Contaminant Transport," presented at the 15th Inter-society Energy Conversion Engineering Conference, Seattle, Washington, August 18-22, 1980.

4. Nelson, R. William, "Evaluating the Environmental Consequences of Groundwater Contamination. 1. An Overview of Contaminant Arrival Distribution as General Evaluation Requirements," Water Resources Res., Vol. 14, 1978, pp. 409.

5. Weinberger, H. F., "A First Course in Partial Differential Equations," Blaisdell Publishing Company, Waltham, MA (1965).

SENSITIVITY OF CALCULATED PORE-FLUID PRESSURE TO REPOSITORY FRACTURE GEOMETRY*

R. R. Eaton, D. E. Larson, and C. M. Korbin

Sandia National Laboratories**
P. O. Box 5800
Albuquerque, NM 87185

ABSTRACT

Calculations have been made to predict the sensitivity of pore-fluid pressure to the geometry of rock fracture patterns in the near vicinity of nuclear waste buried in a tuff repository. The SHAFT 79 code (for multi-dimensional, two-phase flow in porous medium) has been used. The repository considered contains 1.0, 2.1 and 2.5 kW/can, canisters emplaced in welded tuff with a global repository loading of 75 kW/acre. The resulting calculations show that predicted pore fluid pressures vary from 55 bars for a homogeneous, fracture-free rock mass to 5 bars for a well fractured mass. This order of magnitude variation shows the importance of properly modeling fracture patterns in hard rock repository analysis.

INTRODUCTION

Numerous one-and two-dimensional thermal analyses have been made to assess the consequences of radioactive waste burial in various geologic media.[1] These models provide complete temperature fields for a wide range of materials and heat loadings but are incapable of predicting pore-fluid pressures

*This work was supported by the U.S. Department of Energy under contract DE-AC04-76DP00789.

**A U.S. Department of Energy facility.

(pore pressure), boiling-front locations, dry out and convective energy transport. The pore pressures obtained in an earlier study for argillite[2] and for tuff, using the two-phase flow code, SHAFT[3], for a saturated homogeneous (fracture-free) repository with perfect backfill, exceed the local lithostatic load by more than 600 bars. Such overpressures could significantly affect the integrity of a repository since typical tensile strength of these rocks are of the order of 10 bars. These pressure calculations may be unrealistic in that they do not consider volatile release by means of the fractures inherent in welded tuff.

This paper extends this previous study to show the importance of pressure-relieving fractures in the vicinity of a single waste canister at times prior to backfilling the drift. Material properties for the tuff considered in this paper are: conductivity = 3 W/m^oC, density = 2280 kg/m^3, specific heat = 1597 J/kg °C, matrix permeability(k) = 4×10^{-7} darcy, fracture permeability (k_f) = 1 darcy, porosity(ϕ) = 23%. Also assumed are global repository loading = 75 kW/acre, extraction ratio = 20%, and canister loading (Q) = 1.0, 2.1 and 2.5 kW/can. The permeability value is representative of a minimum value measured on the matrix of simple devitrified welded tuff.

ANALYSIS

The geometric nodalization used is based on an effective radius or unit cell concept. This approximation transforms the exact three dimensional configuration into an equal mass axisymmetric shape, Fig. 1. The floor of the repository drift is located 710 m below the ground surface. The drift is held at 1 atmosphere to simulate conditions prior to drift backfill. The welded tuff is assumed saturated. Four near field fracture geometries are considered in this study as shown in Figure 1 (a)-(d). The geometries of the four cases are: (1) homogeneous material (welded tuff) no fractures, base case, (Figure 1a); (2) a single axisymmetric fracture surrounding the canister (at r = radius of canister (.165 m) plus radius of tuff block) which extends to the floor of the drift, (Figure 1b); (3) two horizontal fractures separated by 0.5 m and two vertical fractures extending to drift floor at $r = r_{canister}$ and r = 0.665 m, torus, Figure 1c); and (4) the tuff in vicinity of canister assumed to consist of three dimensional "pie" shaped rock blocks with mean cross sections of 0.5 m surrounded by fractures on all six sides, (Figure 1d). In all cases the fracture width is taken to be 1/100 of the width of the homogeneous tuff block of interest and to have a permeability 10^7 larger than the block permeability. In each case the pore pressure (p*) of interest (see Figure 1) is the maximum calculated pore pressure in the block of interest. Results were obtained using the two-phase lumped

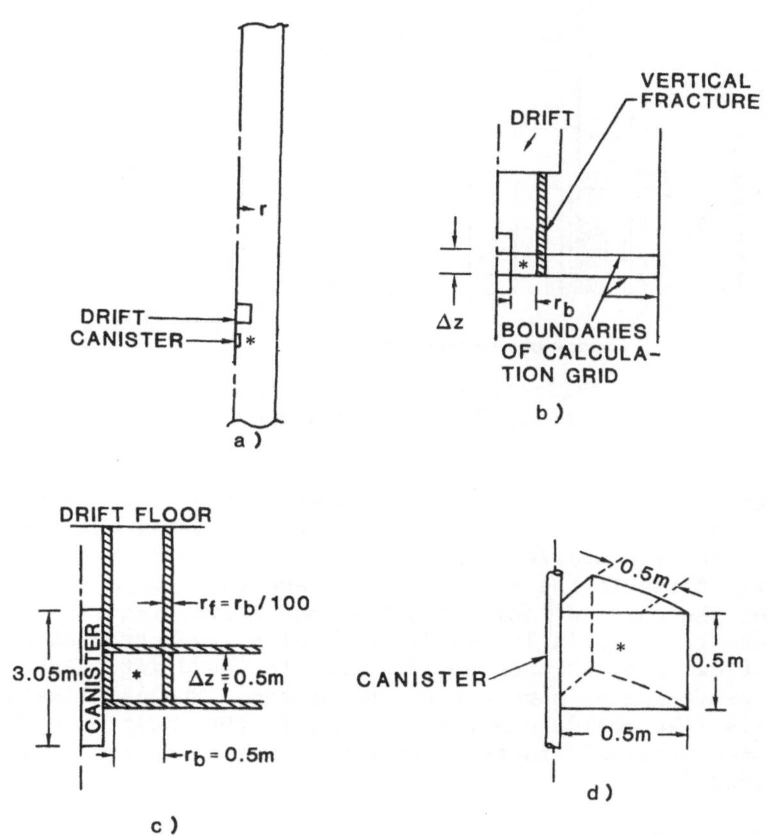

Fig. 1. Fracture Geometries. (Scales are different in each drawing to emphasize important features) (a) Homogeneous Fracture Free Column Extending to Ground Surface (* is point of interest) Figures (b), (c) and (d) Refer to Details of the Three Fracture Cases. (b) Single Axisymmetric Fracture Case (c) Multiple Fractures about Torus Shaped Homogeneous Block (d) Three Dimensional Block with Fractures on all Six Sides. (Drift floor considered but not shown)

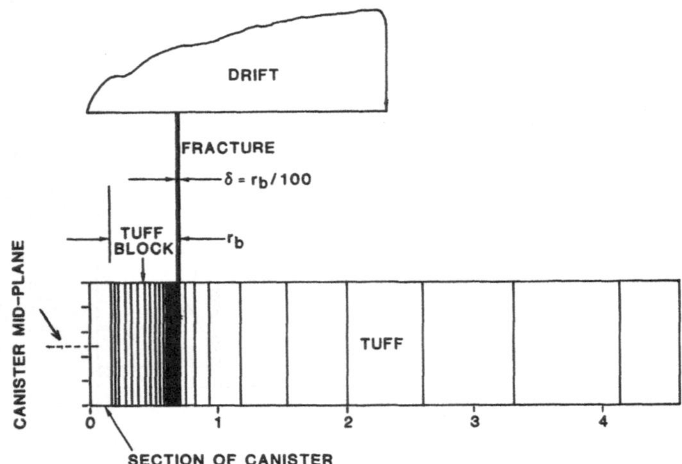

Fig. 2. Nodalization Used for Single Axisymmetric Case (Geometry 2) shown in Figure 1b.

parameter code SHAFT 79. The nodalization was kept moderately simple and therefore the run times were reasonably short (100 to 200 s). The computational grid for case 1 extends from 30 m below the drift floor to the ground surface and uses 102 elements. Details of the computational grid for case 2 are given in Figure 2. This case is essentially one-dimensional axisymmetric except for the vertical flow through the fracture to the drift which is caused by the expansion of pore water when heated. The nodalization for case 3 is like that shown for case 2 with the addition of horizontal and vertical fracture regions. Case 4 has the third dimension added.

SUMMARY OF RESULTS

An initial parametric study determined the sensitivity of pore pressure to matrix permeability (k). Geometry 2 was used for this study with a canister loading (Q) equal to 1.0 kW/can. Figure 3 shows that for k larger than 10^{-5} darcy, pore-pressure increases due to heating are not significant. For k smaller than 10^{-7}d the peak pore pressure becomes a strong function of permeability and can be significant with regards to structural integrity. Figure 4 shows the effect of block size on calculated peak pore pressures for geometry 2 with Q = 2.5kW. A summary of this type of calculation is given in Figure 5 for Q = 1, 2.1 and 2.5 kW/can. It can be seen from Figures 4 and 5 that the calculated peak pressures occurs within one day after canister emplacement and is approximately linear with block size for a given canister output. Figure 6 gives pressure as a function of time for the four geometries, Q = 2.5 kw/can. It shows that fluid

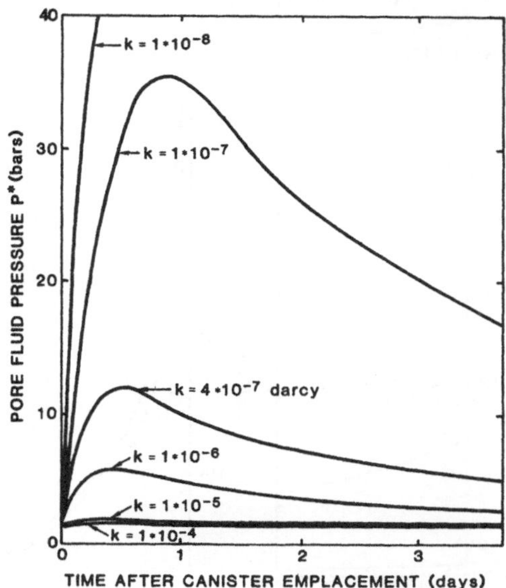

Fig. 3. Effect of Tuff Permeability on Fluid Pressure Increase. Geometry 2, r_b = 0.5, Q = 1.0 kW/can

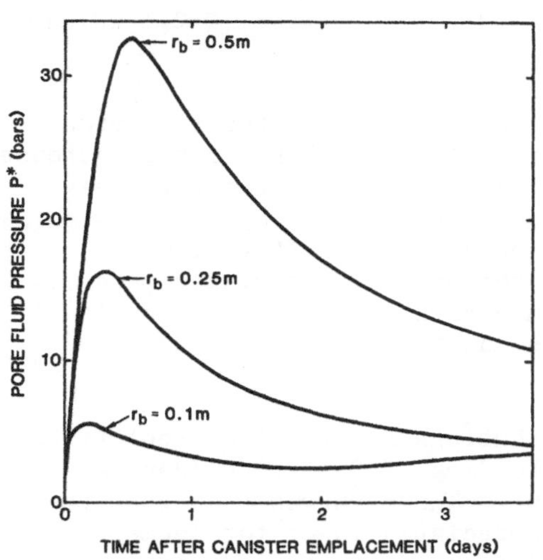

Fig. 4. Effect of Tuff Block Size on P*, Geometry 2, Q=2.5 kW/can, $k=4x10^{-7}$ darcy.

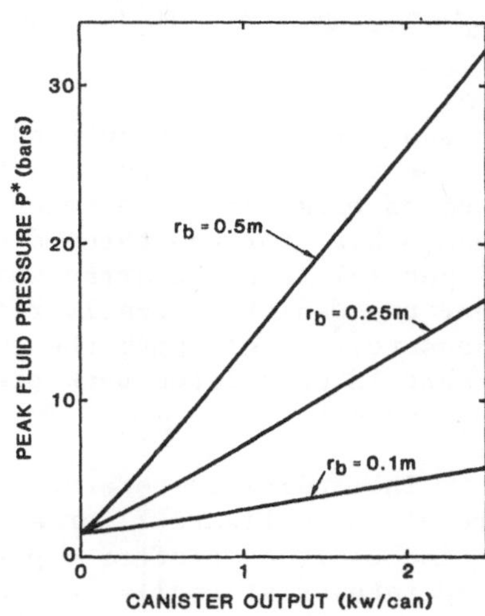

Fig. 5. Peak Pressure as a Function of Block Size and Canister Output, $k = 4x10^{-7}$ darcy Geometry 2.

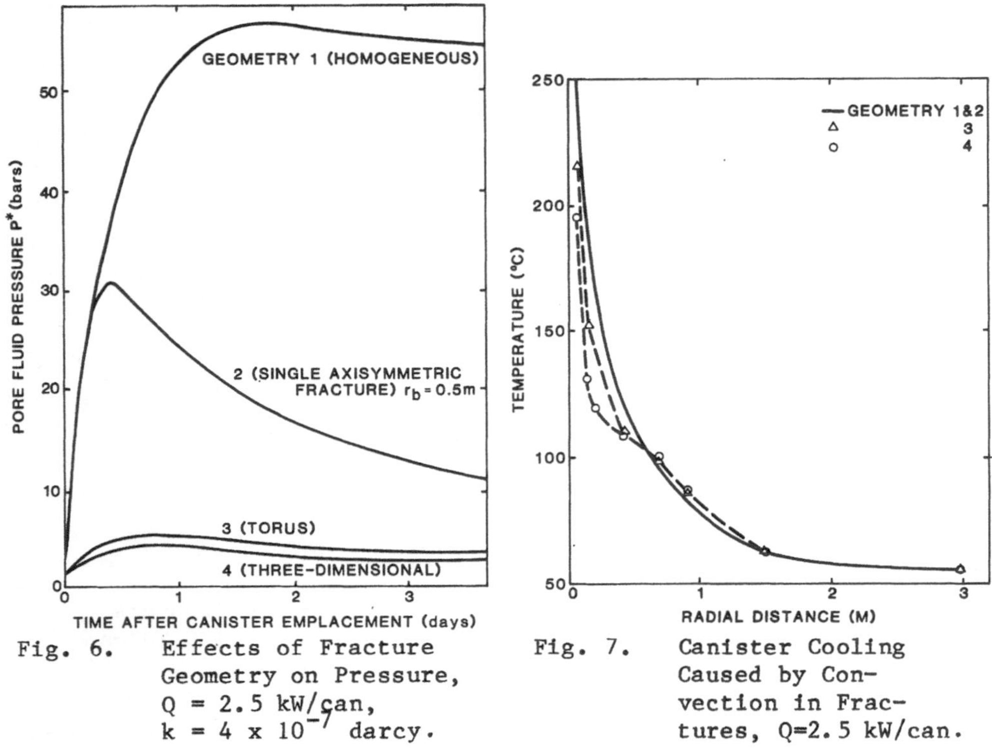

Fig. 6. Effects of Fracture Geometry on Pressure, Q = 2.5 kW/can, k = 4 x 10^{-7} darcy.

Fig. 7. Canister Cooling Caused by Convection in Fractures, Q=2.5 kW/can.

pressure is strongly coupled with the geometry of the fractures in the vicinity of the canister and drift. Peak pressures decrease from 55 bars for the homogeneous fracture free geometry to less than 5 bars for the three-dimensional fracture array. Inclusion of general fracture array which connects the drift to regions near the canister greatly reduces the maximum calculated pore pressures. Neglecting the existence of a fracture array could result in calculated peak pressures up to an order of magnitude too large.

The radial temperature distributions for the four geometries are given in Figure 7 for a heat loading of Q = 2.5 kW/can, and a time of 10 days after emplacement. These curves show that the temperatures at small radii are appreciably lower for geometries 3 and 4. It appears that the additional fractures provide convective pathways sufficient to effect the otherwise conduction dominated temperature distributions. None of these cases contained boiling regions at the time when maximum pressures existed. For Q = 2.5 kW/can, some two-phase and vapor flow existed at later time (3 days). Figure 8 gives the pressure,

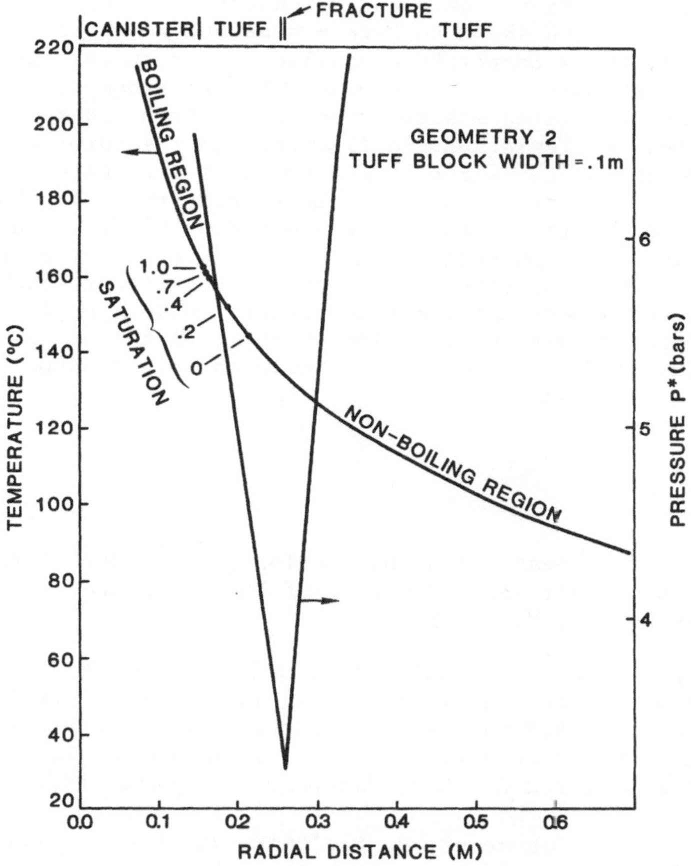

Fig. 8. Temperature and Pressure Near Canister at 10 Days After Burial, Q = 2.5 kW/Can.

temperature and volume fraction of vapor for this boiling case. For the purpose of moedling geologic medium properties, it is often assumed in conduction calculations, that boiling occurs at 100°C and 1 atmosphere.[4] Figure 8 shows that this assumption is not applicable for this case, since the boiling front occurs between 140 and 160°C at a mean pressure of 5.5 bars.

This study shows the effects of fracture network on thermohydrology in the vicinity of a nuclear-waste canister. The presence of fractures has been shown to have a weak but non-negligible influence on resulting temperature profiles and a strong influence on the resulting pore fluid pressures. The presence of fractures decreases the calculated maximum pressure through two mechanisms. First, the reduced resistance to flow through fractures gives the increased volume of the slightly compressible pore water a path through which to migrate away from the hottest regions, thus directly reducing the pressure. Secondly, the movement of fluid in the fractures provides some convective cooling in the conduction dominated problem. It should be emphasized that the details of fracture geometry do not appear extremely important, so long as fractures extending from the very near region of the canister to the drift floor are included, in which case calculated pore pressures are significantly decreased.

References

1. McVey, D. F., "Near Field Interactions," 19th Annual ASME Symposium on Geological Disposal of Nuclear Waste, Albuquerque, NM, March 1979.

2. Eaton, R. R., et al., "Calculated Hydrogeologic Pressures and Temperatures Resulting from Radioactive Waste in the Eleana Argillite," p. 833 in vol. II, Proceedings of International Symposium on the Scientific Basis for Nuclear Waste Management, Materials Research Society, Plenum Press, March 1980.

3. Pruess, K. and Schroeder, R. C., "SHAFT 79, User's Manual," LBL-10861, UC-669, Lawrence Berkeley Laboratory, Berkeley, CA., 1980.

4. Lappin, A. R., "Thermal Conductivity of Silicic Tuffs: Predictive Formulation and Comparison With Preliminary Experimental Results," SAND80-0769, July 1980.

AN EVALUATION OF THE ABILITY TO PREDICT RADIONUCLIDE TRANSPORT IN GROUNDWATER FLOW SYSTEMS

Franklin W. Schwartz and Leslie Smith*

Dept. of Geology
University of Alberta
Edmonton, Canada

*Dept. of Geology and Geophysics
University of Utah
Salt Lake City, Utah 84112

INTRODUCTION

The development of techniques for modeling contaminant transport in groundwater systems has received considerable attention in the past several years. In managing high level radioactive wastes, an ability to predict accurately patterns of contaminant migration would be of significant value in licensing, development and operation of a waste repository in the subsurface. However, an important question that remains unresolved is how much confidence can be placed in the application of transport models to site specific problems.

A variety of factors can contribute to uncertainties in model predictions. These factors are related to both limitations in the transport models themselves and to problems in defining at a site the parameters which control transport. The objective of this paper is to investigate this second source of uncertainty. If one recognizes that porous media are inherently heterogenous, the inability to describe exactly the spatial variability, even with large amounts of field data, gives rise to uncertainty.

Two major differences exist between the model we will describe here and those commonly used in practice. First, macroscopic or large-scale dispersion is accounted for as a mixing process caused primarily by heterogeneities in hydraulic conductivity. The conventional approach assumes that macroscopic dispersion can be approximated as a large-scale diffusion process. Our previous work has shown that such an assumption may not be appropriate for some media and that working with a more realistic model of dispersion overcomes many of the limitations of the conventional approach[1].

Second, the model operates as a stochastic simulator. The parameters of the porous medium which control transport are specified as statistical distributions rather than as a uniquely defined set of values. In effect, the structure of the heterogeneous medium is described as a random process defined by a set of statistical parameters. To form estimates of the probability distributions on model results (predictions), Monte Carlo techniques are used. This approach involves the repetitive generation of many different realizations of the porous medium from the statistical model characterizing the heterogeneity and the modeling of transport in each realization. Although all of the realizations are derived from the same statistical model, each realization will have a unique set and arrangement of hydraulic conductivity values. The variability in transport behavior observed over all of the realizations can be interpreted in terms of prediction uncertainty.

Two distinct sources of uncertainty are related to the heterogeneous nature of the porous medium. Even when the statistical parameters describing the porous medium are assumed known, the inability to describe the heterogeneity exactly creates uncertainty in model prediction. In addition, because the statistical parameters must be estimated from a limited set of field measurements, errors in estimating these parameters will contribute to increased uncertainty. However, field measurements will also contribute to a reduction in uncertainty by constraining the possible patterns of spatial variability at and in the vicinity of the known data points.

DESCRIPTION OF THE MODEL

Figure 1 shows the flow domain for this study. The sides and bottom are no flow boundaries for both water and mass. The water table represents the upper boundary for the region. The porous medium is divided into a set of discrete blocks. Values of hydraulic conductivity and porosity are generated for each block using an algorithm that allows neighboring block values to be correlated[2]. When hydraulic conductivity values are available from field or laboratory measurements, they can be preserved in each of the realizations using the technique of conditional simulation[3]. Details of the modeling procedure are described elsewhere[1,2,4]. Table 1 is a summary of the statistical parameters that are required to generate one or more realizations of a porous medium.

Assuming that hydraulic conductivity is log normally distributed (for justification see 5), μ_y and σ_y characterize its distribution. The integral scales (λ_x, λ_z) define the average distance over which neighboring conductivity values are correlated. Larger integral scales indicate that distinct zones of higher or lower conductivity tend to be more extensive.

Table 1. Statistical parameters necessary to generate a realization

Notation	Definition
μ_y	mean log hydraulic conductivity
σ_y	standard deviation in log conductivity
λ_x, λ_z	integral scales in the x and z direction (meters)
$\bar{Y}_c$	mean log hydraulic conductivity, estimated
S_{yc}	standard deviation in log conductivity, estimated
μ_n	mean porosity
σ_n	standard deviation in porosity
ρ_{yn}	correlation coefficient between porosity and hydraulic conductivity

Taking porosity to be normally distributed[5], μ_n and σ_n characterize this distribution. The correlation coefficient ρ_{yn} expresses the degree of statistical dependence between the hydraulic conductivity and porosity variations. The parameter $\bar{Y}_c$ and S_{yc} represent estimates of μ_y and σ_y formed from available field data.

Having generated an appropriate realization of the porous medium and specified a set of boundary conditions for the flow domain, it is then possible to solve the resulting flow problem numerically to yield values of hydraulic head at node points within the region. The hydraulic head, conductivity and porosity values can be combined in the Darcy equation to yield seepage velocities. The mass transport problem is solved using a hybrid deterministic-probabilitic method[1]. With this technique, mass is tagged to a large number of moving reference particles which are then released in the flow system. By tracing the pattern of advective transport of these particles in the heterogeneous conductivity field, the mass transport problem is in effect solved. For each realization, it is possible to determine the relative concentration of tracer as a function of space and time, concentration distribution along the outflow boundary of the flow domain, and information on various breakthrough times. Such results when averaged over all of the realizations yield estimates of the uncertainty present in a prediction.

EXAMPLE APPLICATIONS

The purpose of this section is to demonstrate the application of the modeling concept. In the examples that follow, contaminants enter the flow system along the line shown in Figure 1a as an instantaneous pulse at the start of the simulation. In all cases, the parameters describing the porous medium are listed on the figures depicting the results. The value of μ_y is representative of a silty sand for hydraulic conductivity measured in m/day. Additional simulation parameters include MC which defines the number of independent realizations in any one Monte Carlo run, NP which defines the number of moving reference particles used to solve the transport equation, and Δ T which indicates the size of the time step used in the solution of the transport equation (in days).

Before looking in detail at the simulation results, we will explain how they can be presented as so-called region probability plots. Consider the flow domain shown in Figure 1b. Contaminants have a finite probability of being located within any cell of the region at a specified time after release. This probability can be estimated by observing in which cells the mass is found as a function of time for each realization, and then averaging over the set of realizations. These probabilities can be contoured as is shown in Figure 1b and interpreted as follows. For the 2% contour, there is a 2% chance of locating a moving reference particle in block A at the given time. However, there is also a 2% chance that a particle could be in block B. There is a much smaller chance that a moving particle could be in blocks C or D at that time. These contours reflect the possibility that in each realization the particular

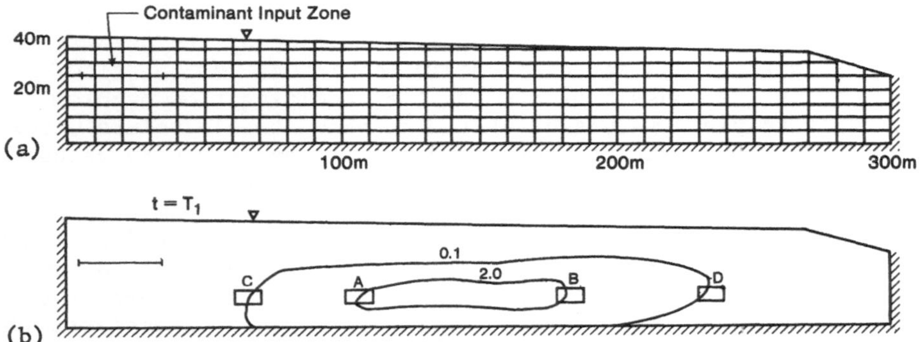

Fig. 1a. Flow domain showing the location of the contaminant input zone.

Fig. 1b. Contours on the regional probability plot define the percent probability of mass being located within each block.

heterogeneity may allow the mass to follow a significantly different flow path through the flow domain.

The area enclosed by the contours reflects the uncertainty in predicting the spatial distribution of contaminants. As uncertainty increases, there is a shrinking of the region enclosed by the higher contours and an expansion of the lower probability region. These region probability plots provide a useful tool to compare uncertainties in the spatial distribution of mass for media with different degrees of heterogeneity.

The first case presented is for a simulation assuming that the statistical parameters of the porous medium are known without uncertainty. Figure 2 compares prediction uncertainty in a medium with $\sigma_y = 0.10$ (a minimal variation in hydraulic conductivity) to that in a medium with a more realistic degree of heterogeneity, $\sigma_y = 0.50$. In this latter case, conductivity variations range over approximately two orders of magnitude. Immediately apparent is the increased prediction uncertainty for the more heterogeneous medium. In this case, there is an increasing possibility that relatively low or relatively high hydraulic conductivity zones will cause rather unexpected directions of mass movement either laterally or even upward away from the source. For $\sigma_y = 0.10$, there is little variability in transport patterns from realization to realization, in addition to the smaller extent of macroscopic dispersion in any one realization. The prediction uncertainty increases in time, very slightly for $\sigma_y = 0.10$ and considerably for $\sigma_y = 0.50$. Because a very large proportion of the flow domain is enclosed by the 0.1% contour for $\sigma_y = 0.50$ at 18500 days, it can be concluded that there is little chance of making accurate predictions in this system until additional information becomes available on the spatial structure of the porous medium.

The results presented in Figure 3 are for the case where the statistical parameters of the medium are estimated from field data. This simulation is carried out by first generating a single realization from a known statistical structure and then assuming that to be the actual repository site. Two regularly spaced sample grids of 24 and 36 data points are tested. By knowing the conductivity values at these measurement points, prediction uncertainty can be reduced. Although the conductivity values are fixed at the measurement points, this only reduces but does not eliminate the uncertainty in the seepage velocity at those points. Results not presented here also show that errors in estimating the statistical parameters ($\bar{Y}_c$, S_{yc}) can contribute significantly to prediction uncertainty. For these reasons, it appears a relatively large amount of information on the spatial distribution of hydraulic conductivity is required to significantly constrain prediction uncertainty.

Note that the integral scales for this simulation are reduced

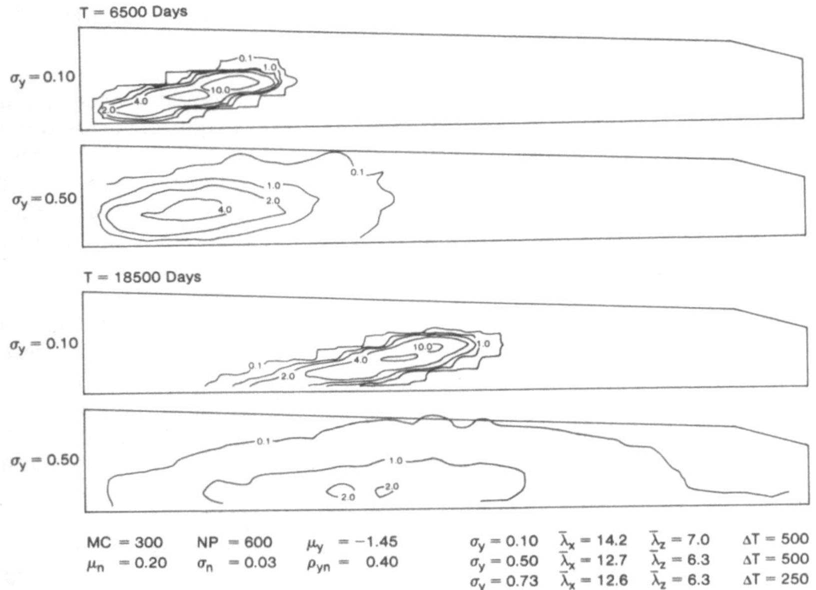

Fig. 2. Influence of the standard deviation in the logarithm of hydraulic conductivity on uncertainty.

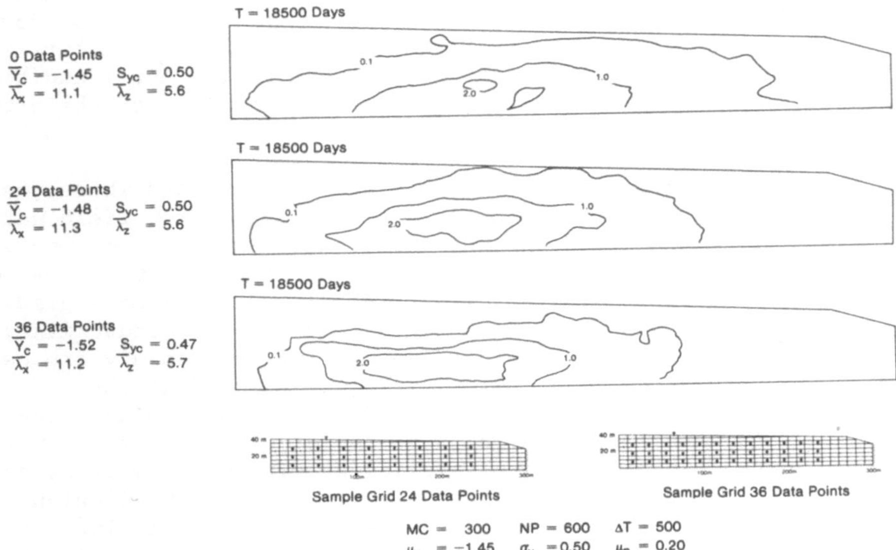

Fig. 3. Influence of the number of known field values of hydraulic conductivity on uncertainty.

from those in Figure 2. Comparison of the case where no data are supplied with the appropriate plot in Figure 2 indicates a general result that prediction uncertainties decrease for reduced spatial continuity in hydraulic conductivity.

CONCLUSIONS

The results of the two sets of simulations demonstrate the potential applicability of this modeling concept. Even from the brief analysis presented here, it is apparent that the certainty with which transport predictions can be made will depend upon the characteristics of the system being modeled, the errors involved in estimating the statistical parameters describing the porous medium and the availability of known values of hydraulic conductivity. The implication of these and similar results from our ongoing studies is that certain kinds of groundwater systems may be more amenable to predictive analysis than others.

The results of the simulations also show how sensitive the pattern of mass transport is to the spatial variation in hydraulic parameters. Thus, even when relatively large quantities of information are available about the porous medium, it appears that predictions still can be uncertain. We conclude that more research on the origin and factors controlling prediction uncertainty is essential to the wide-spread application of modeling techniques.

REFERENCES

1. L. Smith and F. W. Schwartz, Mass transport 1. A stochastic analysis of macroscopic dispersion, Water Resour. Res. 16: 303 (1980).
2. L. Smith and R. A. Freeze, Stochastic Analysis of steady state groundwater flow in a bounded domain, 2, Two - dimensional simulations, Water Resour. Res. 15:1543 (1979).
3. J. P. Delhomme, Spatial variability and uncertainty in groundwater flow patterns, Water Resour. Res. 15:269 (1979).
4. L. Smith and F. W. Schwartz, Mass transport 2. Analysis of uncertainty in prediction, in press, Water Resour. Res. (1980).
5. R. A. Freeze, A stochastic-conceptual analysis of one-dimensional flow in nonuniform homogenious media, Water Resour. Res. 11:725 (1975).

AN ANALYSIS ON THE USE OF ENGINEERED BARRIERS FOR GEOLOGIC ISOLATION OF SPENT FUEL*

M. O. Cloninger

Chemical Technology Department
Pacific Northwest Laboratory
Richland, WA 99352

INTRODUCTION

The development and characterization of waste forms, containers and other engineered barriers destined for use in the isolation of nuclear waste in deep geologic repositories has progressed to the point where there are several options and combinations of barrier systems that are available to help assure safe disposal of nuclear wastes. However, a rigorous basis has not yet developed to define whether various concepts or products are required or desirable, or how durable they should be for how long. This analysis is an attempt to contribute to that basis. The intent of the study is to determine what incentives exist for providing highly effective engineered barriers for the isolation of radioactive waste in a deep geologic repository, based on what effect, if any, the engineered barrier may have on the ultimate potential dose to humans.

Engineered barriers as used here are considered to be anything that man places in the repository to enhance the containment of radioactive material, including the waste form, containers, backfill material, etc.

Past consequences analyses (Burkholder and Cloninger 1978; Burkholder, Cloninger, Baker and Jansen 1975; Burkholder 1976; Cloninger, Cole and Washburn 1980; Cloninger and Cole, 1980) of generic repositories, two containing the reprocessing waste and three containing the spent fuel from the projected U.S. nuclear power economy through the year 2000, developed a methodology for assessing potential consequences of nuclide release from a repository.

*Work supported by the Office of Nuclear Waste Isolation under its contract with the U.S. Department of Energy.

METHODOLOGY

The analysis is based on estimating the potential releases from a deep geologic repository due to a scenario involving groundwater intrusion into the repository, leaching and transport of nuclides to the surface and transport of the nuclides in the biosphere where they interact with humans.

Releases to the biosphere are converted to a dose to a maximally exposed individual ("maximum individual") in order to determine, through parametric analysis, what effects the properties of engineered barriers may have on that resultant dose over a range of geohydrologic conditions.

The Scenario

The highest risk path from the repository to the biosphere is assumed to be the leach incident pathway. While the relative risks of the various pathways from final geologic storage have not been evaluated, the leach incident appears to be the highest probability release event.

The scenario involves the following steps:

- penetration of the repository by groundwater
- failure of the waste container or containment function
- leaching of the radionuclides from the waste form
- transport of the nuclides through the surrounding geology to the biosphere
- transport of the nuclides through the biosphere to man
- interaction of the nuclides with man

The geologic isolation system provides barriers to the first four steps in the process. The transport of nuclides from the repository to the biosphere is limited by several general factors. Some important factors are the solubility of waste form and container at repository conditions, the chemical composition of the groundwater, especially its pH and Eh, ionic strength and the presence in the groundwater of complexing agents, and the sorption capability of the repository media and all geologic media along the groundwater flow path.

These factors are, of course, related to other factors. For example, water velocity and flow direction are a function of aquifer porosity, effective permeability, and hydraulic gradient. They can also be affected by the thermal gradient near the repository. Factors are also interrelated. For example, the solubility of the waste form and the net sorptive capacity of the geologic media can be greatly affected by the pH, Eh and ionic strength of the groundwater.

Through these factors, the geologic repository system provides two general means of protection from radioactive waste. The first means is containment of the waste for a sufficient length of time for the hazardous nuclides to decay to innocuous levels such that unrestricted release to the environment presents no radiological hazard. The second means is to limit the rate of release of nuclides to the biosphere such that their concentration in the constantly renewed local surface water and air do not exceed acceptable limits. Actual repository systems will provide protection using some effective balance of these two means.

The human-caused intrusion water well scenario can essentially short-circuit the protection from nuclides already dissolved in the groundwater that the far-field system provides. The intrusion well scenario was chosen as a limiting case of the leach incident pathway.

The radionuclide inventory used in the analysis is one-fifth of the total for the spent, unreprocessed fuel (no-recycle case) accumulated in the U.S. through the year 2050 that was used in the Commercial Waste Management Generic Impact Statement (CWM/GEIS) (DOE 1979). The assumption was that there would be five operating repositories, each containing an equal fraction of the total U.S. spent fuel accumulation.

The Models

The geosphere transport model is a second-generation Fortran version of the GETOUT code, (DeMier, Cloninger, Burkholder and Liddell 1979) a one-dimensional analytical model that simulates the release of radionuclides from a repository through contact and leaching by flowing groundwater, with simultaneous transport of the nuclides in the groundwater (including the effects of convection, hydrodynamic dispersion, reversible sorption and mineralization, and radioactive decay), and predicts the ultimate time and rate of release of the nuclides to the surface water (lake or river) or water well intake.

In order to simplify the analysis, a multitude of parameters was reduced to terms representing the net effect of several more complex phenomena. The release rate of nuclides into the flowing groundwater (source term) is a constant-rate approximation of what may in reality be a more complex release function. The model parameter representing the rate-limiting capacity of the engineered barriers is the inverse of the release rate (fraction/yr), or the constant-rate release duration in years. A parameter representing the combined effectiveness of the engineered barrier containment function and the stability (or integrity) of the surrounding geology is the delay (years) in release incident initiation, measured from the date of repository closure.

The major hydrologic parameters are the dispersivity, groundwater flow rates and total pathlength to the biosphere, which are functions of geologic media factors such as porosity, permeability, net hydraulic head and gradient, and fracture patterns. For this analysis, these parameters have been reduced to net groundwater flow velocity, total migration pathlength and a net dispersion coefficient. The flow velocity and pathlength, though modeled separately, are combined for graphic presentation into a single parameter, groundwater travel time (pathlength/flow velocity).

Output data of the geosphere model are the input data for the biosphere model, PABLM (Napier, Kennedy and Soldat 1980). The biosphere model predicts the radiation dose to humans from the nuclide release to surface waters via several pathways, including direct ingestion of water, aquatic foods and irrigated food products, as well as irradiation from shoreline deposition and swimming. Food chain reconcentration/accumulation effects are included. For the general scenario, the biosphere parameters are the same as were used in CWM/GEIS, and were held constant through the analysis.

Reference Repository Sites

The models were applied to simulate a leach incident release from a repository model in four reference sites (salt, basalt, granite and shale) having a broad range of characteristics. A 100-year constant rate leach duration was set as a minimum, representing the fastest possible release of nuclides (1%/yr) from the repository due to solubility and mass transfer limitations in a backfilled repository and assumed no credit for any rate-limiting capacity of the engineered system. Actual solubility-limited release rates for most elemental ions of interest are probably much lower. The maximum release duration was one billion yr (one-billionth per yr release rate). The "delay prior to leach initiation" parameter range was arbitrarily chosen in order to cover a very wide range of possibilities without regard to initiating events or processes and was varied from one to one million yr. The remaining site parameters and parameter ranges are based on measured and estimated data from previous analyses of actual sites.

ASSUMPTIONS, RESULTS AND CONCLUSIONS

Several simplifying assumptions that were made to perform the analysis are summarized below. The overall results are then presented in summary form as part of the conclusions discussion.

Assumptions and Approach Summary

1. The radionuclide source is 1/5 of the spent nuclear fuel accumulated in the U.S. through the year 2000 as defined in CWM/GEIS (DOE 1979).
2. The release pathways with the greatest risk to man originate from the waste after ultimate disposal.
3. The highest risk pathway from the site to the biosphere is the leach incident/groundwater transport pathway.
4. All nuclides are soluble in groundwater and the dissolved nuclides do not change speciation during migration.
5. The migrating nuclides are in sorption equilibrium at all points.
6. The water velocity and axial dispersion coefficient are constant for any single parametric calculation.
7. Estimates of the sorption equilibrium constants are used in most cases.
8. Transverse convection and dispersion are neglected.
9. The transporting nuclides are dissolved in the groundwater in trace concentrations, such that there is no inter- or intra-species competition for sorption, nor are solubility limits ever approached.
10. The release of nuclides from the repository is a constant rate band.
11. Doses are calculated for a "maximum" individual.
12. For all dose calculations, the discharging nuclides accumulate in the biosphere for 70 yr; the "maximum individual" is exposed to that accumulation during those 70 yr. and the nuclides are subsequently dispersed into the biosphere with no dose after the exposure period.
13. Humans' dietary and living habits will remain constant throughout geologic time.
14. If the "maximum individual" is adequately protected, all individuals are adequately protected (this tends to ignore any statistical consequences due to integrated population doses based on a linear consequences approach).
15. A mrem dose received by a person living in the year 2051 is of equivalent value in all respects to a mrem dose received by a person living in the year 1002051 (10^6 yr later). There is no value discounting with time of dose occurrence.

Results and Conclusions

Results are completed in considerable detail for a reference salt site repository and in somewhat less detail for reference granite, basalt and shale systems. Based on those results, the following conclusions were reached;

• In nearly all cases, the regional geology is the most effective barrier to release of radionuclides to the biosphere, however, for long-lived isotopes of carbon, technetium and iodine, which were poorly sorbed on the geologic media, the geology is not very effective once a leach incident is initiated.

• While the need for, and the effectiveness of, a release-rate limiting barrier function is somewhat dependent on the sorption properties of the geologic media, generally a release rate of less than $10^{-3}yr^{-1}$ (fractional) is necessary to reduce the potential dose from 14C, 99Tc, and 129I to a baseline level below that of the actinides, and beyond that, a rate of less than 10^{-5} is required before the potential dose from the actinide chains can be further lowered by this mechanism (figure 1, lower two lines). This is because the distribution of actinide chain members in time and space due to their different sorption properties and the characteristics of the groundwater flow field has the same effect as a release rate reduction of between 10^{-5} to $10^{-6}yr^{-1}$. Only in extreme cases of the intrusion water well scenario (upper line, figure 1) is there a direct relationship between release rate from the repository and release to the biosphere.

• Generally, the delay in leach initiation that an engineered delay function (container, sealant backfill, etc.) might provide, 10^4 years or less, is insufficent to accomplish any maximum potential dose reduction. Only the delay provided by a dry, stable geology, 10^5 years or more can accomplish this (bottom two lines, figure 2). However, in the extreme and very unlikely case of the water well scenario (upper line, figure 2, the maximum potential dose is optimally reduced by a containment function lasting about 400 years. If the containment should last beyond 1000 years, the maximum potential dose can actually be increased due to 226Ra accumulation (from uranium decay) in the repository confines.

Primary nuclides of concern, in no particular order are 99Tc, 129I, 14C, 226Ra + daughters (U), 237Np (Am, Np) 234U, and for the extreme case of the well scenario, 90Sr.

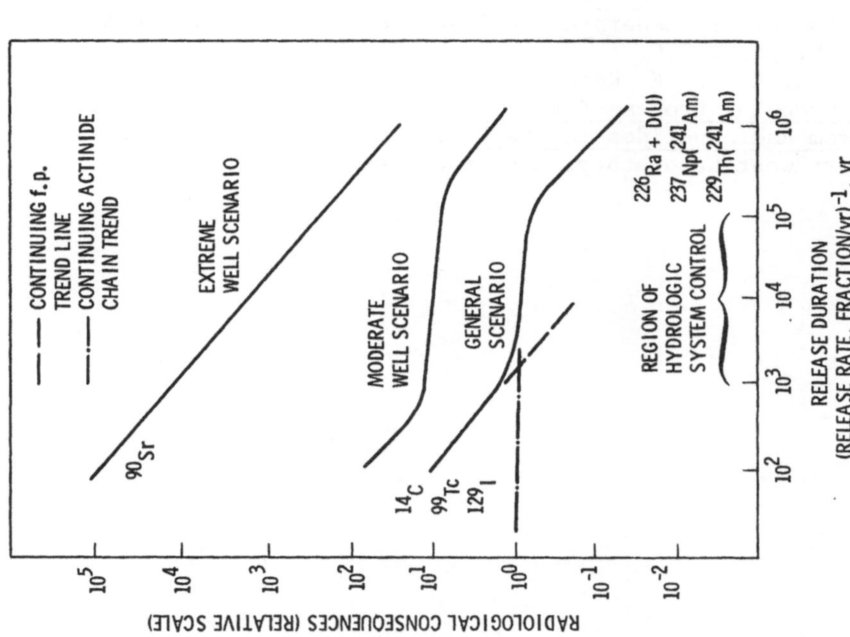

Fig. 1. General effect of rate-limiting barrier function.

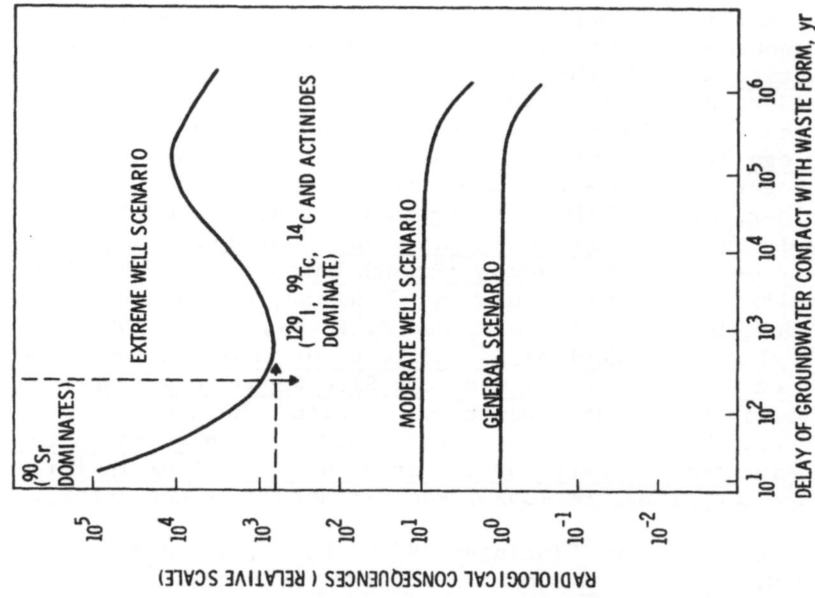

Fig. 2. General effect of engineered delay function.

REFERENCES

Burkholder, H. C., and M. O. Cloninger. 1978. "The Reconcentration Phenomenon of Radionuclide Chain Migration." American Institute of Chemical Engineers Symposium Series, 74(179).

Burkholder, H. C., M. O. Cloninger, D. A. Baker and G. Jansen. 1976. "Incentives for Partitioning High-Level Waste." Nuclear Technology 31:202. (Also, as BNWL-1927. 1975. Pacific Northwest Laboratory, Richland, Washington).

Burkholder, H. C. 1976. "Management Perspectives for Nuclear Fuel Cycle Wastes." Nuclear Waste Management and Transportation Quarterly Report, January through March 1976. BNWL-2029, Pacific Northwest Laboratory, Richland, Washington.

Cloninger, M. O., C. R. Cole, and J. F. Washburn. 1980. An Analysis on the Use of Engineered Barriers for Geologic Isolation of Spent Fuel in a Reference Salt Site Repository, PNL-3356, Pacific Northwest Laboratory, Richland, Washington.

Cloninger, M. O., and C. R. Cole. 1980. An Analysis on the Use of Engineered Barriers for Isolation of Spent Fuel in Three Reference Geologies, PNL-3520, Pacific Northwest Laboratory, Richland, Washington.

Demier, W. V., M. O. Cloninger, H. C. Burkholder and P. J. Liddell. 1979. GETOUT - A Computer Program for Predicting Radionuclide Decay Chain Transport through Geologic Media. PNL-2970, Pacific Northwest Laboratory, Richland, Washington.

DOE. 1979. Draft Environmental Impact Statement: Management of Commercially Generated Radioactive Waste, DOE/EIS-0046-D, DOE Washington, D.C.

Napier, B. A., W. E. Kennedy, Jr. and J. K. Soldat. 1980. PABLM - A Computer Program for Calculating Accumulated Radiation Doses from Radionuclides in the Environment. PNL-3209, Pacific Northwest Laboratory, Richland, Washington.

THERMAL ANALYSIS FOR A SPENT REACTOR FUEL STORAGE TEST IN GRANITE

Donald N. Montan

Lawrence Livermore National Laboratory
P.O. Box 808
Livermore, CA 94550

INTRODUCTION

A test of retrievable geologic storage of spent-fuel assemblies from a commercial nuclear power reactor is underway at the Nevada Test Site of the U.S. Department of Energy (DOE). This project is part of the DOE Nevada Nuclear Waste Storage Investigations, managed by the Nevada Operations Office of the DOE. The Lawrence Livermore National Laboratory (LLNL) is responsible for the technical direction of this test which is located 420 m below the surface in the Climax granite stock. Eleven canisters containing spent-fuel assemblies approximately 2.5 years out of reactor core have been emplaced in the floor of a storage drift along with six electrical simulator canisters for comparison of very near field effects. Two adjacent parallel drifts contain electrical heaters which will simulate within the test array the thermal field of a large repository. During the test, thermomechanical data will be obtained which may ultimately be useful in designing a spent-fuel repository in granite or other hard rock.

CONCEPT AND BASIC DESIGN

The model used in the early phases of the thermal calculations was very simple. The granite was considered to be homogeneous, isotropic and of infinite extent. The model repository, or the spent-fuel test, was modeled as an array of parallel finite line sources whose length was that of the spent-fuel assembly (~ 4 m).

The power vs time history of the sources was taken to be

proportional to that for the PWR fuel with a burnup of 33000 MWD/MTU at a specific power of 37.5 MW/MTU.[1]

This model, of course, ignores the existence of the mined drifts used to emplace the spent fuel and the drill holes in which they are stored. The neglect of the drifts was considered to be conservative in the sense that temperatures calculated would be higher than the "real world" since the openings would be better conductors than the rock they replaced and any ventilation would remove heat from the system. The use of line sources rather than cylindrical sources only affects early transients.

The calculations used in this phase were also simple, namely the superposition in space and time of the continuous point source solution of the diffusion equation.

$$\Delta T = \frac{Q}{4\pi k d} \operatorname{erfc} \sqrt{\frac{d^2}{4\kappa t}} \,.$$

The Model Repository

The model repository is simple in design. It consists of a large number of long parallel drifts with the spent-fuel canisters placed in vertical holes drilled along the center lines of the drifts. The dimensions used in this model are 15 m, center-to-center, for the drifts with the waste canisters emplaced on 3-m centers along the drifts. The rationale for the 15-m drift spacing was an extraction ratio of 33% using a projected drift width of 5 m.

Thermal calculations have been made for the center of such a repository. The parameters used in these calculations were $k = 3$ W/m-K, $\kappa = 1.25$ mm^2/s, $Q = 2$ kW @ 2.5 years out of reactor core, $L = 3.66$ m. Some results of these calculations for the hottest (rock) point in the repository (the rock surface at the fuel assembly midpoint in the center hole of the repository) are shown in Fig. 1. The three curves are for spent fuel that has been out of the reactor core for 2.5, 5, and 10 years, respectively, before emplacement in the rock.

The Spent-Fuel Test

The present spent-fuel-test design calls for fuel that has been out of reactor core for only about 2.5 years. Use of young fuel will allow the assessment of maximum thermal effects in a few years, as opposed to a few decades, and will expose the rock to higher temperatures than might be encountered using older fuel.

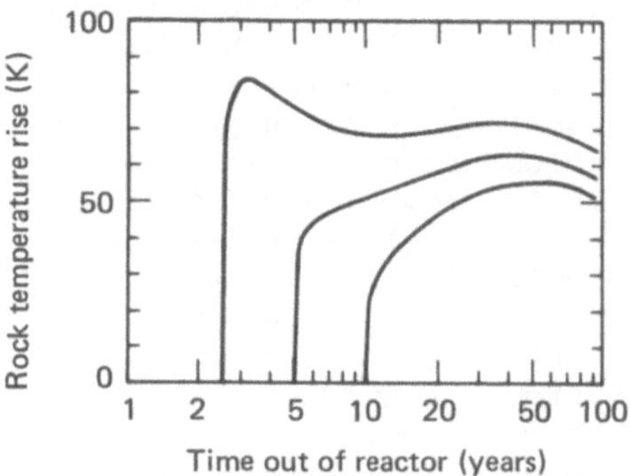

Fig. 1. Temperature history of the rock surface at fuel-assembly midpoint in the center hole of a hypothetical spent-fuel repository, calculated for spent fuel emplaced 2.5, 5, and 10 years after removal from the reactor.

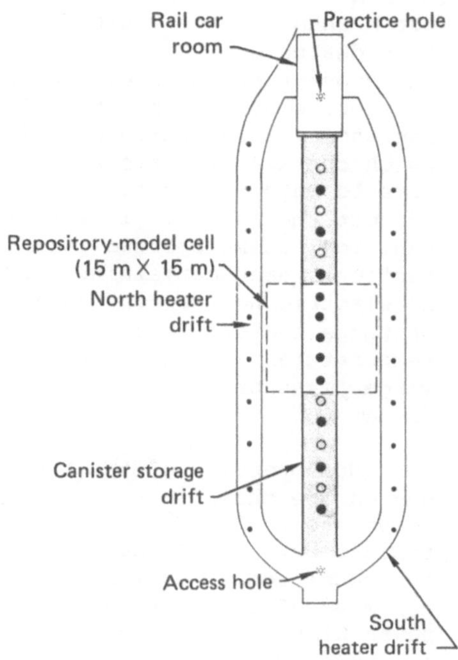

Fig. 2. Plan view of the Climax granite spent-fuel storage test.

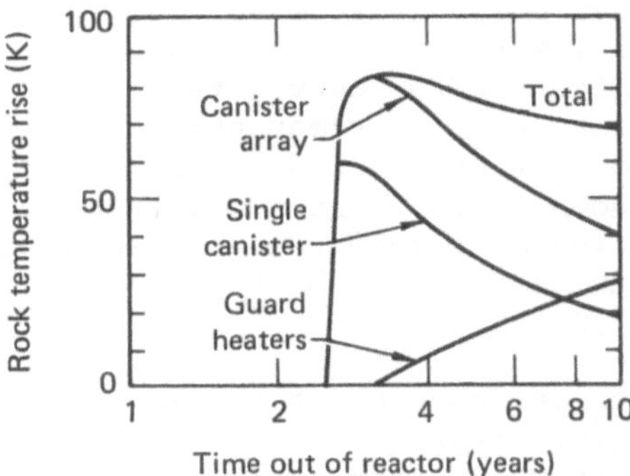

Fig. 3. Calculated temperature history of the rock surface at the fuel-assembly midpoint in the center hole of our 17-canister spent-fuel storage array. The spent fuel, electrical simulators, and heaters are deployed as shown in Fig. 2. The guard heater power is a constant 1739 W; it is turned on at 2.8 years.

After considerable study involving not only thermal effects, but mechanical and mining considerations as well, a spent-fuel-test design evolved whose physical layout is shown in Fig. 2. In the storage drift (5 by 6 m in cross section) is a linear array of 17 storage holes (on 3-m centers) in which are placed storage canisters, 11 of which contain spent-fuel assemblies and 6 of which contain electric heaters whose power level is varied with time to match the decay of the spent fuel. Thus, for the purposes of most thermal calculations the 17 canisters are identical. In addition, there are two parallel side drifts (3 by 3 m in cross section), each containing a row (10 m from the storage array) of 10 electric heaters in small holes on 6-m centers. The function of these heaters is (1) to provide a means of maintaining higher temperatures in the storage canister array, and (2) to heat a larger mass of granite than would be affected by the storage array alone.

Some results of the thermal calculations pertaining to this design are shown in Fig. 3. A single canister gives a peak rock temperature of ~60°C, well below the 85°C calculated for the model repository. The 17-canister array peaks within a few degrees of the model repository, but drops to near half the value in 7 years. A number of calculations led to the design of the guard heater power history. When turned on at 0.3 years after the spent fuel is emplaced and maintained at a constant power of 1732 W each, they produced the contribution shown in the lower curve. When this is added to the canister array curve to give the top curve, the result is a peak rock temperature history that is remarkably close (within 1% for the first 7.5 years) to that of the model repository. Additional calculations have shown that the region between the canister row and the heaters experiences a temperature rise a few degrees higher than the region between rows in the repository design.

Thus, the spent-fuel-storage test, as described here, is expected to subject a substantial granitic mass (~1000 m^2 horizontal area) to thermal effects that match or exceed those that might be expected in an actual repository in a similar medium.

FINAL DESIGN

In order to provide additional information for the final design, the decision was made to do more detailed thermal calculations beyond the capabilities of the calculational techniques used above. These calculations were based on a unit cell of the repository or spent-fuel test. This cell contains one canister and is bounded by two parallel vertical planes perpendicular to the drifts and spaced half-way between canisters. For the model repository calculations two additional vertical planes of symmetry parallel to and midway between the drifts also bound the unit-cell. The four-fold axis of symmetry at the canister center line provides two more vertical planes of symmetry, thus reducing the required calculations to a region comprising one-fourth of the unit-cell.

The mesh typically used contains three basic regions. The innermost region is 20 m wide by 40 m high and is divided into 1600 zones (0.5 by 1 m) to give the desired spatial resolution. To simulate an infinite medium for a long period of time two more regions were added with 5- by 5-m and 20- by 20-m zones giving an overall size of 80 by 160 m. In the regions comprising the drifts the regular zoning was replaced by single zones containing air. For three-dimensional modeling the region beneath the canister drift was three-dimensional with the zoning becoming progressively finer as the source was approached. Additional zones were used to model the canister, liner, shield plugs, railroad tracks, and other details. In all more than 2500 zones were used. For the numerical computations the TRUMP[3] computer program was used.

In some repository calculations that we have seen the drifts were considered as non-conductors (voids) and thus gave a large floor-to-ceiling temperature difference and isotherms that were perpendicular to the drift surfaces. This is quite incorrect. Either thermal radiation <u>or</u> convection will make the drifts better conductors than the rock that was removed to create them. When radiation and convection are properly treated, the drift surfaces are nearly isothermal and the nearby isotherms nearly parallel to the surfaces.

Thermal radiation was modeled by straightforward application of the Stefan-Boltzmann equation. An emittance of 0.8 was normally used. One important consideration (often overlooked) is that in models having plane(s) of symmetry (such as occurs in our model) the adiabatic boundary condition at such planes is equivalent to a mirror. Thus, the "half" main drift in our model with 16 radiating surfaces has 185 radiative connections.

Heat transfer from the drift surfaces to the air in the drift may be expressed by $q = AH\Delta T$, where H is the convection coefficient (typical units W/m^2-K). For the convection coefficient we have used a standard empirical correlation for natural convection in the turbulent regime relating the Nusselt number (N_{Nu}) to the Rayleigh number (N_{Ra}), Ref. 2: $N_{Nu} = 0.13\ N_{Ra}{}^{1/3}$, where the Rayleigh number is the product of the Grashof number and the Prandtl number.

Ventilation

The heat removed from a repository or the spent-fuel test by ventilation can amount to a significant fraction of the heat produced.

Furthermore, the temperature rise of the air as it flows through the drift is nearly a linear function of the flow rate. Thus a careful approach to the calculation of thermal effects of ventilation seems prudent.

One approach is to hold the air, or some external boundary representing it, at the ambient air/rock temperature and then proceed with the calculation using convection coefficients. This ventilation model, while easily incorporated into the unit-cell model, has several serious drawbacks. In the spent-fuel test a fixed amount of air will enter one end of a drift and be removed at the other, picking up heat and rising in temperature as it passes through. The fixed air temperature will overestimate the heat removed, underestimate the rock temperatures (particularly the drift surfaces) and give no estimate of what the actual air temperature will be. On the other hand, physical implementation of the fixed air temperature model would require a very large amount of air.

In order to retain the unit-cell model and still do realistic ventilation effect calculations, we have adopted a "partial-flow" model in which the cell receives only a fraction of the total flow but that partial-flow enters the cell at ambient and leaves at the average air temperature of the cell. The heat removal may then be expressed as: $\Delta q = \rho c \dot{V}' \Delta x T$, where

ρ = air density (kg/m^3)

c = heat capacity (J/kg-K)

$\dot{V}'$ = flow rate per unit length of drift (m^2/s)

Δx = length of calculational cell (m)

T = temperature rise above ambient (K)

The question then arises as to what is the numerical value of V'? The answer is to use several different values. They all represent different locations.

In the situation that we actually wish to model a constant air flow, $\dot{V}_0$ (m^3/s), passes thru the drift and at some point while traveling a distance Δx the temperature increases by an amount ΔT. The heat removal is then: $\Delta q = \rho c V_0 \Delta T$. By equating this equation and the previous one and passing to the limit as Δx becomes infinitesimal we obtain:

$$\frac{dT}{dx} = \frac{\dot{V}'T}{\dot{V}_0} .$$

Thus, by using different values of $\dot{V}'$ in the (complete) model and calculating the resulting air temperatures at a time or times of interest, a functional relationship between $\dot{V}'$ and T may be obtained. The above differential equation may then be integrated to give the desired result, air temperature rise as a function of distance along the drift. In practice we find that the product V'T is adequately approximated as a linear function of T.

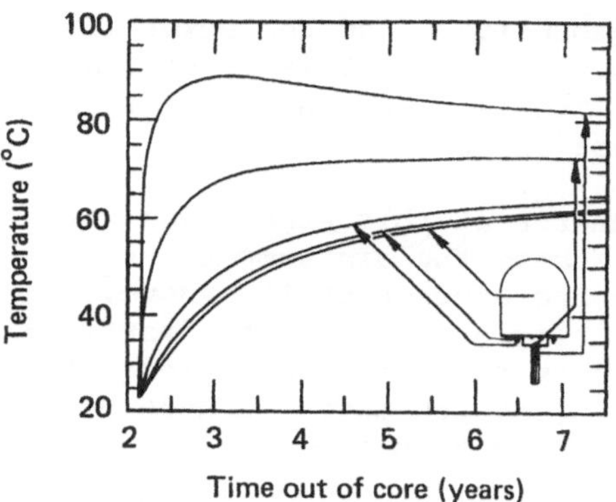

Fig. 4. Three-dimensional TRUMP code calculations of temperature vs time at indicated locations in the storage drift (no ventilation effects are considered).

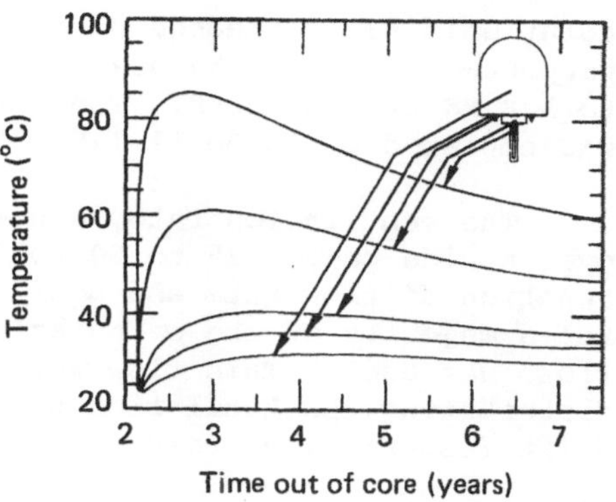

Fig. 5. Three-dimensional TRUMP code calculations of temperature vs time at indicated locations in the storage drift (ventilation effects included). Note the significant cooling compared with Fig. 4.

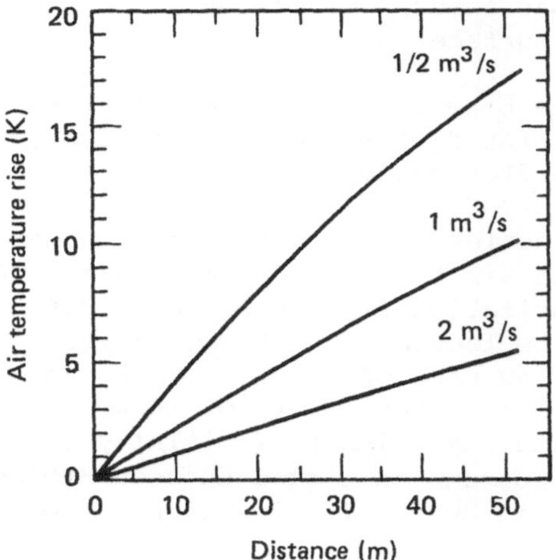

FIG. 6 Temperature for various ventilation rates as a function of distance along the storage drift.

RESULTS

Over a period of about one and one-half years, approximately 100 TRUMP calculations were made in the development and use of the above models. The spent fuel to be stored had been identified (from Turkey

Point Unit #3, discharge date 19 November 1977) and the power level purported to be 2.4 kW at 1.9 years out-of-core. This is equivalent to 1.8 kW at 2.5 years. The first of the 11 spent-fuel canisters was emplaced on 18 April 1980, the last on 28 May.

The ventilation rate, a new parameter at this stage, was varied over a wide range (25 to 50 fold) to assess the air temperature as a function of flow rate and position in the storage drift. However, for a majority of the calculations, a single value of the "partial-flow" was used. This flow was derived by the simple expedient of distributing the total flow uniformly along the active length of the drift (essentially cross-flow as opposed to thru-flow).

The three-dimensional version of our model was developed primarily to study details in the near vicinity of the storage holes. It does, of course, retain all of the features of the two-dimensional version. Shown in Figs. 4 and 5 is an interesting set of results that include items that require 3-D and some that could just as well have been done in 2-D. In both figures we see temperature histories of five locations. They are (from top to bottom)

- the steel shield plug at the top of the canister assembly
- the bottom of the concrete plug directly above
- the railroad tracks
- the floor surface
- the air.

The only difference between the two calculations was ventilation and it makes a significant difference. After five years of storage the air temperatures differ by 30°C.

Results of a series of calculations to determine the spatial distribution of the air temperature are shown in Fig. 6. The heat removed by the ventilation is significant. During most of the test the air will be leaving the storage drift ~10°C hotter than it entered (at a flow rate of 1 m^3/s). With ρc ~1000 J/m^3-K, this amounts to a heat removal rate of 10 kW. When this is compared to the total heat input, (initially 30 kW but decreasing with time) it is apparent that more than one-third of the heat input will be removed by ventilation over a few-year period.

REFERENCES

1. Rockwell International, Spent Unreprocessed Fuel Facility Engineering Studies, Rockwell International, Hanford, Wash., Informal Report RHO-LD-2 (1978).
2. W. H. McAdams, Heat Transmission (McGraw Hill Book Co., New York, 1954), 3rd ed.
3. A. L. Edwards, TRUMP, A Computer Program for Transient and Steady State Temperature Distribution in Multidimensional Systems, Lawrence Livermore Laboratory, Livermore, Calif., UCRL-14754, Rev. 3 (1972).

AUTHOR INDEX

A

B

C

D

E

W

Y

SUBJECT INDEX

Z